Machinery's Handbook Pocket Companion

A REFERENCE BOOK
FOR THE MANUFACTURING AND MECHANICAL ENGINEER, DESIGNER, DRAFTER, METALWORKER, TOOLMAKER, MACHINIST, HOBBYIST, EDUCATOR, AND STUDENT

Machinery's Handbook Pocket Companion

Second Edition

RICHARD P. POHANISH AND CHRISTOPHER J. MCCAULEY

LAURA BRENGELMAN, EDITOR

INDUSTRIAL PRESS, INC.

INDUSTRIAL PRESS, INC.
32 Haviland Street, Suite 3
South Norwalk, Connecticut 06854 U.S.A.
Phone: 203-956-5593
Toll-Free: 888-528-7852
Fax: 203-354-9391
Email: info@industrialpress.com

Title: *Machinery's Handbook Pocket Companion, 2nd Edition*
Authors and Compilers: Richard P. Pohanish and Christopher J. McCauley
Library of Congress Control Number: 2020931274

ISBN PRINT: 978-0-8311-4431-9
ISBN ePDF: 978-0-8311-9567-0
ISBN ePub: 978-0-8311-9568-7
ISBN eMobi: 978-0-8311-9569-4

Printed and bound in the United States of America

**MACHINERY'S HANDBOOK
POCKET COMPANION
2ND EDITION**
First Printing

books.industrialpress.com
ebooks.industrialpress.com

TABLE OF CONTENTS

TABLE OF CONTENTS

STANDARD TAPERS

(Continued)

THREADS

FASTENER INFORMATION

TABLE OF CONTENTS

TABLE OF CONTENTS

MILLING CUTTERS

KEYS AND KEYSEATS

KEYS AND KEYSEATS

(Continued)

BROACHING

CUTTING TOOLS FOR TURNING

MACHINING OPERATIONS

TABLE OF CONTENTS

MACHINING OPERATIONS
(Continued)

COMPUTER NUMERICAL CONTROL

GRINDING WHEELS

GRINDING WHEELS
(Continued)

TABLE OF CONTENTS

GEARING

PROPERTIES OF MATERIALS

PROPERTIES OF MATERIALS

(Continued)

STANDARDS FOR DRAWINGS

SURFACE TEXTURE

TABLE OF CONTENTS

SURFACE TEXTURE

(Continued)

CORROSION

ALLOWANCES AND TOLERANCES

ALLOWANCES AND TOLERANCES

(Continued)

CONVERSION FACTORS

FOREWORD

Twenty years ago, the first edition of the *Machinery's Handbook Pocket Companion* was developed for users of the *Machinery's Handbook* who could benefit by having a smaller, more convenient volume for bench- or desk-side quick reference. Containing key content from the *Machinery's Handbook*, it has evolved into a handy timesaver for anyone in manufacturing, metalworking, and related fields for whom convenient access to fundamental and reliable data is essential. Over the years, the *Pocket Companion* has been perennially popular with practitioners, educators, and students of the machine trades.

A tool designed to provide years of use, this book provides detailed information in a concise package. The presented material has been carefully selected from current and former editions of *Machinery's Handbook*. Some of the subject matter has been reorganized, distilled, or simplified to increase the usefulness of this book without adding to its bulk—though this edition has grown, with replaced and extended material from the 31st edition and a new list of useful online resources (see page 345).

The intention has been to provide information of technical value where only a brief or no introduction and essential data are needed to save time and labor. To obtain the full value of this small handbook, the user must have sufficient knowledge about the subject to apply the tables, formulas, and other data where such information can be used with efficiency. The *Machinery's Handbook Pocket Companion* minimizes explanations of the various subjects, based on the assumption that its users are acquainted with information and procedures necessary for the safe operation and manipulation of machines and tools.

The *Pocket Companion* does not replace the *Machinery's Handbook, 31st Edition*, but instead serves as a handy and more portable distillation of just some of the *Handbook's* vastly larger collection of invaluable text, data, and standards. Readers who require in-depth information, background on manufacturing operations, and theory should refer to discussions in the 31st edition.

This book, like all of the *Machinery's Handbook* product family, is the result of collaborative efforts. Among those credited with the *Pocket Companion* becoming the valuable tool it is today are authors and compilers Richard P. Pohanish and Christopher J. McCauley, as well as Arief Era, John Carleo, Cara Chamberlain, Ken Evans, Robert Green, Steve Heather, Jason Hughes, Kathy McKenzie, Gerald Murray, Julia Phelps, Henry Ryffel, Industrial Press owner Alex Luchars, and the rest of the *Machinery's Handbook, 31st Edition* team.

Many of the American National Standards Institute (ANSI) standards that deal with mechanical engineering, extracts from which are included in the *Pocket Companion*, are published by the American Society of Mechanical Engineers (ASME). The editors thank ASME for its exceptional collaboration in helping to identify and bring essential data up to date, in both the *Machinery's Handbook, 31st Edition,* and this companion volume, according to the latest, definitive industry standards. Information concerning other standards and nomenclature also is included in this book. Official standards and related publications are copyrighted by the issuing organizations; contact them directly for further information regarding standards and to purchase copies. We also thank Carr-Lane Manufacturing, the Norton Company, Sandvik Coromant, and other referenced firms for permission to use their material.

Finally, we wish to thank all of the associations, societies, companies, professionals, hobbyists, scholars, educators, students, and other individuals who have provided invaluable material and input for this book and the 31st edition.

We encourage readers with suggestions for improving or adding to the *Pocket Companion* to send us your thoughts and feedback. We also encourage you to share with us how the *Machinery's Handbook* product family supports and enhances your involvement in this endlessly fascinating field.

Laura Brengelman
Editor

MATHEMATICAL FORMULAS AND TABLES

Dimensions of Plane Figures

Square:

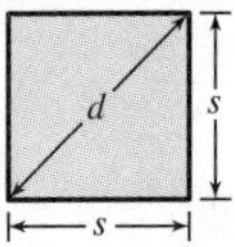

$$A = s^2 = \tfrac{1}{2}d^2$$
$$s = 0.7071d = \sqrt{A}$$
$$d = 1.414s = 1.414\sqrt{A}$$

Example: Side s of a square is 15 in. Find the area of the square and the length of its diagonal.

$$A = s^2 = 15^2 = 225 \text{ in}^2$$
$$d = 1.414s = 1.414 \times 15 = 21.21 \text{ in}$$

Example: The area of a square is 625 cm². Find the length of side s and diagonal d.

$$s = \sqrt{A} = \sqrt{625} = 25 \text{ cm}$$
$$d = 1.414\sqrt{A} = 1.414 \times 25 = 35.35 \text{ cm}$$

Rectangle:

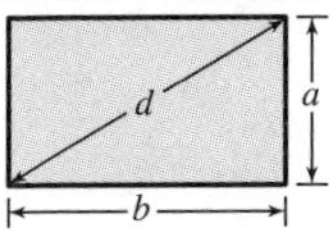

$$A = ab = a\sqrt{d^2 - a^2} = b\sqrt{d^2 - b^2}$$
$$d = \sqrt{a^2 + b^2}$$
$$a = \sqrt{d^2 - b^2} = A/b$$
$$b = \sqrt{d^2 - a^2} = A/a$$

Example: Side a of a rectangle is 12 cm, and the area is 70.5 cm². Find the length of side b and diagonal d.

$$b = A/a = 70.5/12 = 5.875 \text{ cm}$$
$$d = \sqrt{a^2 + b^2} = \sqrt{12^2 + 5.875^2} = \sqrt{178.516} = 13.361 \text{ cm}$$

Example: The sides of a rectangle are 30.5 and 11 cm. Find the area.

$$A = ab = 30.5 \times 11 = 335.5 \text{ cm}^2$$

Parallelogram:

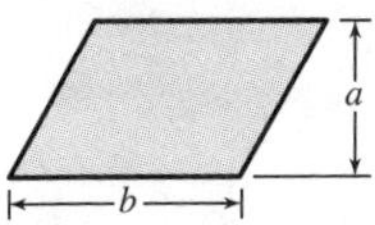

$$A = ab$$
$$a = A/b$$
$$b = A/a$$

Note: The dimension a is the length of the vertical drawn at a right angle to side b. Dimension a is also considered the height of the parallelogram.

Example: Base b of a parallelogram is 16 ft. Height a is 5.5 ft. Find the area.

$$A = ab = 5.5 \times 16 = 88 \text{ ft}^2$$

Example: The area of a parallelogram is 12 in². The height is 1.5 in. Find the length of the base b.

$$b = A/a = 12/1.5 = 8 \text{ in.}$$

Right Triangle (one angle is a 90-degree angle):

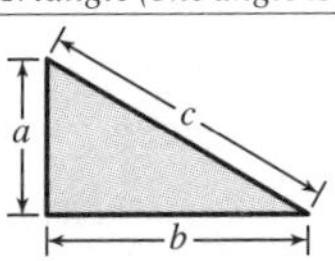

From the Pythagorean theorem, $a^2 + b^2 = c^2$, thus $A = \dfrac{ab}{2}$

$$c = \sqrt{a^2 + b^2} \qquad a = \sqrt{c^2 - b^2} \qquad b = \sqrt{c^2 - a^2}$$

Example: Side a is 6 in. and side b is 8 in. Find side c and area A:

$$c = \sqrt{a^2 + b^2} = \sqrt{6^2 + 8^2} = \sqrt{36 + 64} = \sqrt{100} = 10 \text{ in.}$$
$$A = \frac{ab}{2} = \frac{6 \times 8}{2} = \frac{48}{2} = 24 \text{ in}^2$$

Example: Side $c = 10$ and side $a = 6$. Find side b:

$$b \text{ in.} = \sqrt{c^2 - a^2} = \sqrt{10^2 - 6^2} = \sqrt{100 - 36} = \sqrt{64} = 8 \text{ in.}$$

Acute Triangle (all three angles measure less than 90 degrees):

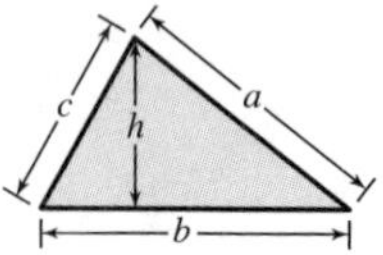

$$A=\frac{bh}{2},\quad h=\sqrt{a^2-\left(\frac{a^2+b^2-c^2}{2b}\right)^2},\quad \text{so}\quad A=\frac{b}{2}\sqrt{a^2-\left(\frac{a^2+b^2-c^2}{2b}\right)^2}$$

$$A=\sqrt{S(S-a)(S-b)(S-c)},\quad \text{where}\quad S=\frac{a+b+c}{2}$$

Example: Side $b = 7$ inches, $h = 4$ inches, so $A = bh/2 = (7\text{ in.}\times 4\text{ in.})/2 = 28\text{ in}^2/2 = 14\text{ in}^2$
Example: Side $a = 10$ cm, $b = 9$ cm, and $c = 8\text{ cm}^2$. Find the area.

$$A=\frac{b}{2}\sqrt{a^2-\left(\frac{a^2+b^2-c^2}{2b}\right)^2}=\frac{9}{2}\sqrt{10^2-\left(\frac{10^2+9^2-8^2}{2\times 9}\right)^2}=4.5\sqrt{100-\left(\frac{117}{18}\right)^2}$$

$$=4.5\sqrt{100-42.25}=4.5\sqrt{57.75}=4.5\times 7.60=34.20\text{ cm}^2$$

Obtuse Triangle (one angle measures greater than 90 degrees):

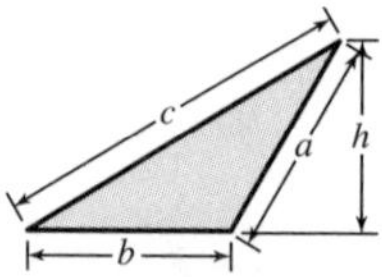

$$A=\frac{bh}{2},\quad h=\sqrt{a^2-\left(\frac{a^2+b^2-c^2}{2b}\right)^2},\quad \text{so}\quad A=\frac{b}{2}\sqrt{a^2-\left(\frac{a^2+b^2-c^2}{2b}\right)^2}$$

$$A=\sqrt{S(S-a)(S-b)(S-c)},\quad \text{where}\quad S=\frac{a+b+c}{2}$$

Example: If $b = 5$ cm and $h = 3$ cm, then $A = bh/2 = (5\text{ cm}\times 3\text{ cm})/2 = 15\text{ cm}^2/2 = 7.5\text{ cm}^2$

Example: Side $a = 5$ in., side $b = 4$ in., and side $c = 8$ in. Find the area.

$$S=(a+b+c)/2=(5+4+8)/2=17/2=8.5$$

$$A=\sqrt{S(S-a)(S-b)(S-c)}=\sqrt{8.5(8.5-5)(8.5-4)(8.5-8)}$$

$$=\sqrt{8.5\times 3.5\times 4.5\times 0.5}=\sqrt{66.937}=8.18\text{ in}^2$$

Trapezoid:

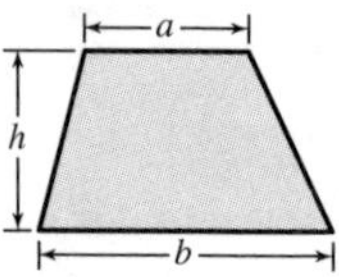

$$\text{Area}=A=\frac{(a+b)h}{2}$$

Note: In Britain, this figure is called a *trapezium* and the figure below it is known as a *trapezoid*, which is the reverse of the US terms.

Example: Side $a = 23$ meters, side $b = 32$ meters, and height $h =$ 12 meters. Find the area.

$$A=\frac{(a+b)h}{2}=\frac{(23+32)\times 12}{2}=\frac{55\times 12}{2}=330\text{ m}^2$$

Trapezium:

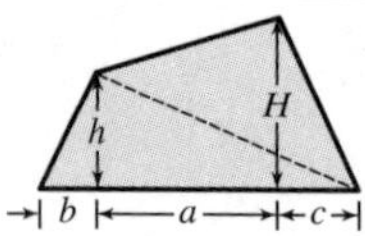

$$\text{Area}=A=\frac{(H+h)a+bh+cH}{2}$$

The area of a trapezium also can be found by dividing it into two triangles, as indicated by the dashed line. Each area is added to give the total area of the trapezium.

Example: Let $a = 10$ in., $b = 2$, $c = 3$ in., $h = 8$ in., and $H = 12$ in. Find the area.

$$A=\frac{(H+h)a+bh+cH}{2}=\frac{(12+8)\times 10+(2\times 8)+(3\times 12)}{2}$$

$$=\frac{(20\times 10)+16+36}{2}=\frac{252}{2}=126\text{ in}^2$$

Regular Hexagon:

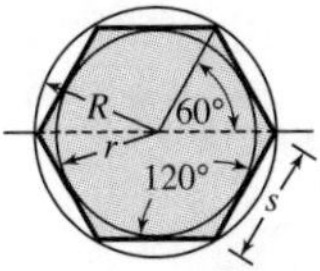

$A = 2.598s^2 = 2.598R^2 = 3.464r^2$

$R = s$ = radius of circumscribed circle $= 1.155r$

r = radius of inscribed circle $= 0.866s = 0.866R$

$s = R = 1.155r$

Example: The side s of a regular hexagon is 40 millimeters. Find the area and the radius r of the inscribed (drawn inside) circle.

$$A = 2.598s^2 = 2.598 \times 40^2 = 2.598 \times 1600 = 4156.8 \text{ mm}^2$$

$$r = 0.866s = 0.866 \times 40 = 34.64 \text{ mm}$$

Example: What is the length of the side of a hexagon circumscribed on (drawn around) a circle of 50 millimeters radius? In this case, because the hexagon is circumscribed on the circle, the circle is *inscribed* (drawn within) the hexagon. Hence, $r = 50$ mm and $s = 1.155r = 1.155 \times 50 = 57.75$ mm

Regular Octagon:

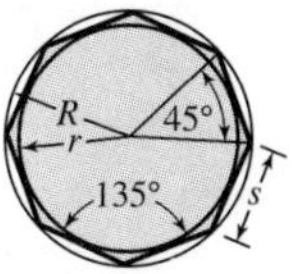

A = area $= 4.828s^2 = 2.828R^2 = 3.314r^2$

R = radius of circumscribed circle $= 1.307s = 1.082r$

r = radius of inscribed circle $= 1.207s = 0.924R$

$s = 0.765R = 0.828r$

Example: Find the area and the length of the side of an octagon inscribed (drawn inside) in a circle of 12 inches diameter.

Diameter of circumscribed (drawn around) circle = 12 inches; hence, $R = 6$ in.

$$A = 2.828R^2 = 2.828 \times 6^2 = 2.828 \times 36 = 101.81 \text{ in}^2$$

$$s = 0.765R = 0.765 \times 6 = 4.590 \text{ in.}$$

Circle:

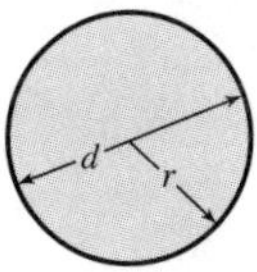

$$\text{Area} = A = \pi r^2 = 3.1416r^2 = 0.7854d^2$$

$$\text{Circumference} = C = 2\pi r = 6.2832r = 3.1416d$$

$$r = C \div 6.2832 = \sqrt{A \div 3.1416} = 0.564\sqrt{A}$$

$$d = C \div 3.1416 = \sqrt{A \div 0.7854} = 1.128\sqrt{A}$$

Length of arc for center angle of 1° $= 0.008727d$

Length of arc for center angle of $n°$ $= 0.008727nd$

Example: Find area A and circumference C of a circle with a diameter of $2\frac{3}{4}$ inches.

$$A = 0.7854d^2 = 0.7854 \times 2.75^2 = 0.7854 \times 2.75 \times 2.75 = 5.9396 \text{ in}^2$$

$$C = 3.1416d = 3.1416 \times 2.75 = 8.6394 \text{ in}$$

Example: The area of a circle is 16.8 in². Find its diameter.

$$d = 1.128\sqrt{A} = 1.128\sqrt{16.8} = 1.128 \times 4.099 = 4.624 \text{ in.}$$

Sector of a Circle:

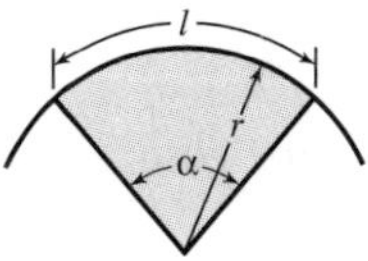

$$\text{Length of arc} = l = \frac{3.1416\, r\, \alpha}{180} = 0.01745r\alpha = \frac{2A}{r}$$

$$\text{Area} = A = \tfrac{1}{2}rl = 0.008727\alpha r^2$$

$$\text{Central angle, in degrees} = \alpha = \frac{57.296\, l}{r}, \qquad r = \frac{2A}{l} = \frac{57.296\, l}{\alpha}$$

Example: The radius of a circle is 35 millimeters, and angle α of a sector of the circle is 60 degrees. Find the area of the sector and the length of arc l.

$$A = 0.008727\alpha r^2 = 0.008727 \times 60 \times 35^2 = 641.41 \text{ mm}^2 = 6.41 \text{ cm}^2$$

$$l = 0.01745r\alpha = 0.01745 \times 35 \times 60 = 36.645 \text{ mm}$$

Segment of a Circle:

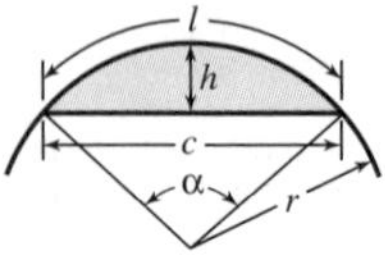

A = area l = length of arc α = angle, in degrees

$$c = 2\sqrt{h(2r-h)} \qquad A = \tfrac{1}{2}[rl - c(r-h)]$$

$$r = \frac{c^2 + 4h^2}{8h} \qquad l = 0.01745 r\alpha$$

$$h = r - \tfrac{1}{2}\sqrt{4r^2 - c^2} = r[1 - \cos(\alpha/2)] \qquad \alpha = \frac{57.296\, l}{r}$$

See also *Segments of Circles* starting on page 7.

Example: The radius r is 60 inches and the height h is 8 inches. Find the length of the chord c.

$$c = 2\sqrt{h(2r-h)} = 2\sqrt{8 \times (2 \times 60 - 8)} = 2\sqrt{896} = 2 \times 29.93 = 59.86 \text{ in.}$$

Example: If $c = 16$, and $h = 6$ inches, what is the radius of the circle of which the segment is a part?

$$r = \frac{c^2 + 4h^2}{8h} = \frac{16^2 + 4 \times 6^2}{8 \times 6} = \frac{256 + 144}{48} = \frac{400}{48} = 8\tfrac{1}{3} \text{ in.}$$

Cycloid:

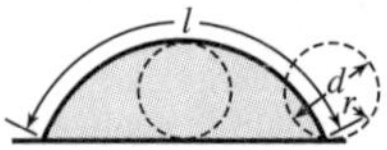

$$\text{Area} = A = 3\pi r^2 = 9.4248 r^2 = 2.3562 d^2$$
$$= 3 \times \text{area of generating circle}$$
$$\text{Length of cycloid} = l = 8r = 4d$$

Example: The diameter of the generating circle of a cycloid is 6 inches. Find the length l of the cycloidal curve and the area enclosed between the curve and the base line.

$$l = 4d = 4 \times 6 = 24 \text{ in.} \qquad A = 2.3562 d^2 = 2.3562 \times 6^2 = 84.82 \text{ in.}^2$$

Circular Ring (Annulus):

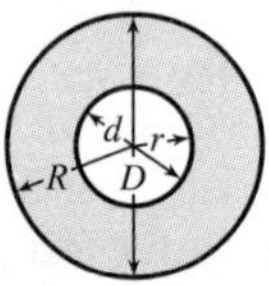

$$\text{Area} = A = \pi(R^2 - r^2) = 3.1416(R^2 - r^2)$$
$$= 3.1416(R + r)(R - r)$$
$$= 0.7854(D^2 - d^2) = 0.7854(D + d)(D - d)$$

Example: Let the outside diameter $D = 12$ centimeters and the inside diameter $d = 8$ centimeters. Find the area of the ring.

$$A = 0.7854(D^2 - d^2) = 0.7854(12^2 - 8^2) = 0.7854(144 - 64) = 0.7854 \times 80$$
$$= 62.83 \text{ cm}^2$$

By the alternative formula:

$$A = 0.7854(D + d)(D - d) = 0.7854(12 + 8)(12 - 8) = 0.7854 \times 20 \times 4$$
$$= 62.83 \text{ cm}^2$$

Sector of Circular Ring:

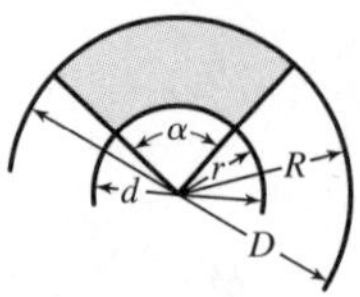

A = area, α = central angle, in degrees

$$A = \frac{\alpha\pi}{360}(R^2 - r^2) = 0.00873\alpha(R^2 - r^2)$$

$$= \frac{\alpha\pi}{4 \times 360}(D^2 - d^2) = 0.00218\alpha(D^2 - d^2)$$

Example: Find the area, if the outside radius $R = 5$ inches, the inside radius $r = 2$ inches, and $\alpha = 72$ degrees.

$$A = 0.00873\alpha(R^2 - r^2) = 0.00873 \times 72(5^2 - 2^2)$$
$$= 0.6286(25 - 4) = 0.6286 \times 21 = 13.2 \text{ in.}^2$$

Ellipse:

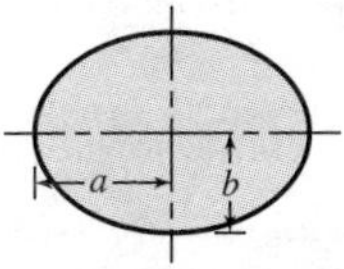

$$\text{Area} = A = \pi ab = 3.1416ab$$

An approximate formula for the perimeter is

$$\text{Perimeter} = P = 3.1416\sqrt{2(a^2+b^2)}$$

A closer approximation is $P = 3.1416\sqrt{2(a^2+b^2) - \frac{(a-b)^2}{2.2}}$

Example: The larger, or major, axis is 200 millimeters. The smaller, or minor, axis is 150 millimeters. Find the area and the approximate circumference. Here, then, $a = 100$, and $b = 75$.

$$A = 3.1416ab = 3.1416 \times 100 \times 75 = 23{,}562 \text{ mm}^2 = 235.62 \text{ cm}^2$$

$$P = 3.1416\sqrt{2(a^2+b^2)} = 3.1416\sqrt{2(100^2+75^2)} = 3.1416\sqrt{2 \times 15{,}625}$$

$$= 3.1416\sqrt{31{,}250} = 3.1416 \times 176.78 = 555.37 \text{ mm} = 55.537 \text{ cm}$$

Spandrel or Fillet:

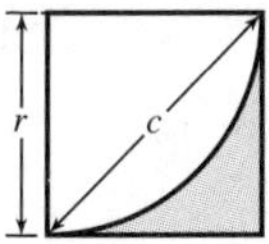

The shaded region is the spandrel (fillet).

$$\text{Area} = A = r^2 - \frac{\pi r^2}{4} = 0.215r^2 = 0.1075c^2$$

Example: Find the area of a spandrel, the radius of which is 0.7 inch.

$$A = 0.215r^2 = 0.215 \times 0.7^2 = 0.105 \text{ in}^2$$

Example: If chord c were given as 2.2 inches, what would be the area?

$$A = 0.1075c^2 = 0.1075 \times 2.2^2 = 0.520 \text{ in}^2$$

Parabola:

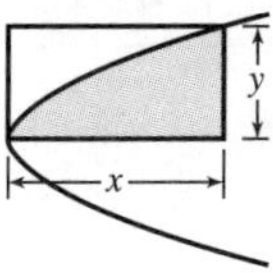

$$\text{Area} = A = \tfrac{2}{3}xy$$

The area of the shaded portion is equal to two-thirds of a rectangle which has x for its base and y for its height.

Example: Let x in the illustration be 15 centimeters, and y be 9 centimeters. Find the area of the shaded portion of the parabola.

$$A = \tfrac{2}{3}xy = \tfrac{2}{3} \times 15 \times 9 = 10 \times 9 = 90 \text{ cm}^2$$

Parabola:

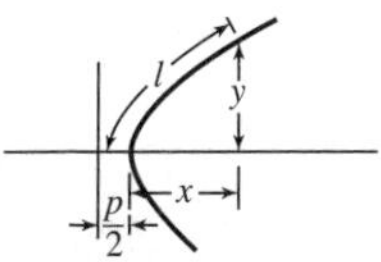

$$l = \text{length of arc} = \frac{p}{2}\left[\sqrt{\frac{2x}{p}}\left(1 + \frac{2x}{p}\right) + \ln\left(\sqrt{\frac{2x}{p}} + \sqrt{1 + \frac{2x}{p}}\right)\right]$$

When x is small in proportion to y, the following is a close approximation:

$$l = y\left[1 + \frac{2}{3}\left(\frac{x}{y}\right)^2 - \frac{2}{5}\left(\frac{x}{y}\right)^4\right] \quad \text{or} \quad l = \sqrt{y^2 + \frac{4}{3}x^2}$$

Example: If $x = 2$ feet and $y = 24$ feet, what is the approximate length l of the parabolic curve?

$$l = y\left[1 + \frac{2}{3}\left(\frac{x}{y}\right)^2 - \frac{2}{5}\left(\frac{x}{y}\right)^4\right] = 24\left[1 + \frac{2}{3}\left(\frac{2}{24}\right)^2 - \frac{2}{5}\left(\frac{2}{24}\right)^4\right]$$

$$= 24 \times 1.0046 = 24.04 \text{ ft}$$

Hyperbola:

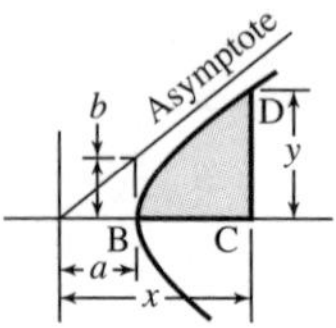

$$\text{Area } BCD = A = \frac{xy}{2} - \frac{ab}{2}\ln\left(\frac{x}{a} + \frac{y}{b}\right)$$

Example: The half-axes a and b are 3 and 2 inches, respectively. Find the area shown shaded in the illustration for $x = 8$ inches and $y = 5$ inches.

Inserting the known values in the formula:

$$\text{Area} = A = \frac{8 \times 5}{2} - \frac{3 \times 2}{2} \times \ln\left(\frac{8}{3} + \frac{5}{2}\right) = 20 - 3 \times \ln(5.167)$$

$$= 20 - 3 \times 1.6423 = 20 - 4.927 = 15.073 \text{ in}^2$$

Formulas and Table for Regular Polygons.—The following formulas and table can be used to calculate the area, length of side, and radii of the inscribed and circumscribed circles of regular polygons (equal sided).

$$A = NS^2 \cot\alpha \div 4 = NR^2 \sin\alpha \cos\alpha = Nr^2 \tan\alpha$$

$$r = R\cos\alpha = (S\cot\alpha) \div 2 = \sqrt{(A\cot\alpha)/N}$$

$$R = S \div (2\sin\alpha) = r \div \cos\alpha = \sqrt{A/(N\sin\alpha\cos\alpha)}$$

$$S = 2R\sin\alpha = 2r\tan\alpha = 2\sqrt{(A\tan\alpha)/N}$$

where N = number of sides
S = length of side
R = radius of circumscribed circle
r = radius of inscribed circle
A = area of polygon
α = 180° ÷ N = one-half center angle of one side

Area, Length of Side, and Inscribed and Circumscribed Radii of Regular Polygons

No. of Sides	$\frac{A}{S^2}$	$\frac{A}{R^2}$	$\frac{A}{r^2}$	$\frac{R}{S}$	$\frac{R}{r}$	$\frac{S}{R}$	$\frac{S}{r}$	$\frac{r}{R}$	$\frac{r}{S}$
3	0.4330	1.2990	5.1962	0.5774	2.0000	1.7321	3.4641	0.5000	0.2887
4	1.0000	2.0000	4.0000	0.7071	1.4142	1.4142	2.0000	0.7071	0.5000
5	1.7205	2.3776	3.6327	0.8507	1.2361	1.1756	1.4531	0.8090	0.6882
6	2.5981	2.5981	3.4641	1.0000	1.1547	1.0000	1.1547	0.8660	0.8660
7	3.6339	2.7364	3.3710	1.1524	1.1099	0.8678	0.9631	0.9010	1.0383
8	4.8284	2.8284	3.3137	1.3066	1.0824	0.7654	0.8284	0.9239	1.2071
9	6.1818	2.8925	3.2757	1.4619	1.0642	0.6840	0.7279	0.9397	1.3737
10	7.6942	2.9389	3.2492	1.6180	1.0515	0.6180	0.6498	0.9511	1.5388
12	11.196	3.0000	3.2154	1.9319	1.0353	0.5176	0.5359	0.9659	1.8660
16	20.109	3.0615	3.1826	2.5629	1.0196	0.3902	0.3978	0.9808	2.5137
20	31.569	3.0902	3.1677	3.1962	1.0125	0.3129	0.3168	0.9877	3.1569
24	45.575	3.1058	3.1597	3.8306	1.0086	0.2611	0.2633	0.9914	3.7979
32	81.225	3.1214	3.1517	5.1011	1.0048	0.1960	0.1970	0.9952	5.0766
48	183.08	3.1326	3.1461	7.6449	1.0021	0.1308	0.1311	0.9979	7.6285
64	325.69	3.1365	3.1441	10.190	1.0012	0.0981	0.0983	0.9988	10.178

Segments of Circles for Radius = 1 (US Customary or Metric Units)

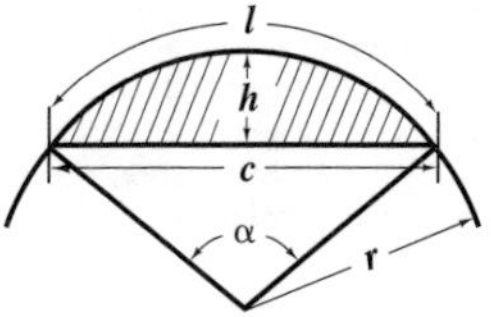

Formulas for segments of circles are given on page 4. When the central angle α and radius *r* are known, the tables on these pages can be used to find the length of arc *l*, height of segment *h*, chord length *c*, and segment area *A*. When angle α and radius *r* are not known, but segment height *h* and chord length *c* are known or can be measured, the ratio *h/c* can be used to enter the table and find α, *l*, and *A* by linear interpolation. Radius *r* is found by the formula on page 4. The value of *l* is then multiplied by the radius *r* and the area *A* by r^2, the square of the radius.

Angle α can be found thus with an accuracy of about 0.001 degree; arc length *l* with an error of about 0.02 percent; and area *A* with an error ranging from about 0.02 percent for the highest entry value of *h/c* to about 1 percent for values of *h/c* of about 0.050. For lower values of *h/c*, and where greater accuracy is required, area *A* should be found by the formula on page 4.

θ, Deg.	*l*	*h*	*c*	*Area A*	*h/c*	θ, Deg.	*l*	*h*	*c*	*Area A*	*h/c*
1	0.01745	0.00004	0.01745	0.0000	0.00218	41	0.71558	0.06333	0.70041	0.0298	0.09041
2	0.03491	0.00015	0.03490	0.0000	0.00436	42	0.73304	0.06642	0.71674	0.0320	0.09267
3	0.05236	0.00034	0.05235	0.0000	0.00655	43	0.75049	0.06958	0.73300	0.0342	0.09493
4	0.06981	0.00061	0.06980	0.0000	0.00873	44	0.76794	0.07282	0.74921	0.0366	0.09719
5	0.08727	0.00095	0.08724	0.0001	0.01091	45	0.78540	0.07612	0.76537	0.0391	0.09946
6	0.10472	0.00137	0.10467	0.0001	0.01309	46	0.80285	0.07950	0.78146	0.0418	0.10173
7	0.12217	0.00187	0.12210	0.0002	0.01528	47	0.82030	0.08294	0.79750	0.0445	0.10400
8	0.13963	0.00244	0.13951	0.0002	0.01746	48	0.83776	0.08645	0.81347	0.0473	0.10628
9	0.15708	0.00308	0.15692	0.0003	0.01965	49	0.85521	0.09004	0.82939	0.0503	0.10856
10	0.17453	0.00381	0.17431	0.0004	0.02183	50	0.87266	0.09369	0.84524	0.0533	0.11085
11	0.19199	0.00460	0.19169	0.0006	0.02402	51	0.89012	0.09741	0.86102	0.0565	0.11314
12	0.20944	0.00548	0.20906	0.0008	0.02620	52	0.90757	0.10121	0.87674	0.0598	0.11543
13	0.22689	0.00643	0.22641	0.0010	0.02839	53	0.92502	0.10507	0.89240	0.0632	0.11773
14	0.24435	0.00745	0.24374	0.0012	0.03058	54	0.94248	0.10899	0.90798	0.0667	0.12004
15	0.26180	0.00856	0.26105	0.0015	0.03277	55	0.95993	0.11299	0.92350	0.0704	0.12235
16	0.27925	0.00973	0.27835	0.0018	0.03496	56	0.97738	0.11705	0.93894	0.0742	0.12466
17	0.29671	0.01098	0.29562	0.0022	0.03716	57	0.99484	0.12118	0.95432	0.0781	0.12698
18	0.31416	0.01231	0.31287	0.0026	0.03935	58	1.01229	0.12538	0.96962	0.0821	0.12931
19	0.33161	0.01371	0.33010	0.0030	0.04155	59	1.02974	0.12964	0.98485	0.0863	0.13164
20	0.34907	0.01519	0.34730	0.0035	0.04374	60	1.04720	0.13397	1.00000	0.0906	0.13397
21	0.36652	0.01675	0.36447	0.0041	0.04594	61	1.06465	0.13837	1.01508	0.0950	0.13632
22	0.38397	0.01837	0.38162	0.0047	0.04814	62	1.08210	0.14283	1.03008	0.0996	0.13866
23	0.40143	0.02008	0.39874	0.0053	0.05035	63	1.09956	0.14736	1.04500	0.1043	0.14101
24	0.41888	0.02185	0.41582	0.0061	0.05255	64	1.11701	0.15195	1.05984	0.1091	0.14337
25	0.43633	0.02370	0.43288	0.0069	0.05476	65	1.13446	0.15661	1.07460	0.1141	0.14574
26	0.45379	0.02563	0.44990	0.0077	0.05697	66	1.15192	0.16133	1.08928	0.1192	0.14811
27	0.47124	0.02763	0.46689	0.0086	0.05918	67	1.16937	0.16611	1.10387	0.1244	0.15048
28	0.48869	0.02970	0.48384	0.0096	0.06139	68	1.18682	0.17096	1.11839	0.1298	0.15287
29	0.50615	0.03185	0.50076	0.0107	0.06361	69	1.20428	0.17587	1.13281	0.1353	0.15525
30	0.52360	0.03407	0.51764	0.0118	0.06583	70	1.22173	0.18085	1.14715	0.1410	0.15765
31	0.54105	0.03637	0.53448	0.0130	0.06805	71	1.23918	0.18588	1.16141	0.1468	0.16005
32	0.55851	0.03874	0.55127	0.0143	0.07027	72	1.25664	0.19098	1.17557	0.1528	0.16246
33	0.57596	0.04118	0.56803	0.0157	0.07250	73	1.27409	0.19614	1.18965	0.1589	0.16488
34	0.59341	0.04370	0.58474	0.0171	0.07473	74	1.29154	0.20136	1.20363	0.1651	0.16730
35	0.61087	0.04628	0.60141	0.0186	0.07696	75	1.30900	0.20665	1.21752	0.1715	0.16973
36	0.62832	0.04894	0.61803	0.0203	0.07919	76	1.32645	0.21199	1.23132	0.1781	0.17216
37	0.64577	0.05168	0.63461	0.0220	0.08143	77	1.34390	0.21739	1.24503	0.1848	0.17461
38	0.66323	0.05448	0.65114	0.0238	0.08367	78	1.36136	0.22285	1.25864	0.1916	0.17706
39	0.68068	0.05736	0.66761	0.0257	0.08592	79	1.37881	0.22838	1.27216	0.1986	0.17952
40	0.69813	0.06031	0.68404	0.0277	0.08816	80	1.39626	0.23396	1.28558	0.2057	0.18199

Segments of Circles for Radius = 1 (US Customary or Metric Units) *(Continued)*

θ, Deg.	*l*	*h*	*c*	*Area A*	*h/c*	θ, Deg.	*l*	*h*	*c*	*Area A*	*h/c*
81	1.41372	0.23959	1.29890	0.2130	0.18446	131	2.28638	0.58531	1.81992	0.7658	0.32161
82	1.43117	0.24529	1.31212	0.2205	0.18694	132	2.30383	0.59326	1.82709	0.7803	0.32470
83	1.44862	0.25104	1.32524	0.2280	0.18943	133	2.32129	0.60125	1.83412	0.7950	0.32781
84	1.46608	0.25686	1.33826	0.2358	0.19193	134	2.33874	0.60927	1.84101	0.8097	0.33094
85	1.48353	0.26272	1.35118	0.2437	0.19444	135	2.35619	0.61732	1.84776	0.8245	0.33409
86	1.50098	0.26865	1.36400	0.2517	0.19696	136	2.37365	0.62539	1.85437	0.8395	0.33725
87	1.51844	0.27463	1.37671	0.2599	0.19948	137	2.39110	0.63350	1.86084	0.8546	0.34044
88	1.53589	0.28066	1.38932	0.2682	0.20201	138	2.40855	0.64163	1.86716	0.8697	0.34364
89	1.55334	0.28675	1.40182	0.2767	0.20456	139	2.42601	0.64979	1.87334	0.8850	0.34686
90	1.57080	0.29289	1.41421	0.2854	0.20711	140	2.44346	0.65798	1.87939	0.9003	0.35010
91	1.58825	0.29909	1.42650	0.2942	0.20967	141	2.46091	0.66619	1.88528	0.9158	0.35337
92	1.60570	0.30534	1.43868	0.3032	0.21224	142	2.47837	0.67443	1.89104	0.9314	0.35665
93	1.62316	0.31165	1.45075	0.3123	0.21482	143	2.49582	0.68270	1.89665	0.9470	0.35995
94	1.64061	0.31800	1.46271	0.3215	0.21741	144	2.51327	0.69098	1.90211	0.9627	0.36327
95	1.65806	0.32441	1.47455	0.3309	0.22001	145	2.53073	0.69929	1.90743	0.9786	0.36662
96	1.67552	0.33087	1.48629	0.3405	0.22261	146	2.54818	0.70763	1.91261	0.9945	0.36998
97	1.69297	0.33738	1.49791	0.3502	0.22523	147	2.56563	0.71598	1.91764	1.0105	0.37337
98	1.71042	0.34394	1.50942	0.3601	0.22786	148	2.58309	0.72436	1.92252	1.0266	0.37678
99	1.72788	0.35055	1.52081	0.3701	0.23050	149	2.60054	0.73276	1.92726	1.0428	0.38021
100	1.74533	0.35721	1.53209	0.3803	0.23315	150	2.61799	0.74118	1.93185	1.0590	0.38366
101	1.76278	0.36392	1.54325	0.3906	0.23582	151	2.63545	0.74962	1.93630	1.0753	0.38714
102	1.78024	0.37068	1.55429	0.4010	0.23849	152	2.65290	0.75808	1.94059	1.0917	0.39064
103	1.79769	0.37749	1.56522	0.4117	0.24117	153	2.67035	0.76655	1.94474	1.1082	0.39417
104	1.81514	0.38434	1.57602	0.4224	0.24387	154	2.68781	0.77505	1.94874	1.1247	0.39772
105	1.83260	0.39124	1.58671	0.4333	0.24657	155	2.70526	0.78356	1.95259	1.1413	0.40129
106	1.85005	0.39818	1.59727	0.4444	0.24929	156	2.72271	0.79209	1.95630	1.1580	0.40489
107	1.86750	0.40518	1.60771	0.4556	0.25202	157	2.74017	0.80063	1.95985	1.1747	0.40852
108	1.88496	0.41221	1.61803	0.4669	0.25476	158	2.75762	0.80919	1.96325	1.1915	0.41217
109	1.90241	0.41930	1.62823	0.4784	0.25752	159	2.77507	0.81776	1.96651	1.2084	0.41585
110	1.91986	0.42642	1.63830	0.4901	0.26028	160	2.79253	0.82635	1.96962	1.2253	0.41955
111	1.93732	0.43359	1.64825	0.5019	0.26306	161	2.80998	0.83495	1.97257	1.2422	0.42328
112	1.95477	0.44081	1.65808	0.5138	0.26585	162	2.82743	0.84357	1.97538	1.2592	0.42704
113	1.97222	0.44806	1.66777	0.5259	0.26866	163	2.84489	0.85219	1.97803	1.2763	0.43083
114	1.98968	0.45536	1.67734	0.5381	0.27148	164	2.86234	0.86083	1.98054	1.2934	0.43464
115	2.00713	0.46270	1.68678	0.5504	0.27431	165	2.87979	0.86947	1.98289	1.3105	0.43849
116	2.02458	0.47008	1.69610	0.5629	0.27715	166	2.89725	0.87813	1.98509	1.3277	0.44236
117	2.04204	0.47750	1.70528	0.5755	0.28001	167	2.91470	0.88680	1.98714	1.3449	0.44627
118	2.05949	0.48496	1.71433	0.5883	0.28289	168	2.93215	0.89547	1.98904	1.3621	0.45020
119	2.07694	0.49246	1.72326	0.6012	0.28577	169	2.94961	0.90415	1.99079	1.3794	0.45417
120	2.09440	0.50000	1.73205	0.6142	0.28868	170	2.96706	0.91284	1.99239	1.3967	0.45817
121	2.11185	0.50758	1.74071	0.6273	0.29159	171	2.98451	0.92154	1.99383	1.4140	0.46220
122	2.12930	0.51519	1.74924	0.6406	0.29452	172	3.00197	0.93024	1.99513	1.4314	0.46626
123	2.14675	0.52284	1.75763	0.6540	0.29747	173	3.01942	0.93895	1.99627	1.4488	0.47035
124	2.16421	0.53053	1.76590	0.6676	0.30043	174	3.03687	0.94766	1.99726	1.4662	0.47448
125	2.18166	0.53825	1.77402	0.6813	0.30341	175	3.05433	0.95638	1.99810	1.4836	0.47865
126	2.19911	0.54601	1.78201	0.6950	0.30640	176	3.07178	0.96510	1.99878	1.5010	0.48284
127	2.21657	0.55380	1.78987	0.7090	0.30941	177	3.08923	0.97382	1.99931	1.5184	0.48708
128	2.23402	0.56163	1.79759	0.7230	0.31243	178	3.10669	0.98255	1.99970	1.5359	0.49135
129	2.25147	0.56949	1.80517	0.7372	0.31548	179	3.12414	0.99127	1.99992	1.5533	0.49566
130	2.26893	0.57738	1.81262	0.7514	0.31854	180	3.14159	1.00000	2.00000	1.5708	0.50000

Diameters of Circles and Sides of Squares of Equal Area
(US Customary or Metric Units)

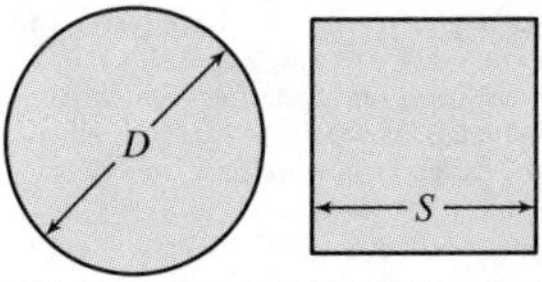

The table below will be found useful for determining the diameter of a circle of an area equal to that of a square, the side of which is known, or for determining the side of a square which has an area equal to that of a circle, the area or diameter of which is known. For example, if the diameter of a circle is 17½ inches, it is found from the table by reading across from the first column that the side of a square of the same area is 15.51 inches. And both have area 240.53 in^2.

Dia. of Circle, *D*	Side of Square, *S*	Area of Circle or Square	Dia. of Circle, *D*	Side of Square, *S*	Area of Circle or Square	Dia. of Circle, *D*	Side of Square, *S*	Area of Circle or Square
½	0.44	0.196	20½	18.17	330.06	40½	35.89	1288.25
1	0.89	0.785	21	18.61	346.36	41	36.34	1320.25
1½	1.33	1.767	21½	19.05	363.05	41½	36.78	1352.65
2	1.77	3.142	22	19.50	380.13	42	37.22	1385.44
2½	2.22	4.909	22½	19.94	397.61	42½	37.66	1418.63
3	2.66	7.069	23	20.38	415.48	43	38.11	1452.20
3½	3.10	9.621	23½	20.83	433.74	43½	38.55	1486.17
4	3.54	12.566	24	21.27	452.39	44	38.99	1520.53
4½	3.99	15.904	24½	21.71	471.44	44½	39.44	1555.28
5	4.43	19.635	25	22.16	490.87	45	39.88	1590.43
5½	4.87	23.758	25½	22.60	510.71	45½	40.32	1625.97
6	5.32	28.274	26	23.04	530.93	46	40.77	1661.90
6½	5.76	33.183	26½	23.49	551.55	46½	41.21	1698.23
7	6.20	38.485	27	23.93	572.56	47	41.65	1734.94
7½	6.65	44.179	27½	24.37	593.96	47½	42.10	1772.05
8	7.09	50.265	28	24.81	615.75	48	42.54	1809.56
8½	7.53	56.745	28½	25.26	637.94	48½	42.98	1847.45
9	7.98	63.617	29	25.70	660.52	49	43.43	1885.74
9½	8.42	70.882	29½	26.14	683.49	49½	43.87	1924.42
10	8.86	78.540	30	26.59	706.86	50	44.31	1963.50
10½	9.31	86.590	30½	27.03	730.62	50½	44.75	2002.96
11	9.75	95.033	31	27.47	754.77	51	45.20	2042.82
11½	10.19	103.87	31½	27.92	779.31	51½	45.64	2083.07
12	10.63	113.10	32	28.36	804.25	52	46.08	2123.72
12½	11.08	122.72	32½	28.80	829.58	52½	46.53	2164.75
13	11.52	132.73	33	29.25	855.30	53	46.97	2206.18
13½	11.96	143.14	33½	29.69	881.41	53½	47.41	2248.01
14	12.41	153.94	34	30.13	907.92	54	47.86	2290.22
14½	12.85	165.13	34½	30.57	934.82	54½	48.30	2332.83
15	13.29	176.71	35	31.02	962.11	55	48.74	2375.83
15½	13.74	188.69	35½	31.46	989.80	55½	49.19	2419.22
16	14.18	201.06	36	31.90	1017.88	56	49.63	2463.01
16½	14.62	213.82	36½	32.35	1046.35	56½	50.07	2507.19
17	15.07	226.98	37	32.79	1075.21	57	50.51	2551.76
17½	15.51	240.53	37½	33.23	1104.47	57½	50.96	2596.72
18	15.95	254.47	38	33.68	1134.11	58	51.40	2642.08
18½	16.40	268.80	38½	34.12	1164.16	58½	51.84	2687.83
19	16.84	283.53	39	34.56	1194.59	59	52.29	2733.97
19½	17.28	298.65	39½	35.01	1225.42	59½	52.73	2780.51
20	17.72	314.16	40	35.45	1256.64	60	53.17	2827.43

Propositions of Geometry

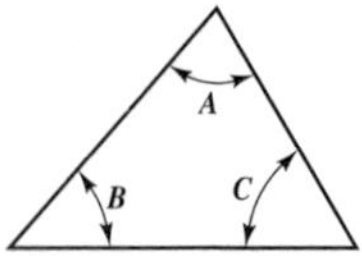

A *triangle* is a three-sided polygon. It is, in fact, the polygon with the least number of sides. The sides of a triangle meet at its *vertices* (singular *vertex*). The sum of the measures of all three angles of a triangle is 180 degrees. Hence, if the measures of any two angles are known, the third angle measure can always be found.

$$A + B + C = 180° \qquad A = 180° - (B + C)$$

$$B = 180° - (A + C) \qquad C = 180° - (A + B)$$

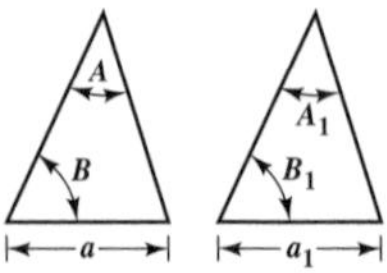

AAS Proposition: If two angles and the non-included side of one triangle are congruent to the corresponding (similarly located) angles and sides of another triangle, the triangles are congruent.

Hence, if $a = a_1$, $A = A_1$, and $B = B_1$, the other corresponding side and angle are equal in measure, and thus the triangles are congruent.

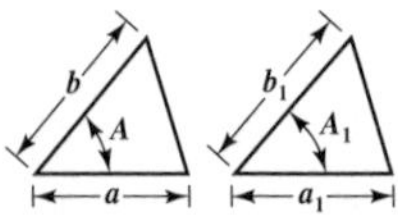

SAS Proposition: If two sides and the included angle (the angle between the sides) of one triangle are congruent (equal in measure) to the corresponding (similarly located) sides and angle of another triangle, then the triangles are congruent.

Hence, in the figure, if $a = a_1$, $b = b_1$, and $A = A_1$, then the remaining side and angles also are equal in measure, and thus the triangles are congruent.

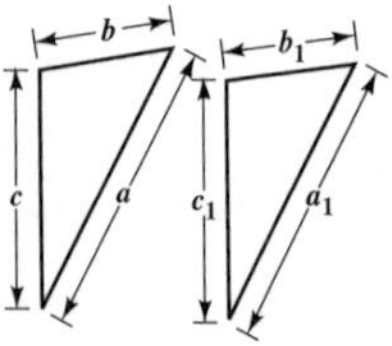

SSS Proposition: If all three sides of one triangle are congruent (equal in measure) to all three sides of another triangle, then the triangles are congruent. If the three sides in one triangle are equal in measure to the three sides of another triangle, then the angles in the two triangles are equal in measure.

If $a = a_1$, $b = b_1$, and $c = c_1$, then the corresponding angles are also equal in measure, and thus the triangles are congruent.

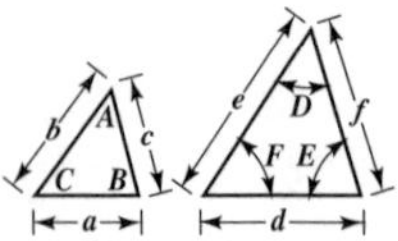

If the three sides of a triangle are proportional to corresponding sides of another triangle, then the triangles are *similar*, and the angles in the one are congruent (equal in measure) to the angles in the other.

Hence, if $a/d = b/e = c/f$ then $A = D$, $B = E$, $C = F$

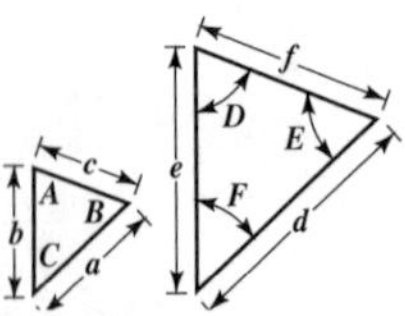

Similar triangles are ones whose corresponding angles are congruent. If this is true then the corresponding sides are proportional. If the angles of one triangle are congruent (equal in measure) to the angles of another triangle, then the triangles are similar and their corresponding sides are proportional.

Hence, if $A = D$, $B = E$, and $C = F$ then $a/d = b/e = c/f$

Propositions of Geometry *(Continued)*

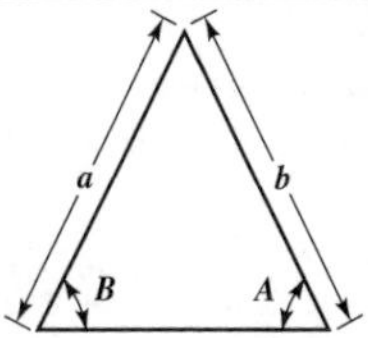	In an *isosceles triangle*, two sides a, b are congruent, hence the two angles opposite them (the base angles) are congruent. An equilateral (equiangular) triangle is also an isosceles triangle.
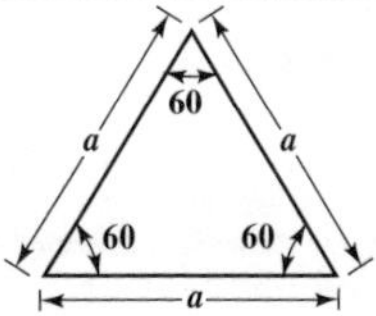	In an *equilateral (equiangular) triangle*, all three sides (angles) are congruent. Since the sum of angle measures in any triangle is 180 degrees, then each angle in an equilateral (equiangular) triangle measures 60 degrees.
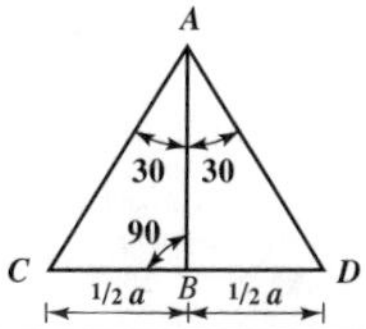	A line in an equilateral triangle that bisects any of the angles (that is, divides it into two 30-degree angles), also bisects the side opposite the bisected angle and is perpendicular (at right angles) to it. Thus, if line *AB* bisects angle *CAD*, it also bisects line *CD* into two equal parts and is perpendicular to it.
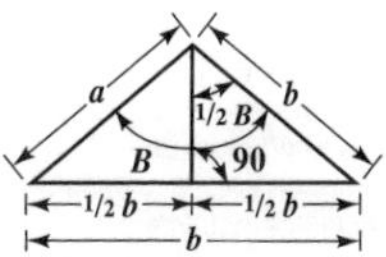	If a line in an isosceles triangle drawn from the vertex where the two congruent sides meet in such a way that is bisects the third side (or base), then it also bisects the angle at the vertex from which it is drawn.
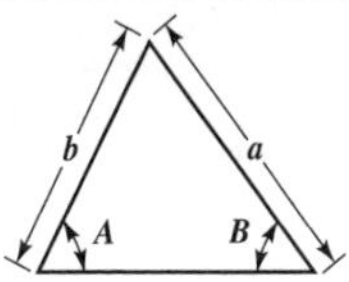	In every triangle, the greatest angle is opposite the longest side. And, the longest side is opposite the greatest angle. Thus, if a is longer than b, then the measure of angle A is greater than that of angle B. And, if angle A measure is greater than B, then side a is longer than side b.
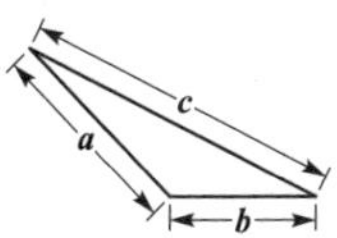	According to the *triangle inequality*, for any triangle, the sum of the lengths of any two sides must be greater than or equal to the length of the remaining side. Thus, $a + b \geq c$.
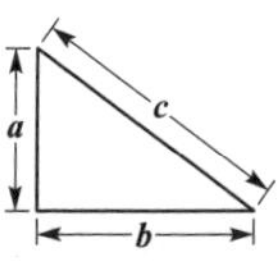	The *Pythagorean theorem* states that in a right-angle triangle, the square of the *hypotenuse*, that is, the side opposite the right angle, is equal to the sum of the squares on the two sides that form the right angle. These sides are the *legs* of the right triangle. Thus, $a^2 + b^2 = c^2$

Propositions of Geometry *(Continued)*

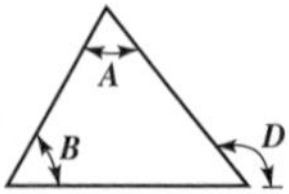

In the figure, angle D is an *exterior angle*. The measure of an exterior angle is equal to the sum of the measures of the non-adjacent interior angles. Thus,

$$A + B = D$$

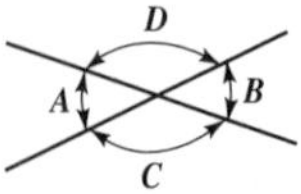

Intersecting lines form congruent *vertical angles*, that is, the angles opposite one another. In the figure, A and B are vertical angles and therefore congruent, and C and D are vertical angles and therefore congruent. That is,

$$A \cong B$$
$$C \cong D$$

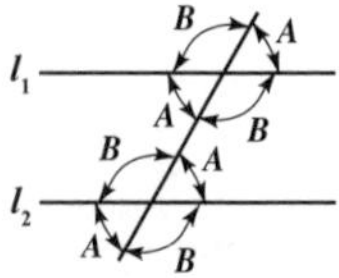

Corresponding angles formed when parallel lines are intersected by another line (a *transversal*) have equal measure (are congruent).

Lines l_1 and l_2 are intersected by s transversal, so all angles A are congruent, and all angles B are congruent.

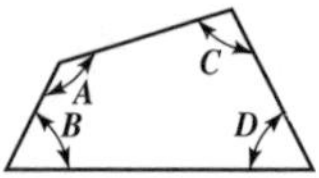

A quadrilateral is a four-sided polygon. In any quadrilateral, the sum of the measures of the interior angles is 360 degrees.

Thus, $A + B + C + D = 360$ degrees

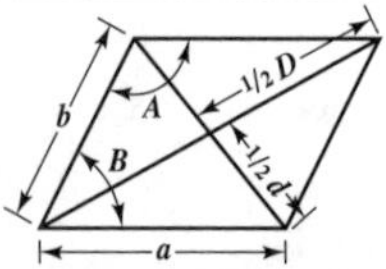

A parallelogram is a quadrilateral in which each pair of opposite sides is parallel and thus congruent. Opposite angles are also congruent. Each diagonal divides the parallelogram into congruent triangles, and the diagonals bisect each other. In the figure, diagonal D bisects diagonal d at their midpoints.

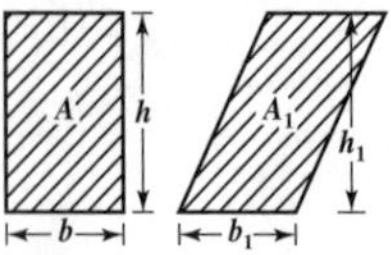

A rectangle is a parallelogram in which all four angles are right angles. In the figure, the rectangle's *base* is marked b; its *height* is marked h; the parallelogram's base is marked b_1, and its height is marked h_1. The area of a parallelogram is $A = bh$. If two parallelograms have a corresponding base and height of equal measure, their areas are equal.

Thus, if $b = b_1$ and h and h_1, then area A = area A_1.

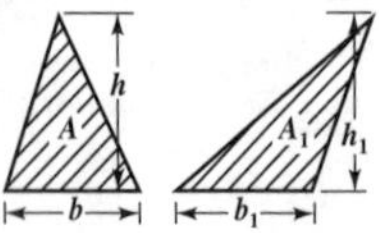

Triangles having equal base and equal height have equal area.

Thus, if $b = b_1$ and h and h_1, then area A = area A_1.

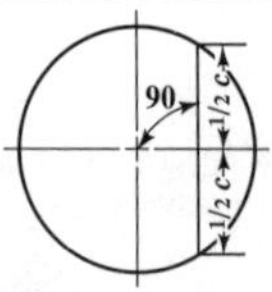

If a diameter of a circle is perpendicular (at right angles) to a chord, then it bisects the chord, dividing it into two equal parts. In the figure, the chord has length c, so its parts have length $c/_2$.

Propositions of Geometry *(Continued)*

<table>
<tr><td>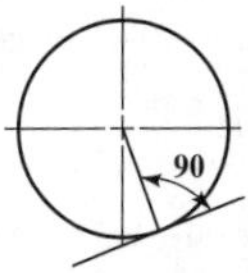</td><td>A line tangent to a circle lies perpendicular (at right angles) to a radius drawn to meet it at the point of tangency.</td></tr>
<tr><td>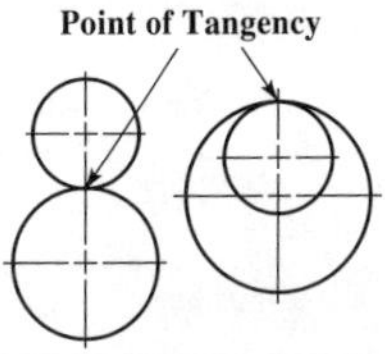</td><td>The figure shows two ways that circles can be tangent. A line drawn through the center of each circle (the diameter) will pass through the point of tangency.</td></tr>
<tr><td>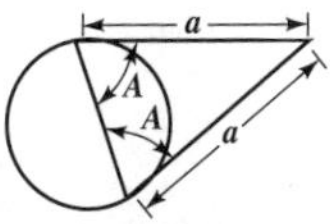</td><td>Two tangents drawn from a single point outside a circle will be equal in length (a), and the angles they make with the chord that connects the points of tangency (A) will be equal in measure. The figure shows this congruency.</td></tr>
<tr><td>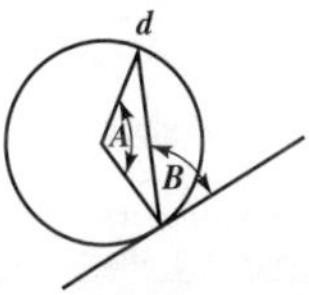</td><td>The angle formed by a tangent and a chord drawn from the point of tangency measures one-half the central angle subtended by that chord.

That is, $B = A/2$</td></tr>
<tr><td>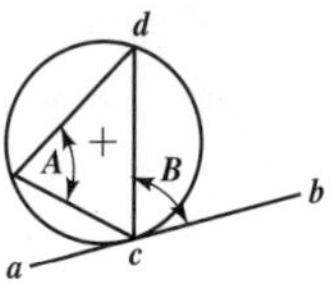</td><td>The angle formed by a tangent and a chord drawn from the point of tangency has measure equal to the inscribed angle subtended by the chord.

That is, $B = A$.</td></tr>
<tr><td>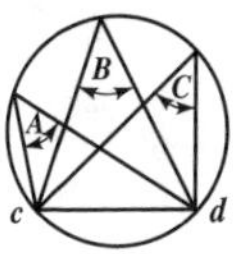</td><td>All inscribed angles subtended by the same chord in a circle are congruent (equal in measure).

In the figure, A, B, and C are subtended by chord cd, and so are equal in measure.</td></tr>
<tr><td></td><td>Referring to the figure, inscribed angle A and central angle B are subtended by the same arc. The measure of inscribed A is half the measure of central angle B. In the figure,

$$A = B/2$$</td></tr>
</table>

Propositions of Geometry *(Continued)*

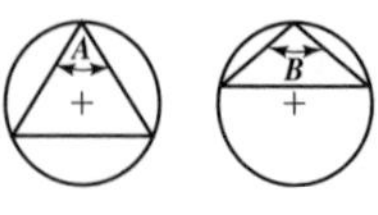	A *major arc* of a circle is one that measures more than half the circumference of the circle. A *minor arc* measures less than half the circumference of the circle. In the figure, angle *A* is subtended by a minor arc, so it is an acute angle (it measures less than 90 degrees). Angle *B* is subtended by a major arc, so it is an obtuse angle (it measures more than 90 degrees).
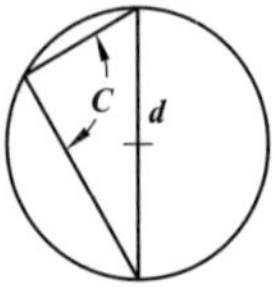	An angle subtended by a circle's diameter is a right angle; the arc described by the diameter is a semicircle. Referring to the figure, right angle *A* is subtended by diameter *d*. Angle $C = 90°$
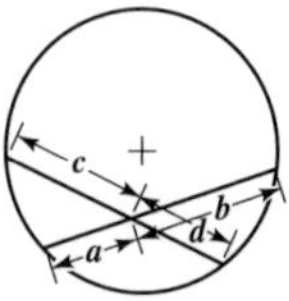	In the figure, the product of line segment lengths formed by intersecting chords in a circle are equal. Thus: $ab = cd$
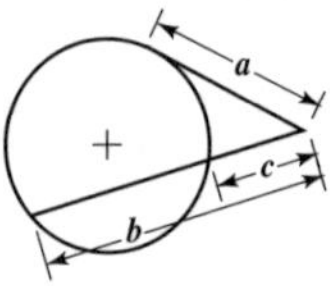	When two lines are drawn from a point outside a circle, one tangent and one through the circle intersecting it at two points, the line segment lengths are such that the square of the tangent segment is equal to the product of the segments formed by the other line. Thus: $a^2 = bc$
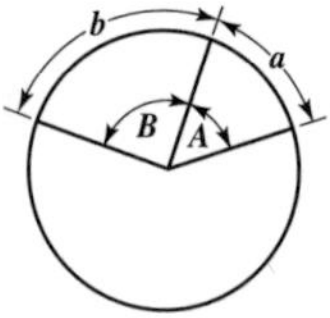	Arc lengths of a circle are proportional to the corresponding central angle measures. Thus: $A{:}B = a{:}b$
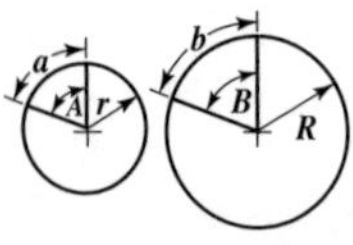	The lengths of circular arcs having the same central angle are proportional to the lengths of the radii. Thus, if $A = B$, then $a/b = r/R$
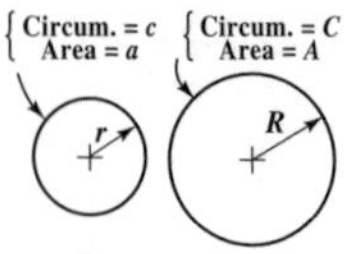	The ratio of the circumferences of two circles is proportional to the ratio of their radii. $c{:}C :: r{:}R$ so $c/C = r/R$ The ratio of the areas of two circles is proportional to the ratio of the squares of their radii. $a{:}A :: r^2{:}R^2$ so $a/A = r^2/R^2$

Useful Trigonometric Relationships

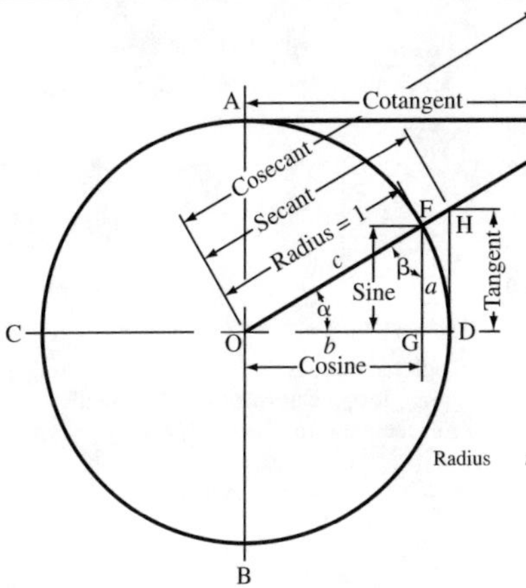

Radius $r = OF$
$= OA$
$= OD$
$= 1$

The following formulas will provide the solutions for most triangular problems in the right-angle triangle FOG, $\overline{OF} = c$, $\overline{FG} = a$, and $\overline{OG} = b$

$$\text{arc FD} = \text{angle } \alpha \cdot \overline{OD}$$

$$\sin\alpha = \overline{FG} \qquad \cos\alpha = \overline{OG} \qquad \tan\alpha = \overline{DH} \qquad \cot\alpha = \overline{AE}$$
$$\csc\alpha = \overline{OE} \qquad \sec\alpha = \overline{OH} \qquad \text{chord } \alpha = \overline{FD}$$

$$\sin\alpha = \frac{a}{c} = \cos\beta \qquad \cos\alpha = \frac{b}{c} = \sin\beta \qquad \tan\alpha = \frac{a}{b} = \cot\beta$$
$$\cot\alpha = \frac{b}{a} = \tan\beta \qquad \sec\alpha = \frac{c}{b} = \csc\beta \qquad \csc\alpha = \frac{c}{a} = \sec\beta$$

$$a = \sin\alpha\, c \qquad b = \cos\alpha\, c \qquad a = \tan\alpha\, b \qquad b = \cot\alpha\, a \qquad c = \sec\alpha\, b \qquad c = \csc\alpha\, a$$

$$c = \frac{a}{\sin\alpha} \qquad c = \frac{b}{\cos\alpha} \qquad b = \frac{a}{\tan\alpha} \qquad a = \frac{b}{\cot\alpha} \qquad b = \frac{c}{\sec\alpha} \qquad a = \frac{c}{\csc\alpha}$$

Signs of Trigonometric Functions

This diagram shows the proper sign (+ or −) for the trigonometric functions of angles in each of the four quadrants of a complete circle.

Examples: The sine of 226° is −0.71934; of 326°, −0.55919.

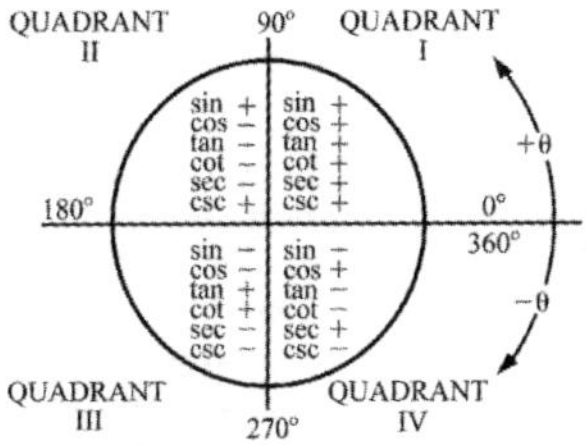

Useful Relationships Among Angles

Angle Function	θ	$-\theta$	$90° \pm \theta$	$180° \pm \theta$	$270° \pm \theta$	$360° \pm \theta$
sine	$\sin\theta$	$-\sin\theta$	$+\cos\theta$	$\mp\sin\theta$	$-\cos\theta$	$\pm\sin\theta$
cosine	$\cos\theta$	$+\cos\theta$	$\mp\sin\theta$	$-\cos\theta$	$\pm\sin\theta$	$+\cos\theta$
tangent	$\tan\theta$	$-\tan\theta$	$\mp\cot\theta$	$\pm\tan\theta$	$\mp\cot\theta$	$\pm\tan\theta$
cotangent	$\cot\theta$	$-\cot\theta$	$\mp\tan\theta$	$\pm\cot\theta$	$\mp\tan\theta$	$\pm\cot\theta$
secant	$\sec\theta$	$+\sec\theta$	$\mp\csc\theta$	$-\sec\theta$	$\pm\csc\theta$	$+\sec\theta$
cosecant	$\csc\theta$	$-\csc\theta$	$+\sec\theta$	$\mp\csc\theta$	$-\sec\theta$	$\pm\csc\theta$

Examples: $\cos(270° - \theta) = -\sin\theta$; $\tan(90° + \theta) = -\cot\theta$.

The Law of Sines.—In any triangle, any side is to the sine of the angle opposite that side as any other side is to the sine of the angle opposite that side. If a, b, and c are the sides, and A, B, and C their opposite angles, respectively, then:

$$\frac{a}{\sin A} = \frac{b}{\sin B} = \frac{c}{\sin C}, \quad \text{so that:}$$

$$a = \frac{b \sin A}{\sin B} \quad \text{or} \quad a = \frac{c \sin A}{\sin C}$$

$$b = \frac{a \sin B}{\sin A} \quad \text{or} \quad b = \frac{c \sin B}{\sin C}$$

$$c = \frac{a \sin C}{\sin A} \quad \text{or} \quad c = \frac{b \sin C}{\sin B}$$

The Law of Cosines.—In any triangle, the square of any side is equal to the sum of the squares of the other two sides minus twice their product times the cosine of the included angle; or if a, b, and c are the sides and A, B, and C are the opposite angles, respectively, then:

$$a^2 = b^2 + c^2 - 2bc\cos A$$

$$b^2 = a^2 + c^2 - 2ac\cos B$$

$$c^2 = a^2 + b^2 - 2ab\cos C$$

These two laws, together with the proposition that the sum of the three angles equals 180 degrees, are the basis of all formulas relating to the solution of triangles.

Formulas for the solution of right-angled and oblique-angled triangles, arranged in tabular form, are given on the following pages.

Trigonometric Functions and Identities.—On page 15, a diagram, *Signs of Trigonometric Functions*, is given. This diagram shows the proper sign (+ or –) for the trigonometric functions of angles in each of the four quadrants, 0 to 90, 90 to 180, 180 to 270, and 270 to 360 degrees. Thus, the cosine of an angle between 90 and 180 degrees is negative; the sine of the same angle is positive.

Trigonometric identities are formulas that show the relationship between different trigonometric functions. They may be used to change the form of some trigonometric expressions to simplify calculations. For example, if a formula has a term, $2\sin A\cos A$, the equivalent but simpler term $\sin 2A$ may be substituted. The identities that follow may themselves be combined or rearranged in various ways to form new identities.

Basic

$$\tan A = \frac{\sin A}{\cos A} = \frac{1}{\cot A} \qquad \sec A = \frac{1}{\cos A} \qquad \csc A = \frac{1}{\sin A}$$

Negative Angle

$$\sin(-A) = -\sin A \qquad \cos(-A) = \cos A \qquad \tan(-A) = -\tan A$$

Pythagorean

$$\sin^2 A + \cos^2 A = 1 \qquad 1 + \tan^2 A = \sec^2 A \qquad 1 + \cot^2 A = \csc^2 A$$

Sum and Difference of Angles

$$\tan(A+B) = \frac{\tan A + \tan B}{1 - \tan A \tan B} \qquad \tan(A-B) = \frac{\tan A - \tan B}{1 + \tan A \tan B}$$

$$\cot(A+B) = \frac{\cot A \cot B - 1}{\cot B + \cot A} \qquad \cot(A-B) = \frac{\cot A \cot B + 1}{\cot B - \cot A}$$

$$\sin(A+B) = \sin A \cos B + \cos A \sin B \qquad \sin(A-B) = \sin A \cos B - \cos A \sin B$$

$$\cos(A+B) = \cos A \cos B - \sin A \sin B \qquad \cos(A-B) = \cos A \cos B + \sin A \sin B$$

Double-Angle

$$\cos 2A = \cos^2 A - \sin^2 A = 2\cos^2 A - 1 = 1 - 2\sin^2 A \qquad \sin 2A = 2\sin A \cos A$$

$$\tan 2A = \frac{2\tan A}{1 - \tan^2 A} = \frac{2}{\cot A - \tan A}$$

Half-Angle

$$\sin \tfrac{1}{2}A = \sqrt{\tfrac{1}{2}(1 - \cos A)} \qquad \cos \tfrac{1}{2}A = \sqrt{\tfrac{1}{2}(1 + \cos A)}$$

$$\tan \tfrac{1}{2}A = \sqrt{\frac{1 - \cos A}{1 + \cos A}} = \frac{1 - \cos A}{\sin A} = \frac{\sin A}{1 + \cos A}$$

Product-to-Sum

$$\sin A \cos B = \tfrac{1}{2}[\sin(A+B) + \sin(A-B)]$$

$$\cos A \cos B = \tfrac{1}{2}[\cos(A+B) + \cos(A-B)]$$

$$\sin A \sin B = \tfrac{1}{2}[\cos(A-B) - \cos(A+B)] \qquad \tan A \tan B = \frac{\tan A + \tan B}{\cot A + \cot B}$$

Sum and Difference of Functions

$$\sin A + \sin B = 2[\sin \tfrac{1}{2}(A+B) \cos \tfrac{1}{2}(A-B)]$$

$$\sin A - \sin B = 2[\sin \tfrac{1}{2}(A-B) \cos \tfrac{1}{2}(A+B)]$$

$$\cos A + \cos B = 2[\cos \tfrac{1}{2}(A+B) \cos \tfrac{1}{2}(A-B)]$$

$$\cos A - \cos B = -2[\sin \tfrac{1}{2}(A+B) \sin \tfrac{1}{2}(A-B)]$$

$$\tan A + \tan B = \frac{\sin(A+B)}{\cos A \cos B} \qquad \tan A - \tan B = \frac{\sin(A-B)}{\cos A \cos B}$$

$$\cot A + \cot B = \frac{\sin(B+A)}{\sin A \sin B} \qquad \cot A - \cot B = \frac{\sin(B-A)}{\sin A \sin B}$$

Solution of Right Triangles

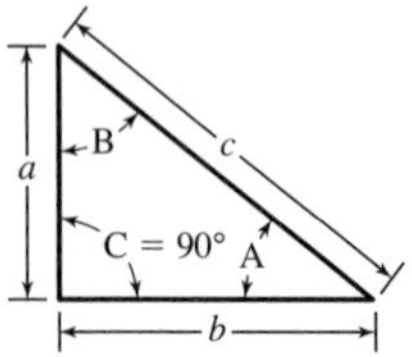

The figure to the left shows right triangle *ABC*. Sides opposite corresponding angles are labeled a, b, c. The formulas in the table are for finding an unknown side or angle from given information and for calculating area. There are several ways to solve for the missing dimension using the three basic trigonometric functions.

Right angle C is always known (90°), thus, $A = 90 - B$ and $B = 90 - A$.

Sides and Angles Known	Formulas for Sides and Angles to be Found			Area
Sides a and b	$c = \sqrt{a^2 + b^2}$	$A = \tan^{-1}(a/b)$	$B = 90° - A$	$\frac{a \times b}{2}$
Side a, hypotenuse c	$b = \sqrt{c^2 - a^2}$	$A = \sin^{-1}(a/c)$ or $A = \cos^{-1}(b/c)$	$B = 90° - A$	$\frac{a \times \sqrt{c^2 - a^2}}{2}$
Side b, hypotenuse c	$a = \sqrt{c^2 - b^2}$	$B = \sin^{-1}(b/c)$ or $B = \cos^{-1}(a/c)$	$A = 90° - B$	$\frac{b \times \sqrt{c^2 - b^2}}{2}$
Hypotenuse c, angle B	$b = c \sin B$ or $a = c \cos B$	$a = c \cos B$ or $b = c \sin B$	$A = 90° - B$	$c^2 \times \sin B \times \cos B$
Hypotenuse c, angle A	$b = c \cos A$ or $a = c \sin A$	$a = c \sin A$ or $b = c \cos A$	$B = 90° - A$	$c^2 \times \sin A \times \cos A$
Side b, angle B	$c = \frac{b}{\sin B}$ or $c = \frac{a}{\cos B}$	$a = \frac{b}{\tan B}$	$A = 90° - B$	$\frac{b^2}{2 \times \tan B}$
Side b, angle A	$c = \frac{b}{\cos A}$ or $c = \frac{a}{\sin A}$	$a = b \tan A$	$B = 90° - A$	$\frac{b^2 \times \tan A}{2}$
Side a, angle B	$c = \frac{a}{\cos B}$	$b = a \tan B$	$A = 90° - B$	$\frac{a^2 \times \tan B}{2}$
Side a, angle A	$c = \frac{a}{\sin A}$	$b = \frac{a}{\tan A}$	$B = 90° - A$	$\frac{a^2}{2 \times \tan A}$

Solution and Examples of Oblique Triangles (US Customary or Metric Units)

One Side and Two Excluded (Not Between) Angles Known (Law of Sines):

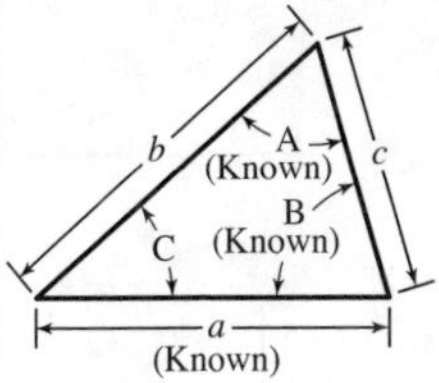

One Side and Two Excluded Angles Known

If side a, angle A opposite it, and angle B, are known:

$$C = 180° - (A + B)$$

$$b = \frac{a \sin B}{\sin A} \qquad c = \frac{a \sin C}{\sin A}$$

$$\text{Area} = \frac{ab \sin C}{2}$$

If angles B and C are known, but not A, then

$$A = 180 - (B + C).$$

Two Sides and Included Angle Known:

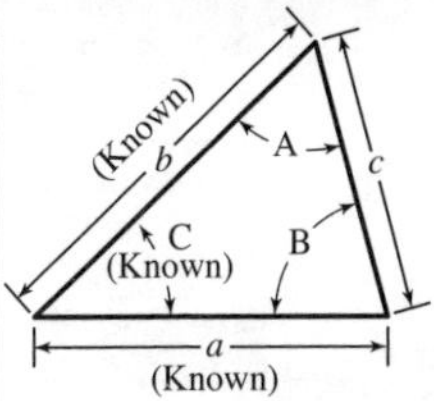

Two Sides and One Included Angle Known

If sides a and b, and angle C between them are known:

$$\tan A = \frac{a \sin C}{b - (a \cos C)}, \text{ so } A = \tan^{-1} \frac{a \sin C}{b - (a \cos C)}$$

$$B = 180° - (A + C) \qquad c = \frac{a \sin C}{\sin A}$$

Side c may also be found directly as below:

$$c = \sqrt{a^2 + b^2 - (2ab \cos C)}$$

$$\text{Area} = \frac{ab \sin C}{2}$$

Two Sides and the Angle Opposite One of the Sides Known:

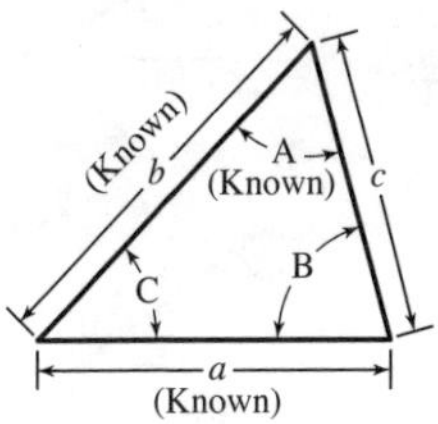

Two Sides and Angle Opposite One Side Known

If angle A, opposite side a, and other side b are known:

$$\sin B = \frac{b \sin A}{a} \qquad C = 180° - (A + B)$$

$$c = \frac{a \sin C}{\sin A} \qquad \text{Area} = \frac{ab \sin C}{2}$$

If $B > A$ but $< 90°$, a second solution, B_2, C_2, c_2, exists:

$$B_2 = 180° - B, \quad C_2 = 180° - (A + B_2)$$

$$c_2 = (a \sin C_2)/\sin A, \quad \text{area} = (ab \sin C_2)/2$$

If $a \geq b \sin A$, then only the first solution exists.

If $a < b \sin A$, then no solution exists.

All Three Sides Known:

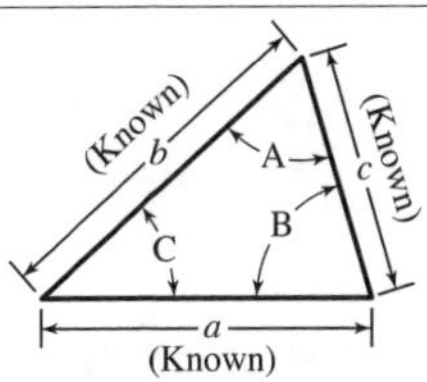

All Three Sides Known

If all three sides a, b, and c are known, then any angle can be found:

$$\cos A = \frac{b^2 + c^2 - a^2}{2bc} \qquad \sin B = \frac{b \sin A}{a}$$

$$C = 180° - (A + B) \qquad \text{Area} = \frac{ab \sin C}{2}$$

Rapid Solution of Right and Oblique Triangles

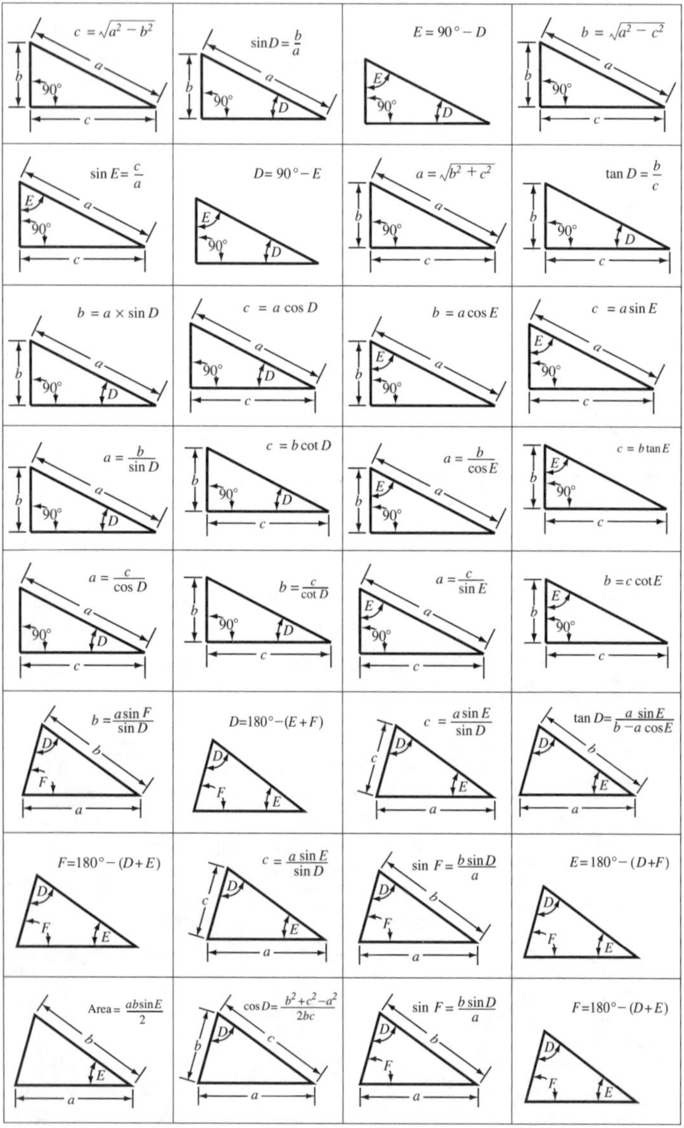

Trigonometric Values of Angles from 0° to 15° and 75° to 90°

Angle	sin	cos	tan	cot		Angle	sin	cos	tan	cot	
0° 0′	0.000000	1.000000	0.000000	—	**90° 0′**	**7° 30′**	0.130526	0.991445	0.131652	7.595754	**82° 30′**
10	0.002909	0.999996	0.002909	343.7737	**50**	**40**	0.133410	0.991061	0.134613	7.428706	**20**
20	0.005818	0.999983	0.005818	171.8854	**40**	**50**	0.136292	0.990669	0.137576	7.268725	**10**
30	0.008727	0.999962	0.008727	114.5887	**30**	**8° 0′**	0.139173	0.990268	0.140541	7.115370	**82° 0′**
40	0.011635	0.999932	0.011636	85.93979	**20**	**10**	0.142053	0.989859	0.143508	6.968234	**50**
50	0.014544	0.999894	0.014545	68.75009	**10**	**20**	0.144932	0.989442	0.146478	6.826944	**40**
1° 0′	0.017452	0.999848	0.017455	57.28996	**89° 0′**	**30**	0.147809	0.989016	0.149451	6.691156	**30**
10	0.020361	0.999793	0.020365	49.10388	**50**	**40**	0.150686	0.988582	0.152426	6.560554	**20**
20	0.023269	0.999729	0.023275	42.96408	**40**	**50**	0.153561	0.988139	0.155404	6.434843	**10**
30	0.026177	0.999657	0.026186	38.18846	**30**	**9° 0′**	0.156434	0.987688	0.158384	6.313752	**81° 0′**
40	0.029085	0.999577	0.029097	34.36777	**20**	**10**	0.159307	0.987229	0.161368	6.197028	**50**
50	0.031992	0.999488	0.032009	31.24158	**10**	**20**	0.162178	0.986762	0.164354	6.084438	**40**
2° 0′	0.034899	0.999391	0.034921	28.63625	**88° 0′**	**30**	0.165048	0.986286	0.167343	5.975764	**30**
10	0.037806	0.999285	0.037834	26.43160	**50**	**40**	0.167916	0.985801	0.170334	5.870804	**20**
20	0.040713	0.999171	0.040747	24.54176	**40**	**50**	0.170783	0.985309	0.173329	5.769369	**10**
30	0.043619	0.999048	0.043661	22.90377	**30**	**10° 0′**	0.173648	0.984808	0.176327	5.671282	**80° 0′**
40	0.046525	0.998917	0.046576	21.47040	**20**	**10**	0.176512	0.984298	0.179328	5.576379	**50**
50	0.049431	0.998778	0.049491	20.20555	**10**	**20**	0.179375	0.983781	0.182332	5.484505	**40**
3° 0′	0.052336	0.998630	0.052408	19.08114	**87° 0′**	**30**	0.182236	0.983255	0.185339	5.395517	**30**
10	0.055241	0.998473	0.055325	18.07498	**50**	**40**	0.185095	0.982721	0.188349	5.309279	**20**
20	0.058145	0.998308	0.058243	17.16934	**40**	**50**	0.187953	0.982178	0.191363	5.225665	**10**
30	0.061049	0.998135	0.061163	16.34986	**30**	**11° 0′**	0.190809	0.981627	0.194380	5.144554	**79° 0′**
40	0.063952	0.997953	0.064083	15.60478	**20**	**10**	0.193664	0.981068	0.197401	5.065835	**50**
50	0.066854	0.997763	0.067004	14.92442	**10**	**20**	0.196517	0.980500	0.200425	4.989403	**40**
4° 0′	0.069756	0.997564	0.069927	14.30067	**86° 0′**	**30**	0.199368	0.979925	0.203452	4.915157	**30**
10	0.072658	0.997357	0.072851	13.72674	**50**	**40**	0.202218	0.979341	0.206483	4.843005	**20**
20	0.075559	0.997141	0.075775	13.19688	**40**	**50**	0.205065	0.978748	0.209518	4.772857	**10**
30	0.078459	0.996917	0.078702	12.70621	**30**	**12° 0′**	0.207912	0.978148	0.212557	4.704630	**78° 0′**
40	0.081359	0.996685	0.081629	12.25051	**20**	**10**	0.210756	0.977539	0.215599	4.638246	**50**
50	0.084258	0.996444	0.084558	11.82617	**10**	**20**	0.213599	0.976921	0.218645	4.573629	**40**
5° 0′	0.087156	0.996195	0.087489	11.43005	**85° 0′**	**30**	0.216440	0.976296	0.221695	4.510709	**30**
10	0.090053	0.995937	0.090421	11.05943	**50**	**40**	0.219279	0.975662	0.224748	4.449418	**20**
20	0.092950	0.995671	0.093354	10.71191	**40**	**50**	0.222116	0.975020	0.227806	4.389694	**10**
30	0.095846	0.995396	0.096289	10.38540	**30**	**13° 0′**	0.224951	0.974370	0.230868	4.331476	**77° 0′**
40	0.098741	0.995113	0.099226	10.07803	**20**	**10**	0.227784	0.973712	0.233934	4.274707	**50**
50	0.101635	0.994822	0.102164	9.788173	**10**	**20**	0.230616	0.973045	0.237004	4.219332	**40**
6° 0′	0.104528	0.994522	0.105104	9.514364	**84° 0′**	**30**	0.233445	0.972370	0.240079	4.165300	**30**
10	0.107421	0.994214	0.108046	9.255304	**50**	**40**	0.236273	0.971687	0.243157	4.112561	**20**
20	0.110313	0.993897	0.110990	9.009826	**40**	**50**	0.239098	0.970995	0.246241	4.061070	**10**
30	0.113203	0.993572	0.113936	8.776887	**30**	**14° 0′**	0.241922	0.970296	0.249328	4.010781	**76° 0′**
40	0.116093	0.993238	0.116883	8.555547	**20**	**10**	0.244743	0.969588	0.252420	3.961652	**50**
50	0.118982	0.992896	0.119833	8.344956	**10**	**20**	0.247563	0.968872	0.255516	3.913642	**40**
7° 0′	0.121869	0.992546	0.122785	8.144346	**83° 0′**	**30**	0.250380	0.968148	0.258618	3.866713	**30**
10	0.124756	0.992187	0.125738	7.953022	**50**	**40**	0.253195	0.967415	0.261723	3.820828	**20**
20	0.127642	0.991820	0.128694	7.770351	**40**	**50**	0.256008	0.966675	0.264834	3.775952	**10**
7° 30′	0.130526	0.991445	0.131652	7.595754	**82° 30′**	**15° 0′**	0.258819	0.965926	0.267949	3.732051	**75° 0′**
	cos	sin	cot	tan		Angle	cos	sin	cot	tan	Angle

For angles 0° to 15° 0′ (angles found in a column to the left of the data), use the column labels at the top of the table; for angles 75° to 90° 0′ (angles found in a column to the right of the data), use the column labels at the bottom of the table.

Trigonometric Values of Angles from 15° to 30° and 60° to 75°

Angle	sin	cos	tan	cot		Angle	sin	cos	tan	cot	
15° 0′	0.258819	0.965926	0.267949	3.732051	**75° 0′**	**22° 30′**	0.382683	0.923880	0.414214	2.414214	**67° 30′**
10	0.261628	0.965169	0.271069	3.689093	**50**	**40**	0.385369	0.922762	0.417626	2.394489	**20**
20	0.264434	0.964404	0.274194	3.647047	**40**	**50**	0.388052	0.921638	0.421046	2.375037	**10**
30	0.267238	0.963630	0.277325	3.605884	**30**	**23° 0′**	0.390731	0.920505	0.424475	2.355852	**67° 0′**
40	0.270040	0.962849	0.280460	3.565575	**20**	**10**	0.393407	0.919364	0.427912	2.336929	**50**
50	0.272840	0.962059	0.283600	3.526094	**10**	**20**	0.396080	0.918216	0.431358	2.318261	**40**
16° 0′	0.275637	0.961262	0.286745	3.487414	**74° 0′**	**30**	0.398749	0.917060	0.434812	2.299843	**30**
10	0.278432	0.960456	0.289896	3.449512	**50**	**40**	0.401415	0.915896	0.438276	2.281669	**20**
20	0.281225	0.959642	0.293052	3.412363	**40**	**50**	0.404078	0.914725	0.441748	2.263736	**10**
30	0.284015	0.958820	0.296213	3.375943	**30**	**24° 0′**	0.406737	0.913545	0.445229	2.246037	**66° 0′**
40	0.286803	0.957990	0.299380	3.340233	**20**	**10**	0.409392	0.912358	0.448719	2.228568	**50**
50	0.289589	0.957151	0.302553	3.305209	**10**	**20**	0.412045	0.911164	0.452218	2.211323	**40**
17° 0′	0.292372	0.956305	0.305731	3.270853	**73° 0′**	**30**	0.414693	0.909961	0.455726	2.194300	**30**
10	0.295152	0.955450	0.308914	3.237144	**50**	**40**	0.417338	0.908751	0.459244	2.177492	**20**
20	0.297930	0.954588	0.312104	3.204064	**40**	**50**	0.419980	0.907533	0.462771	2.160896	**10**
30	0.300706	0.953717	0.315299	3.171595	**30**	**25° 0′**	0.422618	0.906308	0.466308	2.144507	**65° 0′**
40	0.303479	0.952838	0.318500	3.139719	**20**	**10**	0.425253	0.905075	0.469854	2.128321	**50**
50	0.306249	0.951951	0.321707	3.108421	**10**	**20**	0.427884	0.903834	0.473410	2.112335	**40**
18° 0′	0.309017	0.951057	0.324920	3.077684	**72° 0′**	**30**	0.430511	0.902585	0.476976	2.096544	**30**
10	0.311782	0.950154	0.328139	3.047492	**50**	**40**	0.433135	0.901329	0.480551	2.080944	**20**
20	0.314545	0.949243	0.331364	3.017830	**40**	**50**	0.435755	0.900065	0.484137	2.065532	**10**
30	0.317305	0.948324	0.334595	2.988685	**30**	**26° 0′**	0.438371	0.898794	0.487733	2.050304	**64° 0′**
40	0.320062	0.947397	0.337833	2.960042	**20**	**10**	0.440984	0.897515	0.491339	2.035256	**50**
50	0.322816	0.946462	0.341077	2.931888	**10**	**20**	0.443593	0.896229	0.494955	2.020386	**40**
19° 0′	0.325568	0.945519	0.344328	2.904211	**71° 0′**	**30**	0.446198	0.894934	0.498582	2.005690	**30**
10	0.328317	0.944568	0.347585	2.876997	**50**	**40**	0.448799	0.893633	0.502219	1.991164	**20**
20	0.331063	0.943609	0.350848	2.850235	**40**	**50**	0.451397	0.892323	0.505867	1.976805	**10**
30	0.333807	0.942641	0.354119	2.823913	**30**	**27° 0′**	0.453990	0.891007	0.509525	1.962611	**63° 0′**
40	0.336547	0.941666	0.357396	2.798020	**20**	**10**	0.456580	0.889682	0.513195	1.948577	**50**
50	0.339285	0.940684	0.360679	2.772545	**10**	**20**	0.459166	0.888350	0.516875	1.934702	**40**
20° 0′	0.342020	0.939693	0.363970	2.747477	**70° 0′**	**30**	0.461749	0.887011	0.520567	1.920982	**30**
10	0.344752	0.938694	0.367268	2.722808	**50**	**40**	0.464327	0.885664	0.524270	1.907415	**20**
20	0.347481	0.937687	0.370573	2.698525	**40**	**50**	0.466901	0.884309	0.527984	1.893997	**10**
30	0.350207	0.936672	0.373885	2.674621	**30**	**28° 0′**	0.469472	0.882948	0.531709	1.880726	**62° 0′**
40	0.352931	0.935650	0.377204	2.651087	**20**	**10**	0.472038	0.881578	0.535446	1.867600	**50**
50	0.355651	0.934619	0.380530	2.627912	**10**	**20**	0.474600	0.880201	0.539195	1.854616	**40**
21° 0′	0.358368	0.933580	0.383864	2.605089	**69° 0′**	**30**	0.477159	0.878817	0.542956	1.841771	**30**
10	0.361082	0.932534	0.387205	2.582609	**50**	**40**	0.479713	0.877425	0.546728	1.829063	**20**
20	0.363793	0.931480	0.390554	2.560465	**40**	**50**	0.482263	0.876026	0.550513	1.816489	**10**
30	0.366501	0.930418	0.393910	2.538648	**30**	**29° 0′**	0.484810	0.874620	0.554309	1.804048	**61° 0′**
40	0.369206	0.929348	0.397275	2.517151	**20**	**10**	0.487352	0.873206	0.558118	1.791736	**50**
50	0.371908	0.928270	0.400646	2.495966	**10**	**20**	0.489890	0.871784	0.561939	1.779552	**40**
22° 0′	0.374607	0.927184	0.404026	2.475087	**68° 0′**	**30**	0.492424	0.870356	0.565773	1.767494	**30**
10	0.377302	0.926090	0.407414	2.454506	**50**	**40**	0.494953	0.868920	0.569619	1.755559	**20**
20	0.379994	0.924989	0.410810	2.434217	**40**	**50**	0.497479	0.867476	0.573478	1.743745	**10**
22° 30′	0.382683	0.923880	0.414214	2.414214	**67° 30′**	**30° 0′**	0.500000	0.866025	0.577350	1.732051	**60° 0′**
	cos	sin	cot	tan	Angle		cos	sin	cot	tan	Angle

For angles 15° to 30° 0′ (angles found in a column to the left of the data), use the column labels at the top of the table; for angles 60° to 75° 0′ (angles found in a column to the right of the data), use the column labels at the bottom of the table.

Trigonometric Values of Angles from 30° to 60°

Angle	sin	cos	tan	cot		Angle	sin	cos	tan	cot	
30° 0′	0.500000	0.866025	0.577350	1.732051	**60° 0′**	**37° 30′**	0.608761	0.793353	0.767327	1.303225	**52° 30′**
10	0.502517	0.864567	0.581235	1.720474	**50**	**40**	0.611067	0.791579	0.771959	1.295406	**20**
20	0.505030	0.863102	0.585134	1.709012	**40**	**50**	0.613367	0.789798	0.776612	1.287645	**10**
30	0.507538	0.861629	0.589045	1.697663	**30**	**38° 0′**	0.615661	0.788011	0.781286	1.279942	**52° 0′**
40	0.510043	0.860149	0.592970	1.686426	**20**	**10**	0.617951	0.786217	0.785981	1.272296	**50**
50	0.512543	0.858662	0.596908	1.675299	**10**	**20**	0.620235	0.784416	0.790697	1.264706	**40**
31° 0′	0.515038	0.857167	0.600861	1.664279	**59° 0′**	**30**	0.622515	0.782608	0.795436	1.257172	**30**
10	0.517529	0.855665	0.604827	1.653366	**50**	**40**	0.624789	0.780794	0.800196	1.249693	**20**
20	0.520016	0.854156	0.608807	1.642558	**40**	**50**	0.627057	0.778973	0.804979	1.242268	**10**
30	0.522499	0.852640	0.612801	1.631852	**30**	**39° 0′**	0.629320	0.777146	0.809784	1.234897	**51° 0′**
40	0.524977	0.851117	0.616809	1.621247	**20**	**10**	0.631578	0.775312	0.814612	1.227579	**50**
50	0.527450	0.849586	0.620832	1.610742	**10**	**20**	0.633831	0.773472	0.819463	1.220312	**40**
32° 0′	0.529919	0.848048	0.624869	1.600335	**58° 0′**	**30**	0.636078	0.771625	0.824336	1.213097	**30**
10	0.532384	0.846503	0.628921	1.590024	**50**	**40**	0.638320	0.769771	0.829234	1.205933	**20**
20	0.534844	0.844951	0.632988	1.579808	**40**	**50**	0.640557	0.767911	0.834155	1.198818	**10**
30	0.537300	0.843391	0.637070	1.569686	**30**	**40° 0′**	0.642788	0.766044	0.839100	1.191754	**50° 0′**
40	0.539751	0.841825	0.641167	1.559655	**20**	**10**	0.645013	0.764171	0.844069	1.184738	**50**
50	0.542197	0.840251	0.645280	1.549715	**10**	**20**	0.647233	0.762292	0.849062	1.177770	**40**
33° 0′	0.544639	0.838671	0.649408	1.539865	**57° 0′**	**30**	0.649448	0.760406	0.854081	1.170850	**30**
10	0.547076	0.837083	0.653551	1.530102	**50**	**40**	0.651657	0.758514	0.859124	1.163976	**20**
20	0.549509	0.835488	0.657710	1.520426	**40**	**50**	0.653861	0.756615	0.864193	1.157149	**10**
30	0.551937	0.833886	0.661886	1.510835	**30**	**41° 0′**	0.656059	0.754710	0.869287	1.150368	**49° 0′**
40	0.554360	0.832277	0.666077	1.501328	**20**	**10**	0.658252	0.752798	0.874407	1.143633	**50**
50	0.556779	0.830661	0.670284	1.491904	**10**	**20**	0.660439	0.750880	0.879553	1.136941	**40**
34° 0′	0.559193	0.829038	0.674509	1.482561	**56° 0′**	**30**	0.662620	0.748956	0.884725	1.130294	**30**
10	0.561602	0.827407	0.678749	1.473298	**50**	**40**	0.664796	0.747025	0.889924	1.123691	**20**
20	0.564007	0.825770	0.683007	1.464115	**40**	**50**	0.666966	0.745088	0.895151	1.117130	**10**
30	0.566406	0.824126	0.687281	1.455009	**30**	**42° 0′**	0.669131	0.743145	0.900404	1.110613	**48° 0′**
40	0.568801	0.822475	0.691572	1.445980	**20**	**10**	0.671289	0.741195	0.905685	1.104137	**50**
50	0.571191	0.820817	0.695881	1.437027	**10**	**20**	0.673443	0.739239	0.910994	1.097702	**40**
35° 0′	0.573576	0.819152	0.700208	1.428148	**55° 0′**	**30**	0.675590	0.737277	0.916331	1.091309	**30**
10	0.575957	0.817480	0.704551	1.419343	**50**	**40**	0.677732	0.735309	0.921697	1.084955	**20**
20	0.578332	0.815801	0.708913	1.410610	**40**	**50**	0.679868	0.733334	0.927091	1.078642	**10**
30	0.580703	0.814116	0.713293	1.401948	**30**	**43° 0′**	0.681998	0.731354	0.932515	1.072369	**47° 0′**
40	0.583069	0.812423	0.717691	1.393357	**20**	**10**	0.684123	0.729367	0.937968	1.066134	**50**
50	0.585429	0.810723	0.722108	1.384835	**10**	**20**	0.686242	0.727374	0.943451	1.059938	**40**
36° 0′	0.587785	0.809017	0.726543	1.376382	**54° 0′**	**30**	0.688355	0.725374	0.948965	1.053780	**30**
10	0.590136	0.807304	0.730996	1.367996	**50**	**40**	0.690462	0.723369	0.954508	1.047660	**20**
20	0.592482	0.805584	0.735469	1.359676	**40**	**50**	0.692563	0.721357	0.960083	1.041577	**10**
30	0.594823	0.803857	0.739961	1.351422	**30**	**44° 0′**	0.694658	0.719340	0.965689	1.035530	**46° 0′**
40	0.597159	0.802123	0.744472	1.343233	**20**	**10**	0.696748	0.717316	0.971326	1.029520	**50**
50	0.599489	0.800383	0.749003	1.335108	**10**	**20**	0.698832	0.715286	0.976996	1.023546	**40**
37° 0′	0.601815	0.798636	0.753554	1.327045	**53° 0′**	**30**	0.700909	0.713250	0.982697	1.017607	**30**
10	0.604136	0.796882	0.758125	1.319044	**50**	**40**	0.702981	0.711209	0.988432	1.011704	**20**
20	0.606451	0.795121	0.762716	1.311105	**40**	**50**	0.705047	0.709161	0.994199	1.005835	**10**
37° 30′	0.608761	0.793353	0.767327	1.303225	**52° 30′**	**45° 0′**	0.707107	0.707107	1.000000	1.000000	**45° 0′**
	cos	sin	cot	tan	Angle		cos	sin	cot	tan	Angle

For angles 30° to 45°0′ (angles found in a column to the left of the data), use the column labels at the top of the table; for angles 45° to 60° 0′ (angles found in a column to the right of the data), use the column labels at the bottom of the table.

Formulas for Compound Angles

<table>
<tr>
<td>
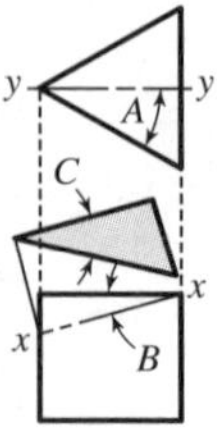

Fig. 1.
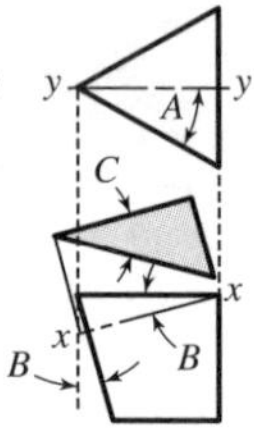

Fig. 2.
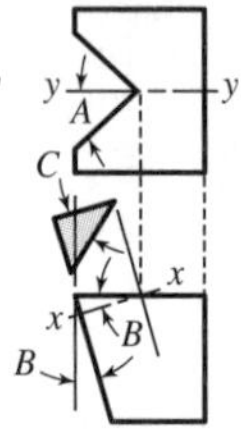

Fig. 3.
</td>
<td>
For given angles A and B, find the resultant angle C in plane x-x. Angle B is measured in vertical plane y-y of midsection.

Fig. 1 $\tan C = \tan A \times \cos B$

Fig. 2 $\tan C = \dfrac{\tan A}{\cos B}$

Fig. 3 (Same formula as for Fig. 2)
</td>
</tr>
<tr>
<td>
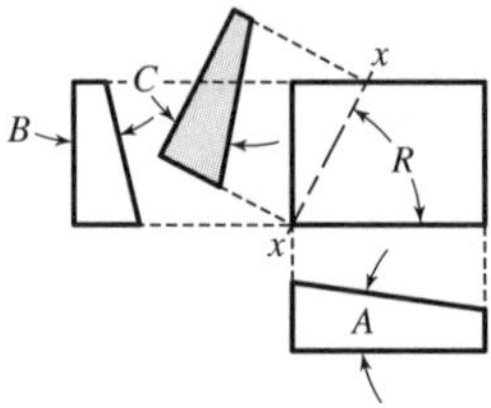

Fig. 4.
</td>
<td>
Fig. 4. In machining a plate to angles A and B, the plate is held at angle C in plane x-x. Angle of rotation R in plane parallel to base (or complement of R) is for locating plate so that plane x-x is perpendicular to axis of pivot on angle-plate or work-holding vise.

$$\tan R = \frac{\tan B}{\tan A}, \quad \tan C = \frac{\tan A}{\cos R}$$
</td>
</tr>
<tr>
<td>
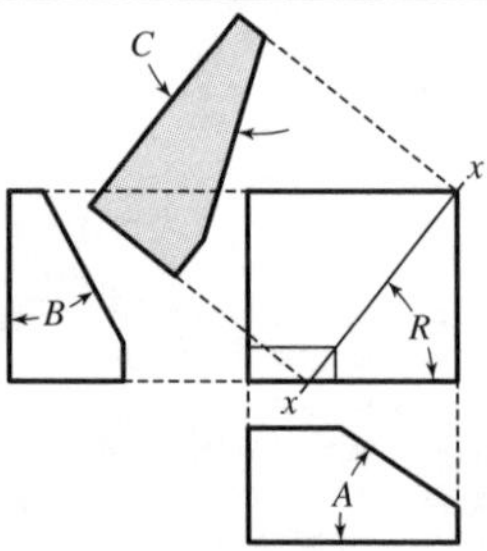

Fig. 5.
</td>
<td>
Fig. 5. Angle R in horizontal plane parallel to base is angle from plane x-x to side having angle A.

$$\tan R = \frac{\tan A}{\tan B}$$

$$\tan C = \tan A \times \cos R = \tan B \times \sin R$$

Compound angle C is angle in plane x-x from base to corner formed by intersection of planes inclined to angles A and B. This formula for C may be used to find cotangent of complement of C_1, Fig. 6.
</td>
</tr>
<tr>
<td>
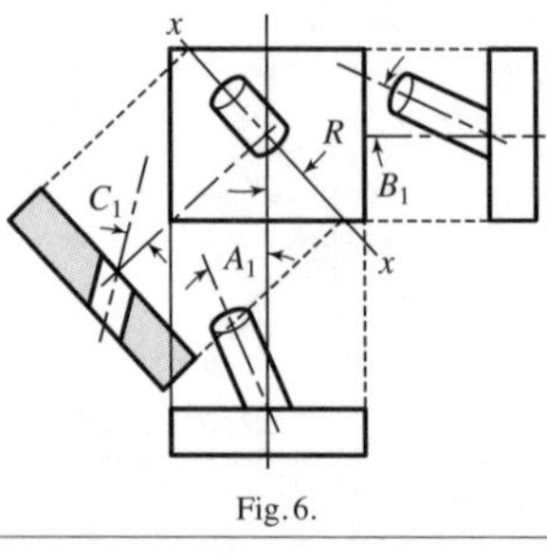

Fig. 6.
</td>
<td>
Fig. 6. Angles A_1 and B_1 are measured in vertical planes of front and side elevations. Plane x-x is located by angle R from center-line or from plane of angle B_1.

$$\tan R = \frac{\tan A_1}{\tan B_1}$$

$$\tan C_1 = \frac{\tan A_1}{\sin R} = \frac{\tan B_1}{\cos R}$$

The resultant angle C_1 would be required in drilling hole for pin.
</td>
</tr>
</table>

Lengths of Chords for Spacing Off the Circumferences of Circles.—The table below, which may be used by toolmakers when setting "buttons" in circular formation, is intended to make possible the division of the periphery into a number of equal parts without trials with the dividers.

Example: Assume that it is required to divide the periphery of a circle of 20 inches diameter into thirty-two equal parts. From the table the length of the chord is found to be 0.098017 inch, if the diameter of the circle were 1 inch. With a diameter of 20 inches the length of the chord for one division would be $20 \times 0.098017 = 1.9603$ inches.

Example, Metric Units: For a 100 millimeter diameter requiring 5 equal divisions, the length of the chord for one division would be $100 \times 0.587785 = 58.7785$ millimeters.

Lengths of Chords for Spacing Off the Circumferences of Circles with a Diameter Equal to 1 (US Customary or Metric Units)

No. of Spaces	Length of Chord	No. of Spaces	Length of Chord	No. of Spaces	Length of Chord	No. of Spaces	Length of Chord
3	0.866025	41	0.076549	79	0.039757	117	0.026848
4	0.707107	42	0.074730	80	0.039260	118	0.026621
5	0.587785	43	0.072995	81	0.038775	119	0.026397
6	0.500000	44	0.071339	82	0.038303	120	0.026177
7	0.433884	45	0.069756	83	0.037841	121	0.025961
8	0.382683	46	0.068242	84	0.037391	122	0.025748
9	0.342020	47	0.066793	85	0.036951	123	0.025539
10	0.309017	48	0.065403	86	0.036522	124	0.025333
11	0.281733	49	0.064070	87	0.036102	125	0.025130
12	0.258819	50	0.062791	88	0.035692	126	0.024931
13	0.239316	51	0.061561	89	0.035291	127	0.024734
14	0.222521	52	0.060378	90	0.034899	128	0.024541
15	0.207912	53	0.059241	91	0.034516	129	0.024351
16	0.195090	54	0.058145	92	0.034141	130	0.024164
17	0.183750	55	0.057089	93	0.033774	131	0.023979
18	0.173648	56	0.056070	94	0.033415	132	0.023798
19	0.164595	57	0.055088	95	0.033063	133	0.023619
20	0.156434	58	0.054139	96	0.032719	134	0.023443
21	0.149042	59	0.053222	97	0.032382	135	0.023269
22	0.142315	60	0.052336	98	0.032052	136	0.023098
23	0.136167	61	0.051479	99	0.031728	137	0.022929
24	0.130526	62	0.050649	100	0.031411	138	0.022763
25	0.125333	63	0.049846	101	0.031100	139	0.022599
26	0.120537	64	0.049068	102	0.030795	140	0.022438
27	0.116093	65	0.048313	103	0.030496	141	0.022279
28	0.111964	66	0.047582	104	0.030203	142	0.022122
29	0.108119	67	0.046872	105	0.029915	143	0.021967
30	0.104528	68	0.046183	106	0.029633	144	0.021815
31	0.101168	69	0.045515	107	0.029356	145	0.021664
32	0.098017	70	0.044865	108	0.029085	146	0.021516
33	0.095056	71	0.044233	109	0.028818	147	0.021370
34	0.092268	72	0.043619	110	0.028556	148	0.021225
35	0.089639	73	0.043022	111	0.028299	149	0.021083
36	0.087156	74	0.042441	112	0.028046	150	0.020942
37	0.084806	75	0.041876	113	0.027798	151	0.020804
38	0.082579	76	0.041325	114	0.027554	152	0.020667
39	0.080467	77	0.040789	115	0.027315	153	0.020532
40	0.078459	78	0.040266	116	0.027079	154	0.020399

The table is calculated for circles having a diameter equal to 1. For circles of other diameters, multiply given length by diameter of circle.

Coordinates for Locating Equally-Spaced Holes (US Customary or Metric Units)

Multiply values shown by diameter of pitch circle.

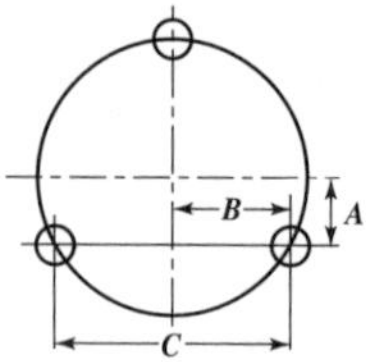

3-Hole
$A = 0.25000$
$B = 0.43301$
$C = 0.86603$

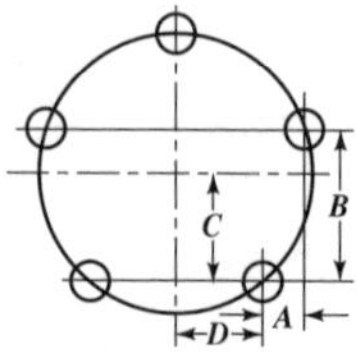

5-Hole
$A = 0.18164$
$B = 0.55902$
$C = 0.40451$
$D = 0.29389$

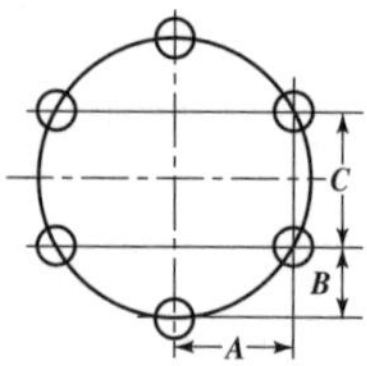

6-Hole
$A = 0.43301$
$B = 0.25000$
$C = 0.50000$

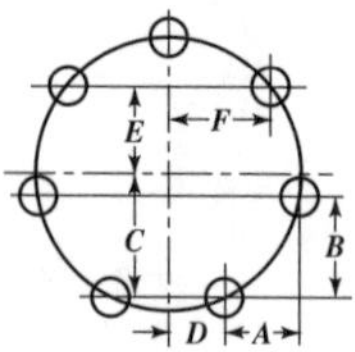

7-Hole
$A = 0.27052$
$B = 0.33922$
$C = 0.45049$
$D = 0.21694$
$E = 0.31175$
$F = 0.39092$

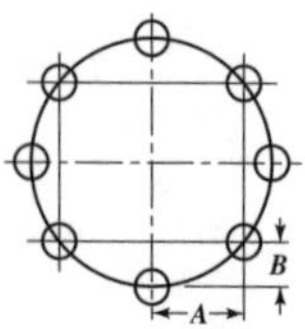

8-Hole
$A = 0.35355$
$B = 0.14645$

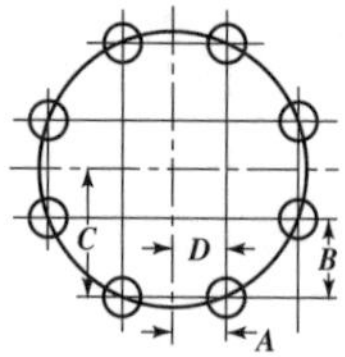

8-Hole
$A = 0.27059$
$B = 0.27059$
$C = 0.46194$
$D = 0.19134$

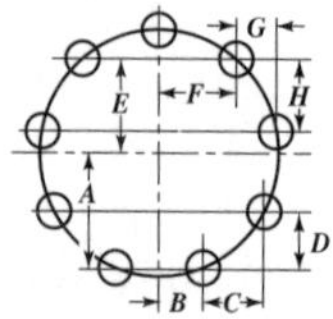

9-Hole
$A = 0.46985$
$B = 0.17101$
$C = 0.26201$
$D = 0.21985$
$E = 0.38302$
$F = 0.32139$
$G = 0.17101$
$H = 0.29620$

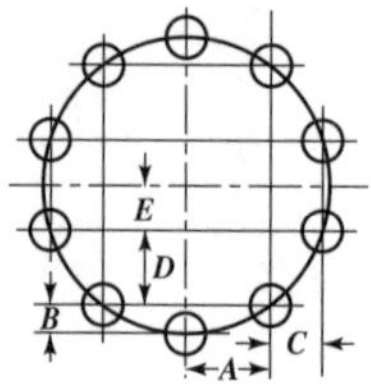

10-Hole
$A = 0.29389$
$B = 0.09549$
$C = 0.18164$
$D = 0.25000$
$E = 0.15451$

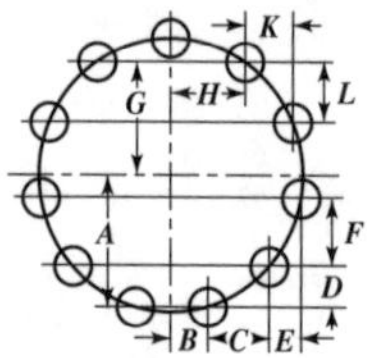

11-Hole
$A = 0.47975$
$B = 0.14087$
$C = 0.23700$
$D = 0.15231$
$E = 0.11704$
$F = 0.25627$
$G = 0.42063$
$H = 0.27032$
$K = 0.18449$
$L = 0.21291$

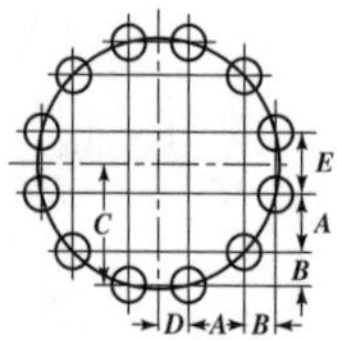

12-Hole
$A = 0.22415$
$B = 0.12941$
$C = 0.48296$
$D = 0.12941$
$E = 0.25882$

The constants in the table are multiplied by the diameter of the bolt hole pitch circle to obtain the longitudinal and lateral adjustments of the right-angle slides of the jig borer in boring equally spaced holes. While holes may be located by these right-angle measurements, an auxiliary rotary table provides a more direct method. With a rotary table, the holes are spaced by precise angular movements after adjustment to the required radius.

Decimal Equivalents, Squares, Cubes, Square Roots, Cube Roots, and Logarithms of Fractions from 1/64 to 1, by 64ths

Fraction	Decimal Equivalent	Log	Square	Log of Square	Cube	Log of Cube	Square Root	Log of Square Root	Cube Root	Log of Cube Root
1/64	0.015625	-1.80618	0.00024	-3.61236	0.00000	-5.41854	0.12500	-0.90309	0.25000	-0.60206
1/32	0.031250	-1.50515	0.00098	-3.01030	0.00003	-4.51545	0.17678	-0.75257	0.31498	-0.50172
3/64	0.046875	-1.32906	0.00220	-2.65812	0.00010	-3.98718	0.21651	-0.66453	0.36056	-0.44302
1/16	0.062500	-1.20412	0.00391	-2.40824	0.00024	-3.61236	0.25000	-0.60206	0.39685	-0.40137
5/64	0.078125	-1.10721	0.00610	-2.21442	0.00048	-3.32163	0.27951	-0.55361	0.42749	-0.36907
3/32	0.093750	-1.02803	0.00879	-2.05606	0.00082	-3.08409	0.30619	-0.51402	0.45428	-0.34268
7/64	0.109375	-0.96108	0.01196	-1.92216	0.00131	-2.88325	0.33072	-0.48054	0.47823	-0.32036
1/8	0.125000	-0.90309	0.01563	-1.80618	0.00195	-2.70927	0.35355	-0.45155	0.50000	-0.30103
9/64	0.140625	-0.85194	0.01978	-1.70388	0.00278	-2.55581	0.37500	-0.42597	0.52002	-0.28398
5/32	0.156250	-0.80618	0.02441	-1.61236	0.00381	-2.41854	0.39529	-0.40309	0.53861	-0.26873
11/64	0.171875	-0.76479	0.02954	-1.52958	0.00508	-2.29436	0.41458	-0.38239	0.55600	-0.25493
3/16	0.187500	-0.72700	0.03516	-1.45400	0.00659	-2.18100	0.43301	-0.36350	0.57236	-0.24233
13/64	0.203125	-0.69224	0.04126	-1.38447	0.00838	-2.07671	0.45069	-0.34612	0.58783	-0.23075
7/32	0.218750	-0.66005	0.04785	-1.32010	0.01047	-1.98016	0.46771	-0.33003	0.60254	-0.22002
15/64	0.234375	-0.63009	0.05493	-1.26018	0.01287	-1.89027	0.48412	-0.31504	0.61655	-0.21003
1/4	0.250000	-0.60206	0.06250	-1.20412	0.01563	-1.80618	0.50000	-0.30103	0.62996	-0.20069
17/64	0.265625	-0.57573	0.07056	-1.15146	0.01874	-1.72719	0.51539	-0.28787	0.64282	-0.19191
9/32	0.281250	-0.55091	0.07910	-1.10182	0.02225	-1.65272	0.53033	-0.27545	0.65519	-0.18364
19/64	0.296875	-0.52743	0.08813	-1.05485	0.02617	-1.58228	0.54486	-0.26371	0.66710	-0.17581
5/16	0.312500	-0.50515	0.09766	-1.01030	0.03052	-1.51545	0.55902	-0.25258	0.67860	-0.16838
21/64	0.328125	-0.48396	0.10767	-0.96792	0.03533	-1.45188	0.57282	-0.24198	0.68973	-0.16132
11/32	0.343750	-0.46376	0.11816	-0.92752	0.04062	-1.39127	0.58630	-0.23188	0.70051	-0.15459
23/64	0.359375	-0.44445	0.12915	-0.88890	0.04641	-1.33336	0.59948	-0.22223	0.71097	-0.14815
3/8	0.375000	-0.42597	0.14063	-0.85194	0.05273	-1.27791	0.61237	-0.21299	0.72113	-0.14199
25/64	0.390625	-0.40824	0.15259	-0.81648	0.05960	-1.22472	0.62500	-0.20412	0.73100	-0.13608
13/32	0.406250	-0.39121	0.16504	-0.78241	0.06705	-1.17362	0.63738	-0.19560	0.74062	-0.13040
27/64	0.421875	-0.37482	0.17798	-0.74963	0.07508	-1.12445	0.64952	-0.18741	0.75000	-0.12494
7/16	0.437500	-0.35902	0.19141	-0.71804	0.08374	-1.07707	0.66144	-0.17951	0.75915	-0.11967
29/64	0.453125	-0.34378	0.20532	-0.68756	0.09304	-1.03135	0.67315	-0.17189	0.76808	-0.11459
15/32	0.468750	-0.32906	0.21973	-0.65812	0.10300	-0.98718	0.68465	-0.16453	0.77681	-0.10969
31/64	0.484375	-0.31482	0.23462	-0.62964	0.11364	-0.94446	0.69597	-0.15741	0.78535	-0.10494
1/2	0.500000	-0.30103	0.25000	-0.60206	0.12500	-0.90309	0.70711	-0.15052	0.79370	-0.10034

Decimal Equivalents, Squares, Cubes, Square Roots, Cube Roots, and Logarithms of Fractions from $^{1}/_{64}$ to 1, by 64ths *(Continued)*

Fraction	Decimal Equivalent	Log	Square	Log of Square	Cube	Log of Cube	Square Root	Log of Square Root	Cube Root	Log of Cube Root
$^{33}/_{64}$	0.515625	-0.28767	0.26587	-0.57533	0.13709	-0.86300	0.71807	-0.14383	0.80188	-0.09589
$^{17}/_{32}$	0.531250	-0.27470	0.28223	-0.54940	0.14993	-0.82410	0.72887	-0.13735	0.80990	-0.09157
$^{35}/_{64}$	0.546875	-0.26211	0.29907	-0.52422	0.16356	-0.78634	0.73951	-0.13106	0.81777	-0.08737
$^{9}/_{16}$	0.562500	-0.24988	0.31641	-0.49976	0.17798	-0.74963	0.75000	-0.12494	0.82548	-0.08329
$^{37}/_{64}$	0.578125	-0.23798	0.33423	-0.47596	0.19323	-0.71394	0.76035	-0.11899	0.83306	-0.07933
$^{19}/_{32}$	0.593750	-0.22640	0.35254	-0.45279	0.20932	-0.67919	0.77055	-0.11320	0.84049	-0.07547
$^{39}/_{64}$	0.609375	-0.21512	0.37134	-0.43023	0.22628	-0.64535	0.78063	-0.10756	0.84780	-0.07171
$^{5}/_{8}$	0.625000	-0.20412	0.39063	-0.40824	0.24414	-0.61236	0.79057	-0.10206	0.85499	-0.06804
$^{41}/_{64}$	0.640625	-0.19340	0.41040	-0.38679	0.26291	-0.58019	0.80039	-0.09670	0.86205	-0.06447
$^{21}/_{32}$	0.656250	-0.18293	0.43066	-0.36586	0.28262	-0.54879	0.81009	-0.09147	0.86901	-0.06098
$^{43}/_{64}$	0.671875	-0.17271	0.45142	-0.34542	0.30330	-0.51814	0.81968	-0.08636	0.87585	-0.05757
$^{11}/_{16}$	0.687500	-0.16273	0.47266	-0.32546	0.32495	-0.48818	0.82916	-0.08136	0.88259	-0.05424
$^{45}/_{64}$	0.703125	-0.15297	0.49438	-0.30594	0.34761	-0.45890	0.83853	-0.07648	0.88922	-0.05099
$^{23}/_{32}$	0.718750	-0.14342	0.51660	-0.28684	0.37131	-0.43027	0.84779	-0.07171	0.89576	-0.04781
$^{47}/_{64}$	0.734375	-0.13408	0.53931	-0.26816	0.39605	-0.40225	0.85696	-0.06704	0.90221	-0.04469
$^{3}/_{4}$	0.750000	-0.12494	0.56250	-0.24988	0.42188	-0.37482	0.86603	-0.06247	0.90856	-0.04165
$^{49}/_{64}$	0.765625	-0.11598	0.58618	-0.23197	0.44880	-0.34795	0.87500	-0.05799	0.91483	-0.03866
$^{25}/_{32}$	0.781250	-0.10721	0.61035	-0.21442	0.47684	-0.32163	0.88388	-0.05361	0.92101	-0.03574
$^{51}/_{64}$	0.796875	-0.09861	0.63501	-0.19722	0.50602	-0.29583	0.89268	-0.04931	0.92711	-0.03287
$^{13}/_{16}$	0.812500	-0.09018	0.66016	-0.18035	0.53638	-0.27053	0.90139	-0.04509	0.93313	-0.03006
$^{53}/_{64}$	0.828125	-0.08190	0.68579	-0.16381	0.56792	-0.24571	0.91001	-0.04095	0.93907	-0.02730
$^{27}/_{32}$	0.843750	-0.07379	0.71191	-0.14757	0.60068	-0.22136	0.91856	-0.03689	0.94494	-0.02460
$^{55}/_{64}$	0.859375	-0.06582	0.73853	-0.13164	0.63467	-0.19745	0.92703	-0.03291	0.95074	-0.02194
$^{7}/_{8}$	0.875000	-0.05799	0.76563	-0.11598	0.66992	-0.17398	0.93541	-0.02900	0.95647	-0.01933
$^{57}/_{64}$	0.890625	-0.05031	0.79321	-0.10061	0.70646	-0.15092	0.94373	-0.02515	0.96213	-0.01677
$^{29}/_{32}$	0.906250	-0.04275	0.82129	-0.08550	0.74429	-0.12826	0.95197	-0.02138	0.96772	-0.01425
$^{59}/_{64}$	0.921875	-0.03533	0.84985	-0.07066	0.78346	-0.10598	0.96014	-0.01766	0.97325	-0.01178
$^{15}/_{16}$	0.937500	-0.02803	0.87891	-0.05606	0.82397	-0.08409	0.96825	-0.01401	0.97872	-0.00934
$^{61}/_{64}$	0.953125	-0.02085	0.90845	-0.04170	0.86586	-0.06255	0.97628	-0.01043	0.98412	-0.00695
$^{31}/_{32}$	0.968750	-0.01379	0.93848	-0.02758	0.90915	-0.04137	0.98425	-0.00689	0.98947	-0.00460
$^{63}/_{64}$	0.984375	-0.00684	0.96899	-0.01368	0.95385	-0.02052	0.99216	-0.00342	0.99476	-0.00228
1	1.000000	0.00000	1.00000	0.00000	1.00000	0.00000	1.00000	0.00000	1.00000	0.00000

Diameter, Circumference, and Area of a Circle

Diameter	Circumference	Area	Diameter	Circumference	Area	Diameter	Circumference	Area
1/64	0.0491	0.0002	2	6.2832	3.1416	5	15.7080	19.635
1/32	0.0982	0.0008	2 1/16	6.4795	3.3410	5 1/16	15.9043	20.129
1/16	0.1963	0.0031	2 1/8	6.6759	3.5466	5 1/8	16.1007	20.629
3/32	0.2945	0.0069	2 3/16	6.8722	3.7583	5 3/16	16.2970	21.135
1/8	0.3927	0.0123	2 1/4	7.0686	3.9761	5 1/4	16.4934	21.648
5/32	0.4909	0.0192	2 5/16	7.2649	4.2000	5 5/16	16.6897	22.166
3/16	0.5890	0.0276	2 3/8	7.4613	4.4301	5 3/8	16.8861	22.691
7/32	0.6872	0.0376	2 7/16	7.6576	4.6664	5 7/16	17.0824	23.221
1/4	0.7854	0.0491	2 1/2	7.8540	4.9087	5 1/2	17.2788	23.758
9/32	0.8836	0.0621	2 9/16	8.0503	5.1572	5 9/16	17.4751	24.301
5/16	0.9817	0.0767	2 5/8	8.2467	5.4119	5 5/8	17.6715	24.850
11/32	1.0799	0.0928	2 11/16	8.4430	5.6727	5 11/16	17.8678	25.406
3/8	1.1781	0.1104	2 3/4	8.6394	5.9396	5 3/4	18.0642	25.967
13/32	1.2763	0.1296	2 13/16	8.8357	6.2126	5 13/16	18.2605	26.535
7/16	1.3744	0.1503	2 7/8	9.0321	6.4918	5 7/8	18.4569	27.109
15/32	1.4726	0.1726	2 15/16	9.2284	6.7771	5 15/16	18.6532	27.688
1/2	1.5708	0.1963	3	9.4248	7.0686	6	18.8496	28.274
17/32	1.6690	0.2217	3 1/16	9.6211	7.3662	6 1/8	19.2423	29.465
9/16	1.7671	0.2485	3 1/8	9.8175	7.6699	6 1/4	19.6350	30.680
19/32	1.8653	0.2769	3 3/16	10.0138	7.9798	6 3/8	20.0277	31.919
5/8	1.9635	0.3068	3 1/4	10.2102	8.2958	6 1/2	20.4204	33.183
21/32	2.0617	0.3382	3 5/16	10.4065	8.6179	6 5/8	20.8131	34.472
11/16	2.1598	0.3712	3 3/8	10.6029	8.9462	6 3/4	21.2058	35.785
23/32	2.2580	0.4057	3 7/16	10.7992	9.2806	6 7/8	21.5984	37.122
3/4	2.3562	0.4418	3 1/2	10.9956	9.6211	7	21.9911	38.485
25/32	2.4544	0.4794	3 9/16	11.1919	9.9678	7 1/8	22.3838	39.871
13/16	2.5525	0.5185	3 5/8	11.388	10.3206	7 1/4	22.7765	41.282
27/32	2.6507	0.5591	3 11/16	11.585	10.6796	7 3/8	23.1692	42.718
7/8	2.7489	0.6013	3 3/4	11.781	11.0447	7 1/2	23.5619	44.179
29/32	2.8471	0.6450	3 13/16	11.977	11.4159	7 5/8	23.9546	45.664
15/16	2.9452	0.6903	3 7/8	12.174	11.7932	7 3/4	24.3473	47.173
31/32	3.0434	0.7371	3 15/16	12.370	12.1767	7 7/8	24.7400	48.707
1	3.1416	0.7854	4	12.566	12.5664	8	25.1327	50.265
1 1/16	3.3379	0.8866	4 1/16	12.763	12.9621	8 1/8	25.5254	51.849
1 1/8	3.5343	0.9940	4 1/8	12.959	13.3640	8 1/4	25.9181	53.456
1 3/16	3.7306	1.1075	4 3/16	13.155	13.7721	8 3/8	26.3108	55.088
1 1/4	3.9270	1.2272	4 1/4	13.352	14.1863	8 1/2	26.7035	56.745
1 5/16	4.1233	1.3530	4 5/16	13.548	14.6066	8 5/8	27.0962	58.426
1 3/8	4.3197	1.4849	4 3/8	13.744	15.0330	8 3/4	27.4889	60.132
1 7/16	4.5160	1.6230	4 7/16	13.941	15.4656	8 7/8	27.8816	61.862
1 1/2	4.7124	1.7671	4 1/2	14.137	15.9043	9	28.2743	63.617
1 9/16	4.9087	1.9175	4 9/16	14.334	16.3492	9 1/8	28.6670	65.397
1 5/8	5.1051	2.0739	4 5/8	14.530	16.8002	9 1/4	29.0597	67.201
1 11/16	5.3014	2.2365	4 11/16	14.726	17.2573	9 3/8	29.4524	69.029
1 3/4	5.4978	2.4053	4 3/4	14.923	17.7205	9 1/2	29.8451	70.882
1 13/16	5.6941	2.5802	4 13/16	15.119	18.1899	9 5/8	30.2378	72.760
1 7/8	5.8905	2.7612	4 7/8	15.315	18.6655	9 3/4	30.6305	74.662
1 15/16	6.0868	2.9483	4 15/16	15.512	19.1471	9 7/8	31.0232	76.589

MEASUREMENT AND INSPECTION

Sine-Bar

The sine-bar is used either for very accurate angular measurements or for locating work at a given angle as, for example, in surface grinding templates, gages, etc. The sine-bar is especially useful in measuring or checking angles when the limit of accuracy is 5 minutes or less. Some bevel protractors are equipped with verniers which read to 5 minutes, but the setting depends upon the alignment of graduations, whereas a sine-bar usually is located by positive contact with precision gage-blocks selected for whatever dimension is required for obtaining a given angle.

Types of Sine-Bars.—A sine-bar consists of a hardened, ground and lapped steel bar with very accurate cylindrical plugs of equal diameter attached to or near each end. The form illustrated by Fig. 1 has notched ends for receiving the cylindrical plugs so that they are held firmly against both faces of the notch. The standard center-to-center distance C between the plugs is either 5 or 10 inches. The upper and lower sides of sine-bars are parallel to the centerline of the plugs within very close limits.

The body of the sine-bar ordinarily has several through holes to reduce the weight. In the making of the sine-bar shown in Fig. 2, if too much material is removed from one locating notch, regrinding the shoulder at the opposite end would make it possible to obtain the correct center distance. That is the reason for this change in form. The type of sine-bar illustrated by Fig. 3 has the cylindrical disks or plugs attached to one side. These differences in form or arrangement do not, of course, affect the principle governing the use of the sine-bar. An accurate surface plate or master flat is always used in conjunction with a sine-bar in order to form the base from which the vertical measurements are made.

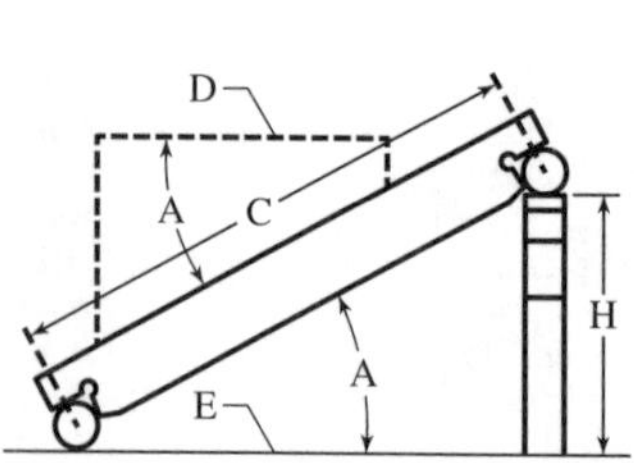

Fig. 1.

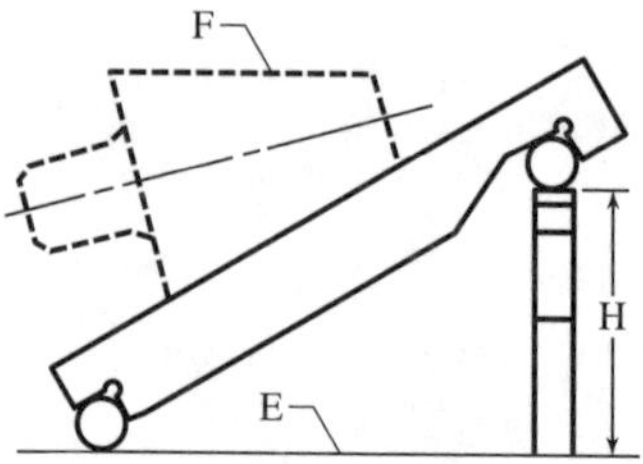

Fig. 2.

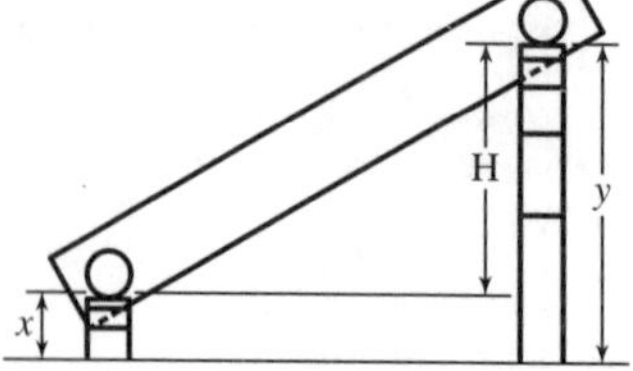

Fig. 3.

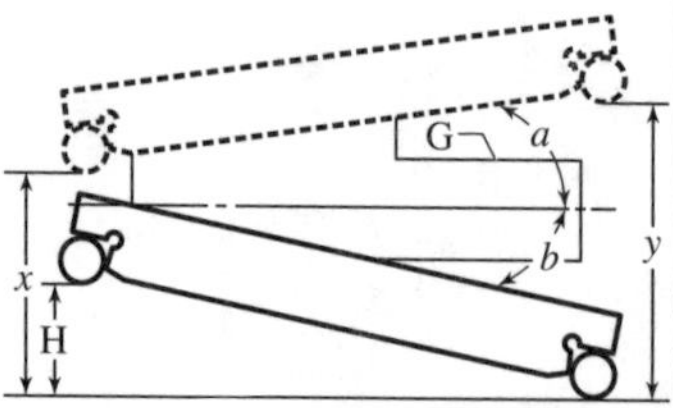

Fig. 4.

Setting a Sine-Bar to a Given Angle.—To find the vertical distance *H*, for setting a sine-bar to the required angle, convert the angle to decimal form on a pocket calculator, take the sine of that angle, and multiply by the distance between the cylinders. For example, if an angle of 31 degrees, 30 minutes is required, the equivalent angle is 31 degrees plus $^{30}/_{60}$ = 31 + 0.5, or 31.5 degrees. The sine of 31.5 degrees is 0.5225 and multiplying this value by the sine-bar length gives 2.613 in. for the height *H*, Fig. 1 and 3, of the gage blocks.

Finding Angle when Height *H* of Sine-Bar is Known.—To find the angle equivalent to a given height *H*, reverse the above procedure. Thus, if the height *H* is 1.4061 in., dividing by 5 gives a sine of 0.28122, which corresponds to an angle of 16.333 degrees, or 16 degrees 20 minutes.

Checking Angle of Templet or Gage by Using Sine-Bar.—Place templet or gage on sine-bar as indicated by dotted lines, Fig. 1. Clamps may be used to hold work in place. Place upper end of sine-bar on gage blocks having total height *H* corresponding to the required angle. If upper edge *D* of work is parallel with surface plate *E*, then angle *A* of work equals angle *A* to which sine-bar is set. Parallelism between edge *D* and surface plate may be tested by checking the height at each end with a dial gage or some type of indicating comparator.

Measuring Angle of Templet or Gage with Sine-Bar.—To measure such an angle, adjust height of gage blocks and sine-bar until edge *D*, Fig. 1, is parallel with surface plate *E*; then find angle corresponding to height *H* of gage blocks. For example, if height *H* is 2.5939 inches when *D* and *E* are parallel, the calculator will show that the angle *A* of the work is 31 degrees, 15 minutes.

Checking Taper per Foot with Sine-Bar.—As an example, assume that the plug gage in Fig. 2 is supposed to have a taper of $6\frac{1}{8}$ inches per foot and taper is to be checked by using a 5-inch sine-bar. The table of *Tapers per Foot and Corresponding Angles* on page 41 shows that the included angle for a taper of $6\frac{1}{8}$ inches per foot is 28 degrees 38 minutes 1 second, or 28.6336 degrees from the calculator. For a 5-inch sine-bar, the calculator gives a value of 2.396 in. for the height *H* of the gage blocks. Using this height, if the upper surface *F* of the plug gage is parallel to the surface plate, the angle corresponds to a taper of $6\frac{1}{8}$ inches per foot.

Setting Sine-Bar Having Plugs Attached to Side.—If the lower plug does not rest directly on the surface plate, as in Fig. 3, the height *H* for the sine-bar is the difference between heights *x* and *y*, or the difference between the heights of the plugs; otherwise, the procedure in setting the sine-bar and checking angles is the same as previously described.

Checking Templets Having Two Angles.—Assume that angle *a* of templet, Fig. 4, is 9 degrees, angle *b* 12 degrees, and that edge *G* is parallel to the surface plate. For an angle *b* of 12 degrees, the calculator shows that the height *H* is 1.03956 inches. For an angle *a* of 9 degrees, the difference between measurements *x* and *y* when the sine-bar is in contact with the upper edge of the templet is 0.78217 inch.

Setting 10-Inch Sine-Bar to Given Angle.—A 10-inch sine-bar may sometimes be preferred because of its longer working surface or because the longer center distance is conducive to greater precision. To obtain the vertical distances *H* for setting a 10-inch sine-bar, multiply the sine of the angle by 10, by shifting the decimal point one place to the right.

For example, the sine of 39 degrees is 0.62932, hence the vertical height *H* for setting a 10-inch sine-bar is 6.2932 inches.

Measuring Tapers with V-Block and Sine-Bar.—The taper on a conical part may be checked or found by placing the part in a V-block which rests on the surface of a sine-plate or sine-bar as shown in the accompanying diagram. The advantage of this method is that the axis of the V-block may be aligned with the sides of the sine-bar. Thus when the tapered part is placed in the V-block it will be aligned perpendicular to the transverse axis of the sine-bar.

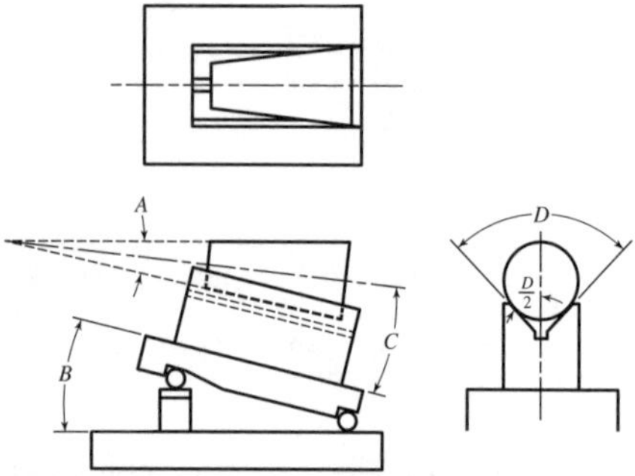

The sine-bar is set to angle $B = (C + A/2)$ where $A/2$ is one-half the included angle of the tapered part. If D is the included angle of the precision V-block, the angle C is calculated from the formula:

$$\sin C = \frac{\sin(A/2)}{\sin(D/2)}$$

If dial indicator readings show no change across all points along the top of the taper surface, then this checks that the angle A of the taper is correct.

If the indicator readings vary, proceed as follows to find the actual angle of taper: 1) Adjust the angle of the sine-bar until the indicator reading is constant. Then find the new angle B' as explained in the paragraph *Measuring Angle of Templet or Gage with Sine-Bar*; and 2) Using the angle B' calculate the actual half-angle $A'/2$ of the taper from the formula:

$$\tan\frac{A'}{2} = \frac{\sin B'}{\csc\frac{D}{2} + \cos B'}$$

Using a Calculator to Determine Sine-Bar Constants for a Given Angle.—The constant required to set a given angle for a sine-bar of any length can be quickly determined by using a scientific calculator. The required formulas are as follows:

a) angle A given in degrees and calculator is set to measure angles in radian

$$H = L \times \sin\left(A \times \frac{\pi}{180}\right)$$

or

a) angle A is given in radian, or
b) angle A is given in degrees and calculator is set to measure angles in degrees

$$H = L \times \sin(A)$$

where L = length of the sine-bar A = angle to which the sine-bar is to be set
H = vertical height to which one end of sine-bar must be set to obtain angle A
$\pi = 3.141592654$

In the previous formulas, the height H and length L must be given in the same units but may be in either metric or US units. Thus, if L is given in mm, then H is in mm; and, if L is given in inches, then H is in inches.

Constants for Setting a 5-Inch Sine-Bar for 1° to 7°

Min.	0°	1°	2°	3°	4°	5°	6°	7°
0	0.00000	0.08726	0.17450	0.26168	0.34878	0.43578	0.52264	0.60935
1	0.00145	0.08872	0.17595	0.26313	0.35023	0.43723	0.52409	0.61079
2	0.00291	0.09017	0.17740	0.26458	0.35168	0.43868	0.52554	0.61223
3	0.00436	0.09162	0.17886	0.26604	0.35313	0.44013	0.52698	0.61368
4	0.00582	0.09308	0.18031	0.26749	0.35459	0.44157	0.52843	0.61512
5	0.00727	0.09453	0.18177	0.26894	0.35604	0.44302	0.52987	0.61656
6	0.00873	0.09599	0.18322	0.27039	0.35749	0.44447	0.53132	0.61801
7	0.01018	0.09744	0.18467	0.27185	0.35894	0.44592	0.53277	0.61945
8	0.01164	0.09890	0.18613	0.27330	0.36039	0.44737	0.53421	0.62089
9	0.01309	0.10035	0.18758	0.27475	0.36184	0.44882	0.53566	0.62234
10	0.01454	0.10180	0.18903	0.27620	0.36329	0.45027	0.53710	0.62378
11	0.01600	0.10326	0.19049	0.27766	0.36474	0.45171	0.53855	0.62522
12	0.01745	0.10471	0.19194	0.27911	0.36619	0.45316	0.54000	0.62667
13	0.01891	0.10617	0.19339	0.28056	0.36764	0.45461	0.54144	0.62811
14	0.02036	0.10762	0.19485	0.28201	0.36909	0.45606	0.54289	0.62955
15	0.02182	0.10907	0.19630	0.28346	0.37054	0.45751	0.54433	0.63099
16	0.02327	0.11053	0.19775	0.28492	0.37199	0.45896	0.54578	0.63244
17	0.02473	0.11198	0.19921	0.28637	0.37344	0.46040	0.54723	0.63388
18	0.02618	0.11344	0.20066	0.28782	0.37489	0.46185	0.54867	0.63532
19	0.02763	0.11489	0.20211	0.28927	0.37634	0.46330	0.55012	0.63677
20	0.02909	0.11634	0.20357	0.29072	0.37779	0.46475	0.55156	0.63821
21	0.03054	0.11780	0.20502	0.29218	0.37924	0.46620	0.55301	0.63965
22	0.03200	0.11925	0.20647	0.29363	0.38069	0.46765	0.55445	0.64109
23	0.03345	0.12071	0.20793	0.29508	0.38214	0.46909	0.55590	0.64254
24	0.03491	0.12216	0.20938	0.29653	0.38360	0.47054	0.55734	0.64398
25	0.03636	0.12361	0.21083	0.29798	0.38505	0.47199	0.55879	0.64542
26	0.03782	0.12507	0.21228	0.29944	0.38650	0.47344	0.56024	0.64686
27	0.03927	0.12652	0.21374	0.30089	0.38795	0.47489	0.56168	0.64830
28	0.04072	0.12798	0.21519	0.30234	0.38940	0.47633	0.56313	0.64975
29	0.04218	0.12943	0.21664	0.30379	0.39085	0.47778	0.56457	0.65119
30	0.04363	0.13088	0.21810	0.30524	0.39230	0.47923	0.56602	0.65263
31	0.04509	0.13234	0.21955	0.30669	0.39375	0.48068	0.56746	0.65407
32	0.04654	0.13379	0.22100	0.30815	0.39520	0.48212	0.56891	0.65551
33	0.04800	0.13525	0.22246	0.30960	0.39665	0.48357	0.57035	0.65696
34	0.04945	0.13670	0.22391	0.31105	0.39810	0.48502	0.57180	0.65840
35	0.05090	0.13815	0.22536	0.31250	0.39954	0.48647	0.57324	0.65984
36	0.05236	0.13961	0.22681	0.31395	0.40099	0.48791	0.57469	0.66128
37	0.05381	0.14106	0.22827	0.31540	0.40244	0.48936	0.57613	0.66272
38	0.05527	0.14252	0.22972	0.31686	0.40389	0.49081	0.57758	0.66417
39	0.05672	0.14397	0.23117	0.31831	0.40534	0.49226	0.57902	0.66561
40	0.05818	0.14542	0.23263	0.31976	0.40679	0.49370	0.58046	0.66705
41	0.05963	0.14688	0.23408	0.32121	0.40824	0.49515	0.58191	0.66849
42	0.06109	0.14833	0.23553	0.32266	0.40969	0.49660	0.58335	0.66993
43	0.06254	0.14979	0.23699	0.32411	0.41114	0.49805	0.58480	0.67137
44	0.06399	0.15124	0.23844	0.32556	0.41259	0.49949	0.58624	0.67281
45	0.06545	0.15269	0.23989	0.32702	0.41404	0.50094	0.58769	0.67425
46	0.06690	0.15415	0.24134	0.32847	0.41549	0.50239	0.58913	0.67570
47	0.06836	0.15560	0.24280	0.32992	0.41694	0.50383	0.59058	0.67714
48	0.06981	0.15705	0.24425	0.33137	0.41839	0.50528	0.59202	0.67858
49	0.07127	0.15851	0.24570	0.33282	0.41984	0.50673	0.59346	0.68002
50	0.07272	0.15996	0.24715	0.33427	0.42129	0.50818	0.59491	0.68146
51	0.07417	0.16141	0.24861	0.33572	0.42274	0.50962	0.59635	0.68290
52	0.07563	0.16287	0.25006	0.33717	0.42419	0.51107	0.59780	0.68434
53	0.07708	0.16432	0.25151	0.33863	0.42564	0.51252	0.59924	0.68578
54	0.07854	0.16578	0.25296	0.34008	0.42708	0.51396	0.60068	0.68722
55	0.07999	0.16723	0.25442	0.34153	0.42853	0.51541	0.60213	0.68866
56	0.08145	0.16868	0.25587	0.34298	0.42998	0.51686	0.60357	0.69010
57	0.08290	0.17014	0.25732	0.34443	0.43143	0.51830	0.60502	0.69154
58	0.08435	0.17159	0.25877	0.34588	0.43288	0.51975	0.60646	0.69298
59	0.08581	0.17304	0.26023	0.34733	0.43433	0.52120	0.60790	0.69443
60	0.08726	0.17450	0.26168	0.34878	0.43578	0.52264	0.60935	0.69587

Constants for Setting a 5-Inch Sine-Bar for 8° to 15°

Min.	8°	9°	10°	11°	12°	13°	14°	15°
0	0.69587	0.78217	0.86824	0.95404	1.03956	1.12476	1.20961	1.29410
1	0.69731	0.78361	0.86967	0.95547	1.04098	1.12617	1.21102	1.29550
2	0.69875	0.78505	0.87111	0.95690	1.04240	1.12759	1.21243	1.29690
3	0.70019	0.78648	0.87254	0.95833	1.04383	1.12901	1.21384	1.29831
4	0.70163	0.78792	0.87397	0.95976	1.04525	1.13042	1.21525	1.29971
5	0.70307	0.78935	0.87540	0.96118	1.04667	1.13184	1.21666	1.30112
6	0.70451	0.79079	0.87683	0.96261	1.04809	1.13326	1.21808	1.30252
7	0.70595	0.79223	0.87827	0.96404	1.04951	1.13467	1.21949	1.30393
8	0.70739	0.79366	0.87970	0.96546	1.05094	1.13609	1.22090	1.30533
9	0.70883	0.79510	0.88113	0.96689	1.05236	1.13751	1.22231	1.30673
10	0.71027	0.79653	0.88256	0.96832	1.05378	1.13892	1.22372	1.30814
11	0.71171	0.79797	0.88399	0.96974	1.05520	1.14034	1.22513	1.30954
12	0.71314	0.79941	0.88542	0.97117	1.05662	1.14175	1.22654	1.31095
13	0.71458	0.80084	0.88686	0.97260	1.05805	1.14317	1.22795	1.31235
14	0.71602	0.80228	0.88829	0.97403	1.05947	1.14459	1.22936	1.31375
15	0.71746	0.80371	0.88972	0.97545	1.06089	1.14600	1.23077	1.31516
16	0.71890	0.80515	0.89115	0.97688	1.06231	1.14742	1.23218	1.31656
17	0.72034	0.80658	0.89258	0.97830	1.06373	1.14883	1.23359	1.31796
18	0.72178	0.80802	0.89401	0.97973	1.06515	1.15025	1.23500	1.31937
19	0.72322	0.80945	0.89544	0.98116	1.06657	1.15166	1.23640	1.32077
20	0.72466	0.81089	0.89687	0.98258	1.06799	1.15308	1.23781	1.32217
21	0.72610	0.81232	0.89830	0.98401	1.06941	1.15449	1.23922	1.32357
22	0.72754	0.81376	0.89973	0.98544	1.07084	1.15591	1.24063	1.32498
23	0.72898	0.81519	0.90117	0.98686	1.07226	1.15732	1.24204	1.32638
24	0.73042	0.81663	0.90260	0.98829	1.07368	1.15874	1.24345	1.32778
25	0.73185	0.81806	0.90403	0.98971	1.07510	1.16015	1.24486	1.32918
26	0.73329	0.81950	0.90546	0.99114	1.07652	1.16157	1.24627	1.33058
27	0.73473	0.82093	0.90689	0.99256	1.07794	1.16298	1.24768	1.33199
28	0.73617	0.82237	0.90832	0.99399	1.07936	1.16440	1.24908	1.33339
29	0.73761	0.82380	0.90975	0.99541	1.08078	1.16581	1.25049	1.33479
30	0.73905	0.82524	0.91118	0.99684	1.08220	1.16723	1.25190	1.33619
31	0.74049	0.82667	0.91261	0.99826	1.08362	1.16864	1.25331	1.33759
32	0.74192	0.82811	0.91404	0.99969	1.08504	1.17006	1.25472	1.33899
33	0.74336	0.82954	0.91547	1.00112	1.08646	1.17147	1.25612	1.34040
34	0.74480	0.83098	0.91690	1.00254	1.08788	1.17288	1.25753	1.34180
35	0.74624	0.83241	0.91833	1.00396	1.08930	1.17430	1.25894	1.34320
36	0.74768	0.83384	0.91976	1.00539	1.09072	1.17571	1.26035	1.34460
37	0.74911	0.83528	0.92119	1.00681	1.09214	1.17712	1.26175	1.34600
38	0.75055	0.83671	0.92262	1.00824	1.09355	1.17854	1.26316	1.34740
39	0.75199	0.83815	0.92405	1.00966	1.09497	1.17995	1.26457	1.34880
40	0.75343	0.83958	0.92547	1.01109	1.09639	1.18136	1.26598	1.35020
41	0.75487	0.84101	0.92690	1.01251	1.09781	1.18278	1.26738	1.35160
42	0.75630	0.84245	0.92833	1.01394	1.09923	1.18419	1.26879	1.35300
43	0.75774	0.84388	0.92976	1.01536	1.10065	1.18560	1.27020	1.35440
44	0.75918	0.84531	0.93119	1.01678	1.10207	1.18702	1.27160	1.35580
45	0.76062	0.84675	0.93262	1.01821	1.10349	1.18843	1.27301	1.35720
46	0.76205	0.84818	0.93405	1.01963	1.10491	1.18984	1.27442	1.35860
47	0.76349	0.84961	0.93548	1.02106	1.10632	1.19125	1.27582	1.36000
48	0.76493	0.85105	0.93691	1.02248	1.10774	1.19267	1.27723	1.36140
49	0.76637	0.85248	0.93834	1.02390	1.10916	1.19408	1.27863	1.36280
50	0.76780	0.85391	0.93976	1.02533	1.11058	1.19549	1.28004	1.36420
51	0.76924	0.85535	0.94119	1.02675	1.11200	1.19690	1.28145	1.36560
52	0.77068	0.85678	0.94262	1.02817	1.11342	1.19832	1.28285	1.36700
53	0.77211	0.85821	0.94405	1.02960	1.11483	1.19973	1.28426	1.36840
54	0.77355	0.85965	0.94548	1.03102	1.11625	1.20114	1.28566	1.36980
55	0.77499	0.86108	0.94691	1.03244	1.11767	1.20255	1.28707	1.37119
56	0.77643	0.86251	0.94833	1.03387	1.11909	1.20396	1.28847	1.37259
57	0.77786	0.86394	0.94976	1.03529	1.12050	1.20538	1.28988	1.37399
58	0.77930	0.86538	0.95119	1.03671	1.12192	1.20679	1.29129	1.37539
59	0.78074	0.86681	0.95262	1.03814	1.12334	1.20820	1.29269	1.37679
60	0.78217	0.86824	0.95404	1.03956	1.12476	1.20961	1.29410	1.37819

Constants for Setting a 5-Inch Sine-Bar for 16° to 23°

Min.	16°	17°	18°	19°	20°	21°	22°	23°
0	1.37819	1.46186	1.54509	1.62784	1.71010	1.79184	1.87303	1.95366
1	1.37958	1.46325	1.54647	1.62922	1.71147	1.79320	1.87438	1.95499
2	1.38098	1.46464	1.54785	1.63059	1.71283	1.79456	1.87573	1.95633
3	1.38238	1.46603	1.54923	1.63197	1.71420	1.79591	1.87708	1.95767
4	1.38378	1.46742	1.55062	1.63334	1.71557	1.79727	1.87843	1.95901
5	1.38518	1.46881	1.55200	1.63472	1.71693	1.79863	1.87977	1.96035
6	1.38657	1.47020	1.55338	1.63609	1.71830	1.79998	1.88112	1.96169
7	1.38797	1.47159	1.55476	1.63746	1.71966	1.80134	1.88247	1.96302
8	1.38937	1.47298	1.55615	1.63884	1.72103	1.80270	1.88382	1.96436
9	1.39076	1.47437	1.55753	1.64021	1.72240	1.80405	1.88516	1.96570
10	1.39216	1.47576	1.55891	1.64159	1.72376	1.80541	1.88651	1.96704
11	1.39356	1.47715	1.56029	1.64296	1.72513	1.80677	1.88786	1.96837
12	1.39496	1.47854	1.56167	1.64433	1.72649	1.80812	1.88920	1.96971
13	1.39635	1.47993	1.56306	1.64571	1.72786	1.80948	1.89055	1.97105
14	1.39775	1.48132	1.56444	1.64708	1.72922	1.81083	1.89190	1.97238
15	1.39915	1.48271	1.56582	1.64845	1.73059	1.81219	1.89324	1.97372
16	1.40054	1.48410	1.56720	1.64983	1.73195	1.81355	1.89459	1.97506
17	1.40194	1.48549	1.56858	1.65120	1.73331	1.81490	1.89594	1.97639
18	1.40333	1.48687	1.56996	1.65257	1.73468	1.81626	1.89728	1.97773
19	1.40473	1.48826	1.57134	1.65394	1.73604	1.81761	1.89863	1.97906
20	1.40613	1.48965	1.57272	1.65532	1.73741	1.81897	1.89997	1.98040
21	1.40752	1.49104	1.57410	1.65669	1.73877	1.82032	1.90132	1.98173
22	1.40892	1.49243	1.57548	1.65806	1.74013	1.82168	1.90266	1.98307
23	1.41031	1.49382	1.57687	1.65943	1.74150	1.82303	1.90401	1.98440
24	1.41171	1.49520	1.57825	1.66081	1.74286	1.82438	1.90535	1.98574
25	1.41310	1.49659	1.57963	1.66218	1.74422	1.82574	1.90670	1.98707
26	1.41450	1.49798	1.58101	1.66355	1.74559	1.82709	1.90804	1.98841
27	1.41589	1.49937	1.58238	1.66492	1.74695	1.82845	1.90939	1.98974
28	1.41729	1.50075	1.58376	1.66629	1.74831	1.82980	1.91073	1.99108
29	1.41868	1.50214	1.58514	1.66766	1.74967	1.83115	1.91207	1.99241
30	1.42008	1.50353	1.58652	1.66903	1.75104	1.83251	1.91342	1.99375
31	1.42147	1.50492	1.58790	1.67041	1.75240	1.83386	1.91476	1.99508
32	1.42287	1.50630	1.58928	1.67178	1.75376	1.83521	1.91610	1.99641
33	1.42426	1.50769	1.59066	1.67315	1.75512	1.83657	1.91745	1.99775
34	1.42565	1.50908	1.59204	1.67452	1.75649	1.83792	1.91879	1.99908
35	1.42705	1.51046	1.59342	1.67589	1.75785	1.83927	1.92013	2.00041
36	1.42844	1.51185	1.59480	1.67726	1.75921	1.84062	1.92148	2.00175
37	1.42984	1.51324	1.59617	1.67863	1.76057	1.84198	1.92282	2.00308
38	1.43123	1.51462	1.59755	1.68000	1.76193	1.84333	1.92416	2.00441
39	1.43262	1.51601	1.59893	1.68137	1.76329	1.84468	1.92550	2.00574
40	1.43402	1.51739	1.60031	1.68274	1.76465	1.84603	1.92685	2.00708
41	1.43541	1.51878	1.60169	1.68411	1.76601	1.84738	1.92819	2.00841
42	1.43680	1.52017	1.60307	1.68548	1.76737	1.84873	1.92953	2.00974
43	1.43820	1.52155	1.60444	1.68685	1.76873	1.85009	1.93087	2.01107
44	1.43959	1.52294	1.60582	1.68821	1.77010	1.85144	1.93221	2.01240
45	1.44098	1.52432	1.60720	1.68958	1.77146	1.85279	1.93355	2.01373
46	1.44237	1.52571	1.60857	1.69095	1.77282	1.85414	1.93490	2.01506
47	1.44377	1.52709	1.60995	1.69232	1.77418	1.85549	1.93624	2.01640
48	1.44516	1.52848	1.61133	1.69369	1.77553	1.85684	1.93758	2.01773
49	1.44655	1.52986	1.61271	1.69506	1.77689	1.85819	1.93892	2.01906
50	1.44794	1.53125	1.61408	1.69643	1.77825	1.85954	1.94026	2.02039
51	1.44934	1.53263	1.61546	1.69779	1.77961	1.86089	1.94160	2.02172
52	1.45073	1.53401	1.61683	1.69916	1.78097	1.86224	1.94294	2.02305
53	1.45212	1.53540	1.61821	1.70053	1.78233	1.86359	1.94428	2.02438
54	1.45351	1.53678	1.61959	1.70190	1.78369	1.86494	1.94562	2.02571
55	1.45490	1.53817	1.62096	1.70327	1.78505	1.86629	1.94696	2.02704
56	1.45629	1.53955	1.62234	1.70463	1.78641	1.86764	1.94830	2.02837
57	1.45769	1.54093	1.62371	1.70600	1.78777	1.86899	1.94964	2.02970
58	1.45908	1.54232	1.62509	1.70737	1.78912	1.87034	1.95098	2.03103
59	1.46047	1.54370	1.62647	1.70873	1.79048	1.87168	1.95232	2.03235
60	1.46186	1.54509	1.62784	1.71010	1.79184	1.87303	1.95366	2.03368

Constants for Setting a 5-Inch Sine-Bar for 24° to 31°

Min.	24°	25°	26°	27°	28°	29°	30°	31°
0	2.03368	2.11309	2.19186	2.26995	2.34736	2.42405	2.50000	2.57519
1	2.03501	2.11441	2.19316	2.27125	2.34864	2.42532	2.50126	2.57644
2	2.03634	2.11573	2.19447	2.27254	2.34993	2.42659	2.50252	2.57768
3	2.03767	2.11704	2.19578	2.27384	2.35121	2.42786	2.50378	2.57893
4	2.03900	2.11836	2.19708	2.27513	2.35249	2.42913	2.50504	2.58018
5	2.04032	2.11968	2.19839	2.27643	2.35378	2.43041	2.50630	2.58142
6	2.04165	2.12100	2.19970	2.27772	2.35506	2.43168	2.50755	2.58267
7	2.04298	2.12231	2.20100	2.27902	2.35634	2.43295	2.50881	2.58391
8	2.04431	2.12363	2.20231	2.28031	2.35763	2.43422	2.51007	2.58516
9	2.04563	2.12495	2.20361	2.28161	2.35891	2.43549	2.51133	2.58640
10	2.04696	2.12626	2.20492	2.28290	2.36019	2.43676	2.51259	2.58765
11	2.04829	2.12758	2.20622	2.28420	2.36147	2.43803	2.51384	2.58889
12	2.04962	2.12890	2.20753	2.28549	2.36275	2.43930	2.51510	2.59014
13	2.05094	2.13021	2.20883	2.28678	2.36404	2.44057	2.51636	2.59138
14	2.05227	2.13153	2.21014	2.28808	2.36532	2.44184	2.51761	2.59262
15	2.05359	2.13284	2.21144	2.28937	2.36660	2.44311	2.51887	2.59387
16	2.05492	2.13416	2.21275	2.29066	2.36788	2.44438	2.52013	2.59511
17	2.05625	2.13547	2.21405	2.29196	2.36916	2.44564	2.52138	2.59635
18	2.05757	2.13679	2.21536	2.29325	2.37044	2.44691	2.52264	2.59760
19	2.05890	2.13810	2.21666	2.29454	2.37172	2.44818	2.52389	2.59884
20	2.06022	2.13942	2.21796	2.29583	2.37300	2.44945	2.52515	2.60008
21	2.06155	2.14073	2.21927	2.29712	2.37428	2.45072	2.52640	2.60132
22	2.06287	2.14205	2.22057	2.29842	2.37556	2.45198	2.52766	2.60256
23	2.06420	2.14336	2.22187	2.29971	2.37684	2.45325	2.52891	2.60381
24	2.06552	2.14468	2.22318	2.30100	2.37812	2.45452	2.53017	2.60505
25	2.06685	2.14599	2.22448	2.30229	2.37940	2.45579	2.53142	2.60629
26	2.06817	2.14730	2.22578	2.30358	2.38068	2.45705	2.53268	2.60753
27	2.06950	2.14862	2.22708	2.30487	2.38196	2.45832	2.53393	2.60877
28	2.07082	2.14993	2.22839	2.30616	2.38324	2.45959	2.53519	2.61001
29	2.07214	2.15124	2.22969	2.30745	2.38452	2.46085	2.53644	2.61125
30	2.07347	2.15256	2.23099	2.30874	2.38579	2.46212	2.53769	2.61249
31	2.07479	2.15387	2.23229	2.31003	2.38707	2.46338	2.53894	2.61373
32	2.07611	2.15518	2.23359	2.31132	2.38835	2.46465	2.54020	2.61497
33	2.07744	2.15649	2.23489	2.31261	2.38963	2.46591	2.54145	2.61621
34	2.07876	2.15781	2.23619	2.31390	2.39091	2.46718	2.54270	2.61745
35	2.08008	2.15912	2.23749	2.31519	2.39218	2.46844	2.54396	2.61869
36	2.08140	2.16043	2.23880	2.31648	2.39346	2.46971	2.54521	2.61993
37	2.08273	2.16174	2.24010	2.31777	2.39474	2.47097	2.54646	2.62117
38	2.08405	2.16305	2.24140	2.31906	2.39601	2.47224	2.54771	2.62241
39	2.08537	2.16436	2.24270	2.32035	2.39729	2.47350	2.54896	2.62364
40	2.08669	2.16567	2.24400	2.32163	2.39857	2.47477	2.55021	2.62488
41	2.08801	2.16698	2.24530	2.32292	2.39984	2.47603	2.55146	2.62612
42	2.08934	2.16830	2.24660	2.32421	2.40112	2.47729	2.55271	2.62736
43	2.09066	2.16961	2.24789	2.32550	2.40239	2.47856	2.55397	2.62860
44	2.09198	2.17092	2.24919	2.32679	2.40367	2.47982	2.55522	2.62983
45	2.09330	2.17223	2.25049	2.32807	2.40494	2.48108	2.55647	2.63107
46	2.09462	2.17354	2.25179	2.32936	2.40622	2.48235	2.55772	2.63231
47	2.09594	2.17485	2.25309	2.33065	2.40749	2.48361	2.55896	2.63354
48	2.09726	2.17616	2.25439	2.33193	2.40877	2.48487	2.56021	2.63478
49	2.09858	2.17746	2.25569	2.33322	2.41004	2.48613	2.56146	2.63602
50	2.09990	2.17877	2.25698	2.33451	2.41132	2.48739	2.56271	2.63725
51	2.10122	2.18008	2.25828	2.33579	2.41259	2.48866	2.56396	2.63849
52	2.10254	2.18139	2.25958	2.33708	2.41386	2.48992	2.56521	2.63972
53	2.10386	2.18270	2.26088	2.33836	2.41514	2.49118	2.56646	2.64096
54	2.10518	2.18401	2.26217	2.33965	2.41641	2.49244	2.56771	2.64219
55	2.10650	2.18532	2.26347	2.34093	2.41769	2.49370	2.56895	2.64343
56	2.10782	2.18663	2.26477	2.34222	2.41896	2.49496	2.57020	2.64466
57	2.10914	2.18793	2.26606	2.34350	2.42023	2.49622	2.57145	2.64590
58	2.11045	2.18924	2.26736	2.34479	2.42150	2.49748	2.57270	2.64713
59	2.11177	2.19055	2.26866	2.34607	2.42278	2.49874	2.57394	2.64836
60	2.11309	2.19186	2.26995	2.34736	2.42405	2.50000	2.57519	2.64960

Constants for Setting a 5-Inch Sine-Bar for 32° to 39°

Min.	32°	33°	34°	35°	36°	37°	38°	39°
0	2.64960	2.72320	2.79596	2.86788	2.93893	3.00908	3.07831	3.14660
1	2.65083	2.72441	2.79717	2.86907	2.94010	3.01024	3.07945	3.14773
2	2.65206	2.72563	2.79838	2.87026	2.94128	3.01140	3.08060	3.14886
3	2.65330	2.72685	2.79958	2.87146	2.94246	3.01256	3.08174	3.14999
4	2.65453	2.72807	2.80079	2.87265	2.94363	3.01372	3.08289	3.15112
5	2.65576	2.72929	2.80199	2.87384	2.94481	3.01488	3.08403	3.15225
6	2.65699	2.73051	2.80319	2.87503	2.94598	3.01604	3.08518	3.15338
7	2.65822	2.73173	2.80440	2.87622	2.94716	3.01720	3.08632	3.15451
8	2.65946	2.73295	2.80560	2.87741	2.94833	3.01836	3.08747	3.15564
9	2.66069	2.73416	2.80681	2.87860	2.94951	3.01952	3.08861	3.15676
10	2.66192	2.73538	2.80801	2.87978	2.95068	3.02068	3.08976	3.15789
11	2.66315	2.73660	2.80921	2.88097	2.95185	3.02184	3.09090	3.15902
12	2.66438	2.73782	2.81042	2.88216	2.95303	3.02300	3.09204	3.16015
13	2.66561	2.73903	2.81162	2.88335	2.95420	3.02415	3.09318	3.16127
14	2.66684	2.74025	2.81282	2.88454	2.95538	3.02531	3.09433	3.16240
15	2.66807	2.74147	2.81402	2.88573	2.95655	3.02647	3.09547	3.16353
16	2.66930	2.74268	2.81523	2.88691	2.95772	3.02763	3.09661	3.16465
17	2.67053	2.74390	2.81643	2.88810	2.95889	3.02878	3.09775	3.16578
18	2.67176	2.74511	2.81763	2.88929	2.96007	3.02994	3.09890	3.16690
19	2.67299	2.74633	2.81883	2.89048	2.96124	3.03110	3.10004	3.16803
20	2.67422	2.74754	2.82003	2.89166	2.96241	3.03226	3.10118	3.16915
21	2.67545	2.74876	2.82123	2.89285	2.96358	3.03341	3.10232	3.17028
22	2.67668	2.74997	2.82243	2.89403	2.96475	3.03457	3.10346	3.17140
23	2.67791	2.75119	2.82364	2.89522	2.96592	3.03572	3.10460	3.17253
24	2.67913	2.75240	2.82484	2.89641	2.96709	3.03688	3.10574	3.17365
25	2.68036	2.75362	2.82604	2.89759	2.96827	3.03803	3.10688	3.17478
26	2.68159	2.75483	2.82723	2.89878	2.96944	3.03919	3.10802	3.17590
27	2.68282	2.75605	2.82843	2.89996	2.97061	3.04034	3.10916	3.17702
28	2.68404	2.75726	2.82963	2.90115	2.97178	3.04150	3.11030	3.17815
29	2.68527	2.75847	2.83083	2.90233	2.97294	3.04265	3.11143	3.17927
30	2.68650	2.75969	2.83203	2.90351	2.97411	3.04381	3.11257	3.18039
31	2.68772	2.76090	2.83323	2.90470	2.97528	3.04496	3.11371	3.18151
32	2.68895	2.76211	2.83443	2.90588	2.97645	3.04611	3.11485	3.18264
33	2.69018	2.76332	2.83563	2.90707	2.97762	3.04727	3.11599	3.18376
34	2.69140	2.76453	2.83682	2.90825	2.97879	3.04842	3.11712	3.18488
35	2.69263	2.76575	2.83802	2.90943	2.97996	3.04957	3.11826	3.18600
36	2.69385	2.76696	2.83922	2.91061	2.98112	3.05073	3.11940	3.18712
37	2.69508	2.76817	2.84042	2.91180	2.98229	3.05188	3.12053	3.18824
38	2.69630	2.76938	2.84161	2.91298	2.98346	3.05303	3.12167	3.18936
39	2.69753	2.77059	2.84281	2.91416	2.98463	3.05418	3.12281	3.19048
40	2.69875	2.77180	2.84401	2.91534	2.98579	3.05533	3.12394	3.19160
41	2.69998	2.77301	2.84520	2.91652	2.98696	3.05648	3.12508	3.19272
42	2.70120	2.77422	2.84640	2.91771	2.98813	3.05764	3.12621	3.19384
43	2.70243	2.77543	2.84759	2.91889	2.98929	3.05879	3.12735	3.19496
44	2.70365	2.77664	2.84879	2.92007	2.99046	3.05994	3.12848	3.19608
45	2.70487	2.77785	2.84998	2.92125	2.99162	3.06109	3.12962	3.19720
46	2.70610	2.77906	2.85118	2.92243	2.99279	3.06224	3.13075	3.19831
47	2.70732	2.78027	2.85237	2.92361	2.99395	3.06339	3.13189	3.19943
48	2.70854	2.78148	2.85357	2.92479	2.99512	3.06454	3.13302	3.20055
49	2.70976	2.78269	2.85476	2.92597	2.99628	3.06568	3.13415	3.20167
50	2.71099	2.78389	2.85596	2.92715	2.99745	3.06683	3.13529	3.20278
51	2.71221	2.78510	2.85715	2.92833	2.99861	3.06798	3.13642	3.20390
52	2.71343	2.78631	2.85834	2.92950	2.99977	3.06913	3.13755	3.20502
53	2.71465	2.78752	2.85954	2.93068	3.00094	3.07028	3.13868	3.20613
54	2.71587	2.78873	2.86073	2.93186	3.00210	3.07143	3.13982	3.20725
55	2.71709	2.78993	2.86192	2.93304	3.00326	3.07257	3.14095	3.20836
56	2.71831	2.79114	2.86311	2.93422	3.00443	3.07372	3.14208	3.20948
57	2.71953	2.79235	2.86431	2.93540	3.00559	3.07487	3.14321	3.21059
58	2.72076	2.79355	2.86550	2.93657	3.00675	3.07601	3.14434	3.21171
59	2.72198	2.79476	2.86669	2.93775	3.00791	3.07716	3.14547	3.21282
60	2.72320	2.79596	2.86788	2.93893	3.00908	3.07831	3.14660	3.21394

Constants for Setting a 5-Inch Sine-Bar for 40° to 47°

Min.	40°	41°	42°	43°	44°	45°	46°	47°
0	3.21394	3.28030	3.34565	3.40999	3.47329	3.53553	3.59670	3.65677
1	3.21505	3.28139	3.34673	3.41106	3.47434	3.53656	3.59771	3.65776
2	3.21617	3.28249	3.34781	3.41212	3.47538	3.53759	3.59872	3.65875
3	3.21728	3.28359	3.34889	3.41318	3.47643	3.53862	3.59973	3.65974
4	3.21839	3.28468	3.34997	3.41424	3.47747	3.53965	3.60074	3.66073
5	3.21951	3.28578	3.35105	3.41531	3.47852	3.54067	3.60175	3.66172
6	3.22062	3.28688	3.35213	3.41637	3.47956	3.54170	3.60276	3.66271
7	3.22173	3.28797	3.35321	3.41743	3.48061	3.54273	3.60376	3.66370
8	3.22284	3.28907	3.35429	3.41849	3.48165	3.54375	3.60477	3.66469
9	3.22395	3.29016	3.35537	3.41955	3.48270	3.54478	3.60578	3.66568
10	3.22507	3.29126	3.35645	3.42061	3.48374	3.54580	3.60679	3.66667
11	3.22618	3.29235	3.35753	3.42168	3.48478	3.54683	3.60779	3.66766
12	3.22729	3.29345	3.35860	3.42274	3.48583	3.54785	3.60880	3.66865
13	3.22840	3.29454	3.35968	3.42380	3.48687	3.54888	3.60981	3.66964
14	3.22951	3.29564	3.36076	3.42486	3.48791	3.54990	3.61081	3.67063
15	3.23062	3.29673	3.36183	3.42592	3.48895	3.55093	3.61182	3.67161
16	3.23173	3.29782	3.36291	3.42697	3.48999	3.55195	3.61283	3.67260
17	3.23284	3.29892	3.36399	3.42803	3.49104	3.55297	3.61383	3.67359
18	3.23395	3.30001	3.36506	3.42909	3.49208	3.55400	3.61484	3.67457
19	3.23506	3.30110	3.36614	3.43015	3.49312	3.55502	3.61584	3.67556
20	3.23617	3.30219	3.36721	3.43121	3.49416	3.55604	3.61684	3.67655
21	3.23728	3.30329	3.36829	3.43227	3.49520	3.55707	3.61785	3.67753
22	3.23838	3.30438	3.36936	3.43332	3.49624	3.55809	3.61885	3.67852
23	3.23949	3.30547	3.37044	3.43438	3.49728	3.55911	3.61986	3.67950
24	3.24060	3.30656	3.37151	3.43544	3.49832	3.56013	3.62086	3.68049
25	3.24171	3.30765	3.37259	3.43649	3.49936	3.56115	3.62186	3.68147
26	3.24281	3.30874	3.37366	3.43755	3.50039	3.56217	3.62286	3.68245
27	3.24392	3.30983	3.37473	3.43861	3.50143	3.56319	3.62387	3.68344
28	3.24503	3.31092	3.37581	3.43966	3.50247	3.56421	3.62487	3.68442
29	3.24613	3.31201	3.37688	3.44072	3.50351	3.56523	3.62587	3.68540
30	3.24724	3.31310	3.37795	3.44177	3.50455	3.56625	3.62687	3.68639
31	3.24835	3.31419	3.37902	3.44283	3.50558	3.56727	3.62787	3.68737
32	3.24945	3.31528	3.38010	3.44388	3.50662	3.56829	3.62887	3.68835
33	3.25056	3.31637	3.38117	3.44494	3.50766	3.56931	3.62987	3.68933
34	3.25166	3.31746	3.38224	3.44599	3.50869	3.57033	3.63087	3.69031
35	3.25277	3.31854	3.38331	3.44704	3.50973	3.57135	3.63187	3.69130
36	3.25387	3.31963	3.38438	3.44810	3.51077	3.57236	3.63287	3.69228
37	3.25498	3.32072	3.38545	3.44915	3.51180	3.57338	3.63387	3.69326
38	3.25608	3.32181	3.38652	3.45020	3.51284	3.57440	3.63487	3.69424
39	3.25718	3.32289	3.38759	3.45126	3.51387	3.57542	3.63587	3.69522
40	3.25829	3.32398	3.38866	3.45231	3.51491	3.57643	3.63687	3.69620
41	3.25939	3.32507	3.38973	3.45336	3.51594	3.57745	3.63787	3.69718
42	3.26049	3.32615	3.39080	3.45441	3.51697	3.57846	3.63886	3.69816
43	3.26159	3.32724	3.39187	3.45546	3.51801	3.57948	3.63986	3.69913
44	3.26270	3.32832	3.39294	3.45651	3.51904	3.58049	3.64086	3.70011
45	3.26380	3.32941	3.39400	3.45757	3.52007	3.58151	3.64186	3.70109
46	3.26490	3.33049	3.39507	3.45862	3.52111	3.58252	3.64285	3.70207
47	3.26600	3.33158	3.39614	3.45967	3.52214	3.58354	3.64385	3.70305
48	3.26710	3.33266	3.39721	3.46072	3.52317	3.58455	3.64484	3.70402
49	3.26820	3.33375	3.39827	3.46177	3.52420	3.58557	3.64584	3.70500
50	3.26930	3.33483	3.39934	3.46281	3.52523	3.58658	3.64683	3.70598
51	3.27040	3.33591	3.40041	3.46386	3.52627	3.58759	3.64783	3.70695
52	3.27150	3.33700	3.40147	3.46491	3.52730	3.58861	3.64882	3.70793
53	3.27260	3.33808	3.40254	3.46596	3.52833	3.58962	3.64982	3.70890
54	3.27370	3.33916	3.40360	3.46701	3.52936	3.59063	3.65081	3.70988
55	3.27480	3.34025	3.40467	3.46806	3.53039	3.59164	3.65181	3.71085
56	3.27590	3.34133	3.40573	3.46910	3.53142	3.59266	3.65280	3.71183
57	3.27700	3.34241	3.40680	3.47015	3.53245	3.59367	3.65379	3.71280
58	3.27810	3.34349	3.40786	3.47120	3.53348	3.59468	3.65478	3.71378
59	3.27920	3.34457	3.40893	3.47225	3.53451	3.59569	3.65578	3.71475
60	3.28030	3.34565	3.40999	3.47329	3.53553	3.59670	3.65677	3.71572

Constants for Setting a 5-Inch Sine-Bar for 48° to 55°

Min.	48°	49°	50°	51°	52°	53°	54°	55°
0	3.71572	3.77355	3.83022	3.88573	3.94005	3.99318	4.04508	4.09576
1	3.71670	3.77450	3.83116	3.88665	3.94095	3.99405	4.04594	4.09659
2	3.71767	3.77546	3.83209	3.88756	3.94184	3.99493	4.04679	4.09743
3	3.71864	3.77641	3.83303	3.88847	3.94274	3.99580	4.04765	4.09826
4	3.71961	3.77736	3.83396	3.88939	3.94363	3.99668	4.04850	4.09909
5	3.72059	3.77831	3.83489	3.89030	3.94453	3.99755	4.04936	4.09993
6	3.72156	3.77927	3.83583	3.89122	3.94542	3.99842	4.05021	4.10076
7	3.72253	3.78022	3.83676	3.89213	3.94631	3.99930	4.05106	4.10159
8	3.72350	3.78117	3.83769	3.89304	3.94721	4.00017	4.05191	4.10242
9	3.72447	3.78212	3.83862	3.89395	3.94810	4.00104	4.05277	4.10325
10	3.72544	3.78307	3.83956	3.89487	3.94899	4.00191	4.05362	4.10409
11	3.72641	3.78402	3.84049	3.89578	3.94988	4.00279	4.05447	4.10492
12	3.72738	3.78498	3.84142	3.89669	3.95078	4.00366	4.05532	4.10575
13	3.72835	3.78593	3.84235	3.89760	3.95167	4.00453	4.05617	4.10658
14	3.72932	3.78688	3.84328	3.89851	3.95256	4.00540	4.05702	4.10741
15	3.73029	3.78783	3.84421	3.89942	3.95345	4.00627	4.05787	4.10823
16	3.73126	3.78877	3.84514	3.90033	3.95434	4.00714	4.05872	4.10906
17	3.73222	3.78972	3.84607	3.90124	3.95523	4.00801	4.05957	4.10989
18	3.73319	3.79067	3.84700	3.90215	3.95612	4.00888	4.06042	4.11072
19	3.73416	3.79162	3.84793	3.90306	3.95701	4.00975	4.06127	4.11155
20	3.73513	3.79257	3.84886	3.90397	3.95790	4.01062	4.06211	4.11238
21	3.73609	3.79352	3.84978	3.90488	3.95878	4.01148	4.06296	4.11320
22	3.73706	3.79446	3.85071	3.90579	3.95967	4.01235	4.06381	4.11403
23	3.73802	3.79541	3.85164	3.90669	3.96056	4.01322	4.06466	4.11486
24	3.73899	3.79636	3.85257	3.90760	3.96145	4.01409	4.06550	4.11568
25	3.73996	3.79730	3.85349	3.90851	3.96234	4.01495	4.06635	4.11651
26	3.74092	3.79825	3.85442	3.90942	3.96322	4.01582	4.06720	4.11733
27	3.74189	3.79919	3.85535	3.91032	3.96411	4.01669	4.06804	4.11816
28	3.74285	3.80014	3.85627	3.91123	3.96500	4.01755	4.06889	4.11898
29	3.74381	3.80109	3.85720	3.91214	3.96588	4.01842	4.06973	4.11981
30	3.74478	3.80203	3.85812	3.91304	3.96677	4.01928	4.07058	4.12063
31	3.74574	3.80297	3.85905	3.91395	3.96765	4.02015	4.07142	4.12145
32	3.74671	3.80392	3.85997	3.91485	3.96854	4.02101	4.07227	4.12228
33	3.74767	3.80486	3.86090	3.91576	3.96942	4.02188	4.07311	4.12310
34	3.74863	3.80581	3.86182	3.91666	3.97031	4.02274	4.07395	4.12392
35	3.74959	3.80675	3.86274	3.91756	3.97119	4.02361	4.07480	4.12475
36	3.75056	3.80769	3.86367	3.91847	3.97207	4.02447	4.07564	4.12557
37	3.75152	3.80863	3.86459	3.91937	3.97296	4.02533	4.07648	4.12639
38	3.75248	3.80958	3.86551	3.92027	3.97384	4.02619	4.07732	4.12721
39	3.75344	3.81052	3.86644	3.92118	3.97472	4.02706	4.07817	4.12803
40	3.75440	3.81146	3.86736	3.92208	3.97560	4.02792	4.07901	4.12885
41	3.75536	3.81240	3.86828	3.92298	3.97649	4.02878	4.07985	4.12967
42	3.75632	3.81334	3.86920	3.92388	3.97737	4.02964	4.08069	4.13049
43	3.75728	3.81428	3.87012	3.92478	3.97825	4.03050	4.08153	4.13131
44	3.75824	3.81522	3.87104	3.92568	3.97913	4.03136	4.08237	4.13213
45	3.75920	3.81616	3.87196	3.92658	3.98001	4.03222	4.08321	4.13295
46	3.76016	3.81710	3.87288	3.92748	3.98089	4.03308	4.08405	4.13377
47	3.76112	3.81804	3.87380	3.92839	3.98177	4.03394	4.08489	4.13459
48	3.76207	3.81898	3.87472	3.92928	3.98265	4.03480	4.08572	4.13540
49	3.76303	3.81992	3.87564	3.93018	3.98353	4.03566	4.08656	4.13622
50	3.76399	3.82086	3.87656	3.93108	3.98441	4.03652	4.08740	4.13704
51	3.76495	3.82179	3.87748	3.93198	3.98529	4.03738	4.08824	4.13785
52	3.76590	3.82273	3.87840	3.93288	3.98616	4.03823	4.08908	4.13867
53	3.76686	3.82367	3.87931	3.93378	3.98704	4.03909	4.08991	4.13949
54	3.76782	3.82461	3.88023	3.93468	3.98792	4.03995	4.09075	4.14030
55	3.76877	3.82554	3.88115	3.93557	3.98880	4.04081	4.09158	4.14112
56	3.76973	3.82648	3.88207	3.93647	3.98967	4.04166	4.09242	4.14193
57	3.77068	3.82742	3.88298	3.93737	3.99055	4.04252	4.09326	4.14275
58	3.77164	3.82835	3.88390	3.93826	3.99143	4.04337	4.09409	4.14356
59	3.77259	3.82929	3.88481	3.93916	3.99230	4.04423	4.09493	4.14437
60	3.77355	3.83022	3.88573	3.94005	3.99318	4.04508	4.09576	4.14519

Accurate Measurement of Angles and Tapers.—When great accuracy is required in the measurement of angles, or when originating tapers, disks are commonly used. The principle of the disk method of taper measurement is that if two disks of unequal diameters are placed either in contact or a certain distance apart, lines tangent to their peripheries will represent an angle or taper, the degree of which depends upon the diameters of the two disks and the distance between them.

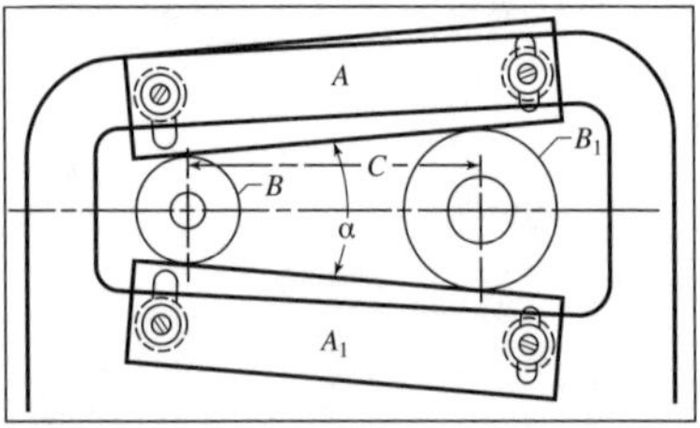

The gage shown in the accompanying illustration, which is a form commonly used for originating tapers or measuring angles accurately, is set by means of disks. This gage consists of two adjustable straight edges A and A_1, which are in contact with disks B and B_1. The angle α or the taper between the straight edges depends, of course, upon the diameters of the disks and the center distance C, and as these three dimensions can be measured accurately, it is possible to set the gage to a given angle within very close limits. Moreover, if a record of the three dimensions is kept, the exact setting of the gage can be reproduced quickly at any time. The following rules may be used for adjusting a gage of this type, and cover all problems likely to arise in practice. Disks are also occasionally used for the setting of parts in angular positions when they are to be machined accurately to a given angle: the rules are applicable to these conditions also.

Rules for Figuring Tapers

D_L = diameter of taper, large end
D_S = diameter of taper, small end
L = length of taper (inches)
TPF = taper per foot
TPI = taper per inch
T = taper in a certain length, in inches

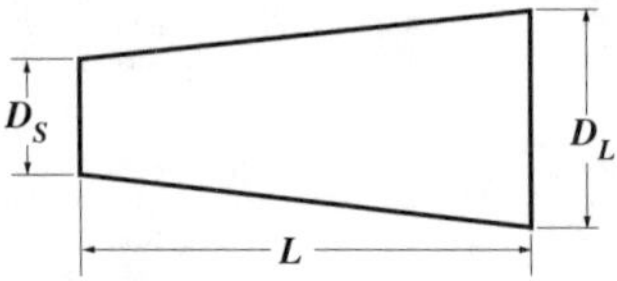

Given	To Find	Rule
TPF	TPI	$TPI = TPF/12$
TPI	TPF	$TPF = TPI \times 12$
D_L, D_S, L	TPF	$TPF = 12\dfrac{D_L - D_S}{L}$
D_L, L, TPF	D_S	$D_S = D_L - \dfrac{TPF}{12}L$
D_S, L, TPF	D_L	$D_L = D_S + \dfrac{TPF}{12}L$
D_L, D_S, TPF	L	$L = (D_L - D_S)\dfrac{12}{TPF}$
L, TPF	T	$T = \dfrac{TPF}{12} \times L$

Tapers per Foot and Corresponding Angles

Taper per Foot	Included Angle				Angle with Center Line			Taper per Foot	Included Angle				Angle with Center Line		
1/64	0.074604°	0°	4′	29″	0°	2′	14″	1 7/8	8.934318°	8°	56′	4″	4°	28′	2″
1/32	0.149208°	0	8	57	0	4	29	1 15/16	9.230863°	9	13	51	4	36	56
1/16	0.298415	0	17	54	0	8	57	2	9.527283	9	31	38	4	45	49
3/32	0.447621	0	26	51	0	13	26	2 1/8	10.119738	10	7	11	5	3	36
1/8	0.596826	0	35	49	0	17	54	2 1/4	10.711650	10	42	42	5	21	21
5/32	0.746028	0	44	46	0	22	23	2 3/8	11.302990	11	18	11	5	39	5
3/16	0.895228	0	53	43	0	26	51	2 1/2	11.893726	11	53	37	5	56	49
7/32	1.044425	1	2	40	0	31	20	2 5/8	12.483829	12	29	2	6	14	31
1/4	1.193619	1	11	37	0	35	49	2 3/4	13.073267	13	4	24	6	32	12
9/32	1.342808	1	20	34	0	40	17	2 7/8	13.662012	13	39	43	6	49	52
5/16	1.491993	1	29	31	0	44	46	3	14.250033	14	15	0	7	7	30
11/32	1.641173	1	38	28	0	49	14	3 1/8	14.837300	14	50	14	7	25	7
3/8	1.790347	1	47	25	0	53	43	3 1/4	15.423785	15	25	26	7	42	43
13/32	1.939516	1	56	22	0	58	11	3 3/8	16.009458	16	0	34	8	0	17
7/16	2.088677	2	5	19	1	2	40	3 1/2	16.594290	16	35	39	8	17	50
15/32	2.237832	2	14	16	1	7	8	3 5/8	17.178253	17	10	42	8	35	21
1/2	2.386979	2	23	13	1	11	37	3 3/4	17.761318	17	45	41	8	52	50
17/32	2.536118	2	32	10	1	16	5	3 7/8	18.343458	18	20	36	9	10	18
9/16	2.685248	2	41	7	1	20	33	4	18.924644	18	55	29	9	27	44
19/32	2.834369	2	50	4	1	25	2	4 1/8	19.504850	19	30	17	9	45	9
5/8	2.983481	2	59	1	1	29	30	4 1/4	20.084047	20	5	3	10	2	31
21/32	3.132582	3	7	57	1	33	59	4 3/8	20.662210	20	39	44	10	19	52
11/16	3.281673	3	16	54	1	38	27	4 1/2	21.239311	21	14	22	10	37	11
23/32	3.430753	3	25	51	1	42	55	4 5/8	21.815324	21	48	55	10	54	28
3/4	3.579821	3	34	47	1	47	24	4 3/4	22.390223	22	23	25	11	11	42
25/32	3.728877	3	43	44	1	51	52	4 7/8	22.963983	22	57	50	11	28	55
13/16	3.877921	3	52	41	1	56	20	5	23.536578	23	32	12	11	46	6
27/32	4.026951	4	1	37	2	0	49	5 1/8	24.107983	24	6	29	12	3	14
7/8	4.175968	4	10	33	2	5	17	5 1/4	24.678175	24	40	41	12	20	21
29/32	4.324970	4	19	30	2	9	45	5 3/8	25.247127	25	14	50	12	37	25
15/16	4.473958	4	28	26	2	14	13	5 1/2	25.814817	25	48	53	12	54	27
31/32	4.622931	4	37	23	2	18	41	5 5/8	26.381221	26	22	52	13	11	26
1	4.771888	4	46	19	2	23	9	5 3/4	26.946316	26	56	47	13	28	23
1 1/16	5.069753	5	4	11	2	32	6	5 7/8	27.510079	27	30	36	13	45	18
1 1/8	5.367550	5	22	3	2	41	2	6	28.072487	28	4	21	14	2	10
1 3/16	5.665275	5	39	55	2	49	57	6 1/8	28.633518	28	38	1	14	19	0
1 1/4	5.962922	5	57	47	2	58	53	6 1/4	29.193151	29	11	35	14	35	48
1 5/16	6.260490	6	15	38	3	7	49	6 3/8	29.751364	29	45	5	14	52	32
1 3/8	6.557973	6	33	29	3	16	44	6 1/2	30.308136	30	18	29	15	9	15
1 7/16	6.855367	6	51	19	3	25	40	6 5/8	30.863447	30	51	48	15	25	54
1 1/2	7.152669	7	9	10	3	34	35	6 3/4	31.417276	31	25	2	15	42	31
1 9/16	7.449874	7	27	0	3	43	30	6 7/8	31.969603	31	58	11	15	59	5
1 5/8	7.746979	7	44	49	3	52	25	7	32.520409	32	31	13	16	15	37
1 11/16	8.043980	8	2	38	4	1	19	7 1/8	33.069676	33	4	11	16	32	5
1 3/4	8.340873	8	20	27	4	10	14	7 1/4	33.617383	33	37	3	16	48	31
1 13/16	8.637654	8	38	16	4	19	8	7 3/8	34.163514	34	9	49	17	4	54

Taper per foot represents inches of taper per foot of length.

To find angle α for given taper *T* in inches per foot.—

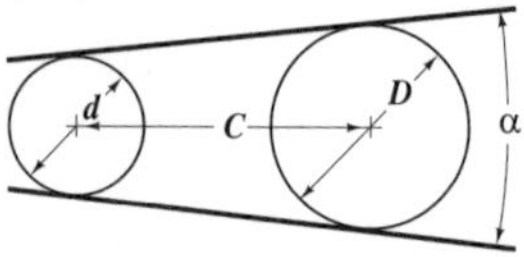

$$\alpha = 2\arctan(T/24)$$

To find taper per foot *T* given angle α in degrees.—

$$T = 24\ \tan(\alpha/2) \text{ inches per foot}$$

To find angle a given dimensions *D*, *d*, and *C*.—Let K be the difference in the disk diameters divided by twice the center distance. $K = (D - d)/(2C)$, then $\alpha = 2\arcsin K$

To find taper *T* measured at right angles to a line through the disk centers given dimensions *D*, *d*, and distance *C*.—Find K using the formula in the previous example, then $T = 24K/\sqrt{1 - K^2}$ inches per foot

To find center distance *C* for a given taper *T* in inches per foot.—

$$C = \frac{D - d}{2} \times \frac{\sqrt{1 + (T/24)^2}}{T/24} \text{ inches}$$

To find center distance *C* for a given angle α and dimensions *D* and *d*.—

$$C = (D - d)/2\sin(\alpha/2) \text{ inches}$$

To find taper *T* measured at right angles to one side.—When one side is taken as a base line and the taper is measured at right angles to that side, calculate K as explained above and use the following formula for determining the taper T:

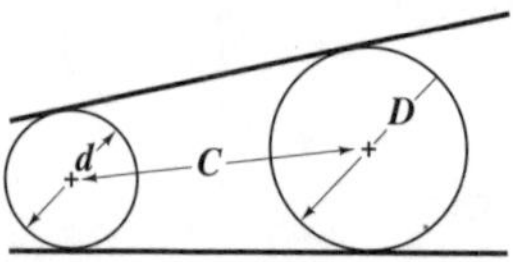

$$T = 24K\frac{\sqrt{1 - K^2}}{1 - 2K^2} \text{ inches per foot}$$

To find center distance *C* when taper *T* is measured from one side.—

$$C = \frac{D - d}{\sqrt{2 - 2/\sqrt{1 + (T/12)^2}}} \text{ inches}$$

To find diameter *D* of a large disk in contact with a small disk of diameter *d* given angle α.—

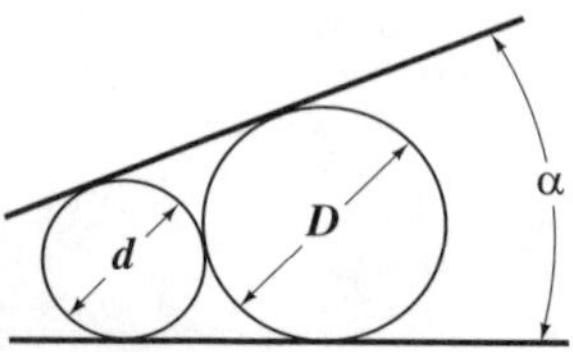

$$D = d \times \frac{1 + \sin(\alpha/2)}{1 - \sin(\alpha/2)} \text{ inches}$$

Gage Block Sets—Inch Sizes *Federal Specification GGG-G-15C*

Set Number 1 (81 Blocks)								
First Series: 0.0001 Inch Increments (9 Blocks)								
0.1001	0.1002	0.1003	0.1004	0.1005	0.1006	0.1007	0.1008	0.1009

Second Series: 0.001 Inch Increments (49 Blocks)									
0.101	0.102	0.103	0.104	0.105	0.106	0.107	0.108	0.109	0.110
0.111	0.112	0.113	0.114	0.115	0.116	0.117	0.118	0.119	0.120
0.121	0.122	0.123	0.124	0.125	0.126	0.127	0.128	0.129	0.130
0.131	0.132	0.133	0.134	0.135	0.136	0.137	0.138	0.139	0.140
0.141	0.142	0.143	0.144	0.145	0.146	0.147	0.148	0.149	

Third Series: 0.050 Inch Increments (19 Blocks)									
0.050	0.100	0.150	0.200	0.250	0.300	0.350	0.400	0.450	0.500
0.550	0.600	0.650	0.700	0.750	0.800	0.850	0.900	0.950	

Fourth Series: 1.000 Inch Increments (4 Blocks)			
1.000	2.000	3.000	4.000

Set number 4 is not shown, and the Specification does not list a set 2 or 3.
Arranged here in incremental series for convenience of use.

Example, Making a Gage Block Stack: Determine the blocks required to obtain a dimension of 3.6742 inch.

1) Use the fewest numbers of blocks for a given dimension. Otherwise the chance of error can be increased by the wringing interval between blocks.

2) Block selection is based on successively eliminating the right-hand figure of the desired dimension.

3) Stacks can be constructed with or without wear blocks.

3.6742
-0.100 Subtract 0.100 for the two 0.050 wear blocks
3.5742
-0.1002 Eliminate the 0.0002 with the 0.1002 block from the first series and subtract
3.4740
-0.124 Eliminate the 0.004 with the 0.124 block from the second series and subtract
3.3500
−0.350 Eliminate the 0.350 with the 0.350 block from the third series and eliminate
-3.0000 the 3.000 with the 3.0000 block from the fourth series and subtract
0.0000

The combined blocks are:	2	0.050 wear blocks
	1	0.124 block
	1	0.350 block
	1	3.0000 block

Measuring Dovetail Slides.—Dovetail slides that must be machined accurately to a given width are commonly gaged by using pieces of cylindrical rod or wire and measuring as indicated by the dimensions *x* and *y* of the accompanying illustrations.

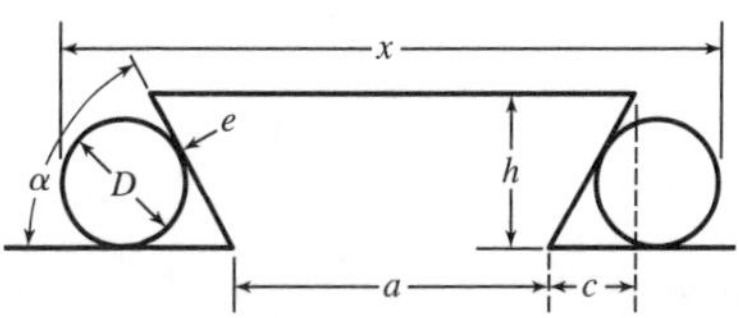

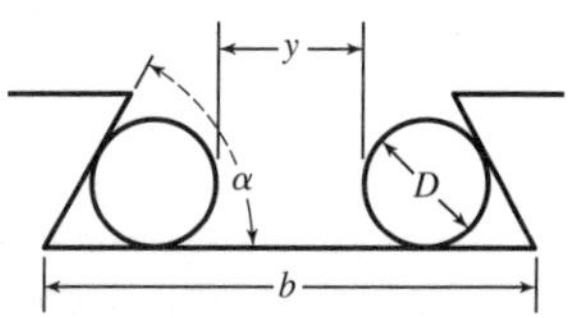

To obtain dimension x for measuring male dovetails, add 1 to the cotangent of one-half the dovetail angle α, multiply by diameter D of the rods used, and add the product to dimension a. To obtain dimension y for measuring a female dovetail, add 1 to the cotangent of one-half the dovetail angle α, multiply by diameter D of the rod used, and subtract the result from dimension b. Expressing these rules as formulas:

$$x = D(1 + \cot \tfrac{1}{2}\alpha) + a$$
$$y = b - D(1 + \cot \tfrac{1}{2}\alpha)$$
$$c = h \times \cot\alpha$$

The rod or wire used should be small enough so that the point of contact e is somewhat below the corner or edge of the dovetail.

Checking a V-Shaped Groove by Measurement Over Pins.—In checking a groove of the shape shown in Fig. 5, it is necessary to measure the dimension X over the pins of radius R. If values for the radius R, dimension Z, and the angles α and β are known, the problem is to determine the distance Y, to arrive at the required overall dimension for X. If a line AC is drawn from the bottom of the V to the center of the pin at the left in Fig. 5, and a line CB from the center of this pin to its point of tangency with the side of the V, a right-angled triangle is formed in which one side, CB, is known and one angle, CAB, can be determined. A line drawn from the center of a circle to the point of intersection of two tangents to the circle bisects the angle made by the tangent lines, and angle CAB therefore equals $\tfrac{1}{2}(\alpha + \beta)$. The length AC and the angle DAC can now be found, and with AC known in the right-angled triangle ADC, AD, which is equal to Y, can be found.

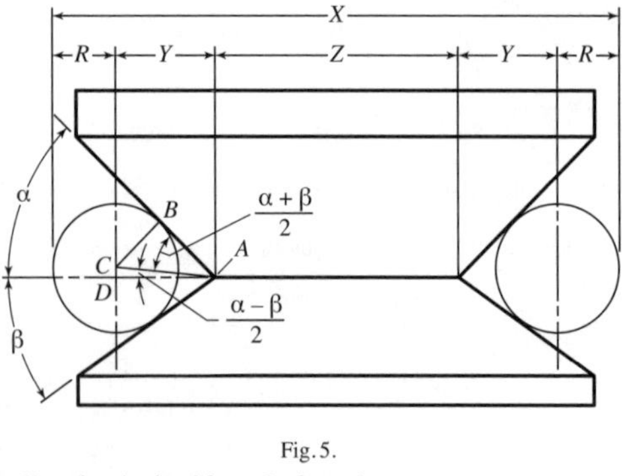

Fig. 5.

The value for X can be obtained from the formula

$$X = Z + 2R\left(\csc\frac{\alpha+\beta}{2}\cos\frac{\alpha-\beta}{2} + 1\right)$$

For example, if $R = 0.500$, $Z = 1.824$, $\alpha = 45$ degrees, and $\beta = 35$ degrees,

$$X = 1.824 + (2 \cdot 0.5)\left(\csc\frac{45° + 35°}{2}\cos\frac{45° - 35°}{2} + 1\right)$$

$$X = 1.824 + \csc 40° \cos 5° + 1$$
$$X = 1.824 + 1.5557 \cdot 0.99619 + 1$$
$$X = 1.824 + 1.550 + 1 = 4.374$$

Diameters of Wires for Measuring American Standard and British Standard Whitworth Screw Threads

Threads per Inch	Pitch, Inch	Wire Diameters for American Standard Threads			Wire Diameters for Whitworth Standard Threads		
		Max.	Min.	Pitch Line Contact	Max.	Min.	Pitch Line Contact
4	0.2500	0.2250	0.1400	0.1443	0.1900	0.1350	0.1409
4½	0.2222	0.2000	0.1244	0.1283	0.1689	0.1200	0.1253
5	0.2000	0.1800	0.1120	0.1155	0.1520	0.1080	0.1127
5½	0.1818	0.1636	0.1018	0.1050	0.1382	0.0982	0.1025
6	0.1667	0.1500	0.0933	0.0962	0.1267	0.0900	0.0939
7	0.1428	0.1283	0.0800	0.0825	0.1086	0.0771	0.0805
8	0.1250	0.1125	0.0700	0.0722	0.0950	0.0675	0.0705
9	0.1111	0.1000	0.0622	0.0641	0.0844	0.0600	0.0626
10	0.1000	0.0900	0.0560	0.0577	0.0760	0.0540	0.0564
11	0.0909	0.0818	0.0509	0.0525	0.0691	0.0491	0.0512
12	0.0833	0.0750	0.0467	0.0481	0.0633	0.0450	0.0470
13	0.0769	0.0692	0.0431	0.0444	0.0585	0.0415	0.0434
14	0.0714	0.0643	0.0400	0.0412	0.0543	0.0386	0.0403
16	0.0625	0.0562	0.0350	0.0361	0.0475	0.0337	0.0352
18	0.0555	0.0500	0.0311	0.0321	0.0422	0.0300	0.0313
20	0.0500	0.0450	0.0280	0.0289	0.0380	0.0270	0.0282
22	0.0454	0.0409	0.0254	0.0262	0.0345	0.0245	0.0256
24	0.0417	0.0375	0.0233	0.0240	0.0317	0.0225	0.0235
28	0.0357	0.0321	0.0200	0.0206	0.0271	0.0193	0.0201
32	0.0312	0.0281	0.0175	0.0180	0.0237	0.0169	0.0176
36	0.0278	0.0250	0.0156	0.0160	0.0211	0.0150	0.0156
40	0.0250	0.0225	0.0140	0.0144	0.0190	0.0135	0.0141

Notation Used in Formulas for Checking Pitch Diameters of Screw Threads by Three-Wire Method

A = one-half included thread angle in the axial plane

A_n = one-half included thread angle in the normal plane or in plane perpendicular to sides of thread = one-half included angle of cutter when thread is milled ($\tan A_n = \tan A \times \cos B$)

(*Note*: Included angle of milling cutter or grinding wheel may equal the nominal included angle of thread, or may be reduced to whatever normal angle is required to make the thread angle standard in the axial plane. In either case, A_n = one-half cutter angle.)

B = lead angle at pitch diameter = helix angle of thread as measured from a plane perpendicular to the axis. Tan $B = L \div 3.1416E$

D = basic major or outside diameter

E = pitch diameter (basic, maximum, or minimum) for which M is required, or pitch diameter corresponding to measurement M

H = helix angle at pitch diameter and measured from axis = $90° - B$ or $\tan H = \cot B$

H_b = helix angle at R_b measured from axis

L = lead of thread = pitch $P \times$ number of threads S

M = dimension over wires

P = pitch = $1 \div$ number of threads per inch

S = number of "starts" or threads on a multiple-threaded worm or screw

T = $0.5\,P$ = width of thread in axial plane at diameter E

T_a = arc thickness on pitch cylinder in plane perpendicular to axis

W = wire or pin diameter

Formulas for Checking Pitch Diameters of Screw Threads

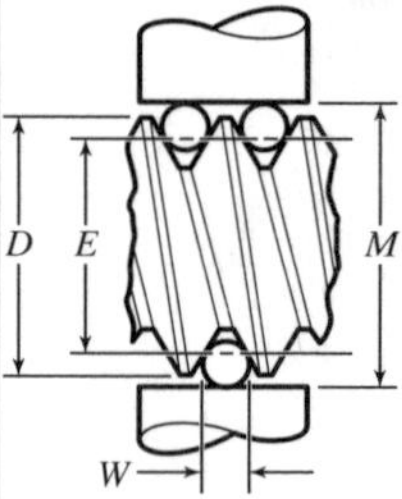

The formulas below do not compensate for the effect of the lead angle upon measurement M, but they are sufficiently accurate for checking standard single-thread screws unless exceptional accuracy is required. See accompanying information on effect of lead angle; also matter relating to measuring wire sizes, accuracy required for such wires, and contact or measuring pressure. The approximate best wire size for pitch-line contact may be obtained by the formula

$$W = 0.5 \times \text{pitch} \times \sec \tfrac{1}{2} \text{ included thread angle}$$

For 60-degree threads, $W = 0.57735 \times \text{pitch}$.

Form of Thread	Formulas for determining measurement M corresponding to correct pitch diameter and the pitch diameter E corresponding to a given measurement over wires.[a]
American National Standard Unified	When measurement M is known. $E = M + 0.86603P - 3W$ When pitch diameter E is used in formula. $M = E - 0.86603P + 3W$ The American Standard formerly was known as US Standard.
British Standard Whitworth	When measurement M is known. $E = M + 0.9605P - 3.1657W$ When pitch diameter E is used in formula. $M = E - 0.9605P + 3.1657W$
British Association Standard	When measurement M is known. $E = M + 1.1363P - 3.4829W$ When pitch diameter E is used in formula. $M = E - 1.1363P + 3.4829W$
Lowenherz Thread	When measurement M is known. $E = M + P - 3.2359W$ When pitch diameter E is used in formula. $M = E - P + 3.2359W$
Sharp V-Thread	When measurement M is known. $E = M + 0.86603P - 3W$ When pitch diameter E is used in formula. $M = E - 0.86603P + 3W$
International Standard	Use the formula given above for the American National Standard Unified Thread.
Buttress Form of Thread	Various forms of buttress threads are used. See paragraph on *Three-Wire Method Applied to Buttress Threads*, on page 47.

[a] The wires must be lapped to a uniform diameter and it is very important to insert in the rule or formula the wire diameter as determined by precise means of measurement. Any error will be multiplied.

Three-Wire Method Applied to Buttress Threads.—The angles of buttress threads vary somewhat, especially on the front or load-resisting side. Formula (1), which follows, may be applied to any angles required. In this formula, M = measurement over wires when *pitch diameter* E is correct; A = included angle of thread and thread groove; a = angle of front face or load-resisting side, measured from a line perpendicular to screw thread axis; P = pitch of thread; and W = wire diameter.

$$M = E - \left[\frac{P}{\tan a + \tan(A-a)}\right] + W\left[1 + \cos\frac{A}{2} - a \times \csc\frac{A}{2}\right] \tag{1}$$

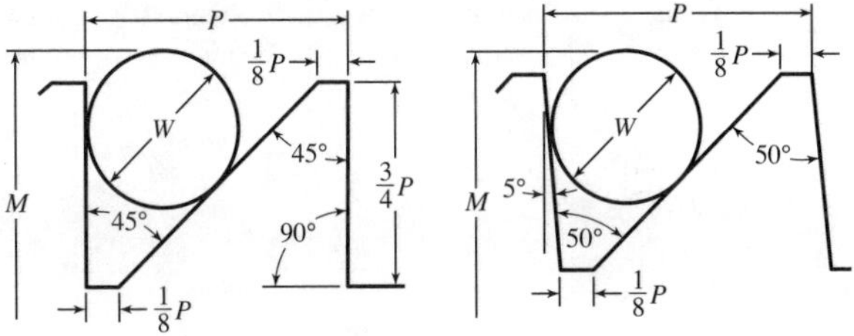

For given angles A and a, this general formula may be simplified as shown by Formulas (3) and (4). These simplified formulas contain constants with values depending upon angles A and a.

Wire Diameter: The wire diameter for obtaining pitch-line contact at the back of a buttress thread may be determined by the following general Formula (2):

$$W = P\left(\frac{\cos a}{1 + \cos A}\right) \tag{2}$$

45-Degree Buttress Thread: The buttress thread shown by the diagram at the left, has a front or load-resisting side that is perpendicular to the axis of the screw. Measurement M equivalent to a correct pitch diameter E may be determined by Formula (3):

$$M = E - P + (W \times 3.4142) \tag{3}$$

Wire diameter W for pitch-line contact at back of thread = 0.586 × pitch.

50-Degree Buttress Thread with Front-face Inclination of 5 Degrees: This buttress thread form is illustrated by the diagram at the right. Measurement M equivalent to the correct pitch diameter E may be determined by Formula (4):

$$M = E - (P \times 0.91955) + (W \times 3.2235) \tag{4}$$

Wire diameter W for pitch-line contact at back of thread = 0.606 × pitch. If the width of flat at crest and root = ⅛ × pitch, depth = 0.69 × pitch.

American National Standard Buttress Threads ANSI/ASME B1.9-1973 (R2017): This buttress screw thread has an included thread angle of 52 degrees and a front face inclination of 7 degrees. Measurements M equivalent to a pitch diameter E may be determined by Formula (5):

$$M = E - 0.89064P + 3.15689W + c \tag{5}$$

The wire angle correction factor c is less than 0.0004 inch for recommended combinations of thread diameters and pitches and may be neglected. Use of wire diameter W = $0.54147P$ is recommended.

Constants Used in Formulas for Measuring Pitch Diameters of Inch Screws by the Three-Wire System

No. of Threads per Inch	American Standard Unified and Sharp V-Thread 0.86603*P*	Whitworth Thread 0.9605*P*	No. of Threads per Inch	American Standard Unified and Sharp V-Thread 0.86603*P*	Whitworth Thread 0.9605*P*
2¼	0.38490	0.42689	18	0.04811	0.05336
2⅜	0.36464	0.40442	20	0.04330	0.04803
2½	0.34641	0.38420	22	0.03936	0.04366
2⅝	0.32992	0.36590	24	0.03608	0.04002
2¾	0.31492	0.34927	26	0.03331	0.03694
2⅞	0.30123	0.33409	28	0.03093	0.03430
3	0.28868	0.32017	30	0.02887	0.03202
3¼	0.26647	0.29554	32	0.02706	0.03002
3½	0.24744	0.27443	34	0.02547	0.02825
4	0.21651	0.24013	36	0.02406	0.02668
4½	0.19245	0.21344	38	0.02279	0.02528
5	0.17321	0.19210	40	0.02165	0.02401
5½	0.15746	0.17464	42	0.02062	0.02287
6	0.14434	0.16008	44	0.01968	0.02183
7	0.12372	0.13721	46	0.01883	0.02088
8	0.10825	0.12006	48	0.01804	0.02001
9	0.09623	0.10672	50	0.01732	0.01921
10	0.08660	0.09605	52	0.01665	0.01847
11	0.07873	0.08732	56	0.01546	0.01715
12	0.07217	0.08004	60	0.01443	0.01601
13	0.06662	0.07388	64	0.01353	0.01501
14	0.06186	0.06861	68	0.01274	0.01412
15	0.05774	0.06403	72	0.01203	0.01334
16	0.05413	0.06003	80	0.01083	0.01201

Constants Used in Formulas for Measuring Pitch Diameters of Metric Screws by the Three-Wire System

Pitch in mm	0.86603*P* in Inches	*W* in Inches	Pitch in mm	0.86603*P* in Inches	*W* in Inches	Pitch in mm	0.86603*P* in Inches	*W* in Inches
0.2	0.00682	0.00455	0.75	0.02557	0.01705	3.5	0.11933	0.07956
0.25	0.00852	0.00568	0.8	0.02728	0.01818	4	0.13638	0.09092
0.3	0.01023	0.00682	1	0.03410	0.02273	4.5	0.15343	0.10229
0.35	0.01193	0.00796	1.25	0.04262	0.02841	5	0.17048	0.11365
0.4	0.01364	0.00909	1.5	0.05114	0.03410	5.5	0.18753	0.12502
0.45	0.01534	0.01023	1.75	0.05967	0.03978	6	0.20457	0.13638
0.5	0.01705	0.01137	2	0.06819	0.04546	8	0.30686	0.18184
0.6	0.02046	0.01364	2.5	0.08524	0.05683	…	…	…
0.7	0.02387	0.01591	3	0.10229	0.06819	…	…	…

This table may be used for American National Standard Metric Threads. The formulas for American Standard Unified Threads on page 47 are used. In the table above, the values of 0.86603P and W are in inches so that the values for E and M calculated from the formulas on page 47 are also in inches.

STANDARD TAPERS

Morse Taper.—Dimensions relating to Morse standard taper shanks and sockets may be found in an accompanying table. The taper for different numbers of Morse tapers is slightly different, but it is approximately $\frac{5}{8}$ inch per foot in most cases. The table gives the actual tapers, accurate to five decimal places. Morse taper shanks are used on a variety of tools, and exclusively on the shanks of twist drills.

Table 1. Morse Standard Taper Shanks

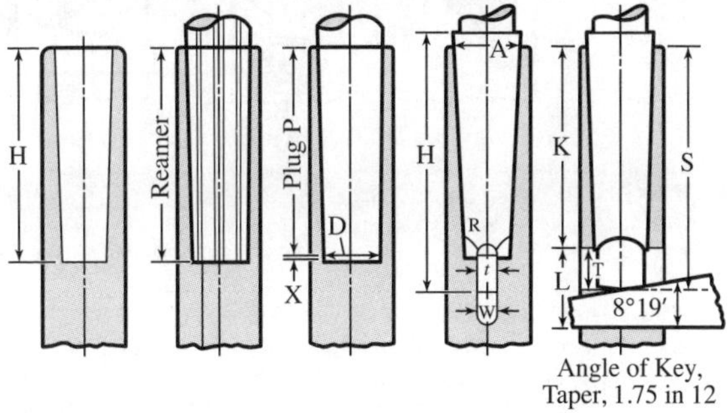

No. of Taper	Taper per Foot	Taper per Inch	Small End of Plug *D*	Diameter End of Socket *A*	Shank Length *B*	Shank Depth *S*	Depth of Hole *H*
0	0.62460	0.05205	0.252	0.3561	$2\frac{11}{32}$	$2\frac{7}{32}$	$2\frac{1}{32}$
1	0.59858	0.04988	0.369	0.475	$2\frac{9}{16}$	$2\frac{7}{16}$	$2\frac{5}{32}$
2	0.59941	0.04995	0.572	0.700	$3\frac{1}{8}$	$2\frac{15}{16}$	$2\frac{39}{64}$
3	0.60235	0.05019	0.778	0.938	$3\frac{7}{8}$	$3\frac{11}{16}$	$3\frac{1}{4}$
4	0.62326	0.05193	1.020	1.231	$4\frac{7}{8}$	$4\frac{5}{8}$	$4\frac{1}{8}$
5	0.63151	0.05262	1.475	1.748	$6\frac{1}{8}$	$5\frac{7}{8}$	$5\frac{1}{4}$
6	0.62565	0.05213	2.116	2.494	$8\frac{9}{16}$	$8\frac{1}{4}$	$7\frac{21}{64}$
7	0.62400	0.05200	2.750	3.270	$11\frac{5}{8}$	$11\frac{1}{4}$	$10\frac{5}{64}$

Plug Depth *P*	Tang or Tongue Thickness *t*	Tang or Tongue Length *T*	Tang or Tongue Radius *R*	Tang or Tongue Diameter	Keyway Width *W*	Keyway Length *L*	Keyway to End *K*
2	0.1562	$\frac{1}{4}$	$\frac{5}{32}$	0.235	$\frac{11}{64}$	$\frac{9}{16}$	$1\frac{15}{16}$
$2\frac{1}{8}$	0.2031	$\frac{3}{8}$	$\frac{3}{16}$	0.343	0.218	$\frac{3}{4}$	$2\frac{1}{16}$
$2\frac{9}{16}$	0.2500	$\frac{7}{16}$	$\frac{1}{4}$	$\frac{17}{32}$	0.266	$\frac{7}{8}$	$2\frac{1}{2}$
$3\frac{3}{16}$	0.3125	$\frac{9}{16}$	$\frac{9}{32}$	$\frac{23}{32}$	0.328	$1\frac{3}{16}$	$3\frac{1}{16}$
$4\frac{1}{16}$	0.4687	$\frac{5}{8}$	$\frac{5}{16}$	$\frac{31}{32}$	0.484	$1\frac{1}{4}$	$3\frac{7}{8}$
$5\frac{3}{16}$	0.6250	$\frac{3}{4}$	$\frac{3}{8}$	$1\frac{13}{32}$	0.656	$1\frac{1}{2}$	$4\frac{15}{16}$
$7\frac{1}{4}$	0.7500	$1\frac{1}{8}$	$\frac{1}{2}$	2	0.781	$1\frac{3}{4}$	7
10	1.1250	$1\frac{3}{8}$	$\frac{3}{4}$	$2\frac{5}{8}$	1.156	$2\frac{5}{8}$	$9\frac{1}{2}$

Table 2. Morse Stub Taper Shanks

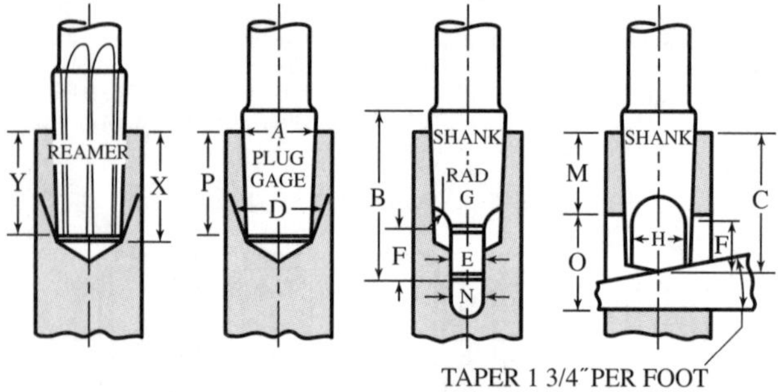

No. of Taper	Taper per Foot[a]	Taper per Inch[b]	Small End of Plug,[b] *D*	Dia. End of Socket,[a] *A*	Shank: Total Length, *B*	Shank: Depth, *C*	Tang: Thickness, *E*	Tang: Length, *F*
1	0.59858	0.049882	0.4314	0.475	1 5/16	1 1/8	13/64	5/16
2	0.59941	0.049951	0.6469	0.700	1 11/16	1 7/16	19/64	7/16
3	0.60235	0.050196	0.8753	0.938	2	1 3/4	25/64	9/16
4	0.62326	0.051938	1.1563	1.231	2 3/8	2 1/16	33/64	11/16
5	0.63151	0.052626	1.6526	1.748	3	2 11/16	3/4	15/16

No. of Taper	Tang: Radius of Mill, *G*	Tang: Diameter, *H*	Socket: Plug Depth, *P*	Socket: Min. Depth of Tapered Hole: Drilled, *X*	Socket: Min. Depth of Tapered Hole: Reamed, *Y*	Socket: Socket End to Tang Slot, *M*	Tang Slot: Width, *N*	Tang Slot: Length, *O*
1	3/16	13/32	7/8	5/16	29/32	25/32	7/32	23/32
2	7/32	39/64	1 1/16	1 5/32	1 7/64	15/16	5/16	15/16
3	9/32	13/16	1 1/4	1 3/8	1 5/16	1 1/16	13/32	1 1/8
4	3/8	1 3/32	1 7/16	1 9/16	1 1/2	1 3/16	17/32	1 3/8
5	9/16	1 19/32	1 13/16	1 15/16	1 7/8	1 7/16	25/32	1 3/4

All dimensions in inches.
Radius *J* is 3/64, 1/16, 5/64, 3/32, and 1/8 inch respectively for Nos. 1, 2, 3, 4, and 5 tapers.

[a] These are basic dimensions.
[b] These dimensions are calculated for reference only.

Table 3. Dimensions of Morse Taper Sleeves

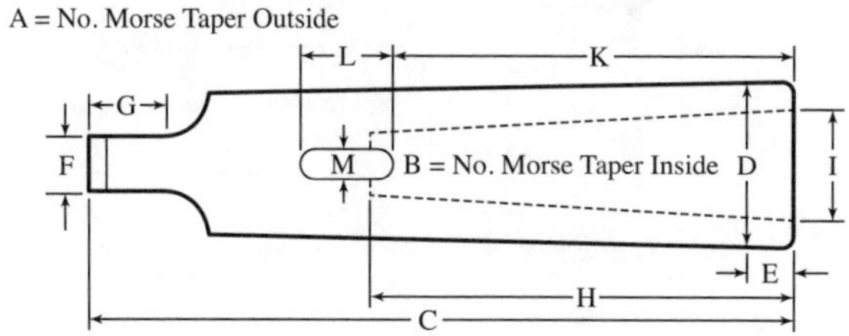

A	*B*	*C*	*D*	*E*	*F*	*G*	*H*	*I*	*K*	*L*	*M*
2	1	3 9/16	0.700	5/8	1/4	7/16	2 3/16	0.475	2 1/16	3/4	0.213
3	1	3 15/16	0.938	1/4	5/16	9/16	2 3/16	0.475	2 1/16	3/4	0.213
3	2	4 7/16	0.938	3/4	5/16	9/16	2 5/8	0.700	2 1/2	7/8	0.260
4	1	4 7/8	1.231	1/4	15/32	5/8	2 3/16	0.475	2 1/16	3/4	0.213
4	2	4 7/8	1.231	1/4	15/32	5/8	2 5/8	0.700	2 1/2	7/8	0.260
4	3	5 3/8	1.231	3/4	15/32	5/8	3 1/4	0.938	3 1/16	1 3/16	0.322
5	1	6 1/8	1.748	1/4	5/8	3/4	2 3/16	0.475	2 1/16	3/4	0.213
5	2	6 1/8	1.748	1/4	5/8	3/4	2 5/8	0.700	2 1/2	7/8	0.260
5	3	6 1/8	1.748	1/4	5/8	3/4	3 1/4	0.938	3 1/16	1 3/16	0.322
5	4	6 5/8	1.748	3/4	5/8	3/4	4 1/8	1.231	3 7/8	1 1/4	0.478
6	1	8 5/8	2.494	3/8	3/4	1 1/8	2 3/16	0.475	2 1/16	3/4	0.213
6	2	8 5/8	2.494	3/8	3/4	1 1/8	2 5/8	0.700	2 1/2	7/8	0.260
6	3	8 5/8	2.494	3/8	3/4	1 1/8	3 1/4	0.938	3 1/16	1 3/16	0.322
6	4	8 5/8	2.494	3/8	3/4	1 1/8	4 1/8	1.231	3 7/8	1 1/4	0.478
6	5	8 5/8	2.494	3/8	3/4	1 1/8	5 1/4	1.748	4 15/16	1 1/2	0.635
7	3	11 5/8	3.270	3/8	1 1/8	1 3/8	3 1/4	0.938	3 1/16	1 3/16	0.322
7	4	11 5/8	3.270	3/8	1 1/8	1 3/8	4 1/8	1.231	3 7/8	1 1/4	0.478
7	5	11 5/8	3.270	3/8	1 1/8	1 3/8	5 1/4	1.748	4 15/16	1 1/2	0.635
7	6	12 1/2	3.270	1 1/4	1 1/8	1 3/8	7 3/8	2.494	7	1 3/4	0.760

Jarno Taper.—The Jarno taper was originally proposed by Oscar J. Beale of the Brown & Sharpe Mfg. Co. This taper is based on such simple formulas that practically no calculations are required when the number of taper is known. The taper per foot of all Jarno taper sizes is 0.600 inch on the diameter. The diameter at the large end is as many eighths, the diameter at the small end is as many tenths, and the length as many half inches as are indicated by the number of the taper. (See Table 4 on page 52.) For example, a No. 7 Jarno taper is $\frac{7}{8}$ inch in diameter at the large end; $\frac{7}{10}$, or 0.700 inch at the small end; and $\frac{7}{2}$, or $3\frac{1}{2}$ inches long; hence, diameter at large end = No. of taper ÷ 8; diameter at small end = No. of taper ÷ 10; length of taper = No. of taper ÷ 2.

The Jarno taper is used on various machine tools, especially profiling machines and die-sinking machines. It has also been used for the headstock and tailstock spindles of some lathes.

Table 4. Jarno Taper Shanks

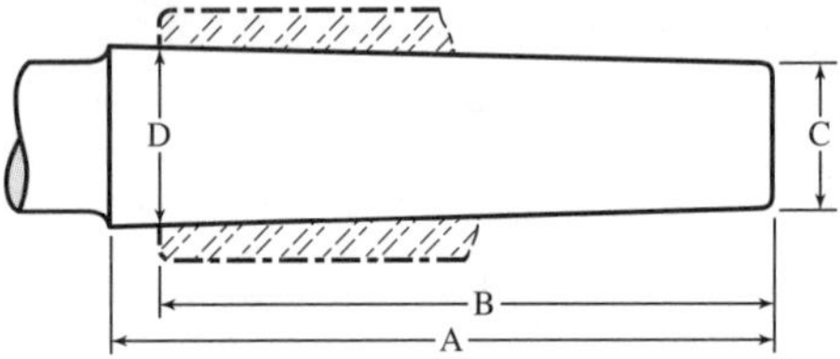

$$D = \frac{\text{No. of Taper}}{8} \quad C = \frac{\text{No. of Taper}}{10} \quad B = \frac{\text{No. of Taper}}{2}$$

Number of Taper	Length *A*	Length *B*	Diameter *C*	Diameter *D*	Taper per foot
2	$1\frac{1}{8}$	1	0.20	0.250	0.600
3	$1\frac{5}{8}$	$1\frac{1}{2}$	0.30	0.375	0.600
4	$2\frac{3}{16}$	2	0.40	0.500	0.600
5	$2\frac{11}{16}$	$2\frac{1}{2}$	0.50	0.625	0.600
6	$3\frac{3}{16}$	3	0.60	0.750	0.600
7	$3\frac{11}{16}$	$3\frac{1}{2}$	0.70	0.875	0.600
8	$4\frac{3}{16}$	4	0.80	1.000	0.600
9	$4\frac{11}{16}$	$4\frac{1}{2}$	0.90	1.125	0.600
10	$5\frac{1}{4}$	5	1.00	1.250	0.600
11	$5\frac{3}{4}$	$5\frac{1}{2}$	1.10	1.375	0.600
12	$6\frac{1}{4}$	6	1.20	1.500	0.600
13	$6\frac{3}{4}$	$6\frac{1}{2}$	1.30	1.625	0.600
14	$7\frac{1}{4}$	7	1.40	1.750	0.600
15	$7\frac{3}{4}$	$7\frac{1}{2}$	1.50	1.875	0.600
16	$8\frac{5}{16}$	8	1.60	2.000	0.600
17	$8\frac{13}{16}$	$8\frac{1}{2}$	1.70	2.125	0.600
18	$9\frac{5}{16}$	9	1.80	2.250	0.600
19	$9\frac{13}{16}$	$9\frac{1}{2}$	1.90	2.375	0.600
20	$10\frac{5}{16}$	10	2.00	2.500	0.600

Brown & Sharpe Taper.—This standard taper is used for taper shanks on tools such as end mills and reamers, the taper being approximately $\frac{1}{2}$ inch per foot for all sizes except for taper No. 10, where the taper is 0.5161 inch per foot. (See Table 5 on page 53.)

Brown & Sharpe taper sockets are used for many arbors, collets, and machine tool spindles, especially milling machines and grinding machines. In many cases there are a number of different lengths of sockets corresponding to the same number of taper; all these tapers, however, are of the same diameter at the small end.

Table 5. Brown & Sharpe Taper Shanks

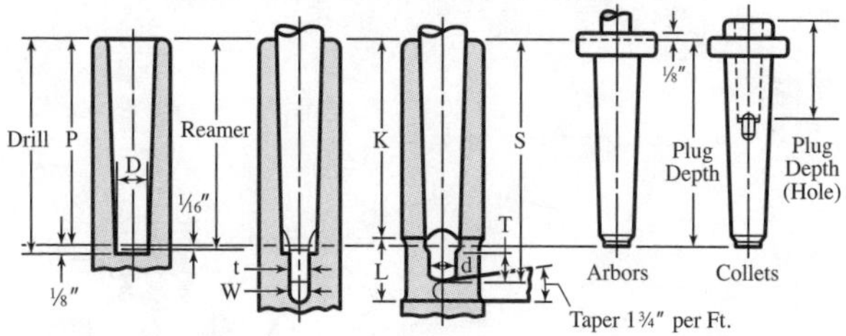

Number of Taper	Taper per Foot (inch)	Dia. of Plug at Small End D	Plug Depth, P: B & S[b] Standard	Plug Depth, P: Mill. Mach. Standard	Plug Depth, P: Miscell.	Keyway from End of Spindle K	Shank Depth S	Length of Keyway[a] L	Width of Keyway W	Length of Arbor Tongue T	Diameter of Arbor Tongue d	Thickness of Arbor Tongue t
1[c]	.50200	.20000	15/16	…	…	15/16	1 3/16	3/8	.135	3/16	.170	1/8
2[c]	.50200	.25000	1 3/16	…	…	1 11/64	1 1/2	1/2	.166	1/4	.220	5/32
3[c]	.50200	.31250	1 1/2	…	…	1 15/32	1 7/8	5/8	.197	5/16	.282	3/16
			…	…	1 3/4	1 23/32	2 1/8	5/8	.197	5/16	.282	3/16
			…	…	2	1 31/32	2 3/8	5/8	.197	5/16	.282	3/16
4	.50240	.35000	…	1 1/4	…	1 13/64	1 21/32	11/16	.228	11/32	.320	7/32
			1 11/16	…	…	1 41/64	2 3/32	11/16	.228	11/32	.320	7/32
5	.50160	.45000	…	1 3/4	…	1 11/16	2 3/16	3/4	.260	3/8	.420	1/4
			…	…	2	1 15/16	2 7/16	3/4	.260	3/8	.420	1/4
			2 1/8	…	…	2 1/16	2 9/16	3/4	.260	3/8	.420	1/4
6	.50329	.50000	2 3/8	…	…	2 19/64	2 7/8	7/8	.291	7/16	.460	9/32
7	.50147	.60000	…	…	2 1/2	2 13/32	3 1/32	15/16	.322	15/32	.560	5/16
			2 7/8	…	…	2 25/32	3 13/32	15/16	.322	15/32	.560	5/16
			…	3	…	2 29/32	3 17/32	15/16	.322	15/32	.560	5/16
8	.50100	.75000	3 9/16	…	…	3 29/64	4 1/8	1	.353	1/2	.710	11/32
9	.50085	.90010	…	4	…	3 7/8	4 5/8	1 1/8	.385	9/16	.860	3/8
			4 1/4	…	…	4 1/8	4 7/8	1 1/8	.385	9/16	.860	3/8
10	.51612	1.04465	5	…	…	4 27/32	5 23/32	1 5/16	.447	21/32	1.010	7/16
			…	5 11/16	…	5 17/32	6 13/32	1 5/16	.447	21/32	1.010	7/16
			…	…	6 7/32	6 1/16	6 15/16	1 5/16	.447	21/32	1.010	7/16
11	.50100	1.24995	5 15/16	…	…	5 25/32	6 21/32	1 5/16	.447	21/32	1.210	7/16
			…	6 3/4	…	6 19/32	7 15/32	1 5/16	.447	21/32	1.210	7/16
12	.49973	1.50010	7 1/8	7 1/8	…	6 15/16	7 15/16	1 1/2	.510	3/4	1.460	1/2
			…	…	6 1/4	…	…	…	…	…	…	…
13	.50020	1.75005	7 3/4	…	…	7 9/16	8 9/16	1 1/2	.510	3/4	1.710	1/2
14	.50000	2.00000	8 1/4	8 1/4	…	8 1/32	9 5/32	1 11/16	.572	27/32	1.960	9/16
15	.5000	2.25000	8 3/4	…	…	8 17/32	9 21/32	1 11/16	.572	27/32	2.210	9/16
16	.50000	2.50000	9 1/4	…	…	9	10 1/4	1 7/8	.635	15/16	2.450	5/8
17	.50000	2.75000	9 3/4	…	…	…	…	…	…	…	…	…
18	.50000	3.00000	10 1/4	…	…	…	…	…	…	…	…	…

[a] Special lengths of keyway are used instead of standard lengths in some places. Standard lengths need not be used when keyway is for driving only and not for admitting key to force out tool.

[b] "B & S Standard" Plug Depths are not used in all cases.

[c] Adopted by American Standards Association.

Table 6. Essential Dimensions of American National Standard Spindle Noses for Milling Machines *ANSI/ASME B5.18-1972 (R2014)*

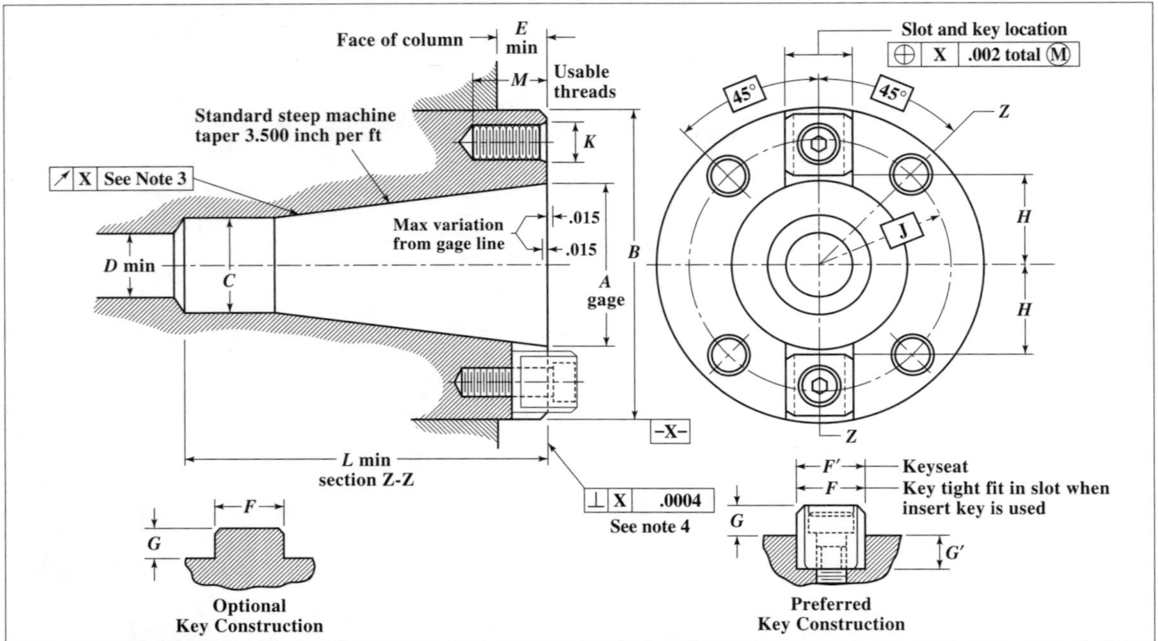

Table 6. *(Continued)* **Essential Dimensions of American National Standard Spindle Noses for Milling Machines** *ANSI/ASME B5.18-1972 (R2014)*

Size No.	Gage Dia. of Taper *A*	Dia. of Spindle *B*	Pilot Dia. *C*	Clearance Hole for Draw-in Bolt Min. *D*	Minimum Dimension Spindle End to Column *E*	Width of Driving Key *F*	Width of Keyseat *F′*	Maximum Height of Driving Key *G*	Minimum Depth of Keyseat *G′*	Distance from Center to Driving Keys *H*	Radius of Bolt Hole Circle *J*	Size of Threads for Bolt Holes UNC-2B *K*	Full Depth of Arbor Hole in Spindle Min. *L*	Depth of Usable Thread for Bolt Hole *M*
30	1.250	2.7493 2.7488	0.692 0.685	0.66	0.50	0.6255 0.6252	0.624 0.625	0.31	0.31	0.660 0.654	1.0625 (Note 1)	0.375-16	2.88	0.62
40	1.750	3.4993 3.4988	1.005 0.997	0.66	0.62	0.6255 0.6252	0.624 0.625	0.31	0.31	0.910 0.904	1.3125 (Note 1)	0.500-13	3.88	0.81
45	2.250	3.9993 3.9988	1.286 1.278	0.78	0.62	0.7505 0.7502	0.749 0.750	0.38	0.38	1.160 1.154	1.500 (Note 1)	0.500-13	4.75	0.81
50	2.750	5.0618 5.0613	1.568 1.559	1.06	0.75	1.0006 1.0002	0.999 1.000	0.50	0.50	1.410 1.404	2.000 (Note 2)	0.625-11	5.50	1.00
60	4.250	8.7180 8.7175	2.381 2.371	1.38	1.50	1.0006 1.0002	0.999 1.000	0.50	0.50	2.420 2.414	3.500 (Note 2)	0.750-10	8.62	1.25

All dimensions are given in inches.
Tolerances:
Two-digit decimal dimensions ±0.010 unless otherwise specified.
A—Taper: Tolerance on rate of taper to be 0.001 inch per foot applied only in direction which decreases rate of taper.
F′—Centrality of keyway with axis of taper 0.002 total at maximum material condition. (0.002 Total indicator variation)
F—Centrality of solid key with axis of taper 0.002 total at maximum material condition. (0.002 Total indicator variation)
Note 1: Holes spaced as shown and located within 0.006 inch diameter of true position.
Note 2: Holes spaced as shown and located within 0.010 inch diameter of true position.
Note 3: Maximum turnout on test plug: 0.0004 at 1 inch projection from gage line. 0.0010 at 12 inch projection from gage line.
Note 4: Squareness of mounting face measured near mounting bolt hole circle.

Table 7. Essential Dimensions for American National Standard Spindle Nose with Large Flange *ANSI/ASME B5.18-1972 (R2014)*

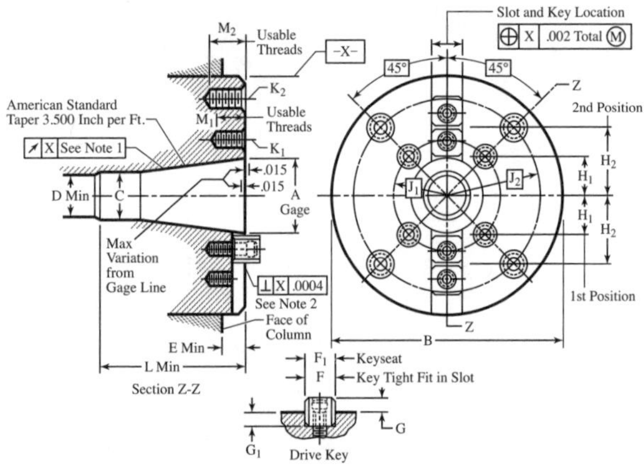

Size No.	Gage Diam. of Taper *A*	Dia. of Spindle Flange *B*	Pilot Dia. *C*	Clearance Hole for Draw-in Bolt Min. *D*	Min. Dim. Spindle End to Column *E*	Width of Driving Key *F*	Height of Driving Key Max. *G*	Depth of Keyseat Min. G_1	Distance from enter to Driving Keys First Position H_1
50A	2.750	8.7180 8.7175	1.568 1.559	1.06	0.75	1.0006 1.0002	0.50	0.50	1.410 1.404

Size No.	Distance from enter to Driving Key second Position	Radius of Bolt Hole Circles (See Note 3)		Size of Threads for Bolt Holes UNC-2B		Full Depth of Arbor Hole in Spindle Min.	Depth of Usable Thread for Bolt Holes		Width of Keyseat
		Inner	Outer						
	H_2	J_1	J_2	K_1	K_2	*L*	M_1	M_2	F_1
50A	2.420 2.410	2.000	3.500	0.625–11	0.750–10	5.50	1.00	1.25	0.999 1.000

All dimensions are given in inches.

Tolerances: Two-digit decimal dimensions ±0.010 unless otherwise specified.

A—Tolerance on rate of taper to be 0.001 inch per foot applied only in direction which decreases rate of taper.

F—Centrality of solid key with axis of taper 0.002 inch total at maximum material condition. (0.002 inch total indicator variation)

F_1—Centrality of keyseat with axis of taper 0.002 inch total at maximum material condition. (0.002 inch total indicator variation)

Note 1: Maximum runout on test plug:
0.0004 at 1 inch projection from gage line.
0.0010 at 12 inch projection from gage line.

Note 2: Squareness of mounting face measured near mounting bolt hole circle.

Note 3: Holes located as shown and within 0.010 inch diameter of true position.

Table 8. Essential Dimensions of American National Standard Tool Shanks for Milling Machines *ANSI/ASME B5.18-1972 (R2014)*

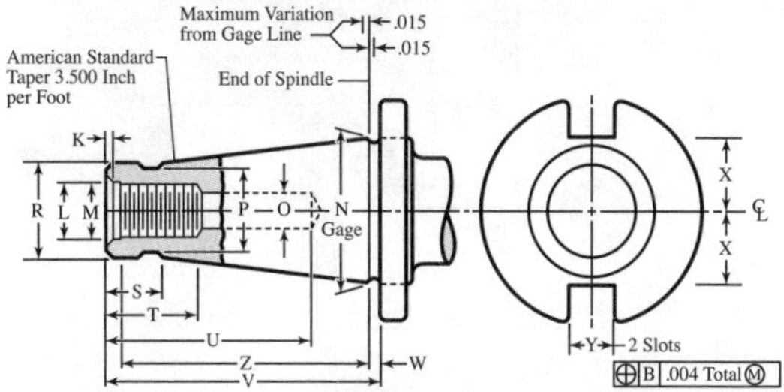

Size No.	Gage Dia. of Taper *N*	Tap Drill Size for Draw-in Thread *O*	Dia. of Neck *P*	Size of Thread for Draw-in Bolt UNC-2B *M*	Pilot Dia. *R*	Length of Pilot *S*	Minimum Length of Usable Thread *T*	Minimum Depth of Clearance Hole *U*
30	1.250	0.422 0.432	0.66 0.65	0.500–13	0.675 0.670	0.81	1.00	2.00
40	1.750	0.531 0.541	0.94 0.93	0.625–11	0.987 0.980	1.00	1.12	2.25
45	2.250	0.656 0.666	1.19 1.18	0.750–10	1.268 1.260	1.00	1.50	2.75
50	2.750	0.875 0.885	1.50 1.49	1.000–8	1.550 1.540	1.00	1.75	3.50
60	4.250	1.109 1.119	2.28 2.27	1.250–7	2.360 2.350	1.75	2.25	4.25

Size No.	Distance from Rear of Flange to End of Arbor *V*	Clearance of Flange from Gage Diameter *W*	Tool Shank Centerline to Driving Slot *X*	Width of Driving Slot *Y*	Distance from Gage Line to Bottom of C'bore *Z*	Depth of 60° Center *K*	Diameter of C'bore *L*
30	2.75	0.045 0.075	0.640 0.625	0.635 0.645	2.50	0.05 0.07	0.525 0.530
40	3.75	0.045 0.075	0.890 0.875	0.635 0.645	3.50	0.05 0.07	0.650 0.655
45	4.38	0.105 0.135	1.140 1.125	0.760 0.770	4.06	0.05 0.07	0.775 0.780
50	5.12	0.105 0.135	1.390 1.375	1.010 1.020	4.75	0.05 0.12	1.025 1.030
60	8.25	0.105 0.135	2.400 2.385	1.010 1.020	7.81	0.05 0.12	1.307 1.312

All dimensions are given in inches.

Tolerances: Two digit decimal dimensions ±0.010 inch unless otherwise specified.

M—Permissible for Class 2B "No Go" gage to enter five threads before interference.

N—Taper tolerance on rate of taper to be 0.001 inch per foot applied only in direction which increases rate of taper.

Y—Centrality of drive slot with axis of taper shank 0.004 inch at maximum material condition. (0.004 inch total indicator variation)

Table 9. Essential Dimensions of V-Flange Tool Shanks *ANSI/ASME B5.50-2015*

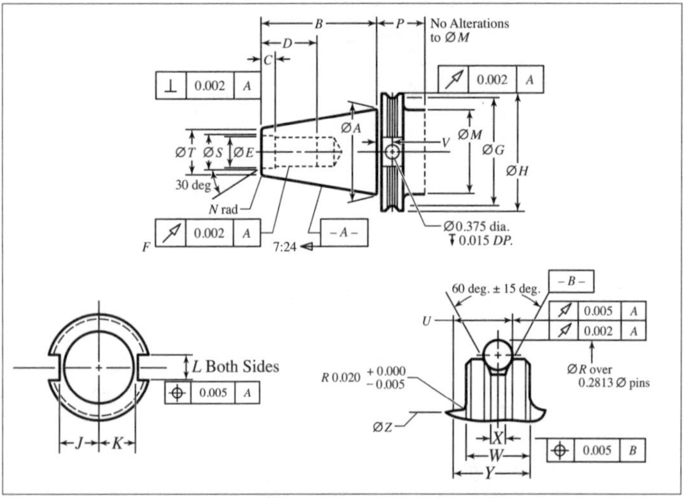

	A	B	C	D	E	F	G	H	J	K
Tolerance		±0.005	±0.010	Min.	+0.015 -0.000	UNC 2B	±0.010	±0.002	+0.000 -0.015	+0.000 -0.015
Size	Gage Dia.									
30	1.250	1.875	0.188	1.00	0.516	0.500-13	1.531	1.812	0.735	0.640
40	1.750	2.687	0.188	1.12	0.641	0.625-11	2.219	2.500	0.985	0.890
45	2.250	3.250	0.188	1.50	0.766	0.750-10	2.969	3.250	1.235	1.140
50	2.750	4.000	0.250	1.75	1.031	1.000-8	3.594	3.875	1.485	1.390
60	4.250	6.375	0.312	2.25	1.281	1.250-7	5.219	5.500	2.235	2.140

	A	L	M	N	P	R	S	T	Z
Tolerance		±0.001	Max.	±0.010	Min.	±0.002	±0.010	Min.	+0.000 -0.005
Size	Gage Dia.								
30	1.250	0.645	1.812	0.020	1.38	2.176	0.590	0.652	1.250
40	1.750	0.645	2.500	0.040	1.38	2.863	0.720	0.880	1.750
45	2.250	0.770	3.250	0.040	1.38	3.613	0.850	1.233	2.250
50	2.750	1.020	3.875	0.040	1.38	4.238	1.125	1.427	2.750
60	4.250	1.020	5.500	0.040	1.500	5.863	1.375	2.309	4.250

Notes: Taper tolerance to be 0.001 in. in 12 in. applied in direction that increases rate of taper. Geometric dimensions symbols are to ANSI/ASME Y14.5-2018. Dimensions are in inches. Deburr all sharp edges. Unspecified fillets and radii to be 0.03 ± 0.010*R*, or 0.03 ± 0.010 × 45 degrees. Data for size 60 are not part of Standard. For all sizes, the values for dimensions *U* (tol. ± 0.005) are 0.579: for *V* (tol. ±0.010), 0.440; for *W* (tol. ±0.002), 0.625; for *X* (tol. ±0.005), 0.152; and for *Y* (tol. ±0.002), 0.750.

Table 10. Essential Dimensions of V-Flange Tool Shank Retention Knobs
ANSI/ASME B5.50-2015

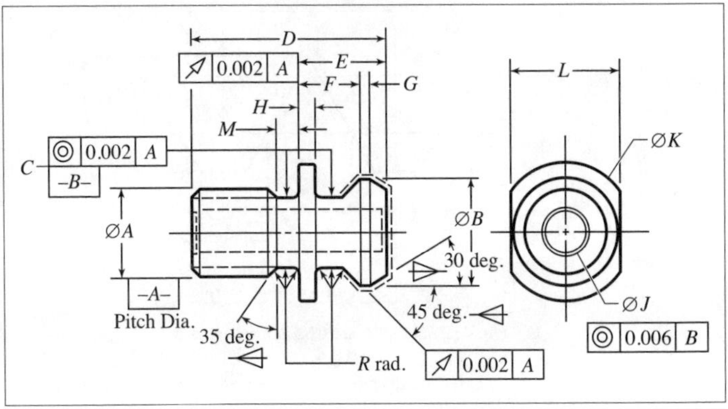

Size/ Totals	A	B	C	D	E	F
	UNC 2A	±0.005	±0.005	±0.040	±0.005	±0.005
30	0.500-13	0.520	0.385	1.10	0.460	0.320
40	0.625-11	0.740	0.490	1.50	0.640	0.440
45	0.750-10	0.940	0.605	1.80	0.820	0.580
50	1.000-8	1.140	0.820	2.30	1.000	0.700
60	1.250-7	1.460	1.045	3.20	1.500	1.080

Size / Totals	G	H	J	K	L	M	R
	±0.010	±0.010	±0.010		+0.000 −0.010	±0.040	+0.010 −0.005
30	0.04	0.10	0.187	0.65 0.64	0.53	0.19	0.094
40	0.06	0.12	0.281	0.94 0.92	0.75	0.22	0.094
45	0.08	0.16	0.375	1.20 1.18	1.00	0.22	0.094
50	0.10	0.20	0.468	1.44 1.42	1.25	0.25	0.125
60	0.14	0.30	0.500	2.14 2.06	1.50	0.31	0.125

Notes: Dimensions are in inches. Material: low-carbon steel. Heat treatment: carburize and harden to 0.016 to 0.028 in. effective case depth. Hardness of noted surfaces to be Rockwell 56-60; core hardness Rockwell C35-45. Hole *J* shall not be carburized. Surfaces *C* and *R* to be free from tool marks. Deburr all sharp edges. Geometric dimension symbols are to ANSI/ASME Y14.5-2018. Data for size 60 are not part of Standard.

THREADS

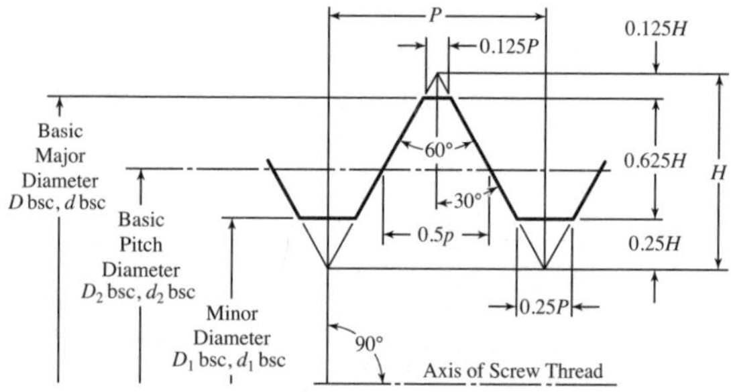

Fig. 1. Basic Profile of UN and UNF Screw Threads

Thread Classes.—Thread classes are distinguished from each other by the amounts of tolerance and allowance. Classes identified by a numeral followed by the letters A and B are derived from certain Unified formulas (not shown here) in which the pitch diameter tolerances are based on increments of the basic major (nominal) diameter, the pitch, and the length of engagement. These formulas and the class identification or symbols apply to all of the Unified threads.

Classes 1A, 2A, and 3A apply to external threads only, and Classes 1B, 2B, and 3B apply to internal threads only. The disposition of the tolerances, allowances, and crest clearances for the various classes is illustrated on pages 61 and 62.

Classes 2A and 2B: Classes 2A and 2B are the most commonly used for general applications, including production of bolts, screws, nuts, and similar fasteners.

The maximum diameters of Class 2A (external) uncoated threads are less than basic by the amount of the allowance. The allowance minimizes galling and seizing in high-cycle wrench assembly, or it can be used to accommodate plated finishes or other coating. However, for threads with additive finish, the maximum diameters of Class 2A may be exceeded by the amount of the allowance. For example, the 2A maximum diameters apply to an unplated part or to a part before plating whereas the basic diameters (the 2A maximum diameter plus allowance) apply to a part after plating. The minimum diameters of Class 2B (internal) threads, whether or not plated or coated, are basic, affording no allowance or clearance in assembly at maximum metal limits.

Class 2AG: Certain applications require an allowance for rapid assembly to permit application of the proper lubricant or for residual growth due to high-temperature expansion. In these applications, when the thread is coated and the 2A allowance is not permitted to be consumed by such coating, the thread class symbol is qualified by G following the class symbol.

Classes 3A and 3B: Classes 3A and 3B may be used if closer tolerances are desired than those provided by Classes 2A and 2B. The maximum diameters of Class 3A (external) threads and the minimum diameters of Class 3B (internal) threads, whether or not plated or coated, are basic, affording no allowance or clearance for assembly of maximum metal components.

Classes 1A and 1B: Classes 1A and 1B threads replaced American National Class 1. These classes are intended for ordnance and other special uses. They are used on threaded components where quick and easy assembly is necessary and where a liberal allowance is required to permit ready assembly, even with slightly bruised or dirty threads.

Maximum diameters of Class 1A (external) threads are less than basic by the amount of the same allowance as applied to Class 2A. For the intended applications in American practice the allowance is not available for plating or coating. Where the thread is plated or coated, special provisions are necessary. The minimum diameters of Class 1B (internal) threads, whether or not plated or coated, are basic, affording no allowance or clearance for assembly with maximum metal external thread components having maximum diameters which are basic.

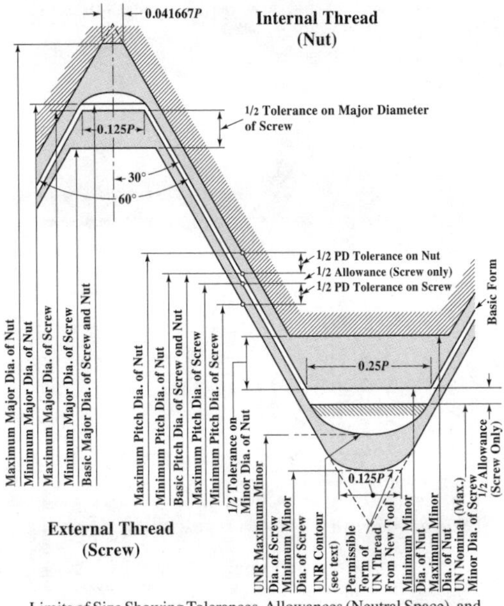

Limits of Size Showing Tolerances, Allowances (Neutral Space), and Crest Clearances for Unified Classes 1A, 2A, 1B, and 2B

Sharp V-Thread.—The sides of the thread form an angle of 60 degrees with each other. The top and bottom of the thread are, theoretically, sharp, but in practice it is necessary to make the thread with a slight flat, owing to the difficulty of producing a perfectly sharp edge and because of the tendency of such an edge to wear away or become battered. There is no standard adopted for this flat, but it is usually made about one-twenty-fifth of the pitch. If p = pitch of thread, and d = depth of thread, then:

$$d = p \times \cos 30 \text{ deg.} = p \times 0.866 = \frac{0.866}{\text{no. of threads per inch}}$$

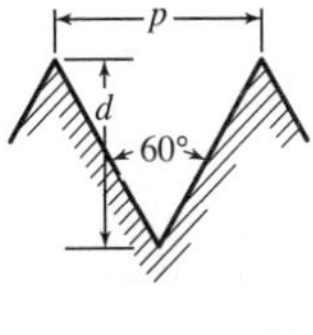

Some modified V-threads, for locomotive boiler taps particularly, have a depth of 0.8 × pitch.

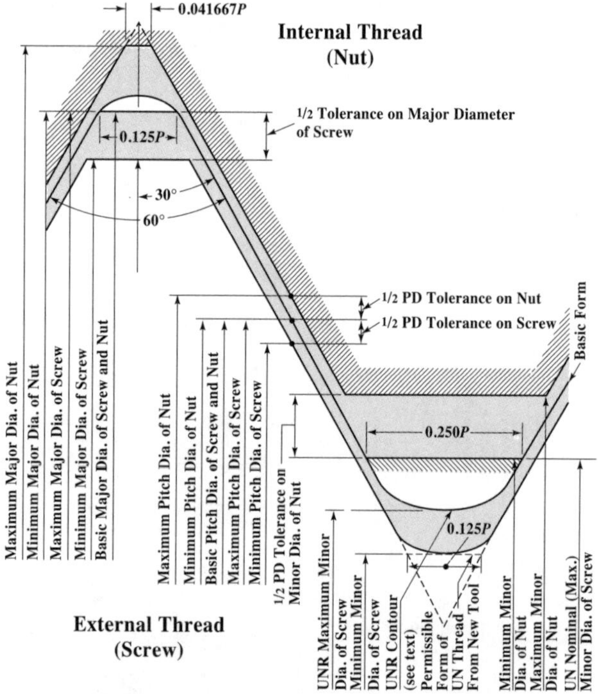

Limits of Size Showing Tolerances and Crest Clearances for Unified Classes 3A and 3B and American National Classes 2 and 3

UN External Screw Threads: A flat root contour is specified, but it is necessary to provide for some threading tool crest wear, hence a rounded root contour cleared beyond the 0.25*P* flat width of the Basic Profile is optional.

UNR External Screw Threads: To reduce the rate of threading tool crest wear and to improve fatigue strength of a flat root thread, the Design Profile of the UNR thread has a smooth, continuous, non-reversing contour with a radius of curvature not less than 0.108*P* at any point and blends tangentially into the flanks and any straight segment. At the maximum material condition, the point of tangency is specified to be at a distance not less than 0.625*H* (where *H* is the height of a sharp V-thread) below the basic major diameter.

UN and UNR External Screw Threads: The design profiles of both UN and UNR external screw threads have flat crests. However, in practice, product threads are produced with partially or completely rounded crests. A rounded crest tangent at 0.125*P* flat is shown as an option on page .

UN Internal Screw Thread: In practice it is necessary to provide for some threading tool crest wear; therefore the root of the design profile is rounded and cleared beyond the 0.125*p* flat width of the basic profile. There is no internal UNR screw thread.

Table 1. American National Standard Unified Internal and External Screw Thread Design Profiles (Maximum Material Condition)

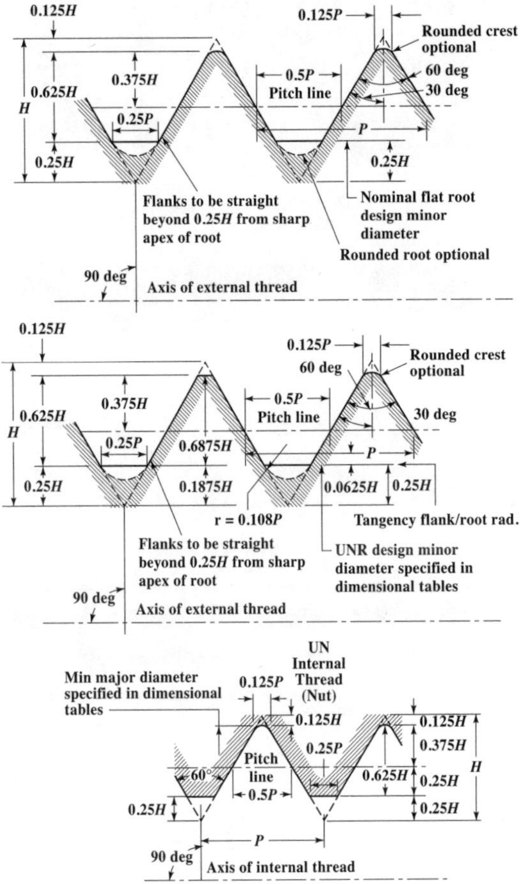

(*H* = height of sharp V-thread = 0.86603 × pitch)

Fine-Thread Series.—This series, UNF/UNRF, is suitable for the production of bolts, screws, and nuts and for other applications where the Coarse series is not applicable. External threads of this series have greater tensile stress area than comparable sizes of the Coarse series. The Fine series is suitable when the resistance to stripping of both external and mating internal threads equals or exceeds the tensile load carrying capacity of the externally threaded member. It is also used where the length of engagement is short, where a smaller lead angle is desired, where the wall thickness demands a fine pitch, or where finer adjustment is needed.

Table 2. Standard Series and Selected Combinations—Unified Screw Threads

Nominal Size, Threads per Inch, and Series Designation[a]	External[b]								Internal[b]					
	Class	Allowance	Major Diameter			Pitch Diameter		UNR Minor Dia.,[c] Max (Ref.)	Class	Minor Diameter		Pitch Diameter		Major Diameter
			Max[d]	Min	Min[e]	Max[d]	Min			Min	Max	Min	Max	Min
0-80 UNF	2A	0.0005	0.0595	0.0563	—	0.0514	0.0496	0.0446	2B	0.0465	0.0514	0.0519	0.0542	0.0600
	3A	0.0000	0.0600	0.0568	—	0.0519	0.0506	0.0451	3B	0.0465	0.0514	0.0519	0.0536	0.0600
1-64 UNC	2A	0.0006	0.0724	0.0686	—	0.0623	0.0603	0.0538	2B	0.0561	0.0622	0.0629	0.0655	0.0730
	3A	0.0000	0.0730	0.0692	—	0.0629	0.0614	0.0544	3B	0.0561	0.0622	0.0629	0.0648	0.0730
1-72 UNF	2A	0.0006	0.0724	0.0689	—	0.0634	0.0615	0.0559	2B	0.0580	0.0634	0.0640	0.0665	0.0730
	3A	0.0000	0.0730	0.0695	—	0.0640	0.0626	0.0565	3B	0.0580	0.0634	0.0640	0.0659	0.0730
2-56 UNC	2A	0.0006	0.0854	0.0813	—	0.0738	0.0717	0.0641	2B	0.0667	0.0737	0.0744	0.0772	0.0860
	3A	0.0000	0.0860	0.0819	—	0.0744	0.0728	0.0647	3B	0.0667	0.0737	0.0744	0.0765	0.0860
2-64 UNF	2A	0.0006	0.0854	0.0816	—	0.0753	0.0733	0.0668	2B	0.0691	0.0752	0.0759	0.0786	0.0860
	3A	0.0000	0.0860	0.0822	—	0.0759	0.0744	0.0674	3B	0.0691	0.0752	0.0759	0.0779	0.0860
3-48 UNC	2A	0.0007	0.0983	0.0938	—	0.0848	0.0825	0.0735	2B	0.0764	0.0845	0.0855	0.0885	0.0990
	3A	0.0000	0.0990	0.0945	—	0.0855	0.0838	0.0742	3B	0.0764	0.0845	0.0855	0.0877	0.0990
3-56 UNF	2A	0.0007	0.0983	0.0942	—	0.0867	0.0845	0.0770	2B	0.0797	0.0865	0.0874	0.0902	0.0990
	3A	0.0000	0.0990	0.0949	—	0.0874	0.0858	0.0777	3B	0.0797	0.0865	0.0874	0.0895	0.0990
4-40 UNC	2A	0.0008	0.1112	0.1061	—	0.0950	0.0925	0.0814	2B	0.0849	0.0939	0.0958	0.0991	0.1120
	3A	0.0000	0.1120	0.1069	—	0.0958	0.0939	0.0822	3B	0.0849	0.0939	0.0958	0.0982	0.1120
4-48 UNF	2A	0.0007	0.1113	0.1068	—	0.0978	0.0954	0.0865	2B	0.0894	0.0968	0.0985	0.1016	0.1120
	3A	0.0000	0.1120	0.1075	—	0.0985	0.0967	0.0872	3B	0.0894	0.0968	0.0985	0.1008	0.1120
5-40 UNC	2A	0.0008	0.1242	0.1191	—	0.1080	0.1054	0.0944	2B	0.0979	0.1062	0.1088	0.1121	0.1250
	3A	0.0000	0.1250	0.1199	—	0.1088	0.1069	0.0952	3B	0.0979	0.1062	0.1088	0.1113	0.1250
5-44 UNF	2A	0.0007	0.1243	0.1195	—	0.1095	0.1070	0.0972	2B	0.1004	0.1079	0.1102	0.1134	0.1250
	3A	0.0000	0.1250	0.1202	—	0.1102	0.1083	0.0979	3B	0.1004	0.1079	0.1102	0.1126	0.1250
6-32 UNC	2A	0.0008	0.1372	0.1312	—	0.1169	0.1141	0.1000	2B	0.104	0.114	0.1177	0.1214	0.1380
	3A	0.0000	0.1380	0.1320	—	0.1177	0.1156	0.1008	3B	0.1040	0.1139	0.1177	0.1204	0.1380
6-40 UNF	2A	0.0008	0.1372	0.1321	—	0.1210	0.1184	0.1074	2B	0.111	0.119	0.1218	0.1252	0.1380
	3A	0.0000	0.1380	0.1329	—	0.1218	0.1198	0.1082	3B	0.1110	0.1186	0.1218	0.1243	0.1380
8-32 UNC	2A	0.0009	0.1631	0.1571	—	0.1428	0.1399	0.1259	2B	0.130	0.139	0.1437	0.1475	0.1640
	3A	0.0000	0.1640	0.1580	—	0.1437	0.1415	0.1268	3B	0.1300	0.1388	0.1437	0.1465	0.1640
8-36 UNF	2A	0.0008	0.1632	0.1577	—	0.1452	0.1424	0.1301	2B	0.134	0.142	0.1460	0.1496	0.1640
	3A	0.0000	0.1640	0.1585	—	0.1460	0.1439	0.1309	3B	0.1340	0.1416	0.1460	0.1487	0.1640

Table 2. *(Continued)* **Standard Series and Selected Combinations — Unified Screw Threads**

Nominal Size, Threads per Inch, and Series Designation[a]	External[b]								Internal[b]					
	Class	Allowance	Major Diameter			Pitch Diameter		UNR Minor Dia.,[c] Max (Ref.)	Class	Minor Diameter		Pitch Diameter		Major Diameter
			Max[d]	Min	Min[e]	Max[d]	Min			Min	Max	Min	Max	Min
10-24 UNC	2A	0.0010	0.1890	0.1818	—	0.1619	0.1586	0.1394	2B	0.145	0.155	0.1629	0.1672	0.1900
	3A	0.0000	0.1900	0.1828	—	0.1629	0.1604	0.1404	3B	0.1450	0.1555	0.1629	0.1661	0.1900
10-28 UNS	2A	0.0010	0.1890	0.1825	—	0.1658	0.1625	0.1465	2B	0.151	0.160	0.1668	0.1711	0.1900
10-32 UNF	2A	0.0009	0.1891	0.1831	—	0.1688	0.1658	0.1519	2B	0.156	0.164	0.1697	0.1736	0.1900
	3A	0.0000	0.1900	0.1840	—	0.1697	0.1674	0.1528	3B	0.1560	0.1641	0.1697	0.1726	0.1900
10-36 UNS	2A	0.0009	0.1891	0.1836	—	0.1711	0.1681	0.1560	2B	0.160	0.166	0.1720	0.1759	0.1900
10-40 UNS	2A	0.0009	0.1891	0.1840	—	0.1729	0.1700	0.1593	2B	0.163	0.169	0.1738	0.1775	0.1900
10-48 UNS	2A	0.0008	0.1892	0.1847	—	0.1757	0.1731	0.1644	2B	0.167	0.172	0.1765	0.1799	0.1900
10-56 UNS	2A	0.0007	0.1893	0.1852	—	0.1777	0.1752	0.1680	2B	0.171	0.175	0.1784	0.1816	0.1900
12-24 UNC	2A	0.0010	0.2150	0.2078	—	0.1879	0.1845	0.1654	2B	0.171	0.181	0.1889	0.1933	0.2160
	3A	0.0000	0.2160	0.2088	—	0.1889	0.1863	0.1664	3B	0.1710	0.1807	0.1889	0.1922	0.2160
12-28 UNF	2A	0.0010	0.2150	0.2085	—	0.1918	0.1886	0.1725	2B	0.177	0.186	0.1928	0.1970	0.2160
	3A	0.0000	0.2160	0.2095	—	0.1928	0.1904	0.1735	3B	0.1770	0.1857	0.1928	0.1959	0.2160
12-32 UNEF	2A	0.0010	0.2150	0.2090	—	0.1947	0.1915	0.1778	2B	0.182	0.190	0.1957	0.1998	0.2160
	3A	0.0000	0.2160	0.2100	—	0.1957	0.1933	0.1788	3B	0.1820	0.1895	0.1957	0.1988	0.2160
12-36 UNS	2A	0.0009	0.2151	0.2096	—	0.1971	0.1941	0.1820	2B	0.186	0.193	0.1980	0.2019	0.2160
12-40 UNS	2A	0.0009	0.2151	0.2100	—	0.1989	0.1960	0.1853	2B	0.189	0.195	0.1998	0.2036	0.2160
12-48 UNS	2A	0.0008	0.2152	0.2107	—	0.2017	0.1990	0.1904	2B	0.193	0.198	0.2025	0.2060	0.2160
12-56 UNS	2A	0.0008	0.2152	0.2111	—	0.2036	0.2011	0.1939	2B	0.197	0.201	0.2044	0.2077	0.2160
$\frac{1}{4}$-20 UNC	1A	0.0011	0.2489	0.2367	—	0.2164	0.2108	0.1894	1B	0.196	0.207	0.2175	0.2248	0.2500
	2A	0.0011	0.2489	0.2408	0.2367	0.2164	0.2127	0.1894	2B	0.196	0.207	0.2175	0.2224	0.2500
	3A	0.0000	0.2500	0.2419	—	0.2175	0.2147	0.1905	3B	0.1960	0.2067	0.2175	0.2211	0.2500
$\frac{1}{4}$-24 UNS	2A	0.0011	0.2489	0.2417	—	0.2218	0.2181	0.1993	2B	0.205	0.215	0.2229	0.2277	0.2500
$\frac{1}{4}$-27 UNS	2A	0.0010	0.2490	0.2423	—	0.2249	0.2214	0.2049	2B	0.210	0.219	0.2259	0.2304	0.2500
$\frac{1}{4}$-28 UNF	1A	0.0010	0.2490	0.2392	—	0.2258	0.2208	0.2065	1B	0.211	0.220	0.2268	0.2333	0.2500
	2A	0.0010	0.2490	0.2425	—	0.2258	0.2225	0.2065	2B	0.211	0.220	0.2268	0.2311	0.2500
	3A	0.0000	0.2500	0.2435	—	0.2268	0.2243	0.2075	3B	0.2110	0.2190	0.2268	0.2300	0.2500
$\frac{1}{4}$-32 UNEF	2A	0.0010	0.2490	0.2430	—	0.2287	0.2255	0.2118	2B	0.216	0.224	0.2297	0.2339	0.2500
	3A	0.0000	0.2500	0.2440	—	0.2297	0.2273	0.2128	3B	0.2160	0.2229	0.2297	0.2328	0.2500

Table 2. *(Continued)* **Standard Series and Selected Combinations — Unified Screw Threads**

Nominal Size, Threads per Inch, and Series Designation[a]	External[b]								Internal[b]					
	Class	Allowance	Major Diameter			Pitch Diameter		UNR Minor Dia.,[c] Max (Ref.)	Class	Minor Diameter		Pitch Diameter		Major Diameter
			Max[d]	Min	Min[e]	Max[d]	Min			Min	Max	Min	Max	Min
$^1/_4$-36 UNS	2A	0.0009	0.2491	0.2436	—	0.2311	0.2280	0.2160	2B	0.220	0.227	0.2320	0.2360	0.2500
$^1/_4$-40 UNS	2A	0.0009	0.2491	0.2440	—	0.2329	0.2300	0.2193	2B	0.223	0.229	0.2338	0.2376	0.2500
$^1/_4$-48 UNS	2A	0.0008	0.2492	0.2447	—	0.2357	0.2330	0.2244	2B	0.227	0.232	0.2365	0.2401	0.2500
$^1/_4$-56 UNS	2A	0.0008	0.2492	0.2451	—	0.2376	0.2350	0.2279	2B	0.231	0.235	0.2384	0.2417	0.2500
$^5/_{16}$-18 UNC	1A	0.0012	0.3113	0.2982	—	0.2752	0.2691	0.2451	1B	0.252	0.265	0.2764	0.2843	0.3125
	2A	0.0012	0.3113	0.3026	0.2982	0.2752	0.2712	0.2451	2B	0.252	0.265	0.2764	0.2817	0.3125
	3A	0.0000	0.3125	0.3038	—	0.2764	0.2734	0.2463	3B	0.2520	0.2630	0.2764	0.2803	0.3125
$^5/_{16}$-20 UN	2A	0.0012	0.3113	0.3032	—	0.2788	0.2747	0.2518	2B	0.258	0.270	0.2800	0.2853	0.3125
	3A	0.0000	0.3125	0.3044	—	0.2800	0.2770	0.2530	3B	0.2580	0.2680	0.2800	0.2840	0.3125
$^5/_{16}$-24 UNF	1A	0.0011	0.3114	0.3006	—	0.2843	0.2788	0.2618	1B	0.267	0.277	0.2854	0.2925	0.3125
	2A	0.0011	0.3114	0.3042	—	0.2843	0.2806	0.2618	2B	0.267	0.277	0.2854	0.2902	0.3125
	3A	0.0000	0.3125	0.3053	—	0.2854	0.2827	0.2629	3B	0.2670	0.2754	0.2854	0.2890	0.3125
$^5/_{16}$-27 UNS	2A	0.0011	0.3114	0.3047	—	0.2873	0.2837	0.2673	2B	0.272	0.281	0.2884	0.2930	0.3125
$^5/_{16}$-28 UN	2A	0.0010	0.3115	0.3050	—	0.2883	0.2848	0.2690	2B	0.274	0.282	0.2893	0.2938	0.3125
	3A	0.0000	0.3125	0.3060	—	0.2893	0.2867	0.2700	3B	0.2740	0.2807	0.2893	0.2927	0.3125
$^5/_{16}$-32 UNEF	2A	0.0010	0.3115	0.3055	—	0.2912	0.2879	0.2743	2B	0.279	0.286	0.2922	0.2965	0.3125
	3A	0.0000	0.3125	0.3065	—	0.2922	0.2897	0.2753	3B	0.2790	0.2846	0.2922	0.2954	0.3125
$^5/_{16}$-36 UNS	2A	0.0009	0.3116	0.3061	—	0.2936	0.2905	0.2785	2B	0.282	0.289	0.2945	0.2986	0.3125
$^5/_{16}$-40 UNS	2A	0.0009	0.3116	0.3065	—	0.2954	0.2924	0.2818	2B	0.285	0.291	0.2963	0.3002	0.3125
$^5/_{16}$-48 UNS	2A	0.0008	0.3117	0.3072	—	0.2982	0.2954	0.2869	2B	0.290	0.295	0.2990	0.3026	0.3125
$^3/_8$-16 UNC	1A	0.0013	0.3737	0.3595	—	0.3331	0.3266	0.2993	1B	0.307	0.321	0.3344	0.3429	0.3750
	2A	0.0013	0.3737	0.3643	0.3595	0.3331	0.3287	0.2993	2B	0.307	0.321	0.3344	0.3401	0.3750
	3A	0.0000	0.3750	0.3656	—	0.3344	0.3311	0.3006	3B	0.3070	0.3182	0.3344	0.3387	0.3750
$^3/_8$-18 UNS	2A	0.0013	0.3737	0.3650	—	0.3376	0.3333	0.3075	2B	0.315	0.328	0.3389	0.3445	0.3750
$^3/_8$-20 UN	2A	0.0012	0.3738	0.3657	—	0.3413	0.3372	0.3143	2B	0.321	0.332	0.3425	0.3479	0.3750
	3A	0.0000	0.3750	0.3669	—	0.3425	0.3394	0.3155	3B	0.3210	0.3297	0.3425	0.3465	0.3750

Table 2. *(Continued)* **Standard Series and Selected Combinations — Unified Screw Threads**

Nominal Size, Threads per Inch, and Series Designation[a]	External[b]								Internal[b]					
	Class	Allowance	Major Diameter			Pitch Diameter		UNR Minor Dia.,[c] Max (Ref.)	Class	Minor Diameter		Pitch Diameter		Major Diameter
			Max[d]	Min	Min[e]	Max[d]	Min			Min	Max	Min	Max	Min
3/8-24 UNF	1A	0.0011	0.3739	0.3631	—	0.3468	0.3411	0.3243	1B	0.330	0.340	0.3479	0.3553	0.3750
	2A	0.0011	0.3739	0.3667	—	0.3468	0.3430	0.3243	2B	0.330	0.340	0.3479	0.3528	0.3750
	3A	0.0000	0.3750	0.3678	—	0.3479	0.3450	0.3254	3B	0.3300	0.3372	0.3479	0.3516	0.3750
3/8-27 UNS	2A	0.0011	0.3739	0.3672	—	0.3498	0.3462	0.3298	2B	0.335	0.344	0.3509	0.3556	0.3750
3/8-28 UN	2A	0.0011	0.3739	0.3674	—	0.3507	0.3471	0.3314	2B	0.336	0.345	0.3518	0.3564	0.3750
	3A	0.0000	0.3750	0.3685	—	0.3518	0.3491	0.3325	3B	0.3360	0.3426	0.3518	0.3553	0.3750
3/8-32 UNEF	2A	0.0010	0.3740	0.3680	—	0.3537	0.3503	0.3368	2B	0.341	0.349	0.3547	0.3591	0.3750
	3A	0.0000	0.3750	0.3690	—	0.3547	0.3522	0.3378	3B	0.3410	0.3469	0.3547	0.3580	0.3750
3/8-36 UNS	2A	0.0010	0.3740	0.3685	—	0.3560	0.3528	0.3409	2B	0.345	0.352	0.3570	0.3612	0.3750
3/8-40 UNS	2A	0.0009	0.3741	0.3690	—	0.3579	0.3548	0.3443	2B	0.348	0.354	0.3588	0.3628	0.3750
0.390-27 UNS	2A	0.0011	0.3889	0.3822	—	0.3648	0.3612	0.3448	2B	0.350	0.359	0.3659	0.3706	0.3900
7/16-14 UNC	1A	0.0014	0.4361	0.4206	—	0.3897	0.3826	0.3510	1B	0.360	0.376	0.3911	0.4003	0.4375
	2A	0.0014	0.4361	0.4258	0.4206	0.3897	0.3850	0.3510	2B	0.360	0.376	0.3911	0.3972	0.4375
	3A	0.0000	0.4375	0.4272	—	0.3911	0.3876	0.3524	3B	0.3600	0.3717	0.3911	0.3957	0.4375
7/16-16 UN	2A	0.0014	0.4361	0.4267	—	0.3955	0.3909	0.3617	2B	0.370	0.384	0.3969	0.4029	0.4375
	3A	0.0000	0.4375	0.4281	—	0.3969	0.3934	0.3631	3B	0.3700	0.3800	0.3969	0.4014	0.4375
7/16-18 UNS	2A	0.0013	0.4362	0.4275	—	0.4001	0.3957	0.3700	2B	0.377	0.390	0.4014	0.4071	0.4375
7/16-20 UNF	1A	0.0013	0.4362	0.4240	—	0.4037	0.3974	0.3767	1B	0.383	0.395	0.4050	0.4131	0.4375
	2A	0.0013	0.4362	0.4281	—	0.4037	0.3995	0.3767	2B	0.383	0.395	0.4050	0.4104	0.4375
	3A	0.0000	0.4375	0.4294	—	0.4050	0.4019	0.3780	3B	0.3830	0.3916	0.4050	0.4091	0.4375
7/16-24 UNS	2A	0.0012	0.4363	0.4291	—	0.4092	0.4053	0.3867	2B	0.392	0.402	0.4104	0.4154	0.4375
7/16-27 UNS	2A	0.0011	0.4364	0.4297	—	0.4123	0.4086	0.3923	2B	0.397	0.406	0.4134	0.4182	0.4375
7/16-28 UNEF	2A	0.0011	0.4364	0.4299	—	0.4132	0.4096	0.3939	2B	0.399	0.407	0.4143	0.4190	0.4375
	3A	0.0000	0.4375	0.4310	—	0.4143	0.4116	0.3950	3B	0.3990	0.4051	0.4143	0.4178	0.4375
7/16-32 UN	2A	0.0010	0.4365	0.4305	—	0.4162	0.4128	0.3993	2B	0.404	0.411	0.4172	0.4216	0.4375
	3A	0.0000	0.4375	0.4315	—	0.4172	0.4146	0.4003	3B	0.4040	0.4094	0.4172	0.4205	0.4375

Table 2. *(Continued)* **Standard Series and Selected Combinations—Unified Screw Threads**

Nominal Size, Threads per Inch, and Series Designation[a]	External[b]								Internal[b]					
	Class	Allowance	Major Diameter			Pitch Diameter		UNR Minor Dia.,[c] Max (Ref.)	Class	Minor Diameter		Pitch Diameter		Major Diameter
			Max[d]	Min	Min[e]	Max[d]	Min			Min	Max	Min	Max	Min
$^1/_2$-12 UNS	2A	0.0016	0.4984	0.4870	—	0.4443	0.4389	0.3992	2B	0.410	0.428	0.4459	0.4529	0.5000
	3A	0.0000	0.5000	0.4886	—	0.4459	0.4419	0.4008	3B	0.4100	0.4185	0.4459	0.4511	0.5000
$^1/_2$-13 UNC	1A	0.0015	0.4985	0.4822	—	0.4485	0.4411	0.4069	1B	0.417	0.434	0.4500	0.4597	0.5000
	2A	0.0015	0.4985	0.4876	0.4822	0.4485	0.4435	0.4069	2B	0.417	0.434	0.4500	0.4565	0.5000
	3A	0.0000	0.5000	0.4891	—	0.4500	0.4463	0.4084	3B	0.4170	0.4284	0.4500	0.4548	0.5000
$^1/_2$-14 UNS	2A	0.0015	0.4985	0.4882	—	0.4521	0.4471	0.4134	2B	0.423	0.438	0.4536	0.4601	0.5000
$^1/_2$-16 UN	2A	0.0014	0.4986	0.4892	—	0.4580	0.4533	0.4242	2B	0.432	0.446	0.4594	0.4655	0.5000
	3A	0.0000	0.5000	0.4906	—	0.4594	0.4559	0.4256	3B	0.4320	0.4420	0.4594	0.4640	0.5000
$^1/_2$-18 UNS	2A	0.0013	0.4987	0.4900	—	0.4626	0.4582	0.4325	2B	0.440	0.453	0.4639	0.4697	0.5000
$^1/_2$-20 UNF	1A	0.0013	0.4987	0.4865	—	0.4662	0.4598	0.4392	1B	0.446	0.457	0.4675	0.4759	0.5000
	2A	0.0013	0.4987	0.4906	—	0.4662	0.4619	0.4392	2B	0.446	0.457	0.4675	0.4731	0.5000
	3A	0.0000	0.5000	0.4919	—	0.4675	0.4643	0.4405	3B	0.4460	0.4537	0.4675	0.4717	0.5000
$^1/_2$-24 UNS	2A	0.0012	0.4988	0.4916	—	0.4717	0.4678	0.4492	2B	0.455	0.465	0.4729	0.4780	0.5000
$^1/_2$-27 UNS	2A	0.0011	0.4989	0.4922	—	0.4748	0.4711	0.4548	2B	0.460	0.469	0.4759	0.4807	0.5000
$^1/_2$-28 UNEF	2A	0.0011	0.4989	0.4924	—	0.4757	0.4720	0.4564	2B	0.461	0.470	0.4768	0.4816	0.5000
	3A	0.0000	0.5000	0.4935	—	0.4768	0.4740	0.4575	3B	0.4610	0.4676	0.4768	0.4804	0.5000
$^1/_2$-32 UN	2A	0.0010	0.4990	0.4930	—	0.4787	0.4752	0.4618	2B	0.466	0.474	0.4797	0.4842	0.5000
	3A	0.0000	0.5000	0.4940	—	0.4797	0.4771	0.4628	3B	0.4660	0.4719	0.4797	0.4831	0.5000
$^9/_{16}$-12 UNC	1A	0.0016	0.5609	0.5437	—	0.5068	0.4990	0.4617	1B	0.472	0.490	0.5084	0.5186	0.5625
	2A	0.0016	0.5609	0.5495	0.5437	0.5068	0.5016	0.4617	2B	0.472	0.490	0.5084	0.5152	0.5625
	3A	0.0000	0.5625	0.5511	—	0.5084	0.5045	0.4633	3B	0.4720	0.4843	0.5084	0.5135	0.5625
$^9/_{16}$-14 UNS	2A	0.0015	0.5610	0.5507	—	0.5146	0.5096	0.4759	2B	0.485	0.501	0.5161	0.5226	0.5625
$^9/_{16}$-16 UN	2A	0.0014	0.5611	0.5517	—	0.5205	0.5158	0.4867	2B	0.495	0.509	0.5219	0.5280	0.5625
	3A	0.0000	0.5625	0.5531	—	0.5219	0.5184	0.4881	3B	0.4950	0.5041	0.5219	0.5265	0.5625
$^9/_{16}$-18 UNF	1A	0.0014	0.5611	0.5480	—	0.5250	0.5182	0.4949	1B	0.502	0.515	0.5264	0.5353	0.5625
	2A	0.0014	0.5611	0.5524	—	0.5250	0.5205	0.4949	2B	0.502	0.515	0.5264	0.5323	0.5625
	3A	0.0000	0.5625	0.5538	—	0.5264	0.5230	0.4963	3B	0.5020	0.5106	0.5264	0.5308	0.5625

Table 2. *(Continued)* **Standard Series and Selected Combinations — Unified Screw Threads**

Nominal Size, Threads per Inch, and Series Designation[a]	External[b]								Internal[b]					
	Class	Allowance	Major Diameter			Pitch Diameter		UNR Minor Dia.,[c] Max (Ref.)	Class	Minor Diameter		Pitch Diameter		Major Diameter
			Max[d]	Min	Min[e]	Max[d]	Min			Min	Max	Min	Max	Min
$^9/_{16}$-20 UN	2A	0.0013	0.5612	0.5531	—	0.5287	0.5244	0.5017	2B	0.508	0.520	0.5300	0.5356	0.5625
	3A	0.0000	0.5625	0.5544	—	0.5300	0.5268	0.5030	3B	0.5080	0.5161	0.5300	0.5342	0.5625
$^9/_{16}$-24 UNEF	2A	0.0012	0.5613	0.5541	—	0.5342	0.5302	0.5117	2B	0.517	0.527	0.5354	0.5405	0.5625
	3A	0.0000	0.5625	0.5553	—	0.5354	0.5324	0.5129	3B	0.5170	0.5244	0.5354	0.5393	0.5625
$^9/_{16}$-27 UNS	2A	0.0011	0.5614	0.5547	—	0.5373	0.5335	0.5173	2B	0.522	0.531	0.5384	0.5433	0.5625
$^9/_{16}$-28 UN	2A	0.0011	0.5614	0.5549	—	0.5382	0.5345	0.5189	2B	0.524	0.532	0.5393	0.5441	0.5625
	3A	0.0000	0.5625	0.5560	—	0.5393	0.5365	0.5200	3B	0.5240	0.5301	0.5393	0.5429	0.5625
$^9/_{16}$-32 UN	2A	0.0011	0.5614	0.5554	—	0.5411	0.5376	0.5242	2B	0.529	0.536	0.5422	0.5468	0.5625
	3A	0.0000	0.5625	0.5565	—	0.5422	0.5396	0.5253	3B	0.5290	0.5344	0.5422	0.5456	0.5625
$^5/_8$-11 UNC	1A	0.0017	0.6233	0.6051	—	0.5643	0.5560	0.5150	1B	0.527	0.546	0.5660	0.5767	0.6250
	2A	0.0017	0.6233	0.6112	0.6052	0.5643	0.5588	0.5150	2B	0.527	0.546	0.5660	0.5732	0.6250
	3A	0.0000	0.6250	0.6129	—	0.5660	0.5619	0.5167	3B	0.5270	0.5391	0.5660	0.5714	0.6250
$^5/_8$-12 UN	2A	0.0016	0.6234	0.6120	—	0.5693	0.5639	0.5242	2B	0.535	0.553	0.5709	0.5780	0.6250
	3A	0.0000	0.6250	0.6136	—	0.5709	0.5668	0.5258	3B	0.5350	0.5463	0.5709	0.5762	0.6250
$^5/_8$-14 UNS	2A	0.0015	0.6235	0.6132	—	0.5771	0.5720	0.5384	2B	0.548	0.563	0.5786	0.5852	0.6250
$^5/_8$-16 UN	2A	0.0014	0.6236	0.6142	—	0.5830	0.5782	0.5492	2B	0.557	0.571	0.5844	0.5906	0.6250
	3A	0.0000	0.6250	0.6156	—	0.5844	0.5808	0.5506	3B	0.5570	0.5662	0.5844	0.5890	0.6250
$^5/_8$-18 UNF	1A	0.0014	0.6236	0.6105	—	0.5875	0.5805	0.5574	1B	0.565	0.578	0.5889	0.5980	0.6250
	2A	0.0014	0.6236	0.6149	—	0.5875	0.5828	0.5574	2B	0.565	0.578	0.5889	0.5949	0.6250
	3A	0.0000	0.6250	0.6163	—	0.5889	0.5854	0.5588	3B	0.5650	0.5730	0.5889	0.5934	0.6250
$^5/_8$-20 UN	2A	0.0013	0.6237	0.6156	—	0.5912	0.5869	0.5642	2B	0.571	0.582	0.5925	0.5981	0.6250
	3A	0.0000	0.6250	0.6169	—	0.5925	0.5893	0.5655	3B	0.5710	0.5786	0.5925	0.5967	0.6250
$^5/_8$-24 UNEF	2A	0.0012	0.6238	0.6166	—	0.5967	0.5927	0.5742	2B	0.580	0.590	0.5979	0.6031	0.6250
	3A	0.0000	0.6250	0.6178	—	0.5979	0.5949	0.5754	3B	0.5800	0.5869	0.5979	0.6018	0.6250
$^5/_8$-27 UNS	2A	0.0011	0.6239	0.6172	—	0.5998	0.5960	0.5798	2B	0.585	0.594	0.6009	0.6059	0.6250
$^5/_8$-28 UN	2A	0.0011	0.6239	0.6174	—	0.6007	0.5969	0.5814	2B	0.586	0.595	0.6018	0.6067	0.6250
	3A	0.0000	0.6250	0.6185	—	0.6018	0.5990	0.5825	3B	0.5860	0.5926	0.6018	0.6055	0.6250

Table 2. *(Continued)* **Standard Series and Selected Combinations — Unified Screw Threads**

Nominal Size, Threads per Inch, and Series Designation[a]	External[b]								Internal[b]					
	Class	Allowance	Major Diameter			Pitch Diameter		UNR Minor Dia.,[c] Max (Ref.)	Class	Minor Diameter		Pitch Diameter		Major Diameter
			Max[d]	Min	Min[e]	Max[d]	Min			Min	Max	Min	Max	Min
$^{5}/_{8}$-32 UN	2A	0.0011	0.6239	0.6179	—	0.6036	0.6000	0.5867	2B	0.591	0.599	0.6047	0.6093	0.6250
	3A	0.0000	0.6250	0.6190	—	0.6047	0.6020	0.5878	3B	0.5910	0.5969	0.6047	0.6082	0.6250
$^{11}/_{16}$-12 UN	2A	0.0016	0.6859	0.6745	—	0.6318	0.6263	0.5867	2B	0.597	0.615	0.6334	0.6405	0.6875
	3A	0.0000	0.6875	0.6761	—	0.6334	0.6293	0.5883	3B	0.5970	0.6085	0.6334	0.6387	0.6875
$^{11}/_{16}$-16 UN	2A	0.0014	0.6861	0.6767	—	0.6455	0.6407	0.6117	2B	0.620	0.634	0.6469	0.6532	0.6875
	3A	0.0000	0.6875	0.6781	—	0.6469	0.6433	0.6131	3B	0.6200	0.6284	0.6469	0.6516	0.6875
$^{11}/_{16}$-20 UN	2A	0.0013	0.6862	0.6781	—	0.6537	0.6493	0.6267	2B	0.633	0.645	0.6550	0.6607	0.6875
	3A	0.0000	0.6875	0.6794	—	0.6550	0.6517	0.6280	3B	0.6330	0.6411	0.6550	0.6593	0.6875
$^{11}/_{16}$-24 UNEF	2A	0.0012	0.6863	0.6791	—	0.6592	0.6552	0.6367	2B	0.642	0.652	0.6604	0.6657	0.6875
	3A	0.0000	0.6875	0.6803	—	0.6604	0.6574	0.6379	3B	0.6420	0.6494	0.6604	0.6643	0.6875
$^{11}/_{16}$-28 UN	2A	0.0011	0.6864	0.6799	—	0.6632	0.6594	0.6439	2B	0.649	0.657	0.6643	0.6692	0.6875
	3A	0.0000	0.6875	0.6810	—	0.6643	0.6614	0.6450	3B	0.6490	0.6551	0.6643	0.6680	0.6875
$^{11}/_{16}$-32 UN	2A	0.0011	0.6864	0.6804	—	0.6661	0.6625	0.6492	2B	0.654	0.661	0.6672	0.6719	0.6875
	3A	0.0000	0.6875	0.6815	—	0.6672	0.6645	0.6503	3B	0.6540	0.6594	0.6672	0.6707	0.6875
$^{3}/_{4}$-10 UNC	1A	0.0018	0.7482	0.7288	—	0.6832	0.6744	0.6291	1B	0.642	0.663	0.6850	0.6965	0.7500
	2A	0.0018	0.7482	0.7353	0.7288	0.6832	0.6773	0.6291	2B	0.642	0.663	0.6850	0.6927	0.7500
	3A	0.0000	0.7500	0.7371	—	0.6850	0.6806	0.6309	3B	0.6420	0.6545	0.6850	0.6907	0.7500
$^{3}/_{4}$-12 UN	2A	0.0017	0.7483	0.7369	—	0.6942	0.6887	0.6491	2B	0.660	0.678	0.6959	0.7031	0.7500
	3A	0.0000	0.7500	0.7386	—	0.6959	0.6918	0.6508	3B	0.6600	0.6707	0.6959	0.7013	0.7500
$^{3}/_{4}$-14 UNS	2A	0.0015	0.7485	0.7382	—	0.7021	0.6970	0.6634	2B	0.673	0.688	0.7036	0.7103	0.7500
$^{3}/_{4}$-16 UNF	1A	0.0015	0.7485	0.7343	—	0.7079	0.7004	0.6741	1B	0.682	0.696	0.7094	0.7192	0.7500
	2A	0.0015	0.7485	0.7391	—	0.7079	0.7029	0.6741	2B	0.682	0.696	0.7094	0.7159	0.7500
	3A	0.0000	0.7500	0.7406	—	0.7094	0.7056	0.6756	3B	0.6820	0.6909	0.7094	0.7143	0.7500
$^{3}/_{4}$-18 UNS	2A	0.0014	0.7486	0.7399	—	0.7125	0.7079	0.6824	2B	0.690	0.703	0.7139	0.7199	0.7500
$^{3}/_{4}$-20 UNEF	2A	0.0013	0.7487	0.7406	—	0.7162	0.7118	0.6892	2B	0.696	0.707	0.7175	0.7232	0.7500
	3A	0.0000	0.7500	0.7419	—	0.7175	0.7142	0.6905	3B	0.6960	0.7036	0.7175	0.7218	0.7500
$^{3}/_{4}$-24 UNS	2A	0.0012	0.7488	0.7416	—	0.7217	0.7176	0.6992	2B	0.705	0.715	0.7229	0.7282	0.7500
$^{3}/_{4}$-27 UNS	2A	0.0012	0.7488	0.7421	—	0.7247	0.7208	0.7047	2B	0.710	0.719	0.7259	0.7310	0.7500

Table 2. *(Continued)* **Standard Series and Selected Combinations — Unified Screw Threads**

Nominal Size, Threads per Inch, and Series Designation[a]	External[b]								Internal[b]					
	Class	Allowance	Major Diameter			Pitch Diameter		UNR Minor Dia.,[c] Max (Ref.)	Class	Minor Diameter		Pitch Diameter		Major Diameter
			Max[d]	Min	Min[e]	Max[d]	Min			Min	Max	Min	Max	Min
3/4-28 UN	2A	0.0012	0.7488	0.7423	—	0.7256	0.7218	0.7063	2B	0.711	0.720	0.7268	0.7318	0.7500
	3A	0.0000	0.7500	0.7435	—	0.7268	0.7239	0.7075	3B	0.7110	0.7176	0.7268	0.7305	0.7500
3/4-32 UN	2A	0.0011	0.7489	0.7429	—	0.7286	0.7250	0.7117	2B	0.716	0.724	0.7297	0.7344	0.7500
	3A	0.0000	0.7500	0.7440	—	0.7297	0.7270	0.7128	3B	0.7160	0.7219	0.7297	0.7333	0.7500
13/16-12 UN	2A	0.0017	0.8108	0.7994	—	0.7567	0.7511	0.7116	2B	0.722	0.740	0.7584	0.7656	0.8125
	3A	0.0000	0.8125	0.8011	—	0.7584	0.7542	0.7133	3B	0.7220	0.7329	0.7584	0.7638	0.8125
13/16-16 UN	2A	0.0015	0.8110	0.8016	—	0.7704	0.7655	0.7366	2B	0.745	0.759	0.7719	0.7783	0.8125
	3A	0.0000	0.8125	0.8031	—	0.7719	0.7682	0.7381	3B	0.7450	0.7534	0.7719	0.7767	0.8125
13/16-20 UNEF	2A	0.0013	0.8112	0.8031	—	0.7787	0.7743	0.7517	2B	0.758	0.770	0.7800	0.7858	0.8125
	3A	0.0000	0.8125	0.8044	—	0.7800	0.7767	0.7530	3B	0.7580	0.7661	0.7800	0.7843	0.8125
13/16-28 UN	2A	0.0012	0.8113	0.8048	—	0.7881	0.7842	0.7688	2B	0.774	0.782	0.7893	0.7943	0.8125
	3A	0.0000	0.8125	0.8060	—	0.7893	0.7864	0.7700	3B	0.7740	0.7801	0.7893	0.7931	0.8125
13/16-32 UN	2A	0.0011	0.8114	0.8054	—	0.7911	0.7874	0.7742	2B	0.779	0.786	0.7922	0.7970	0.8125
	3A	0.0000	0.8125	0.8065	—	0.7922	0.7894	0.7753	3B	0.7790	0.7844	0.7922	0.7958	0.8125
7/8-9 UNC	1A	0.0019	0.8731	0.8523	—	0.8009	0.7914	0.7408	1B	0.755	0.778	0.8028	0.8151	0.8750
	2A	0.0019	0.8731	0.8592	0.8523	0.8009	0.7946	0.7408	2B	0.755	0.778	0.8028	0.8110	0.8750
	3A	0.0000	0.8750	0.8611	—	0.8028	0.7981	0.7427	3B	0.7550	0.7681	0.8028	0.8089	0.8750
7/8-10 UNS	2A	0.0018	0.8732	0.8603	—	0.8082	0.8021	0.7541	2B	0.767	0.788	0.8100	0.8179	0.8750
7/8-12 UN	2A	0.0017	0.8733	0.8619	—	0.8192	0.8136	0.7741	2B	0.785	0.803	0.8209	0.8282	0.8750
	3A	0.0000	0.8750	0.8636	—	0.8209	0.8167	0.7758	3B	0.7850	0.7952	0.8209	0.8264	0.8750
7/8-14 UNF	1A	0.0016	0.8734	0.8579	—	0.8270	0.8189	0.7883	1B	0.798	0.813	0.8286	0.8392	0.8750
	2A	0.0016	0.8734	0.8631	—	0.8270	0.8216	0.7883	2B	0.798	0.813	0.8286	0.8356	0.8750
	3A	0.0000	0.8750	0.8647	—	0.8286	0.8245	0.7899	3B	0.7980	0.8067	0.8286	0.8339	0.8750
7/8-16 UN	2A	0.0015	0.8735	0.8641	—	0.8329	0.8280	0.7991	2B	0.807	0.821	0.8344	0.8408	0.8750
	3A	0.0000	0.8750	0.8656	—	0.8344	0.8307	0.8006	3B	0.8070	0.8159	0.8344	0.8392	0.8750
7/8-18 UNS	2A	0.0014	0.8736	0.8649	—	0.8375	0.8328	0.8074	2B	0.815	0.828	0.8389	0.8450	0.8750
7/8-20 UNEF	2A	0.0013	0.8737	0.8656	—	0.8412	0.8367	0.8142	2B	0.821	0.832	0.8425	0.8483	0.8750
	3A	0.0000	0.8750	0.8669	—	0.8425	0.8391	0.8155	3B	0.8210	0.8286	0.8425	0.8469	0.8750

Table 2. *(Continued)* **Standard Series and Selected Combinations—Unified Screw Threads**

Nominal Size, Threads per Inch, and Series Designation[a]	External[b]								Internal[b]					
	Class	Allow-ance	Major Diameter			Pitch Diameter		UNR Minor Dia.,[c] Max (Ref.)	Class	Minor Diameter		Pitch Diameter		Major Diameter
			Max[d]	Min	Min[e]	Max[d]	Min			Min	Max	Min	Max	Min
$^7/_8$-24 UNS	2A	0.0012	0.8738	0.8666	—	0.8467	0.8425	0.8242	2B	0.830	0.840	0.8479	0.8533	0.8750
$^7/_8$-27 UNS	2A	0.0012	0.8738	0.8671	—	0.8497	0.8457	0.8297	2B	0.835	0.844	0.8509	0.8561	0.8750
$^7/_8$-28 UN	2A	0.0012	0.8738	0.8673	—	0.8506	0.8467	0.8313	2B	0.836	0.845	0.8518	0.8569	0.8750
	3A	0.0000	0.8750	0.8685	—	0.8518	0.8489	0.8325	3B	0.8360	0.8426	0.8518	0.8556	0.8750
$^7/_8$-32 UN	2A	0.0011	0.8739	0.8679	—	0.8536	0.8499	0.8367	2B	0.841	0.849	0.8547	0.8595	0.8750
	3A	0.0000	0.8750	0.8690	—	0.8547	0.8519	0.8378	3B	0.8410	0.8469	0.8547	0.8583	0.8750
$^{15}/_{16}$-12 UN	2A	0.0017	0.9358	0.9244	—	0.8817	0.8761	0.8366	2B	0.847	0.865	0.8834	0.8907	0.9375
	3A	0.0000	0.9375	0.9261	—	0.8834	0.8792	0.8383	3B	0.8470	0.8575	0.8834	0.8889	0.9375
$^{15}/_{16}$-16 UN	2A	0.0015	0.9360	0.9266	—	0.8954	0.8904	0.8616	2B	0.870	0.884	0.8969	0.9033	0.9375
	3A	0.0000	0.9375	0.9281	—	0.8969	0.8932	0.8631	3B	0.8700	0.8784	0.8969	0.9017	0.9375
$^{15}/_{16}$-20 UNEF	2A	0.0014	0.9361	0.9280	—	0.9036	0.8991	0.8766	2B	0.883	0.895	0.9050	0.9109	0.9375
	3A	0.0000	0.9375	0.9294	—	0.9050	0.9016	0.8780	3B	0.8830	0.8911	0.9050	0.9094	0.9375
$^{15}/_{16}$-28 UN	2A	0.0012	0.9363	0.9298	—	0.9131	0.9092	0.8938	2B	0.899	0.907	0.9143	0.9194	0.9375
	3A	0.0000	0.9375	0.9310	—	0.9143	0.9113	0.8950	3B	0.8990	0.9051	0.9143	0.9181	0.9375
$^{15}/_{16}$-32 UN	2A	0.0011	0.9364	0.9304	—	0.9161	0.9123	0.8992	2B	0.904	0.911	0.9172	0.9221	0.9375
	3A	0.0000	0.9375	0.9315	—	0.9172	0.9144	0.9003	3B	0.9040	0.9094	0.9172	0.9209	0.9375
1-8 UNC	1A	0.0020	0.9980	0.9755	—	0.9168	0.9067	0.8492	1B	0.865	0.890	0.9188	0.9320	1.0000
	2A	0.0020	0.9980	0.9830	0.9755	0.9168	0.9101	0.8492	2B	0.865	0.890	0.9188	0.9276	1.0000
	3A	0.0000	1.0000	0.9850	—	0.9188	0.9137	0.8512	3B	0.8650	0.8797	0.9188	0.9254	1.0000
1-10 UNS	2A	0.0018	0.9982	0.9853	—	0.9332	0.9270	0.8791	2B	0.892	0.913	0.9350	0.9430	1.0000
1-12 UNF	1A	0.0018	0.9982	0.9810	—	0.9441	0.9353	0.8990	1B	0.910	0.928	0.9459	0.9573	1.0000
	2A	0.0018	0.9982	0.9868	—	0.9441	0.9382	0.8990	2B	0.910	0.928	0.9459	0.9535	1.0000
	3A	0.0000	1.0000	0.9886	—	0.9459	0.9415	0.9008	3B	0.9100	0.9198	0.9459	0.9516	1.0000

[a] Use UNR designation instead of UN wherever UNR thread form is desired for external use.
[b] Thread classes may be combined; for example, a Class 2A external thread may be used with a Class 1B, 2B, or 3B internal thread.
[c] UN series external thread maximum minor diameter is basic for Class 3A and basic minus allowance for Classes 1A and 2A.
[d] For Class 2A threads having an additive finish the maximum is increased, by the allowance, to the basic size, the value being the same as for Class 3A.
[e] For unfinished hot-rolled material not including standard fasteners with rolled threads.
All dimensions are in inches.
For UNS threads and sizes above 1 inch see ASME/ANSI B1.1-1989 (R2001). Use UNS threads only if Standard Series do not meet requirements.

Table 3. External Inch Screw Thread Calculations for $^1/_2$-28 UNEF-2A

Characteristic Description	Calculation	Notes
Basic major diameter, d_{bsc}	$d_{bsc} = \frac{1}{2} = 0.5 = 0.5000$	d_{bsc} is rounded to four decimal places
Pitch, P	$P = \frac{1}{28} = 0.035714285714 = 0.03571429$	P is rounded to eight decimal places
Maximum external major diameter (d_{max}) = basic major diameter (d_{bsc}) – allowance (es)	$d_{max} = d_{bsc} - es$	es is the basic allowance
Basic major diameter (d_{bsc})	$d_{bsc} = 0.5000$	d_{bsc} is rounded to four decimal places
Allowance (es)	$es = 0.300 \times Td_2$ for Class 2A	Td_2 is the pitch diameter tolerance for Class 2A
External pitch diameter tolerance Td_2	$Td_2 = 0.0015D^{\frac{1}{3}} + 0.0015\sqrt{LE} + 0.015P^{\frac{2}{3}}$ $= 0.0015 \times 0.5^{\frac{1}{3}} + 0.0015\sqrt{9 \times 0.03571429} + 0.015(0.03571429)^{\frac{2}{3}}$ $= 0.001191 + 0.000850 + 0.001627 = 0.003668$	$LE = 9P$ (length of engagement) Td_2 is rounded to six decimal places
Allowance (es)	$es = 0.300 \times 0.003668 = 0.0011004 = 0.0011$	es is rounded to four decimal places
Maximum external major diameter (d_{max})	$d_{max} = d_{base} - es = 0.5000 - 0.0011 = 0.4989$	d_{max} is rounded to four decimal places
Minimum external major diameter (d_{min}) = maximum external major diameter (d_{max}) – major diameter tolerance (Td)	$d_{min} = d_{max} - Td$	Td is the major diameter tolerance
Major diameter tolerance (Td)	$Td = 0.060\sqrt[3]{P^2} = 0.060 \times \sqrt[3]{0.03571429^2}$ $= 0.060 \times \sqrt[3]{0.001276} = 0.060 \times 0.108463$ $= 0.00650778 = 0.0065$	Td is rounded to four decimal places

Table 3. *(Continued)* **External Inch Screw Thread Calculations for $^1/_2$-28 UNEF-2A**

Characteristic Description	Calculation	Notes
Minimum external major diameter (d_{min})	$d_{min} = d_{max} - Td = 0.4989 - 0.006508$ $= 0.492392 = 0.4924$	d_{min} is rounded to four decimal places
Maximum external pitch diameter (d_{2max}) = maximum external major diameter (d_{max}) – twice the external thread addendum (h_{as})	$d_{2max} = d_{max} - 2 \times h_{as}$	h_{as} = external thread addendum
External thread addendum	$h_{as} = \frac{0.64951905P}{2}$ $\quad 2h_{as} = 0.64951905P$ $2h_{as} = 0.64951905 \times 0.03571429 = 0.02319711$ $= 0.023197$	$2h_{as}$ is rounded to six decimal places
Maximum external pitch diameter (d_{2max})	$d_{2max} = d_{max} - 2 \times h_{as} = 0.4989 - 0.23197$ $= 0.475703 = 0.4757$	d_{2max} is rounded to four decimal places
Minimum external pitch diameter (d_{2min}) = maximum external pitch diameter (d_{2max}) – external pitch diameter tolerance (Td_2)	$d_{2min} = d_{2max} - Td_2$	Td_2 = external pitch diameter tolerance (see previous Td_2 calculation in this table)
Minimum external pitch diameter (d_{2min})	$d_{2min} = d_{2max} - Td_2 = 0.4757 - 0.003668$ $= 0.472032 = 0.4720$	d_{2min} is rounded to four decimal places
Maximum external UNR minor diameter (d_{3max}) = maximum external major diameter (d_{max}) – double height of external UNR thread $2h_s$	$d_{3max} = d_{max} - 2 \times h_s$	h_s = external UNR thread height
External UNR thread height ($2h_s$)	$2h_s = 1.19078493P = 1.19078493 \times 0.03571429$ $= 0.042528$	$2h_s$ rounded to six decimal places
Maximum external UNR minor diameter (d_{3max})	$d_{3max} = d_{max} - 2 \times h_s = 0.4989 - 0.042528$ $= 0.456372 = 0.4564$	d_{3max} is rounded to four decimal places

Table 3. *(Continued)* **External Inch Screw Thread Calculations for $^1/_2$-28 UNEF-2A**

Characteristic Description	Calculation	Notes
Maximum external UN minor diameter (d_{1max}) = maximum external major diameter (d_{max}) – double height of external UN thread $2h_s$	$d_{1\max} = d_{max} - 2 \times h_s$	For UN threads, $2h_s = 2h_n$
Double height of external UN thread $2h_s$	$2h_s = 1.08253175P$ $= 1.08253175 \times 0.03571429 = 0.03866185$ $= 0.038662$	$2h_s$ is rounded to six decimal places
Maximum external UN minor diameter (d_{1max})	$d_{1max} = d_{max} - 2 \times h_s$ $= 0.4989 - 0.038662 = 0.460238 = 0.4602$	d_{1max} is rounded to four decimal places

Table 4. Internal Inch Screw Thread Calculations for $^1/_2$-28 UNEF-2B

Characteristic Description	Calculation	Notes
Basic major diameter, d_{bsc}	$d_{bsc} = \frac{1}{2} = 0.5 = 0.5000$	d_{bsc} is rounded to four decimal places
Pitch, P	$P = \frac{1}{28} = 0.035714285714 = 0.03571429$	P is rounded to eight decimal places
Minimum internal minor diameter (D_{1min}) = basic major diameter (D_{bsc}) – double height of external UN thread $2h_n$	$D_{1min} = D_{bsc} - 2h_n$	$2h_n$ is the double height of external UN thread
Double height of external UN thread $2h_s$	$2h_n = 1.08253175P = 1.08253175 \times 0.03571429$ $= 0.03866185 = 0.038662$	$2h_n$ is rounded to six decimal places
Minimum internal major diameter (D_{1min})	$D_{1min} = D_{bsc} - 2 \times h_n = 0.5000 - 0.038662$ $= 0.461338 = 0.461$	For class 2B the value is rounded to three decimal places to obtain the final values

Table 4. *(Continued)* **Internal Inch Screw Thread Calculations for $^1/_2$-28 UNEF-2B**

Characteristic Description	Calculation	Notes
Maximum internal minor diameter (D_{1max}) = minimum internal minor diameter (D_{1min}) + internal minor diameter tolerance TD_1	$D_{1max} = D_{1min} + TD_1$	D_{1min} is rounded to six decimal places
Internal minor diameter tolerance TD_1	$TD_1 = 0.25P - 0.40P^2$ $= 0.25 \times 0.03571429 - 0.40 \times 0.03571429^2$ $= 0.008929 - 0.000510 = 0.008419 = 0.003127$	TD_1 is rounded to four decimal places
Maximum internal minor diameter (D_{1max})	$D_{1max} = D_{1min} + TD_1 = 0.461338 + 0.008419$ $= 0.469757 = 0.470$	For the Class 2B thread D_{1max} is rounded to three decimal places to obtain final values. Other sizes and classes are expressed in four decimal places
Minimum internal pitch diameter (D_{2min}) = basic major diameter (D_{bsc}) – twice the external thread addendum (h_b)	$D_{2min} = D_{bsc} - h_b$	h_b = external thread addendum
External thread addendum (h_b)	$h_b = 0.64951905P = 0.64951905 \times 0.03571429$ $= 0.02319711 = 0.023197$	h_b is rounded to six decimal places
Minimum internal pitch diameter (D_{2min})	$D_{2min} = D_{bsc} - h_b = 0.5000 - 0.023197$ $= 0.476803 = 0.4768$	D_{2min} is rounded to four decimal places
Maximum internal pitch diameter (D_{2max}) = minimum internal pitch diameter (D_{2min}) + internal pitch diameter tolerance (TD_2)	$D_{2max} = D_{2min} + TD_2$	TD_2 = external pitch diameter tolerance
External pitch diameter tolerance TD_2	$TD_2 = 1.30 \times (Td_2 \text{ for Class 2A}) = 1.30 \times 0.003668$ $= 0.0047684 = 0.0048$	Constant 1.30 is for this Class 2B example, and will be different for Classes 1B and 3B. Td_2 for Class 2A (see Table 3) is rounded to six decimal places. TD_2 is rounded to four places
Maximum internal pitch diameter (D_{2max})	$D_{2max} = D_{2min} + TD_2 = 0.4768 + 0.0048 = 0.4816$	D_{2max} is rounded to four decimal places

Table 4. *(Continued)* **Internal Inch Screw Thread Calculations for 1/2 -28 UNEF-2B**

Characteristic Description	Calculation	Notes
Minimum internal major diameter (D_{min}) = basic major diameter (D_{bsc})	$D_{min} = D_{bsc} = 0.5000$	D_{min} is rounded to four decimal places

Table 5. External Inch Screw Thread Calculations for 19/64 -36 UNS-2A

Characteristic Description	Calculation	Notes
Basic major diameter, d_{bsc}	$d_{bsc} = \frac{19}{64} = 0.296875 = 0.2969$	d_{bsc} is rounded to four decimal places
Pitch, P	$P = \frac{1}{36} = 0.0277777777778 = 0.02777778$	P is rounded to eight decimal places
Maximum external major diameter (d_{max}) = basic major diameter (d_{bsc}) – allowance (es)	$d_{max} = d_{bsc} - es$	
Allowance (es)	$es = 0.300 \times Td_2$ for Class 2A	Td_2 is Pitch diameter tolerance for Class 2A
External pitch diameter tolerance, Td_2	$Td_2 = 0.0015D^{\frac{1}{3}} + 0.0015\sqrt{LE} + 0.015P^{\frac{2}{3}}$ $= 0.0015 \times 0.2969^{\frac{1}{3}} + 0.0015\sqrt{9 \times 0.02777778} + 0.015(0.02777778)^{\frac{2}{3}}$ $= 0.001000679 + 0.00075 + 0.001375803 = 0.003126482$ $= 0.003127$	$LE = 9P$ (length of engagement) Td_2 is rounded to six decimal places
Allowance (es)	$es = 0.300 \times 0.003127 = 0.0009381 = 0.0009$	es is rounded to four decimal places
Maximum external major diameter (d_{max})	$d_{max} = d_{bsc} - es = 0.2969 - 0.0009 = 0.2960$	d_{max} is rounded to four decimal places

Table 5. *(Continued)* **External Inch Screw Thread Calculations for $^{19}/_{64}$-36 UNS-2A**

Characteristic Description	Calculation	Notes
Minimum external major diameter (d_{min}) = maximum external major diameter (d_{max}) − major diameter tolerance (Td)	$d_{min} = d_{max} - Td$	Td is the major diameter tolerance
Major diameter tolerance (Td)	$Td = 0.060\sqrt[3]{P^2} = 0.060 \times \sqrt[3]{0.02777778^2}$ $= 0.060 \times \sqrt[3]{0.000772} = 0.060 \times 0.091736$ $= 0.00550416 = 0.0055$	Td is rounded to four decimal places
Minimum external major diameter (d_{min})	$d_{min} = d_{max} - Td = 0.2960 - 0.0055 = 0.2905$	d_{min} is rounded to four decimal places
Maximum external pitch diameter (d_{2max}) = maximum external major diameter (d_{max}) − twice the external thread addendum	$d_{2max} = d_{max} - 2 \times h_{as}$	h_{as} = external thread addendum
External thread addendum	$h_{as} = \dfrac{0.64951905P}{2}$ $\quad 2h_{as} = 0.64951905P$ $2h_{as} = 0.64951905 \times 0.02777778 = 0.0180421972$ $= 0.018042$	h_{as} is rounded to six decimal places
Maximum external pitch diameter (d_{2max})	$d_{2max} = d_{max} - 2h_{as} = 0.2960 - 0.018042$ $= 0.277958 = 0.2780$	d_{2max} is rounded to four decimal places
Minimum external pitch diameter (d_{2min}) = maximum external pitch diameter (d_{2max}) − external pitch diameter tolerance (Td_2)	$d_{2min} = d_{2max} - Td_2$	Td_2 = external pitch diameter tolerance (see previous Td_2 calculation in this table)
Minimum external pitch diameter (d_{2min})	$d_{2min} = d_{2max} - Td_2 = 0.2780 - 0.003127$ $= 0.274873 = 0.2749$	d_{2min} is rounded to four decimal places

Table 5. *(Continued)* **External Inch Screw Thread Calculations for $^{19}/_{64}$-36 UNS-2A**

Characteristic Description	Calculation	Notes
Maximum external UNR minor diameter (d_{3max}) = maximum external major diameter (d_{max}) – double height of external UNR thread $2hs$	$d_{3max} = d_{max} - 2h_s$	h_s = external UNR thread height
External UNR thread height	$2h_s = 1.19078493P = 1.19078493 \times 0.02777778$ $= 0.033077362 = 0.033077$	$2h_s$ is rounded to six decimal places
Maximum external UNR minor diameter (d_{3max})	$d_{3max} = d_{max} - 2h_s = 0.2960 - 0.033077$ $= 0.262923 = 0.2629$	d_{3max} is rounded to four decimal places
Maximum external UN minor diameter (d_{1max}) = maximum external major diameter (d_{max}) – double height of external UN thread $2h_s$	$d_{1max} = d_{max} - 2 \times h_s$	For UN threads, $2h_s = 2h_n$
Double height of external UN thread $2h_s$	$2h_s = 1.08253175P = 1.08253175 \times 0.02777778$ $= 0.030070329 = 0.030070$	For UN threads, $2h_s = 2h_n$ $2h_s$ is rounded to six decimal places
Maximum external UN minor diameter (d_{1max})	$d_{1max} = d_{max} - 2h_s = 0.2960 - 0.030070$ $= 0.265930 = 0.2659$	Maximum external UN minor diameter is rounded to four decimal places

Table 6. Internal Inch Screw Thread Calculations for $^{19}/_{64}$-28 UNS-2B

Characteristic Description	Calculation	Notes
Minimum internal minor diameter (D_{1min}) = basic major diameter (D_{bsc}) – double height of external UN thread $2h_n$	$D_{1min} = D_{bsc} - 2h_n$	$2h_n$ is the double height of external UN threads
Basic major diameter (D_{bsc})	$D_{bsc} = \frac{19}{64} = 0.296875 = 0.2969$	This is the final value of basic major diameter (given) and rounded to four decimal places

Table 6. *(Continued)* **Internal Inch Screw Thread Calculations for $^{19}/_{64}$-28 UNS-2B**

Characteristic Description	Calculation	Notes
Double height of external UN thread $2h_s$	$2h_n = 1.08253175P = 1.08253175 \times 0.02777778$ $= 0.030070329 = 0.030070$	P is rounded to eight decimal places
Minimum internal major diameter (D_{1min})	$D_{1min} = D_{bsc} - 2h_n = 0.2969 - 0.030070$ $= 0.266830 = 0.267$	For class 2B the value is rounded to three decimal places to obtain the final value; other sizes and classes are expressed in a four place decimal
Maximum internal minor diameter (D_{1max}) = minimum internal minor diameter (D_{1min}) + internal minor diameter tolerance TD_1	$D_{1max} = D_{1min} + TD_1$	D_{1min} is rounded to six decimal places
Internal minor diameter tolerance TD_1	$TD_1 = 0.25P - 0.40P^2$ $= 0.25 \times 0.02777778 - 0.40 \times 0.02777778^2$ $= 0.006944 - 0.000309 = 0.006635 = 0.0066$	TD_1 is rounded to four decimal places.
Maximum internal minor diameter (D_{1max})	$D_{1max} = D_{1min} + TD_1 = 0.266830 + 0.006635$ $= 0.273465 = 0.273$	For Class 2B thread the value is rounded to three decimal places to obtain the final values. Other sizes and classes are expressed to four decimal places
Minimum internal pitch diameter (D_{2min}) = basic major diameter (D_{bsc}) − twice the external thread addendum (h_b)	$D_{2min} = D_{1max} - h_b$	h_b = external thread addendum
External thread addendum	$h_b = 0.64951905P = 0.64951905 \times 0.02777778$ $= 0.018042197 = 0.018042$	h_b is rounded to six decimal places
Minimum internal pitch diameter (D_{2min})	$D_{2min} = D_{bsc} - h_b = 0.2969 - 0.018042$ $= 0.278858 = 0.2789$	$D_{2\,min}$ is rounded to four decimal places

Table 6. *(Continued)* **Internal Inch Screw Thread Calculations for $^{19}/_{64}$-28 UNS-2B**

Characteristic Description	Calculation	Notes
Maximum internal pitch diameter (D_{2max}) = minimum internal pitch diameter (D_{2min}) + internal pitch diameter tolerance (TD_2)	$D_{2max} = D_{2min} + TD_2$	TD_2 = external pitch diameter tolerance
External pitch diameter tolerance TD_2	$TD_2 = 1.30 \times (Td_2 \text{ for Class } 2A)$ $= 1.30 \times 0.003127 = 0.0040651 = 0.0041$	The constant 1.30 is for this Class 2B example, and will be different for Classes 1B and 3B. Td_2 for Class 2A (see calculation, Table 5) is rounded to six decimal places
Maximum internal pitch diameter (D_{2max})	$D_{2max} = D_{2min} + TD_2 = 0.2789 + 0.0041 = 0.2830$	D_{2max} is rounded to four decimal places
Minimum internal major diameter (D_{min}) = basic major diameter (D_{bsc})	$D_{min} = D_{bsc} = 0.2969$	D_{min} is rounded to four decimal places

Table 7. Number of Decimal Places for Intermediate and Final Calculations of Thread Characteristics

Symbol	Dimensions	Final		Symbol	Dimensions	Intermediate		Final	
		Inch	Metric			Inch	Metric	Inch	Metric
d	Major diameter, external thread	4	3	LE	Length of thread engagement	6	N/A	…	…
D	Major diameter, internal thread	4	3	P	Pitch	…	…	8	Note [a]
d_2	Pitch diameter, external thread	4	3	Td	Major diameter tolerance	…	…	4	3
D_2	Pitch diameter, internal thread	4	3	Td_2	Pitch diameter tolerance, external thread	…	…	6	3
d_1	Minor diameter, external thread	4	3	TD_2	Pitch diameter tolerance, internal thread	…	…	4	3
d_3	Minor diameter, rounded root external thread	4	3	TD_1	Minor diameter tolerance, internal thread	…	…	4	3
D_1	Minor diameter, internal threads for sizes 0.138 and larger for Classes 1B and 2B only	3	N/A	$h_b = 2h_{as}$	Twice the external thread addendum	6	N/A	…	…
D_1	Minor diameter, internal threads for sizes smaller than 0.138 for Classes 1B and 2B, and all sizes for Class 3B	4	N/A	$2h_s$	Double height of UNR external thread	6	N/A	…	…
D_1	Minor diameter, internal metric thread	N/A	3	$2h_n$	Double height of internal thread and UN external thread	6	N/A	…	…
es	Allowance at major pitch and minor diameters of external thread		3		Twice the external thread addendum	6	N/A	…	…

[a] Metric pitches are not calculated. They are stated in the screw thread designation and are to be used out to the number of decimal places as stated.
Note: Constants based on a function of P are rounded to an 8-place decimal for inch threads and a 7-place decimal for metric threads.

Table 8. Basic Dimensions, American National Standard Taper Pipe Threads, NPT
ANSI/ASME B1.20.1-2013 (R2018)

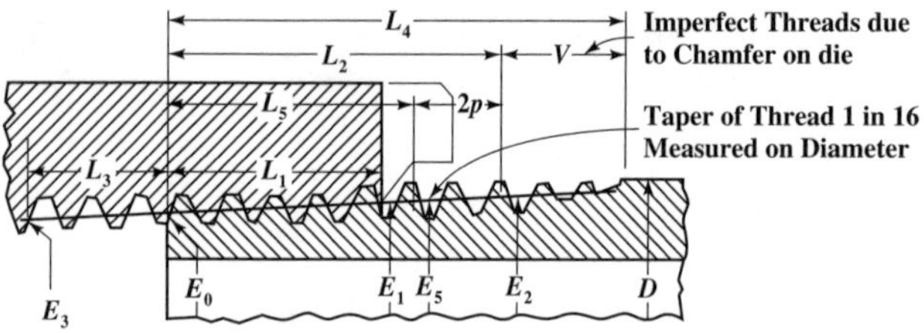

For all dimensions, see corresponding reference letter in table.

Angle between sides of thread is 60 degrees. Taper of thread, on diameter, is ¾ inch per foot. Angle of taper with center line is 1°47′.

The basic maximum thread height, h, of the truncated thread is 0.8 × pitch of thread. The crest and root are truncated a minimum of 0.033 × pitch for all pitches.

Nominal Pipe Size	Outside Dia. of Pipe, D	Threads per Inch, n	Pitch of Thread, p	Pitch Diameter at Beginning of External Thread, E_0	Handtight Engagement: Length,[a] L_1 (Inch)	Handtight Engagement: Dia.,[b] E_1 (Inch)	Effective Thread, External: Length,[c] L_2 (Inch)	Effective Thread, External: Dia., E_2 (Inch)
1/16	0.3125	27	0.03704	0.27118	0.160	0.28118	0.2611	0.28750
1/8	0.405	27	0.03704	0.36351	0.1615	0.37360	0.2639	0.38000
1/4	0.540	18	0.05556	0.47739	0.2278	0.49163	0.4018	0.50250
3/8	0.675	18	0.05556	0.61201	0.240	0.62701	0.4078	0.63750
1/2	0.840	14	0.07143	0.75843	0.320	0.77843	0.5337	0.79178
3/4	1.050	14	0.07143	0.96768	0.339	0.98887	0.5457	1.00178
1	1.315	11½	0.08696	1.21363	0.400	1.23863	0.6828	1.25631
1¼	1.660	11½	0.08696	1.55713	0.420	1.58338	0.7068	1.60131
1½	1.900	11½	0.08696	1.79609	0.420	1.82234	0.7235	1.84131
2	2.375	11½	0.08696	2.26902	0.436	2.29627	0.7565	2.31630
2½	2.875	8	0.12500	2.71953	0.682	2.76216	1.1375	2.79063
3	3.500	8	0.12500	3.34062	0.766	3.38850	1.2000	3.41563
3½	4.000	8	0.12500	3.83750	0.821	3.88881	1.2500	3.91563
4	4.500	8	0.12500	4.33438	0.844	4.38712	1.3000	4.41563
5	5.563	8	0.12500	5.39073	0.937	5.44929	1.4063	5.47863
6	6.625	8	0.12500	6.44609	0.958	6.50597	1.5125	6.54063
8	8.625	8	0.12500	8.43359	1.063	8.50003	1.7125	8.54063
10	10.750	8	0.12500	10.54531	1.210	10.62094	1.9250	10.66563
12	12.750	8	0.12500	12.53281	1.360	12.61781	2.1250	12.66563
14 OD	14.000	8	0.12500	13.77500	1.562	13.87262	2.2500	13.91563
16 OD	16.000	8	0.12500	15.76250	1.812	15.87575	2.4500	15.91563
18 OD	18.000	8	0.12500	17.75000	2.000	17.87500	2.6500	17.91563
20 OD	20.000	8	0.12500	19.73750	2.125	19.87031	2.8500	19.91563
24 OD	24.000	8	0.12500	23.71250	2.375	23.86094	3.2500	23.91563

[a] Also length of thin ring gage and length from gaging notch to small end of plug gage.
[b] Also pitch diameter at gaging notch (handtight plane).
[c] Also length of plug gage.

Table 9. Basic Dimensions, American National Standard Taper Pipe Threads, NPT
ANSI/ASME B1.20.1-2013 (R2018)

Nominal Pipe Size	Wrench Makeup Length for Internal Thread: Length,[c] L_3	Wrench Makeup Length for Internal Thread: Dia., E_3	Vanish Thread, (3.47 thds.), V	Overall Length External Thread, L_4	Nominal Perfect External Threads[a]: Length, L_5	Nominal Perfect External Threads[a]: Dia., E_5	Height of Thread, h	Basic Minor Dia. at Small End of Pipe,[b] K_0
1/16	0.1111	0.26424	0.1285	0.3896	0.1870	0.28287	0.02963	0.2415
1/8	0.1111	0.35656	0.1285	0.3924	0.1898	0.37537	0.02963	0.3338
1/4	0.1667	0.46697	0.1928	0.5946	0.2907	0.49556	0.04444	0.4329
3/8	0.1667	0.60160	0.1928	0.6006	0.2967	0.63056	0.04444	0.5675
1/2	0.2143	0.74504	0.2479	0.7815	0.3909	0.78286	0.05714	0.7014
3/4	0.2143	0.95429	0.2479	0.7935	0.4029	0.99286	0.05714	0.9106
1	0.2609	1.19733	0.3017	0.9845	0.5089	1.24543	0.06957	1.1441
1 1/4	0.2609	1.54083	0.3017	1.0085	0.5329	1.59043	0.06957	1.4876
1 1/2	0.2609	1.77978	0.3017	1.0252	0.5496	1.83043	0.06957	1.7266
2	0.2609	2.25272	0.3017	1.0582	0.5826	2.30543	0.06957	2.1995
2 1/2	0.2500[d]	2.70391	0.4338	1.5712	0.8875	2.77500	0.100000	2.6195
3	0.2500[d]	3.32500	0.4338	1.6337	0.9500	3.40000	0.100000	3.2406
3 1/2	0.2500	3.82188	0.4338	1.6837	1.0000	3.90000	0.100000	3.7374
4	0.2500	4.31875	0.4338	1.7337	1.0500	4.40000	0.100000	4.2343
5	0.2500	5.37511	0.4338	1.8400	1.1563	5.46300	0.100000	5.2907
6	0.2500	6.43047	0.4338	1.9462	1.2625	6.52500	0.100000	6.3460
8	0.2500	8.41797	0.4338	2.1462	1.4625	8.52500	0.100000	8.3335
10	0.2500	10.52969	0.4338	2.3587	1.6750	10.65000	0.100000	10.4453
12	0.2500	12.51719	0.4338	2.5587	1.8750	12.65000	0.100000	12.4328
14 OD	0.2500	13.75938	0.4338	2.6837	2.0000	13.90000	0.100000	13.6749
16 OD	0.2500	15.74688	0.4338	2.8837	2.2000	15.90000	0.100000	15.6624
18 OD	0.2500	17.73438	0.4338	3.0837	2.4000	17.90000	0.100000	17.6499
20 OD	0.2500	19.72188	0.4338	3.2837	2.6000	19.90000	0.100000	19.6374
24 OD	0.2500	23.69688	0.4338	3.6837	3.0000	23.90000	0.100000	23.6124

[a] The length L_5 from the end of the pipe determines the plane beyond which the thread form is imperfect at the crest. The next two threads are perfect at the root. At this plane the cone formed by the crests of the thread intersects the cylinder forming the external surface of the pipe. $L_5 = L_2 - 2p$.

[b] Given as information for use in selecting tap drills.

[c] Three threads for 2-inch size and smaller; two threads for larger sizes.

[d] Military Specification MIL-P-7105 gives the wrench makeup as three threads for 3 in. and smaller. The E_3 dimensions are then as follows: Size 2 1/2 in., 2.69609 and size 3 in., 3.31719.

All dimensions given in inches.

Increase in diameter per thread is equal to $0.0625/n$.

The basic dimensions of the ANSI Standard Taper Pipe Thread are given in inches to four or five decimal places. While this implies a greater degree of precision than is ordinarily attained, these dimensions are the basis of gage dimensions and are so expressed for the purpose of eliminating errors in computations.

Metric Screw Threads—M Profile

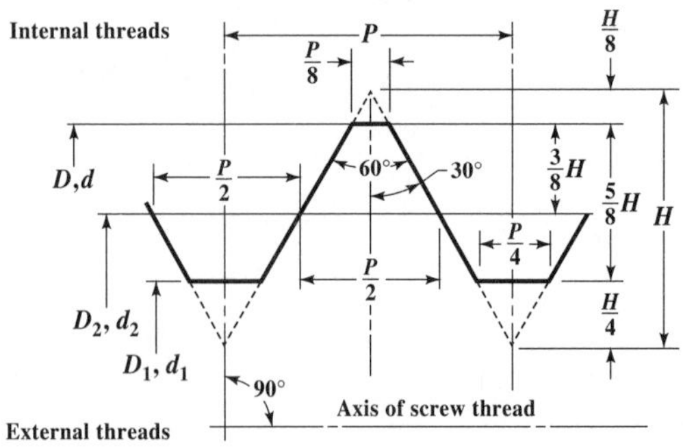

$$H = \frac{\sqrt{3}}{2} \times P \quad = 0.866025P$$

$$0.125H = 0.108253P \quad 0.250H = 0.216506P \quad 0.375H = 0.324760P \quad 0.625H = 0.541266P$$

Fig. 2. Basic M Thread Profile (ISO 68 Basic Profile)

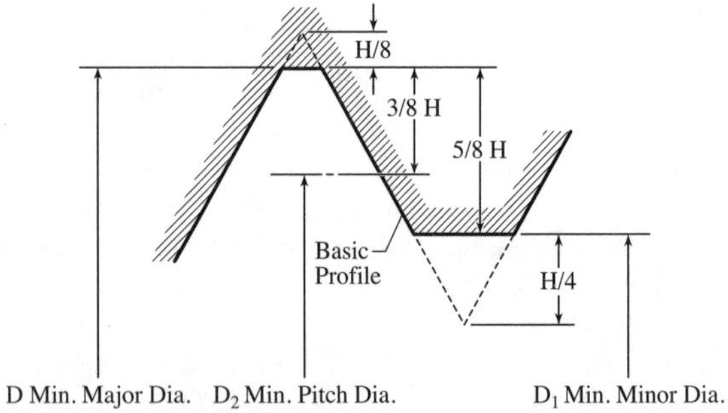

Fig. 3. Internal Thread Design M Profile with No Allowance (Fundamental Deviation) (Maximum Material Condition). For Dimensions, see Table 10

Definitions.—The following definitions apply to metric screw threads—M profile.

Basic Thread Profile: The cyclical outline in an axial plane of the permanently established boundary between the provinces of the external and internal threads. All deviations are with respect to this boundary. (See Fig. 2 and 5.)

Design Profiles: The maximum material profiles permitted for external and internal threads for a specified tolerance class. (See Fig. 3 and 4.)

Fundamental Deviation: For Standard threads, the deviation (upper or lower) closer to the basic size. It is the upper deviation, *es*, for an external thread and the lower deviation, *EI*, for an internal thread. (See Fig. 5.)

Limiting Profiles: The limiting M profile for internal threads is shown in Fig. 6. The limiting M profile for external threads is shown in Fig. 7.

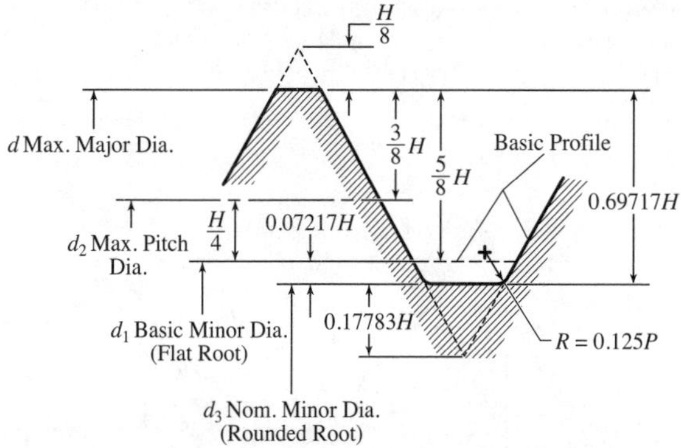

Fig. 4. External Thread Design M Profile with No Allowance (Fundamental Deviation) (Flanks at Maximum Material Condition). For Dimensions, see Table 10

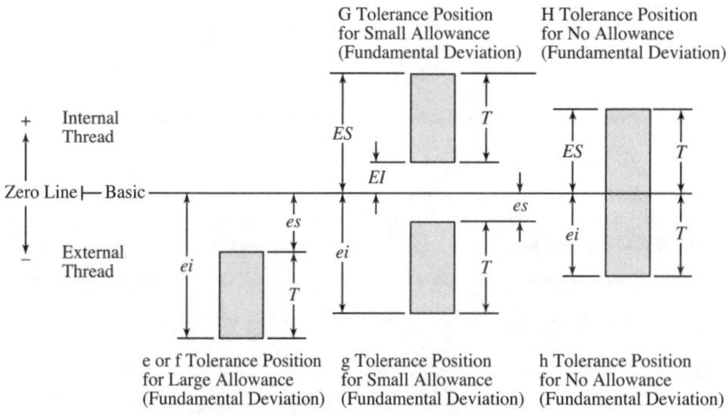

Fig. 5. Metric Tolerance System for Screw Threads

Formulas for M Profile Screw Thread Limiting Dimensions.—The limiting dimensions for M profile screw threads are calculated from the following formulas.

Internal Threads:

Min major dia. = basic major dia. + EI

Min pitch dia. = basic major dia. − 0.6495191P + EI for D_2

Max pitch dia. = min pitch dia. + TD_2

Max major dia. = max pitch dia. + 0.7938566P

Min minor dia. = min major dia. − 1.0825318P

Max minor dia. = min minor dia. + TD_1

External Threads:

Max major dia. = basic major dia. − es (Note that es is an absolute value.)

Min major dia. = max major dia. − Td

Max pitch dia. = basic major dia. − 0.6495191P − es for d_2

Min pitch dia. = max pitch dia. − Td_2

Max flat form minor dia. = max pitch dia. − 0.433013*P*
Max rounded root minor dia. = max pitch dia. − 2 × max trunc.
Min rounded root minor dia. = min pitch dia. − 0.616025*P*
Min root radius = 0.125*P*

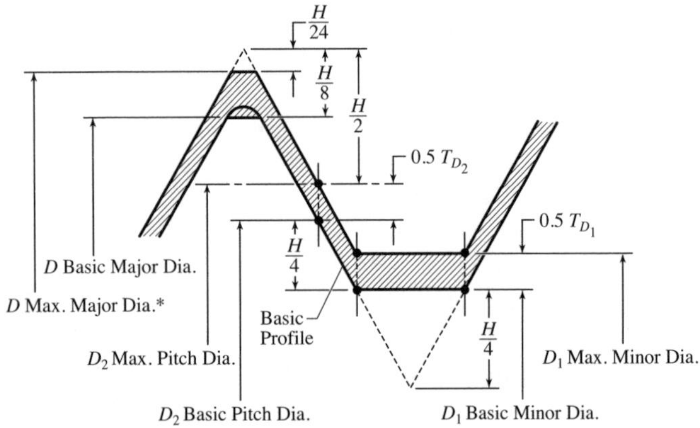

Fig. 6. Internal Thread—Limiting M Profile. Tolerance Position H

*This dimension is used in the design of tools, etc. In dimensioning internal threads it is not normally specified. Generally, major diameter acceptance is based on maximum material condition gaging.

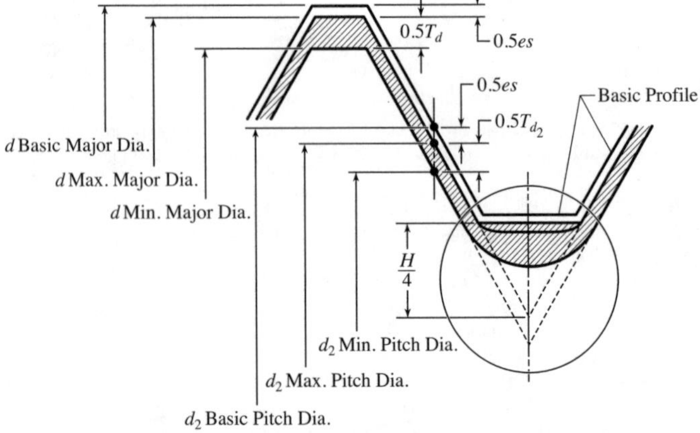

Fig. 7. External Thread—Limiting M Profile. Tolerance Position g

Table 10. American National Standard Metric Thread — M Profile Data *ANSI/ASME B1.13M-2005 (R2015)*

Pitch P	Truncation of Internal Thread Root and External Thread Crest $\frac{H}{8}$ 0.1082532P	Addendum of Internal Thread and Truncation of Internal Thread $\frac{H}{4}$ 0.2165064P	Dedendum of Internal Thread and Addendum External Thread $\frac{3}{8}H$ 0.3247595P	Difference[a] $\frac{H}{2}$ 0.4330127P	Height of Internal Thread and Depth of Thread Engagement $\frac{5}{8}H$ 0.5412659P	Difference[b] 0.711325H 0.6160254P	Twice the External Thread Addendum $\frac{3}{4}H$ 0.6495191P	Difference[c] $\frac{11}{12}H$ 0.7938566P	Height of Sharp V-Thread H 0.8660254P	Double Height of Internal Thread $\frac{5}{4}H$ 1.0825318P
0.2	0.02165	0.04330	0.06495	0.08660	0.10825	0.12321	0.12990	0.15877	0.17321	0.21651
0.25	0.02706	0.05413	0.08119	0.10825	0.13532	0.15401	0.16238	0.19846	0.21651	0.27063
0.3	0.03248	0.06495	0.09743	0.12990	0.16238	0.18481	0.19486	0.23816	0.25981	0.32476
0.35	0.03789	0.07578	0.11367	0.15155	0.18944	0.21561	0.22733	0.27785	0.30311	0.37889
0.4	0.04330	0.08660	0.12990	0.17321	0.21651	0.24541	0.25981	0.31754	0.34641	0.43301
0.45	0.04871	0.09743	0.14614	0.19486	0.24357	0.27721	0.29228	0.35724	0.38971	0.48714
0.5	0.05413	0.10825	0.16238	0.21651	0.27063	0.30801	0.32476	0.39693	0.43301	0.54127
0.6	0.06495	0.12990	0.19486	0.25981	0.32476	0.36962	0.38971	0.47631	0.51962	0.64952
0.7	0.07578	0.15155	0.22733	0.30311	0.37889	0.43122	0.45466	0.55570	0.60622	0.75777
0.75	0.08119	0.16238	0.24357	0.32476	0.40595	0.46202	0.48714	0.59539	0.64952	0.81190
0.8	0.08660	0.17321	0.25981	0.34641	0.43301	0.49282	0.51962	0.63509	0.69282	0.86603
1	0.10825	0.21651	0.32476	0.43301	0.54127	0.61603	0.64952	0.79386	0.86603	1.08253
1.25	0.13532	0.27063	0.40595	0.54127	0.67658	0.77003	0.81190	0.99232	1.08253	1.35316
1.5	0.16238	0.32476	0.48714	0.64952	0.81190	0.92404	0.97428	1.19078	1.29904	1.62380
1.75	0.18944	0.37889	0.56833	0.75777	0.94722	1.07804	1.13666	1.38925	1.51554	1.89443
2	0.21651	0.43301	0.64952	0.86603	1.08253	1.23205	1.29904	1.58771	1.73205	2.16506
2.5	0.27063	0.54127	0.81190	1.08253	1.35316	1.54006	1.62380	1.98464	2.16506	2.70633
3	0.32476	0.64652	0.97428	1.29904	1.62380	1.84808	1.94856	2.38157	2.59808	3.24760
3.5	0.37889	0.75777	1.13666	1.51554	1.89443	2.15609	2.27332	2.77850	3.03109	3.78886
4	0.43301	0.86603	1.29904	1.73205	2.16506	2.46410	2.59808	3.17543	3.46410	4.33013
4.5	0.48714	0.97428	1.46142	1.94856	2.43570	2.77211	2.92284	3.57235	3.89711	4.87139
5	0.54127	1.08253	1.62380	2.16506	2.70633	3.08013	3.24760	3.96928	4.33013	5.41266
5.5	0.59539	1.19079	1.78618	2.38157	2.97696	3.38814	3.57236	4.36621	4.76314	5.95392
6	0.64952	1.29904	1.94856	2.59808	3.24760	3.69615	3.89711	4.76314	5.19615	6.49519
8	0.86603	1.73205	2.59808	3.46410	4.33013	4.92820	5.19615	6.35085	6.92820	8.66025

[a] Difference between max theoretical pitch diameter and max minor diameter of external thread and between min theoretical pitch diameter and min minor diameter of internal thread.

[b] Difference between min theoretical pitch diameter and min design minor diameter of external thread for 0.125P root radius.

[c] Difference between max major diameter and max theoretical pitch diameter of internal thread.

All dimensions are in millimeters.

Table 11. Internal Metric Thread—M Profile Limiting Dimensions *ANSI/ASME B1.13M-2005 (R2015)*

Basic Thread Designation	Toler. Class	Minor Diameter D_1		Pitch Diameter D_2			Major Diameter D	
		Min	Max	Min	Max	Tol	Min	Max[a]
M1.6 × 0.35	6H	1.221	1.321	1.373	1.458	0.085	1.600	1.736
M2 × 0.4	6H	1.567	1.679	1.740	1.830	0.090	2.000	2.148
M2.5 × 0.45	6H	2.013	2.138	2.208	2.303	0.095	2.500	2.660
M3 × 0.5	6H	2.459	2.599	2.675	2.775	0.100	3.000	3.172
M3.5 × 0.6	6H	2.850	3.010	3.110	3.222	0.112	3.500	3.698
M4 × 0.7	6H	3.242	3.422	3.545	3.663	0.118	4.000	4.219
M5 × 0.8	6H	4.134	4.334	4.480	4.605	0.125	5.000	5.240
M6 × 1	6H	4.917	5.153	5.350	5.500	0.150	6.000	6.294
M8 × 1.25	6H	6.647	6.912	7.188	7.348	0.160	8.000	8.340
M8 × 1	6H	6.917	7.153	7.350	7.500	0.150	8.000	8.294
M10 × 0.75	6H	9.188	9.378	9.513	9.645	0.132	10.000	10.240
M10 × 1	6H	8.917	9.153	9.350	9.500	0.150	10.000	10.294
M10 × 1.5	6H	8.376	8.676	9.026	9.206	0.180	10.000	10.397
M10 × 1.25	6H	8.647	8.912	9.188	9.348	0.160	10.000	10.340
M12 × 1.75	6H	10.106	10.441	10.863	11.063	0.200	12.000	12.452
M12 × 1.5	6H	10.376	10.676	11.026	11.216	0.190	12.000	12.407
M12 × 1.25	6H	10.647	10.912	11.188	11.368	0.180	12.000	12.360
M12 × 1	6H	10.917	11.153	11.350	11.510	0.160	12.000	12.304
M14 × 2	6H	11.835	12.210	12.701	12.913	0.212	14.000	14.501
M14 × 1.5	6H	12.376	12.676	13.026	13.216	0.190	14.000	14.407
M15 × 1	6H	13.917	14.153	14.350	14.510	0.160	15.000	15.304
M16 × 2	6H	13.835	14.210	14.701	14.913	0.212	16.000	16.501
M16 × 1.5	6H	14.376	14.676	15.026	15.216	0.190	16.000	16.407
M17 × 1	6H	15.917	16.153	16.350	16.510	0.160	17.000	17.304
M18 × 1.5	6H	16.376	16.676	17.026	17.216	0.190	18.000	18.407
M20 × 2.5	6H	17.294	17.744	18.376	18.600	0.224	20.000	20.585
M20 × 1.5	6H	18.376	18.676	19.026	19.216	0.190	20.000	20.407
M20 × 1	6H	18.917	19.153	19.350	19.510	0.160	20.000	20.304
M22 × 2.5	6H	19.294	19.744	20.376	20.600	0.224	22.000	22.585
M22 × 1.5	6H	20.376	20.676	21.026	21.216	0.190	22.000	22.407
M24 × 3	6H	20.752	21.252	22.051	22.316	0.265	24.000	24.698
M24 × 2	6H	21.835	22.210	22.701	22.925	0.224	24.000	24.513
M25 × 1.5	6H	23.376	23.676	24.026	24.226	0.200	25.000	25.417
M27 × 3	6H	23.752	24.252	25.051	25.316	0.265	27.000	27.698
M27 × 2	6H	24.835	25.210	25.701	25.925	0.224	27.000	27.513
M30 × 3.5	6H	26.211	26.771	27.727	28.007	0.280	30.000	30.786
M30 × 2	6H	27.835	28.210	28.701	28.925	0.224	30.000	30.513
M30 × 1.5	6H	28.376	28.676	29.026	29.226	0.200	30.000	30.417
M33 × 2	6H	30.835	31.210	31.701	31.925	0.224	33.000	33.513
M35 × 1.5	6H	33.376	33.676	34.026	34.226	0.200	35.000	35.417
M36 × 4	6H	31.670	32.270	33.402	33.702	0.300	36.000	36.877
M36 × 2	6H	33.835	34.210	34.701	34.925	0.224	36.000	36.513

Table 11. *(Continued)* **Internal Metric Thread—M Profile Limiting Dimensions** *ANSI/ASME B1.13M-2005 (R2015)*

Basic Thread Designation	Toler. Class	Minor Diameter D_1		Pitch Diameter D_2			Major Diameter D	
		Min	Max	Min	Max	Tol	Min	Max[a]
M39 × 2	6H	36.835	37.210	37.701	37.925	0.224	39.000	39.513
M40 × 1.5	6H	38.376	38.676	39.026	39.226	0.200	40.000	40.417
M42 × 4.5	6H	37.129	37.799	39.077	39.392	0.315	42.000	42.964
M42 × 2	6H	39.835	40.210	40.701	40.925	0.224	42.000	42.513
M45 × 1.5	6H	43.376	43.676	44.026	44.226	0.200	45.000	45.417
M48 × 5	6H	42.587	43.297	44.752	45.087	0.335	48.000	49.056
M48 × 2	6H	45.835	46.210	46.701	46.937	0.236	48.000	48.525
M50 × 1.5	6H	48.376	48.676	49.026	49.238	0.212	50.000	50.429
M55 × 1.5	6H	53.376	53.676	54.026	54.238	0.212	55.000	55.429
M56 × 5.5	6H	50.046	50.796	52.428	52.783	0.355	56.000	57.149
M56 × 2	6H	53.835	54.210	54.701	54.937	0.236	56.000	56.525
M60 × 1.5	6H	58.376	58.676	59.026	59.238	0.212	60.000	60.429
M64 × 6	6H	57.505	58.305	60.103	60.478	0.375	64.000	65.241
M64 × 2	6H	61.835	62.210	62.701	62.937	0.236	64.000	64.525
M65 × 1.5	6H	63.376	63.676	64.026	64.238	0.212	65.000	65.429
M70 × 1.5	6H	68.376	68.676	69.026	69.238	0.212	70.000	70.429
M72 × 6	6H	65.505	66.305	68.103	68.478	0.375	72.000	73.241
M72 × 2	6H	69.835	70.210	70.701	70.937	0.236	72.000	72.525
M75 × 1.5	6H	73.376	73.676	74.026	74.238	0.212	75.000	75.429
M80 × 6	6H	73.505	74.305	76.103	76.478	0.375	80.000	81.241
M80 × 2	6H	77.835	78.210	78.701	78.937	0.236	80.000	80.525
M80 × 1.5	6H	78.376	78.676	79.026	79.238	0.212	80.000	80.429
M85 × 2	6H	82.835	83.210	83.701	83.937	0.236	85.000	85.525
M90 × 6	6H	83.505	84.305	86.103	86.478	0.375	90.000	91.241
M90 × 2	6H	87.835	88.210	88.701	88.937	0.236	90.000	90.525
M95 × 2	6H	92.835	93.210	93.701	93.951	0.250	95.000	95.539
M100 × 6	6H	93.505	94.305	96.103	96.503	0.400	100.000	101.266
M100 × 2	6H	97.835	98.210	98.701	98.951	0.250	100.000	100.539
M105 × 2	6H	102.835	103.210	103.701	103.951	0.250	105.000	105.539
M110 × 2	6H	107.835	108.210	108.701	108.951	0.250	110.000	110.539
M120 × 2	6H	117.835	118.210	118.701	118.951	0.250	120.000	120.539
M130 × 2	6H	127.835	128.210	128.701	128.951	0.250	130.000	130.539
M140 × 2	6H	137.835	138.210	138.701	138.951	0.250	140.000	140.539
M150 × 2	6H	147.835	148.210	148.701	148.951	0.250	150.000	150.539
M160 × 3	6H	156.752	157.252	158.051	158.351	0.300	160.000	160.733
M170 × 3	6H	166.752	167.252	168.051	168.351	0.300	170.000	170.733
M180 × 3	6H	176.752	177.252	178.051	178.351	0.300	180.000	180.733
M190 × 3	6H	186.752	187.252	188.051	188.386	0.335	190.000	190.768
M200 × 3	6H	196.752	197.252	198.051	198.386	0.335	200.000	200.768

[a] This reference dimension is used in design of tools, etc., and is not normally specified. Generally, major diameter acceptance is based upon maximum material condition gaging.

All dimensions are in millimeters.

Table 12. External Metric Thread—M Profile Limiting Dimensions *ANSI/ASME B1.13M-2005 (R2015)*

Basic Thread Designation	Tol. Class	Allowance[a] es	Major Diameter[b] d		Pitch Diameter[b,c] d_2			Minor Dia.[b] d_1	Minor Dia.[d] d_3
			Max.	Min.	Max.	Min.	Tol.	Max.	Min.
M1.6 × 0.35	6g	0.019	1.581	1.496	1.354	1.291	0.063	1.202	1.075
M1.6 × 0.35	6h	0.000	1.600	1.515	1.373	1.310	0.063	1.221	1.094
M1.6 × 0.35	4g6g	0.019	1.581	1.496	1.354	1.314	0.040	1.202	1.098
M2 × 0.4	6g	0.019	1.981	1.886	1.721	1.654	0.067	1.548	1.408
M2 × 0.4	6h	0.000	2.000	1.905	1.740	1.673	0.067	1.567	1.427
M2 × 0.4	4g6g	0.019	1.981	1.886	1.721	1.679	0.042	1.548	1.433
M2.5 × 0.45	6g	0.020	2.480	2.380	2.188	2.117	0.071	1.993	1.840
M2.5 × 0.45	6h	0.000	2.500	2.400	2.208	2.137	0.071	2.013	1.860
M2.5 × 0.45	4g6g	0.020	2.480	2.380	2.188	2.143	0.045	1.993	1.866
M3 × 0.5	6g	0.020	2.980	2.874	2.655	2.580	0.075	2.438	2.272
M3 × 0.5	6h	0.000	3.000	2.894	2.675	2.600	0.075	2.458	2.292
M3 × 0.5	4g6g	0.020	2.980	2.874	2.655	2.607	0.048	2.438	2.299
M3.5 × 0.6	6g	0.021	3.479	3.354	3.089	3.004	0.085	2.829	2.634
M3.5 × 0.6	6h	0.000	3.500	3.375	3.110	3.025	0.085	2.850	2.655
M3.5 × 0.6	4g6g	0.021	3.479	3.354	3.089	3.036	0.053	2.829	2.666
M4 × 0.7	6g	0.022	3.978	3.838	3.523	3.433	0.090	3.220	3.002
M4 × 0.7	6h	0.000	4.000	3.860	3.545	3.455	0.090	3.242	3.024
M4 × 0.7	4g6g	0.022	3.978	3.838	3.523	3.467	0.056	3.220	3.036
M5 × 0.8	6g	0.024	4.976	4.826	4.456	4.361	0.095	4.110	3.868
M5 × 0.8	6h	0.000	5.000	4.850	4.480	4.385	0.095	4.134	3.892
M5 × 0.8	4g6g	0.024	4.976	4.826	4.456	4.396	0.060	4.110	3.903
M6 × 1	6g	0.026	5.974	5.794	5.324	5.212	0.112	4.891	4.596
M6 × 1	6h	0.000	6.000	5.820	5.350	5.238	0.112	4.917	4.622
M6 × 1	4g6g	0.026	5.974	5.794	5.324	5.253	0.071	4.891	4.637
M8 × 1.25	6g	0.028	7.972	7.760	7.160	7.042	0.118	6.619	6.272
M8 × 1.25	6h	0.000	8.000	7.788	7.188	7.070	0.118	6.647	6.300
M8 × 1.25	4g6g	0.028	7.972	7.760	7.160	7.085	0.075	6.619	6.315
M8 × 1	6g	0.026	7.974	7.794	7.324	7.212	0.112	6.891	6.596
M8 × 1	6h	0.000	8.000	7.820	7.350	7.238	0.112	6.917	6.622
M8 × 1	4g6g	0.026	7.974	7.794	7.324	7.253	0.071	6.891	6.637
M10 × 1.5	6g	0.032	9.968	9.732	8.994	8.862	0.132	8.344	7.938
M10 × 1.5	6h	0.000	10.000	9.764	9.026	8.894	0.132	8.376	7.970
M10 × 1.5	4g6g	0.032	9.968	9.732	8.994	8.909	0.085	8.344	7.985
M10 × 1.25	6g	0.028	9.972	9.760	9.160	9.042	0.118	8.619	8.272
M10 × 1.25	6h	0.000	10.000	9.788	9.188	9.070	0.118	8.647	8.300
M10 × 1.25	4g6g	0.028	9.972	9.760	9.160	9.085	0.075	8.619	8.315
M10 × 1	6g	0.026	9.974	9.794	9.324	9.212	0.112	8.891	8.596
M10 × 1	6h	0.000	10.000	9.820	9.350	9.238	0.112	8.917	8.622
M10 × 1	4g6g	0.026	9.974	9.794	9.324	9.253	0.071	8.891	8.637
M10 × 0.75	6g	0.022	9.978	9.838	9.491	9.391	0.100	9.166	8.929
M10 × 0.75	6h	0.000	10.000	9.860	9.513	9.413	0.100	9.188	8.951
M10 × 0.75	4g6g	0.022	9.978	9.838	9.491	9.428	0.063	9.166	8.966
M12 × 1.75	6g	0.034	11.966	11.701	10.829	10.679	0.150	10.071	9.601
M12 × 1.75	6h	0.000	12.000	11.735	10.863	10.713	0.150	10.105	9.635
M12 × 1.75	4g6g	0.034	11.966	11.701	10.829	10.734	0.095	10.071	9.656
M12 × 1.5	6g	0.032	11.968	11.732	10.994	10.854	0.140	10.344	9.930
M12 × 1.5	6h	0.000	12.000	11.764	11.026	10.886	0.140	10.376	9.962
M12 × 1.5	4g6g	0.032	11.968	11.732	10.994	10.904	0.090	10.344	9.980
M12 × 1.25	6g	0.028	11.972	11.760	11.160	11.028	0.132	10.619	10.258
M12 × 1.25	6h	0.000	12.000	11.788	11.188	11.056	0.132	10.647	10.286
M12 × 1.25	4g6g	0.028	11.972	11.760	11.160	11.075	0.085	10.619	10.305
M12 × 1	6g	0.026	11.974	11.794	11.324	11.206	0.118	10.891	10.590
M12 × 1	6h	0.000	12.000	11.820	11.350	11.232	0.118	10.917	10.616
M12 × 1	4g6g	0.026	11.974	11.794	11.324	11.249	0.075	10.891	10.633
M14 × 2	6g	0.038	13.962	13.682	12.663	12.503	0.160	11.797	11.271
M14 × 2	6h	0.000	14.000	13.720	12.701	12.541	0.160	11.835	11.309
M14 × 2	4g6g	0.038	13.962	13.682	12.663	12.563	0.100	11.797	11.331
M14 × 1.5	6g	0.032	13.968	13.732	12.994	12.854	0.140	12.344	11.930
M14 × 1.5	6h	0.000	14.000	13.764	13.026	12.886	0.140	12.376	11.962
M14 × 1.5	4g6g	0.032	13.968	13.732	12.994	12.904	0.090	12.344	11.980
M15 × 1	6g	0.026	14.974	14.794	14.324	14.206	0.118	13.891	13.590
M15 × 1	6h	0.000	15.000	14.820	14.350	14.232	0.118	13.917	13.616
M15 × 1	4g6g	0.026	14.974	14.794	14.324	14.249	0.075	13.891	13.633

Table 12. *(Continued)* **External Metric Thread — M Profile Limiting Dimensions** *ANSI/ASME B1.13M-2005 (R2015)*

Basic Thread Designation	Tol. Class	Allowance[a] *es*	Major Diameter[b] *d* Max.	Major Diameter[b] *d* Min.	Pitch Diameter[b,c] d_2 Max.	Pitch Diameter[b,c] d_2 Min.	Pitch Diameter[b,c] d_2 Tol.	Minor Dia.[b] d_1 Max.	Minor Dia.[d] d_3 Min.
M16 × 2	6g	0.038	15.962	15.682	14.663	14.503	0.160	13.797	13.271
M16 × 2	6h	0.000	16.000	15.720	14.701	14.541	0.160	13.835	13.309
M16 × 2	4g6g	0.038	15.962	15.682	14.663	14.563	0.100	13.797	13.331
M16 × 1.5	6g	0.032	15.968	15.732	14.994	14.854	0.140	14.344	13.930
M16 × 1.5	6h	0.000	16.000	15.764	15.026	14.886	0.140	14.376	13.962
M16 × 1.5	4g6g	0.032	15.968	15.732	14.994	14.904	0.090	14.344	13.980
M17 × 1	6g	0.026	16.974	16.794	16.324	16.206	0.118	15.891	15.590
M17 × 1	6h	0.000	17.000	16.820	16.350	16.232	0.118	15.917	15.616
M17 × 1	4g6g	0.026	16.974	16.794	16.324	16.249	0.075	15.891	15.633
M18 × 1.5	6g	0.032	17.968	17.732	16.994	16.854	0.140	16.344	15.930
M18 × 1.5	6h	0.000	18.000	17.764	17.026	16.886	0.140	16.376	15.962
M18 × 1.5	4g6g	0.032	17.968	17.732	16.994	16.904	0.090	16.344	15.980
M20 × 2.5	6g	0.042	19.958	19.623	18.334	18.164	0.170	17.251	16.624
M20 × 2.5	6h	0.000	20.000	19.665	18.376	18.206	0.170	17.293	16.666
M20 × 2.5	4g6g	0.042	19.958	19.623	18.334	18.228	0.106	17.251	16.688
M20 × 1.5	6g	0.032	19.968	19.732	18.994	18.854	0.140	18.344	17.930
M20 × 1.5	6h	0.000	20.000	19.764	19.026	18.886	0.140	18.376	17.962
M20 × 1.5	4g6g	0.032	19.968	19.732	18.994	18.904	0.090	18.344	17.980
M20 × 1	6g	0.026	19.974	19.794	19.324	19.206	0.118	18.891	18.590
M20 × 1	6h	0.000	20.000	19.820	19.350	19.232	0.118	18.917	18.616
M20 × 1	4g6g	0.026	19.974	19.794	19.324	19.249	0.075	18.891	18.633
M22 × 2.5	6g	0.042	21.958	21.623	20.334	20.164	0.170	19.251	18.624
M22 × 2.5	6h	0.000	22.000	21.665	20.376	20.206	0.170	19.293	18.666
M22 × 1.5	6g	0.032	21.968	21.732	20.994	20.854	0.140	20.344	19.930
M22 × 1.5	6h	0.000	22.000	21.764	21.026	20.886	0.140	20.376	19.962
M22 × 1.5	4g6g	0.032	21.968	21.732	20.994	20.904	0.090	20.344	19.980
M24 × 3	6g	0.048	23.952	23.577	22.003	21.803	0.200	20.704	19.955
M24 × 3	6h	0.000	24.000	23.625	22.051	21.851	0.200	20.752	20.003
M24 × 3	4g6g	0.048	23.952	23.577	22.003	21.878	0.125	20.704	20.030
M24 × 2	6g	0.038	23.962	23.682	22.663	22.493	0.170	21.797	21.261
M24 × 2	6h	0.000	24.000	23.720	22.701	22.531	0.170	21.835	21.299
M24 × 2	4g6g	0.038	23.962	23.682	22.663	22.557	0.106	21.797	21.325
M25 × 1.5	6g	0.032	24.968	24.732	23.994	23.844	0.150	23.344	22.920
M25 × 1.5	6h	0.000	25.000	24.764	24.026	23.876	0.150	23.376	22.952
M25 × 1.5	4g6g	0.032	24.968	24.732	23.994	23.899	0.095	23.344	22.975
M27 × 3	6g	0.048	26.952	26.577	25.003	24.803	0.200	23.704	22.955
M27 × 3	6h	0.000	27.000	26.625	25.051	24.851	0.200	23.752	23.003
M27 × 2	6g	0.038	26.962	26.682	25.663	25.493	0.170	24.797	24.261
M27 × 2	6h	0.000	27.000	26.720	25.701	25.531	0.170	24.835	24.299
M27 × 2	4g6g	0.038	26.962	26.682	25.663	25.557	0.106	24.797	24.325
M30 × 3.5	6g	0.053	29.947	29.522	27.674	27.462	0.212	26.158	25.306
M30 × 3.5	6h	0.000	30.000	29.575	27.727	27.515	0.212	26.211	25.359
M30 × 3.5	4g6g	0.053	29.947	29.522	27.674	27.542	0.132	26.158	25.386
M30 × 2	6g	0.038	29.962	29.682	28.663	28.493	0.170	27.797	27.261
M30 × 2	6h	0.000	30.000	29.720	28.701	28.531	0.170	27.835	27.299
M30 × 2	4g6g	0.038	29.962	29.682	28.663	28.557	0.106	27.797	27.325
M30 × 1.5	6g	0.032	29.968	29.732	28.994	28.844	0.150	28.344	27.920
M30 × 1.5	6h	0.000	30.000	29.764	29.026	28.876	0.150	28.376	27.952
M30 × 1.5	4g6g	0.032	29.968	29.732	28.994	28.899	0.095	28.344	27.975
M33 × 2	6g	0.038	32.962	32.682	31.663	31.493	0.170	30.797	30.261
M33 × 2	6h	0.000	33.000	32.720	31.701	31.531	0.170	30.835	30.299
M33 × 2	4g6g	0.038	32.962	32.682	31.663	31.557	0.106	30.797	30.325
M35 × 1.5	6g	0.032	34.968	34.732	33.994	33.844	0.150	33.344	32.920
M35 × 1.5	6h	0.000	35.000	34.764	34.026	33.876	0.150	33.376	32.952
M36 × 4	6g	0.060	35.940	35.465	33.342	33.118	0.224	31.610	30.654
M36 × 4	6h	0.000	36.000	35.525	33.402	33.178	0.224	31.670	30.714
M36 × 4	4g6g	0.060	35.940	35.465	33.342	33.202	0.140	31.610	30.738
M36 × 2	6g	0.038	35.962	35.682	34.663	34.493	0.170	33.797	33.261
M36 × 2	6h	0.000	36.000	35.720	34.701	34.531	0.170	33.835	33.299
M36 × 2	4g6g	0.038	35.962	35.682	34.663	34.557	0.106	33.797	33.325
M39 × 2	6g	0.038	38.962	38.682	37.663	37.493	0.170	36.797	36.261
M39 × 2	6h	0.000	39.000	38.720	37.701	37.531	0.170	36.835	36.299
M39 × 2	4g6g	0.038	38.962	38.682	37.663	37.557	0.106	36.797	36.325
M40 × 1.5	6g	0.032	39.968	39.732	38.994	38.844	0.150	38.344	37.920

Table 12. *(Continued)* **External Metric Thread—M Profile Limiting Dimensions** *ANSI/ASME B1.13M-2005 (R2015)*

Basic Thread Designation	Tol. Class	Allowance[a] es	Major Diameter[b] d		Pitch Diameter[b,c] d_2			Minor Dia.[b] d_1	Minor Dia.[d] d_3
			Max.	Min.	Max.	Min.	Tol.	Max.	Min.
M40 × 1.5	6h	0.000	40.000	39.764	39.026	38.876	0.150	38.376	37.952
M40 × 1.5	4g6g	0.032	39.968	39.732	38.994	38.899	0.095	38.344	37.975
M42 × 4.5	6g	0.063	41.937	41.437	39.014	38.778	0.236	37.065	36.006
M42 × 4.5	6h	0.000	42.000	41.500	39.077	38.841	0.236	37.128	36.069
M42 × 4.5	4g6g	0.063	41.937	41.437	39.014	38.864	0.150	37.065	36.092
M42 × 2	6g	0.038	41.962	41.682	40.663	40.493	0.170	39.797	39.261
M42 × 2	6h	0.000	42.000	41.720	40.701	40.531	0.170	39.835	39.299
M42 × 2	4g6g	0.038	41.962	41.682	40.663	40.557	0.106	39.797	39.325
M45 × 1.5	6g	0.032	44.968	44.732	43.994	43.844	0.150	43.344	42.920
M45 × 1.5	6h	0.000	45.000	44.764	44.026	43.876	0.150	43.376	42.952
M45 × 1.5	4g6g	0.032	44.968	44.732	43.994	43.899	0.095	43.344	42.975
M48 × 5	6g	0.071	47.929	47.399	44.681	44.431	0.250	42.516	41.351
M48 × 5	6h	0.000	48.000	47.470	44.752	44.502	0.250	42.587	41.422
M48 × 5	4g6g	0.071	47.929	47.399	44.681	44.521	0.160	42.516	41.441
M48 × 2	6g	0.038	47.962	47.682	46.663	46.483	0.180	45.797	45.251
M48 × 2	6h	0.000	48.000	47.720	46.701	46.521	0.180	45.835	45.289
M48 × 2	4g6g	0.038	47.962	47.682	46.663	46.551	0.112	45.797	45.319
M50 × 1.5	6g	0.032	49.968	49.732	48.994	48.834	0.160	48.344	47.910
M50 × 1.5	6h	0.000	50.000	49.764	49.026	48.866	0.160	48.376	47.942
M50 × 1.5	4g6g	0.032	49.968	49.732	48.994	48.894	0.100	48.344	47.970
M55 × 1.5	6g	0.032	54.968	54.732	53.994	53.834	0.160	53.344	52.910
M55 × 1.5	6h	0.000	55.000	54.764	54.026	53.866	0.160	53.376	52.942
M55 × 1.5	4g6g	0.032	54.968	54.732	53.994	53.894	0.100	53.344	52.970
M56 × 5.5	6g	0.075	55.925	55.365	52.353	52.088	0.265	49.971	48.700
M56 × 5.5	6h	0.000	56.000	55.440	52.428	52.163	0.265	50.046	48.775
M56 × 5.5	4g6g	0.075	55.925	55.365	52.353	52.183	0.170	49.971	48.795
M56 × 2	6g	0.038	55.962	55.682	54.663	54.483	0.180	53.797	53.251
M56 × 2	6h	0.000	56.000	55.720	54.701	54.521	0.180	53.835	53.289
M56 × 2	4g6g	0.038	55.962	55.682	54.663	54.551	0.112	53.797	53.319
M60 × 1.5	6g	0.032	59.968	59.732	58.994	58.834	0.160	58.344	57.910
M60 × 1.5	6h	0.000	60.000	59.764	59.026	58.866	0.160	58.376	57.942
M60 × 1.5	4g6g	0.032	59.968	59.732	58.994	58.894	0.100	58.344	57.970
M64 × 6	6g	0.080	63.920	63.320	60.023	59.743	0.280	57.425	56.047
M64 × 6	6h	0.000	64.000	63.400	60.103	59.823	0.280	57.505	56.127
M64 × 6	4g6g	0.080	63.920	63.320	60.023	59.843	0.180	57.425	56.147
M64 × 2	6g	0.038	63.962	63.682	62.663	62.483	0.180	61.797	61.251
M64 × 2	6h	0.000	64.000	63.720	62.701	62.521	0.180	61.835	61.289
M64 × 2	4g6g	0.038	63.962	63.682	62.663	62.551	0.112	61.797	61.319
M65 × 1.5	6g	0.032	64.968	64.732	63.994	63.834	0.160	63.344	62.910
M65 × 1.5	6h	0.000	65.000	64.764	64.026	63.866	0.160	63.376	62.942
M65 × 1.5	4g6g	0.032	64.968	64.732	63.994	63.894	0.100	63.344	62.970
M70 × 1.5	6g	0.032	69.968	69.732	68.994	68.834	0.160	68.344	67.910
M70 × 1.5	6h	0.000	70.000	69.764	69.026	68.866	0.160	68.376	67.942
M70 × 1.5	4g6g	0.032	69.968	69.732	68.994	68.894	0.100	68.344	67.970
M72 × 6	6g	0.080	71.920	71.320	68.023	67.743	0.280	65.425	64.047
M72 × 6	6h	0.000	72.000	71.400	68.103	67.823	0.280	65.505	64.127
M72 × 6	4g6g	0.080	71.920	71.320	68.023	67.843	0.180	65.425	64.147
M72 × 2	6g	0.038	71.962	71.682	70.663	70.483	0.180	69.797	69.251
M72 × 2	6h	0.000	72.000	71.720	70.701	70.521	0.180	69.835	69.289
M72 × 2	4g6g	0.038	71.962	71.682	70.663	70.551	0.112	69.797	69.319
M75 × 1.5	6g	0.032	74.968	74.732	73.994	73.834	0.160	73.344	72.910
M75 × 1.5	6h	0.000	75.000	74.764	74.026	73.866	0.160	73.376	72.942
M75 × 1.5	4g6g	0.032	74.968	74.732	73.994	73.894	0.100	73.344	72.970
M80 × 6	6g	0.080	79.920	79.320	76.023	75.743	0.280	73.425	72.047
M80 × 6	6h	0.000	80.000	79.400	76.103	75.823	0.280	73.505	72.127
M80 × 6	4g6g	0.080	79.920	79.320	76.023	75.843	0.180	73.425	72.147
M80 × 2	6g	0.038	79.962	79.682	78.663	78.483	0.180	77.797	77.251
M80 × 2	6h	0.000	80.000	79.720	78.701	78.521	0.180	77.835	77.289
M80 × 2	4g6g	0.038	79.962	79.682	78.663	78.551	0.112	77.797	77.319
M80 × 1.5	6g	0.032	79.968	79.732	78.994	78.834	0.160	78.344	77.910
M80 × 1.5	6h	0.000	80.000	79.764	79.026	78.866	0.160	78.376	77.942
M80 × 1.5	4g6g	0.032	79.968	79.732	78.994	78.894	0.100	78.344	77.970
M85 × 2	6g	0.038	84.962	84.682	83.663	83.483	0.180	82.797	82.251
M85 × 2	6h	0.000	85.000	84.720	83.701	83.521	0.180	82.835	82.289

Table 12. *(Continued)* External Metric Thread—M Profile Limiting Dimensions *ANSI/ASME B1.13M-2005 (R2015)*

Basic Thread Designation	Tol. Class	Allowance[a] es	Major Diameter[b] d		Pitch Diameter[b,c] d_2			Minor Dia.[b] d_1	Minor Dia.[d] d_3
			Max.	Min.	Max.	Min.	Tol.	Max.	Min.
M85 × 2	4g6g	0.038	84.962	84.682	83.663	83.551	0.112	82.797	82.319
M90 × 6	6g	0.080	89.920	89.320	86.023	85.743	0.280	83.425	82.047
M90 × 6	6h	0.000	90.000	89.400	86.103	85.823	0.280	83.505	82.127
M90 × 6	4g6g	0.080	89.920	89.320	86.023	85.843	0.180	83.425	82.147
M90 × 2	6g	0.038	89.962	89.682	88.663	88.483	0.180	87.797	87.251
M90 × 2	6h	0.000	90.000	89.720	88.701	88.521	0.180	87.835	87.289
M90 × 2	4g6g	0.038	89.962	89.682	88.663	88.551	0.112	87.797	87.319
M95 × 2	6g	0.038	94.962	94.682	93.663	93.473	0.190	92.797	92.241
M95 × 2	6h	0.000	95.000	94.720	93.701	93.511	0.190	92.835	92.279
M95 × 2	4g6g	0.038	94.962	94.682	93.663	93.545	0.118	92.797	92.313
M100 × 6	6g	0.080	99.920	99.320	96.023	95.723	0.300	93.425	92.027
M100 × 6	6h	0.000	100.000	99.400	96.103	95.803	0.300	93.505	92.107
M100 × 6	4g6g	0.080	99.920	99.320	96.023	95.833	0.190	93.425	92.137
M100 × 2	6g	0.038	99.962	99.682	98.663	98.473	0.190	97.797	97.241
M100 × 2	6h	0.000	100.000	99.720	98.701	98.511	0.190	97.835	97.279
M100 × 2	4g6g	0.038	99.962	99.682	98.663	98.545	0.118	97.797	97.313
M105 × 2	6g	0.038	104.962	104.682	103.663	103.473	0.190	102.797	102.241
M105 × 2	6h	0.000	105.000	104.720	103.701	103.511	0.190	102.835	102.279
M105 × 2	4g6g	0.038	104.962	104.682	103.663	103.545	0.118	102.797	102.313
M110 × 2	6g	0.038	109.962	109.682	108.663	108.473	0.190	107.797	107.241
M110 × 2	6h	0.000	110.000	109.720	108.701	108.511	0.190	107.835	107.279
M110 × 2	4g6g	0.038	109.962	109.682	108.663	108.545	0.118	107.797	107.313
M120 × 2	6g	0.038	119.962	119.682	118.663	118.473	0.190	117.797	117.241
M120 × 2	6h	0.000	120.000	119.720	118.701	118.511	0.190	117.835	117.279
M120 × 2	4g6g	0.038	119.962	119.682	118.663	118.545	0.118	117.797	117.313
M130 × 2	6g	0.038	129.962	129.682	128.663	128.473	0.190	127.797	127.241
M130 × 2	6h	0.000	130.000	129.720	128.701	128.511	0.190	127.835	127.279
M130 × 2	4g6g	0.038	129.962	129.682	128.663	128.545	0.118	127.797	127.313
M140 × 2	6g	0.038	139.962	139.682	138.663	138.473	0.190	137.797	137.241
M140 × 2	6h	0.000	140.000	139.720	138.701	138.511	0.190	137.835	137.279
M140 × 2	4g6g	0.038	139.962	139.682	138.663	138.545	0.118	137.797	137.313
M150 × 2	6g	0.038	149.962	149.682	148.663	148.473	0.190	147.797	147.241
M150 × 2	6h	0.000	150.000	149.720	148.701	148.511	0.190	147.835	147.279
M150 × 2	4g6g	0.038	149.962	149.682	148.663	148.545	0.118	147.797	147.313
M160 × 3	6g	0.048	159.952	159.577	158.003	157.779	0.224	156.704	155.931
M160 × 3	6h	0.000	160.000	159.625	158.051	157.827	0.224	156.752	155.979
M160 × 3	4g6g	0.048	159.952	159.577	158.003	157.863	0.140	156.704	156.015
M170 × 3	6g	0.048	169.952	169.577	168.003	167.779	0.224	166.704	165.931
M170 × 3	6h	0.000	170.000	169.625	168.051	167.827	0.224	166.752	165.979
M170 × 3	4g6g	0.048	169.952	169.577	168.003	167.863	0.140	166.704	166.015
M180 × 3	6g	0.048	179.952	179.577	178.003	177.779	0.224	176.704	175.931
M180 × 3	6h	0.000	180.000	179.625	178.051	177.827	0.224	176.752	175.979
M180 × 3	4g6g	0.048	179.952	179.577	178.003	177.863	0.140	176.704	176.015
M190 × 3	6g	0.048	189.952	189.577	188.003	187.753	0.250	186.704	185.905
M190 × 3	6h	0.000	190.000	189.625	188.051	187.801	0.250	186.752	185.953
M190 × 3	4g6g	0.048	189.952	189.577	188.003	187.843	0.160	186.704	185.995
M200 × 3	6g	0.048	199.952	199.577	198.003	197.753	0.250	196.704	195.905
M200 × 3	6h	0.000	200.000	199.625	198.051	197.801	0.250	196.752	195.953
M200 × 3	4g6g	0.048	199.952	199.577	198.003	197.843	0.160	196.704	195.995

[a] *es* is an absolute value.

[b] Coated threads with tolerance classes 6g or 4g6g.

[c] Functional diameter size includes the effects of all variations in pitch diameter, thread form, and profile. The variations in the individual thread characteristics such as flank angle, lead, taper, and roundness on a given thread, cause the measurements of the pitch diameter and functional diameter to vary from one another on most threads. The pitch diameter and the functional diameter on a given thread are equal to one another only when the thread form is perfect. When required to inspect either the pitch diameter, the functional diameter, or both, for thread acceptance, use the same limits of size for the appropriate thread size and class.

[d] Dimension used in the design of tools, etc.; in dimensioning external threads it is not normally specified. Generally, minor diameter acceptance is based on maximum material condition gaging.

All dimensions are in millimeters.

Metric Spark Plug Threads

British Standard for Spark Plugs BS 45:1972 (withdrawn).—This revised British Standard refers to spark plugs used in automobiles and industrial spark ignition internal combustion engines. The basic thread form is that of the ISO metric. In assigning tolerances, consideration has been given to the desirability of achieving the closest possible measure of interchangeability between British spark plugs and engines, and those made to standards of other ISO Member Bodies. The dimensions below are given in millimeters.

Basic Thread Dimensions for Spark Plug and Tapped Hole in Cylinder Head

Nom. Size	Pitch	Thread	Major Dia.		Pitch Dia.		Minor Dia.	
			Max.	Min.	Max.	Min.	Max.	Min.
14	1.25	Plug	13.937[a]	13.725	13.125	12.993	12.402	12.181
14	1.25	Hole		14.00	13.368	13.188	12.912	12.647
18	1.5	Plug	17.933[a]	17.697	16.959	16.819	16.092	15.845
18	1.5	Hole		18.00	17.216	17.026	16.676	16.376

[a] Not specified

The tolerance grades for finished spark plugs and corresponding tapped holes in the cylinder head are: for 14 mm size, 6e for spark plugs and 6H for tapped holes which gives a minimum clearance of 0.063 mm; and for 18 mm size, 6e for spark plugs and 6H for tapped holes which gives a minimum clearance of 0.067 mm. These minimum clearances help prevent seizure due to combustion deposits on the bare threads, when removing the spark plugs; this applies to both ferrous and nonferrous materials. They also should enable spark plugs with threads in accordance with this standard to fit into existing holes.

SAE Spark Plug Screw Threads.—The SAE Standard includes the following sizes: $\frac{7}{8}$-inch nominal diameter with 18 threads per inch: 18-millimeter nominal diameter with a 18-millimeter nominal diameter with 1.5-millimeter pitch; 14-millimeter nominal diameter with a 1.25-millimeter pitch; 10-millimeter nominal diameter with a 1.0 millimeter pitch; $\frac{3}{8}$-inch nominal diameter with 24 threads per inch; and $\frac{1}{4}$-inch nominal diameter with 32 threads per inch. During manufacture, in order to keep the wear on the threading tools within permissible limits, the threads in the spark plug GO (ring) gage should be truncated to the maximum minor diameter of the spark plug; and in the tapped hole GO (plug) gage to the minimum major diameter of the tapped hole.

SAE Standard Threads for Spark Plugs

Size[a] Nom. × Pitch	Major Diameter		Pitch Diameter		Minor Diameter	
	Max.	Min.	Max.	Min.	Max.	Min.
Spark Plug Threads, mm (inches)						
M18 × 1.5	17.933	17.803	16.959	16.853	16.053	…
	(0.7060)	(0.7009)	(0.6677)	(0.6635)	(0.6320)	…
M14 × 1.25	13.868	13.741	13.104	12.997	12.339	…
	(0.5460)	(0.5410)	(0.5159)	(0.5117)	(0.4858)	…
M12 × 1.25	11.862	11.735	11.100	10.998	10.211	…
	(0.4670)	(0.4620)	(0.4370)	(0.4330)	(0.4020)	…
M10 × 1.0	9.974	9.794	9.324	9.212	8.747	…
	(0.3927)	(0.3856)	(0.3671)	(0.3627)	(0.3444)	…
Tapped Hole Threads, mm (inches)						
M18 × 1.5	…	18.039	17.153	17.026	16.426	16.266
	…	(0.7102)	(0.6753)	(0.6703)	(0.6467)	(0.6404)
M14 × 1.25	…	14.034	13.297	13.188	12.692	12.499
	…	(0.5525)	(0.5235)	(0.5192)	(0.4997)	(0.4921)
M12 × 1.25	…	12.000	11.242	11.188	10.559	10.366
	…	(0.4724)	(0.4426)	(0.4405)	(0.4157)	(0.4081)
M10 × 1.0	…	10.000	9.500	9.350	9.153	8.917
	…	(0.3937)	(0.3740)	(0.3681)	(0,3604)	(0.3511)

[a] M14 and M18 are preferred for new applications.

FASTENER INFORMATION

Table 1. Grade Identification Marks and Mechanical Properties of Bolts and Screws

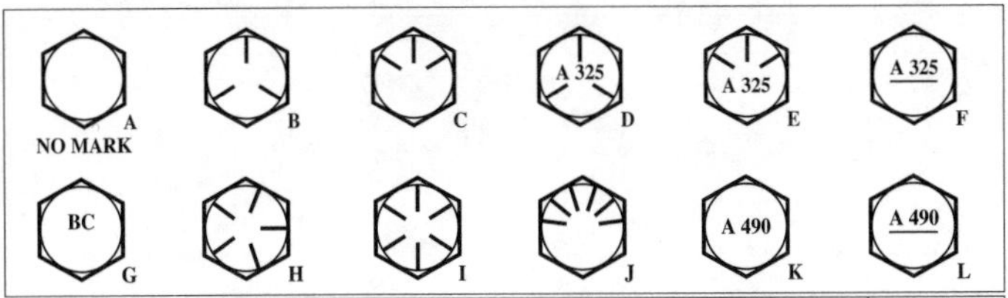

Identifier	Grade	Size (in.)	Min. Strength (10^3 psi) Proof	Tensile	Yield	Material & Treatment
A	SAE Grade 1	$\frac{1}{4}$ to $1\frac{1}{2}$	33	60	36	1
	ASTM A307	$\frac{1}{4}$ to $1\frac{1}{2}$	33	60	36	3
	SAE Grade 2	$\frac{1}{4}$ to $\frac{3}{4}$	55	74	57	1
		$\frac{7}{8}$ to $1\frac{1}{2}$	33	60	36	
	SAE Grade 4	$\frac{1}{4}$ to $1\frac{1}{2}$	65	115	100	2, a
B	SAE Grade 5, ASTM A449	$\frac{1}{4}$ to 1	85	120	92	2, b
	ASTM A449	$1\frac{1}{8}$ to $1\frac{1}{2}$	74	105	81	
	ASTM A449	$1\frac{3}{4}$ to 3	55	90	58	
C	SAE Grade 5.2	$\frac{1}{4}$ to 1	85	120	92	4, b
D	ASTM A325, Type 1	$\frac{1}{2}$ to 1	85	120	92	2, b
		$1\frac{1}{8}$ to $1\frac{1}{2}$	74	105	81	
E	ASTM A325, Type 2	$\frac{1}{2}$ to 1	85	120	92	4, b
		$1\frac{1}{8}$ to $1\frac{1}{2}$	74	105	81	
F	ASTM A325, Type 3	$\frac{1}{2}$ to 1	85	120	92	5, b
		$1\frac{1}{8}$ to $1\frac{1}{2}$	74	105	81	
G	ASTM A354, Grade BC	$\frac{1}{4}$ to $2\frac{1}{2}$	105	125	109	5, b
		$2\frac{5}{8}$ to 4	95	115	99	
H	SAE Grade 7	$\frac{1}{4}$ to $1\frac{1}{2}$	105	133	115	7, b
I	SAE Grade 8	$\frac{1}{4}$ to $1\frac{1}{2}$	120	150	130	7, b
	ASTM A354, Grade BD	$\frac{1}{4}$ to $2\frac{1}{2}$	120	150	130	6, b
		$2\frac{5}{8}$ to 4	105	140	115	
J	SAE Grade 8.2	$\frac{1}{4}$ to 1	120	150	130	4, b
K	ASTM A490, Type 1	$\frac{1}{2}$ to $1\frac{1}{2}$	120	150	130	6, b
L	ASTM A490, Type 3					5, b

Material Steel: 1—low or medium carbon; 2—medium carbon; 3—low carbon; 4—low-carbon martensite; 5—weathering steel; 6—alloy steel; 7—medium-carbon alloy. Treatment: a—cold drawn; b—quench and temper.

Table 2. Applicability of Hexagon and Spline Keys and Bits

Nominal Key or Bit Size		Cap Screws 1960 Series	Flat Countersunk Head Cap Screws	Button Head Cap Screws	Shoulder Screws	Set Screws
		Nominal Screw Sizes				
HEXAGON KEYS AND BITS						
0.028		…	…	…	…	0
0.035		…	0	0	…	1 & 2
0.050		0	1 & 2	1 & 2	…	3 & 4
1/16	0.062	1	3 & 4	3 & 4	…	5 & 6
5/64	0.078	2 & 3	5 & 6	5 & 6	…	8
3/32	0.094	4 & 5	8	8	…	10
7/64	0.109	6	…	…	…	…
1/8	0.125	…	10	10	1/4	1/4
9/64	0.141	8	…	…	…	…
5/32	0.156	10	1/4	1/4	5/16	5/16
3/16	0.188	1/4	5/16	5/16	3/8	3/8
7/32	0.219	…	3/8	3/8	…	7/16
1/4	0.250	5/16	7/16	…	1/2	1/2
5/16	0.312	3/8	1/2	1/2	5/8	5/8
3/8	0.375	7/16 & 1/2	5/8	5/8	3/4	3/4
7/16	0.438	…	…	…	…	…
1/2	0.500	5/8	3/4	…	1	7/8
9/16	0.562	…	7/8	…	…	1 & 1 1/8
5/8	0.625	3/4	1	…	1 1/4	1 1/4 & 1 3/8
3/4	0.750	7/8 & 1	1 1/8	…	…	1 1/2
7/8	0.875	1 1/8 & 1 1/4	1 1/4 & 1 3/8	…	1 1/2	…
1	1.000	1 3/8 & 1 1/2	1 1/2	…	1 3/4	1 3/4 & 2
1 1/4	1.250	1 3/4	…	…	2	…
1 1/2	1.500	2	…	…	…	…
1 3/4	1.750	2 1/4 & 2 1/2	…	…	…	…
2	2.000	2 3/4	…	…	…	…
2 1/4	2.250	3 & 3 1/4	…	…	…	…
2 3/4	2.750	3 1/2 & 3 3/4	…	…	…	…
3	3.000	4	…	…	…	…
SPLINE KEYS AND BITS						
0.033		…	…	…	…	0 & 1
0.048		…	0	0	…	2 & 3
0.060		0	1 & 2	1 & 2	…	4
0.072		1	3 & 4	3 & 4	…	5 & 6
0.096		2 & 3	5 & 6	5 & 6	…	8
0.111		4 & 5	8	8	…	10
0.133		6	…	…	…	…
0.145		…	10	10	…	1/4
0.168		8	…	…	…	…
0.183		10	1/4	1/4	…	5/16
0.216		1/4	5/16	5/16	…	3/8
0.251		…	3/8	3/8	…	7/16
0.291		5/16	7/16	…	…	1/2
0.372		3/8	1/2	1/2	…	5/8
0.454		7/16 & 1/2	5/8 & 3/4	5/8	…	3/4
0.595		5/8	…	…	…	7/8
0.620		3/4	…	…	…	…
0.698		7/8	…	…	…	…
0.790		1	…	…	…	…

Source: Appendix to American National Standard ANSI/ASME B18.3-2012.

Table 3. American National Standard Hexagon and Spline Socket Head Cap Screws *ANSI/ASME B18.3-2012*

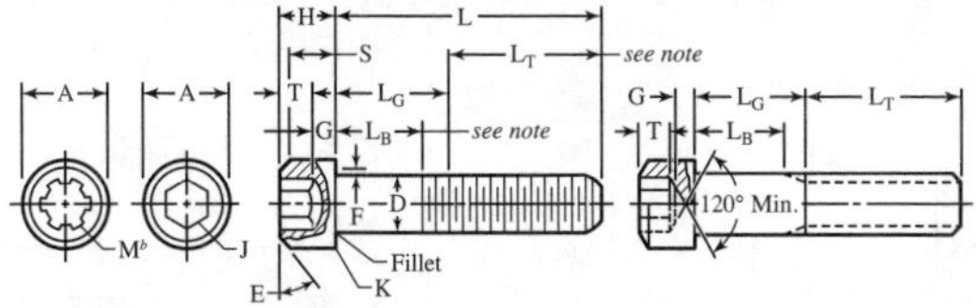

Nominal Size	Body Diameter, *D*		Head Diameter, *A*		Head Height, *H*		Spline Socket[a] Size, *M*[b]	Nom. Hex. Socket Size, *J*	Fillet Ext., *F*	Key Engage-ment,[a] *T*
	Max.	Min.	Max.	Min.	Max.	Min.			Max.	
0	0.0600	0.0568	0.096	0.091	0.060	0.057	0.060	0.050	0.007	0.025
1	0.0730	0.0695	0.118	0.112	0.073	0.070	0.072	1/16 0.062	0.007	0.031
2	0.0860	0.0822	0.140	0.134	0.086	0.083	0.096	5/64 0.078	0.008	0.038
3	0.0990	0.0949	0.161	0.154	0.099	0.095	0.096	5/64 0.078	0.008	0.044
4	0.1120	0.1075	0.183	0.176	0.112	0.108	0.111	3/32 0.094	0.009	0.051
5	0.1250	0.1202	0.205	0.198	0.125	0.121	0.111	3/32 0.094	0.010	0.057
6	0.1380	0.1329	0.226	0.218	0.138	0.134	0.133	7/64 0.109	0.010	0.064
8	0.1640	0.1585	0.270	0.262	0.164	0.159	0.168	9/64 0.141	0.012	0.077
10	0.1900	0.1840	0.312	0.303	0.190	0.185	0.183	5/32 0.156	0.014	0.090
1/4	0.2500	0.2435	0.375	0.365	0.250	0.244	0.216	3/16 0.188	0.014	0.120
5/16	0.3125	0.3053	0.469	0.457	0.312	0.306	0.291	1/4 0.250	0.017	0.151
3/8	0.3750	0.3678	0.562	0.550	0.375	0.368	0.372	5/16 0.312	0.020	0.182
7/16	0.4375	0.4294	0.656	0.642	0.438	0.430	0.454	3/8 0.375	0.023	0.213
1/2	0.5000	0.4919	0.750	0.735	0.500	0.492	0.454	3/8 0.375	0.026	0.245
5/8	0.6250	0.6163	0.938	0.921	0.625	0.616	0.595	1/2 0.500	0.032	0.307
3/4	0.7500	0.7406	1.125	1.107	0.750	0.740	0.620	5/8 0.625	0.039	0.370
7/8	0.8750	0.8647	1.312	1.293	0.875	0.864	0.698	3/4 0.750	0.044	0.432
1	1.0000	0.9886	1.500	1.479	1.000	0.988	0.790	3/4 0.750	0.050	0.495
1 1/8	1.1250	1.1086	1.688	1.665	1.125	1.111	…	7/8 0.875	0.055	0.557
1 1/4	1.2500	1.2336	1.875	1.852	1.250	1.236	…	7/8 0.875	0.060	0.620
1 3/8	1.3750	1.3568	2.062	2.038	1.375	1.360	…	1 1.000	0.065	0.682
1 1/2	1.5000	1.4818	2.250	2.224	1.500	1.485	…	1 1.000	0.070	0.745
1 3/4	1.7500	1.7295	2.625	2.597	1.750	1.734	…	1 1/4 1.250	0.080	0.870
2	2.0000	1.9780	3.000	2.970	2.000	1.983	…	1 1/2 1.500	0.090	0.995
2 1/4	2.2500	2.2280	3.375	3.344	2.250	2.232	…	1 3/4 1.750	0.100	1.120
2 1/2	2.5000	2.4762	3.750	3.717	2.500	2.481	…	1 3/4 1.750	0.110	1.245
2 3/4	2.7500	2.7262	4.125	4.090	2.750	2.730	…	2 2.000	0.120	1.370
3	3.0000	2.9762	4.500	4.464	3.000	2.979	…	2 1/4 2.250	0.130	1.495
3 1/4	3.2500	3.2262	4.875	4.837	3.250	3.228	…	2 1/4 2.250	0.140	1.620
3 1/2	3.5000	3.4762	5.250	5.211	3.500	3.478	…	2 3/4 2.750	0.150	1.745
3 3/4	3.7500	3.7262	5.625	5.584	3.750	3.727	…	2 3/4 2.750	0.160	1.870
4	4.0000	3.9762	6.000	5.958	4.000	3.976	…	3 3.000	0.170	1.995

[a] Key engagement depths are minimum. Spline socket sizes are nominal.

[b] Spline driven fasteners are removed from the latest revision. For additional information on these fasteners, see ANSI/ASME B18.3-1998.

All dimensions in inches. The body length L_B of the screw is the length of the unthreaded cylindrical portion of the shank. The length of thread, L_T, is the distance from the extreme point to the last complete (full form) thread. Standard length increments for screw diameters up to 1 inch are 1/16 inch for lengths 1/8 through 1/4 inch, 1/8 inch for lengths 1/4 through 1 inch, 1/4 inch for lengths 1 through 3 1/2 inches, 1/2 inch for lengths 3 1/2 through 7 inches, 1 inch for lengths 7 through 10 inches and for diameters over 1 inch are 1/2 inch for lengths 1 through 7 inches, 1 inch for lengths 7 through 10 inches, and 2 inches for lengths over 10 inches.

Heads may be plain or knurled, and chamfered to an angle *E* of 30 to 45 degrees with the surface of the flat. The thread conforms to the Unified Standard with radius root, Class 3A UNRC and UNRF for screw sizes No. 0 through 1 inch inclusive, Class 2A UNRC and UNRF for over 1 inch through 1 1/2 inches inclusive, and Class 2A UNRC for larger sizes. For details not shown, including materials, see ANSI/ASME B18.3-2012 and ANSI/ASME B18.3-1998 (for Spline driven fasteners).

Table 4. American National Standard Socket Head Cap Screws—Metric Series
ANSI/ASME B18.3.1M-1986 (Withdrawn)

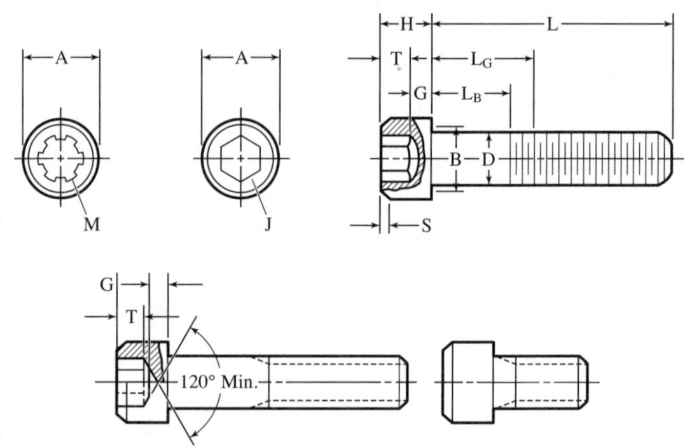

Nom. Size and Thread Pitch	Body Diameter, *D*		Head Diameter *A*		Head Height *H*		Chamfer or Radius *S*	Hexagon Socket Size[a] *J*	Spline Socket Size[a] *M*	Key Engagement *T*	Transition Dia. *B*[a]
	Max	Min	Max	Min	Max	Min	Max	Nom.	Nom.	Min	Max
M1.6 × 0.35	1.60	1.46	3.00	2.87	1.60	1.52	0.16	1.5	1.829	0.80	2.0
M2 × 0.4	2.00	1.86	3.80	3.65	2.00	1.91	0.20	1.5	1.829	1.00	2.6
M2.5 × 0.45	2.50	2.36	4.50	4.33	2.50	2.40	0.25	2.0	2.438	1.25	3.1
M3 × 0.5	3.00	2.86	5.50	5.32	3.00	2.89	0.30	2.5	2.819	1.50	3.6
M4 × 0.7	4.00	3.82	7.00	6.80	4.00	3.88	0.40	3.0	3.378	2.00	4.7
M5 × 0.8	5.00	4.82	8.50	8.27	5.00	4.86	0.50	4.0	4.648	2.50	5.7
M6 × 1	6.00	5.82	10.00	9.74	6.00	5.85	0.60	5.0	5.486	3.00	6.8
M8 × 1.25	8.00	7.78	13.00	12.70	8.00	7.83	0.80	6.0	7.391	4.00	9.2
M10 × 1.5	10.00	9.78	16.00	15.67	10.00	9.81	1.00	8.0	…	5.00	11.2
M12 × 1.75	12.00	11.73	18.00	17.63	12.00	11.79	1.20	10.0	…	6.00	14.2
M14 × 2[b]	14.00	13.73	21.00	20.60	14.00	13.77	1.40	12.0	…	7.00	16.2
M16 × 2	16.00	15.73	24.00	23.58	16.00	15.76	1.60	14.0	…	8.00	18.2
M20 × 2.5	20.00	19.67	30.00	29.53	20.00	19.73	2.00	17.0	…	10.00	22.4
M24 × 3	24.00	23.67	36.00	35.48	24.00	23.70	2.40	19.0	…	12.00	26.4
M30 × 3.5	30.00	29.67	45.00	44.42	30.00	29.67	3.00	22.0	…	15.00	33.4
M36 × 4	36.00	35.61	54.00	53.37	36.00	35.64	3.60	27.0	…	18.00	39.4
M42 × 4.5	42.00	41.61	63.00	62.31	42.00	41.61	4.20	32.0	…	21.00	45.6
M48 × 5	48.00	47.61	72.00	71.27	48.00	47.58	4.80	36.0	…	24.00	52.6

[a] See also Table 2.

[b] The M14 × 2 size is not recommended for use in new designs.

All dimensions are in millimeters. *LG* is grip length and *LB* is body length. For additional manufacturing and acceptance specifications, see ASME B18.3.1M (Withdrawn).

Table 5. Hexagon and Spline Socket Set Screws

ANSI/ASME B18.3-2012 and ANSI/ASME B18.3-1998 (for Spline-Driven Fasteners)

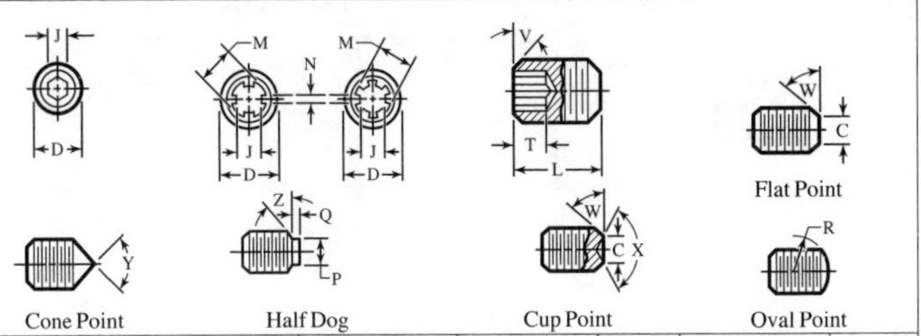

Nominal Size or Basic Screw Diameter		Socket Size: Hex. Nom.	Socket Size: Spl. Nom.	Cup and Flat Point Diameters: Max.	Cup and Flat Point Diameters: Min.	Half Dog Point: Dia. Max.	Half Dog Point: Lgth. Max.	Oval Point Radius Basic	Min. Key Engagement Depth: Hex.	Min. Key Engagement Depth: Spl.	Lgth. Limit for Angle
		J	M	C		P	Q	R	T_H[a]	T_S[a]	Y[b]
0	0.0600	0.028	0.033	0.033	0.027	0.040	0.017	0.045	0.050	0.026	0.09
1	0.0730	0.035	0.033	0.040	0.033	0.049	0.021	0.055	0.060	0.035	0.09
2	0.0860	0.035	0.048	0.047	0.039	0.057	0.024	0.064	0.060	0.040	0.13
3	0.0990	0.050	0.048	0.054	0.045	0.066	0.027	0.074	0.070	0.040	0.13
4	0.1120	0.050	0.060	0.061	0.051	0.075	0.030	0.084	0.070	0.045	0.19
5	0.1250	1/16	0.072	0.067	0.057	0.083	0.033	0.094	0.080	0.055	0.19
6	0.1380	1/16	0.072	0.074	0.064	0.092	0.038	0.104	0.080	0.055	0.19
8	0.1640	5/64	0.096	0.087	0.076	0.109	0.043	0.123	0.090	0.080	0.25
10	0.1900	3/32	0.111	0.102	0.088	0.127	0.049	0.142	0.100	0.080	0.25
1/4	0.2500	1/8	0.145	0.132	0.118	0.156	0.067	0.188	0.125	0.125	0.31
5/16	0.3125	5/32	0.183	0.172	0.156	0.203	0.082	0.234	0.156	0.156	0.38
3/8	0.3750	3/16	0.216	0.212	0.194	0.250	0.099	0.281	0.188	0.188	0.44
7/16	0.4375	7/32	0.251	0.252	0.232	0.297	0.114	0.328	0.219	0.219	0.50
1/2	0.5000	1/4	0.291	0.291	0.270	0.344	0.130	0.375	0.250	0.250	0.57
5/8	0.6250	5/16	0.372	0.371	0.347	0.469	0.164	0.469	0.312	0.312	0.75
3/4	0.7500	3/8	0.454	0.450	0.425	0.562	0.196	0.562	0.375	0.375	0.88
7/8	0.8750	1/2	0.595	0.530	0.502	0.656	0.227	0.656	0.500	0.500	1.00
1	1.0000	9/16	…	0.609	0.579	0.750	0.260	0.750	0.562	…	1.13
1 1/8	1.1250	9/16	…	0.689	0.655	0.844	0.291	0.844	0.562	…	1.25
1 1/4	1.2500	5/8	…	0.767	0.733	0.938	0.323	0.938	0.625	…	1.50
1 3/8	1.3750	5/8	…	0.848	0.808	1.031	0.354	1.031	0.625	…	1.63
1 1/2	1.5000	3/4	…	0.926	0.886	1.125	0.385	1.125	0.750	…	1.75
1 3/4	1.7500	1	…	1.086	1.039	1.312	0.448	1.321	1.000	…	2.00
2	2.0000	1	…	1.244	1.193	1.500	0.510	1.500	1.000	…	2.25

[a] Reference should be made to the Standard for shortest optimum nominal lengths to which the minimum key engagement depths T_H and T_S apply.

[b] Cone point angle Y is 90 degrees plus or minus 2 degrees for these nominal lengths or longer and 118 degrees plus or minus 2 degrees for shorter nominal lengths.

All dimensions are in inches. The thread conforms to the Unified Standard, Class 3A, UNC and UNF series. The socket depth T is included in the Standard and some are shown here. The nominal length L of all socket-type set screws is the total or overall length. For nominal screw lengths of 1/16 through 3/16 inch (0 through 3 sizes incl.) the standard length increment is 0.06 inch; for lengths 1/8 through 1 inch the increment is 1/8 inch; for lengths 1 through 2 inches the increment is 1/4 inch; for lengths 2 through 6 inches the increment is 1/2 inch; for lengths 6 inches and longer the increment is 1 inch.

Length Tolerance: The allowable tolerance on length L for all set screws of the socket type is ±0.01 inch for set screws up to 5/8 inch long; ±0.02 inch for screws over 5/8 to 2 inches long; ±0.03 inch for screws over 2 to 6 inches long and ± 0.06 inch for screws over 6 inches long. For manufacturing details, including materials, not shown, see American National Standard ANSI/ASME B18.3-2 and ASNI/ASME B18.3-1998 (for Spline driven fasteners).

Table 6. Drill and Counterbore Sizes for Socket Head Cap Screws (1960 Series)

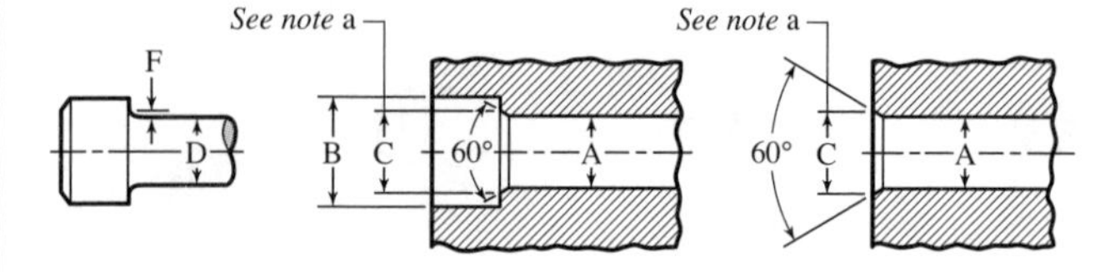

Nominal Size or Basic Screw Diameter		Nominal Drill Size				Counterbore Diameter	Countersink Diameter[a]
		Close Fit[b]		Normal Fit[c]			
		Number or Fractional Size	Decimal Size	Number or Fractional Size	Decimal Size		
		A				B	C
0	0.0600	51	0.067	49	0.073	1/8	0.074
1	0.0730	46	0.081	43	0.089	5/32	0.087
2	0.0860	3/32	0.094	36	0.106	3/16	0.102
3	0.0990	36	0.106	31	0.120	7/32	0.115
4	0.1120	1/8	0.125	29	0.136	7/32	0.130
5	0.1250	9/64	0.141	23	0.154	1/4	0.145
6	0.1380	23	0.154	18	0.170	9/32	0.158
8	0.1640	15	0.180	10	0.194	5/16	0.188
10	0.1900	5	0.206	2	0.221	3/8	0.218
1/4	0.2500	17/64	0.266	9/32	0.281	7/16	0.278
5/16	0.3125	21/64	0.328	11/32	0.344	17/32	0.346
3/8	0.3750	25/64	0.391	13/32	0.406	5/8	0.415
7/16	0.4375	29/64	0.453	15/32	0.469	23/32	0.483
1/2	0.5000	33/64	0.516	17/32	0.531	13/16	0.552
5/8	0.6250	41/64	0.641	21/32	0.656	1	0.689
3/4	0.7500	49/64	0.766	25/32	0.781	1 3/16	0.828
7/8	0.8750	57/64	0.891	29/32	0.906	1 3/8	0.963
1	1.0000	1 1/64	1.016	1 1/32	1.031	1 5/8	1.100
1 1/4	1.2500	1 9/32	1.281	1 5/16	1.312	2	1.370
1 1/2	1.5000	1 17/32	1.531	1 9/16	1.562	2 3/8	1.640
1 3/4	1.7500	1 25/32	1.781	1 13/16	1.812	2 3/4	1.910
2	2.0000	2 1/32	2.031	2 1/16	2.062	3 1/8	2.180

[a] *Countersink:* It is considered good practice to countersink or break the edges of holes smaller than (D Max + $2F$ Max) in parts having a hardness which approaches, equals, or exceeds the screw hardness. If such holes are not countersunk, the heads of screws may not seat properly or the sharp edges on holes may deform the fillets on screws, thereby making them susceptible to fatigue in applications involving dynamic loading. The countersink or corner relief, however, should not be larger than is necessary to insure that the fillet on the screw is cleared.

[b] *Close Fit:* The close fit is normally limited to holes for those lengths of screws which are threaded to the head in assemblies where only one screw is to be used or where two or more screws are to be used and the mating holes are to be produced either at assembly or by matched and coordinated tooling.

[c] *Normal Fit:* The normal fit is intended for screws of relatively long length or for assemblies involving two or more screws where the mating holes are to be produced by conventional tolerancing methods. It provides for the maximum allowable eccentricity of the longest standard screws and for certain variations in the parts to be fastened, such as: deviations in hole straightness, angularity between the axis of the tapped hole and that of the hole for the shank, differences in center distances of the mating holes, etc.

All dimensions in inches.

Source: Appendix to American National Standard ANSI/ASME B18.3-2012 and ANSI/ASME B18.3-1998 (for Spline driven fasteners).

Table 7. Drill and Counterbore Sizes for Metric Socket Head Cap Screws

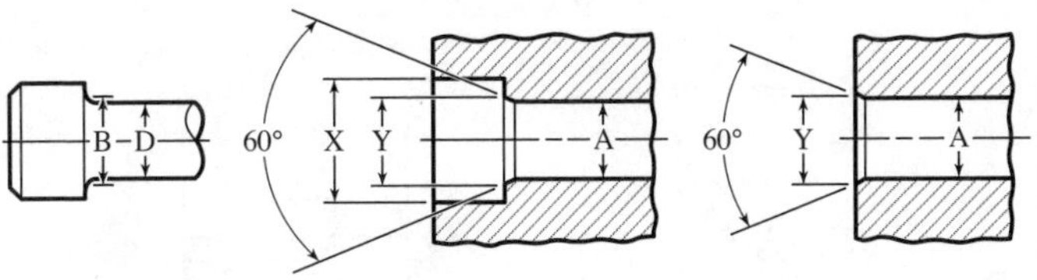

Nominal Size or Basic Screw Diameter	Nominal Drill Size, *A*		Counterbore Diameter, *X*	Countersink Diameter,[a] *Y*
	Close Fit[b]	Normal Fit[c]		
M1.6	1.80	1.95	3.50	2.0
M2	2.20	2.40	4.40	2.6
M2.5	2.70	3.00	5.40	3.1
M3	3.40	3.70	6.50	3.6
M4	4.40	4.80	8.25	4.7
M5	5.40	5.80	9.75	5.7
M6	6.40	6.80	11.25	6.8
M8	8.40	8.80	14.25	9.2
M10	10.50	10.80	17.25	11.2
M12	12.50	12.80	19.25	14.2
M14	14.50	14.75	22.25	16.2
M16	16.50	16.75	25.50	18.2
M20	20.50	20.75	31.50	22.4
M24	24.50	24.75	37.50	26.4
M30	30.75	31.75	47.50	33.4
M36	37.00	37.50	56.50	39.4
M42	43.00	44.00	66.00	45.6
M48	49.00	50.00	75.00	52.6

[a] *Countersink:* It is considered good practice to countersink or break the edges of holes which are smaller than *B* Max. (see Table 21, page 117) in parts having a hardness which approaches, equals, or exceeds the screw hardness. If such holes are not countersunk, the heads of screws may not seat properly or the sharp edges on holes may deform the fillets on screws, thereby making them susceptible to fatigue in applications involving dynamic loading. The countersink or corner relief, however, should not be larger than is necessary to ensure that the fillet on the screw is cleared. Normally, the diameter of countersink does not have to exceed *B* Max. Countersinks or corner reliefs in excess of this diameter reduce the effective bearing area and introduce the possibility of embedment where the parts to be fastened are softer than the screws or of brinnelling or flaring the heads of the screws where the parts to be fastened are harder than the screws.

[b] *Close Fit:* The close fit is normally limited to holes for those lengths of screws which are threaded to the head in assemblies where only one screw is to be used or where two or more screws are to be used and the mating holes are to be produced either at assembly or by matched and coordinated tooling.

[c] *Normal Fit:* The normal fit is intended for screws of relatively long length or for assemblies involving two or more screws where the mating holes are to be produced by conventional tolerancing methods. It provides for the maximum allowable eccentricity of the longest standard screws and for certain variations in the parts to be fastened, such as: deviations in hole straightness, angularity between the axis of the tapped hole and that of the hole for shank, differences in center distances of the mating holes, etc.

All dimensions are in millimeters.

Table 8. American National Standard Hexagon and Spline Socket Flat Countersunk Head Cap Screws *ANSI/ASME B18.3-2012*

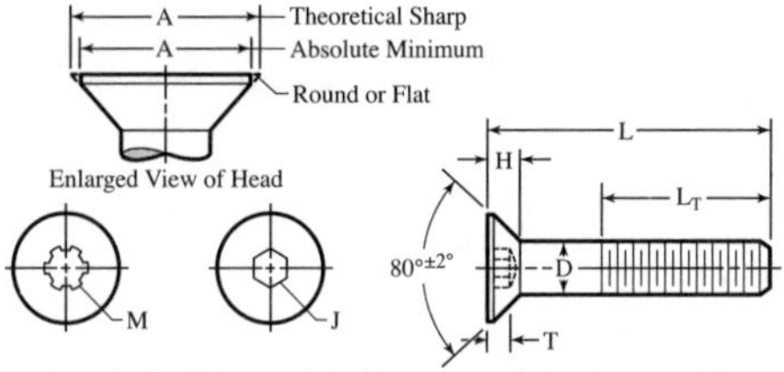

Nominal Size	Body Diameter Max.	Body Diameter Min.	Head Diameter Theoretical Sharp Max.	Head Diameter Abs. Min.	Head Height Reference	Spline Socket Size	Hexagon Socket Size Nom.	Key Engagement Min.
	D		*A*		*H*	*M*	*J*	*T*
0	0.0600	0.0568	0.138	0.117	0.044	0.048	0.035	0.025
1	0.0730	0.0695	0.168	0.143	0.054	0.060	0.050	0.031
2	0.0860	0.0822	0.197	0.168	0.064	0.060	0.050	0.038
3	0.0990	0.0949	0.226	0.193	0.073	0.072	1/16	0.044
4	0.1120	0.1075	0.255	0.218	0.083	0.072	1/16	0.055
5	0.1250	0.1202	0.281	0.240	0.090	0.096	5/64	0.061
6	0.1380	0.1329	0.307	0.263	0.097	0.096	5/64	0.066
8	0.1640	0.1585	0.359	0.311	0.112	0.111	3/32	0.076
10	0.1900	0.1840	0.411	0.359	0.127	0.145	1/8	0.087
1/4	0.2500	0.2435	0.531	0.480	0.161	0.183	5/32	0.111
5/16	0.3125	0.3053	0.656	0.600	0.198	0.216	3/16	0.135
3/8	0.3750	0.3678	0.781	0.720	0.234	0.251	7/32	0.159
7/16	0.4375	0.4294	0.844	0.781	0.234	0.291	1/4	0.159
1/2	0.5000	0.4919	0.938	0.872	0.251	0.372	5/16	0.172
5/8	0.6250	0.6163	1.188	1.112	0.324	0.454	3/8	0.220
3/4	0.7500	0.7406	1.438	1.355	0.396	0.454	1/2	0.220
7/8	0.8750	0.8647	1.688	1.604	0.468	…	9/16	0.248
1	1.0000	0.9886	1.938	1.841	0.540	…	5/8	0.297
1 1/8	1.1250	1.1086	2.188	2.079	0.611	…	3/4	0.325
1 1/4	1.2500	1.2336	2.438	2.316	0.683	…	7/8	0.358
1 3/8	1.3750	1.3568	2.688	2.553	0.755	…	7/8	0.402
1 1/2	1.5000	1.4818	2.938	2.791	0.827	…	1	0.435

All dimensions in inches.

The body of the screw is the unthreaded cylindrical portion of the shank where not threaded to the head, the shank being the portion of the screw from the point of juncture of the conical bearing surface and the body to the flat of the point. The length of thread *LT* is the distance measured from the extreme point to the last complete (full form) thread.

Standard length increments of No. 0 through 1-inch sizes are as follows: 1/16 inch for nominal screw lengths of 1/8 through 1/4 inch; 1/8 inch for lengths of 1/4 through 1 inch; 1/4 inch for lengths of 1 inch through 3 1/2 inches; 1/2 inch for lengths of 3 1/2 through 7 inches; and 1 inch for lengths of 7 through 10 inches, incl. For screw sizes over 1 inch, length increments are: 1/2 inch for nominal screw lengths of 1 inch through 7 inches; 1 inch for lengths of 7 through 10 inches; and 2 inches for lengths over 10 inches.

Threads shall be Unified external threads with radius root; Class 3A UNRC and UNRF series for sizes No. 0 through 1 inch and Class 2A UNRC and UNRF series for sizes over 1 inch to 1 1/2 inches, incl.

For manufacturing details not shown, including materials, see ANSI/ASME B18.3-2012 and ANSI/ASME B18.3-1998 (for Spline driven fasteners).

Table 9. American National Standard Slotted Flat Countersunk Head Cap Screws *ANSI/ASME B18.6.2-1998 (R2010)*

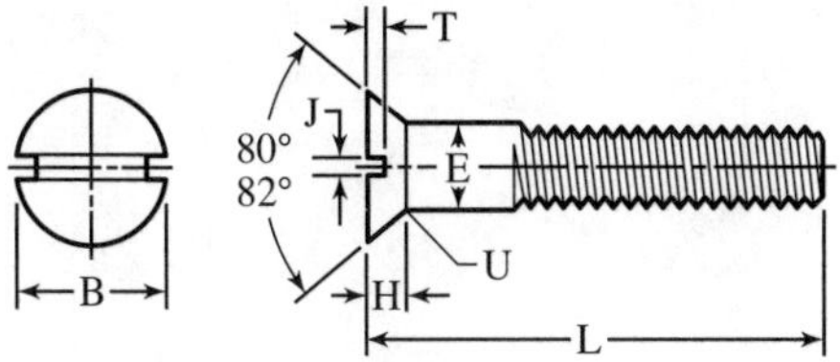

Nominal Size[a] or Basic Screw Dia.		Body Dia., E		Head Dia., A Edge Sharp	Head Dia., A Edge Rnd'd. or Flat	Head Hgt., H	Slot Width, J		Slot Depth, T		Fillet Rad., U
		Max.	Min.	Max.	Min.	Ref.	Max.	Min.	Max.	Min.	Max.
1/4	0.2500	.2500	.2450	.500	.452	.140	.075	.064	.068	.045	.100
5/16	0.3125	.3125	.3070	.625	.567	.177	.084	.072	.086	.057	.125
3/8	0.3750	.3750	.3690	.750	.682	.210	.094	.081	.103	.068	.150
7/16	0.4375	.4375	.4310	.812	.736	.210	.094	.081	.103	.068	.175
1/2	0.5000	.5000	.4930	.875	.791	.210	.106	.091	.103	.068	.200
9/16	0.5625	.5625	.5550	1.000	.906	.244	.118	.102	.120	.080	.225
5/8	0.6250	.6250	.6170	1.125	1.020	.281	.133	.116	.137	.091	.250
3/4	0.7500	.7500	.7420	1.375	1.251	.352	.149	.131	.171	.115	.300
7/8	0.8750	.8750	.8660	1.625	1.480	.423	.167	.147	.206	.138	.350
1	1.0000	1.0000	.9900	1.875	1.711	.494	.188	.166	.240	.162	.400
1 1/8	1.1250	1.1250	1.1140	2.062	1.880	.529	.196	.178	.257	.173	.450
1 1/4	1.2500	1.2500	1.2390	2.312	2.110	.600	.211	.193	.291	.197	.500
1 3/8	1.3750	1.3750	1.3630	2.562	2.340	.665	.226	.208	.326	.220	.550
1 1/2	1.5000	1.5000	1.4880	2.812	2.570	.742	.258	.240	.360	.244	.600

[a] When specifying a nominal size in decimals, the zero preceding the decimal point is omitted as is any zero in the fourth decimal place.

All dimensions are in inches. *Threads:* Threads are Unified Standard Class 2A; UNC, UNF and 8 UN Series or UNRC, UNRF, and 8 UNR Series.

Table 10. American National Standard Hardened Ground Machine Dowel Pins *ANSI/ASME B18.8.2-2000 (R2010)*

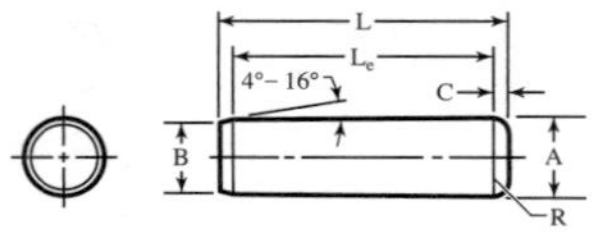

Nominal Size[a] or Nominal Pin Diameter		Pin Diameter, *A*						Point Diameter, *B*		Crown Height, *C*	Crown Radius, *R*	Range of Preferred Lengths,[b] *L*	Single Shear Load, for Carbon or Alloy Steel, Calculated, lb	Suggested Hole Diameter[c]	
		Standard Series Pins			Oversize Series Pins										
		Basic	Max	Min	Basic	Max	Min	Max	Min	Max	Min			Max	Min
1/16	0.0625	0.0627	0.0628	0.0626	0.0635	0.0636	0.0634	0.058	0.048	0.020	0.008	3/16-3/4	400	0.0625	0.0620
5/64[d]	0.0781	0.0783	0.0784	0.0782	0.0791	0.0792	0.0790	0.074	0.064	0.026	0.010	…	620	0.0781	0.0776
3/32	0.0938	0.0940	0.0941	0.0939	0.0948	0.0949	0.0947	0.089	0.079	0.031	0.012	5/16-1	900	0.0937	0.0932
1/8	0.1250	0.1252	0.1253	0.1251	0.1260	0.1261	0.1259	0.120	0.110	0.041	0.016	3/8-2	1,600	0.1250	0.1245
5/32[d]	0.1562	0.1564	0.1565	0.1563	0.1572	0.1573	0.1571	0.150	0.140	0.052	0.020	…	2,500	0.1562	0.1557
3/16	0.1875	0.1877	0.1878	0.1876	0.1885	0.1886	0.1884	0.180	0.170	0.062	0.023	1/2-2	3,600	0.1875	0.1870
1/4	0.2500	0.2502	0.2503	0.2501	0.2510	0.2511	0.2509	0.240	0.230	0.083	0.031	1/2-2 1/2	6,400	0.2500	0.2495
5/16	0.3125	0.3127	0.3128	0.3126	0.3135	0.3136	0.3134	0.302	0.290	0.104	0.039	1/2-2 1/2	10,000	0.3125	0.3120
3/8	0.3750	0.3752	0.3753	0.3751	0.3760	0.3761	0.3759	0.365	0.350	0.125	0.047	1/2-3	14,350	0.3750	0.3745
7/16	0.4375	0.4377	0.4378	0.4376	0.4385	0.4386	0.4384	0.424	0.409	0.146	0.055	7/8-3	19,550	0.4375	0.4370
1/2	0.5000	0.5002	0.5003	0.5001	0.5010	0.5011	0.5009	0.486	0.471	0.167	0.063	3/4, 1-4	25,500	0.5000	0.4995
5/8	0.6250	0.6252	0.6253	0.6251	0.6260	0.6261	0.6259	0.611	0.595	0.208	0.078	1 1/4-5	39,900	0.6250	0.6245
3/4	0.7500	0.7502	0.7503	0.7501	0.7510	0.7511	0.7509	0.735	0.715	0.250	0.094	1 1/2-6	57,000	0.7500	0.7495
7/8	0.8750	0.8752	0.8753	0.8751	0.8760	0.8761	0.8759	0.860	0.840	0.293	0.109	2, 2 1/2-6	78,000	0.8750	0.8745
1	1.0000	1.0002	1.0003	1.0001	1.0010	1.0011	1.0009	0.980	0.960	0.333	0.125	2, 2 1/2-5, 6	102,000	1.0000	0.9995

[a] Where specifying nominal size as basic diameter, zeros preceding decimal and in the fourth decimal place are omitted.

[b] Lengths increase in 1/16-inch steps up to 3/8 inch, in 1/8-inch steps from 3/8 inch to 1 inch, in 1/4-inch steps from 1 inch to 2 1/2 inches, and in 1/2-inch steps above 2 1/2 inches. Tolerance on length is ±0.010 inch.

[c] These hole sizes have been commonly used for press fitting Standard Series machine dowel pins into materials such as mild steels and cast iron. In soft materials such as aluminum or zinc die castings, hole size limits are usually decreased by 0.0005 inch to increase the press fit.

[d] Nonpreferred sizes, not recommended for use in new designs.

All dimensions are in inches.

American National Standard Hardened Ground Production Dowel Pins.—Hardened ground production dowel pins have basic diameters that are 0.0002 inch over the nominal pin diameter.

Preferred Lengths and Sizes: The preferred lengths and sizes of these pins are given in Table 10. Other sizes and lengths are produced as required by the purchaser.

Table 11. American National Standard Hardened Ground Production Dowel Pins *ANSI/ASME B18.8.2-2000 (R2010)*

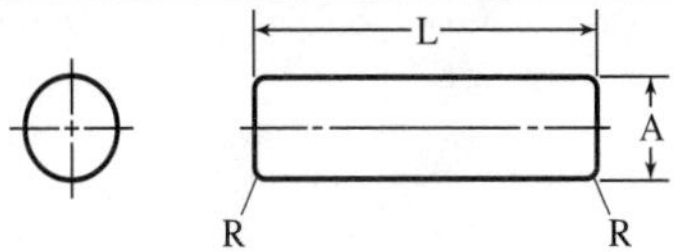

Nominal Size[a] or Nominal Pin Diameter	Pin Diameter, *A*			Corner Radius, *R*		Range of Preferred Lengths,[b] *L*	Single Shear Load, Carbon Steel	Suggested Hole Diameter[c]	
	Basic	Max	Min	Max	Min			Max	Min
$^{1}/_{16}$ 0.0625	0.0627	0.0628	0.0626	0.020	0.010	$^{3}/_{16}$–1	395	0.0625	0.0620
$^{3}/_{32}$ 0.0938	0.0939	0.0940	0.0938	0.020	0.010	$^{3}/_{16}$–2	700	0.0937	0.0932
$^{7}/_{64}$ 0.1094	0.1095	0.1096	0.1094	0.020	0.010	$^{3}/_{16}$–2	950	0.1094	0.1089
$^{1}/_{8}$ 0.1250	0.1252	0.1253	0.1251	0.020	0.010	$^{3}/_{16}$–2	1,300	0.1250	0.1245
$^{5}/_{32}$ 0.1562	0.1564	0.1565	0.1563	0.020	0.010	$^{3}/_{16}$–2	2,050	0.1562	0.1557
$^{3}/_{16}$ 0.1875	0.1877	0.1878	0.1876	0.020	0.010	$^{3}/_{16}$–2	2,950	0.1875	0.1870
$^{7}/_{32}$ 0.2188	0.2189	0.2190	0.2188	0.020	0.010	$^{1}/_{4}$–2	3,800	0.2188	0.2183
$^{1}/_{4}$ 0.2500	0.2502	0.2503	0.2501	0.020	0.010	$^{1}/_{4}$–1$^{1}/_{2}$, 1$^{3}/_{4}$, 2–2$^{1}/_{2}$	5,000	0.2500	0.2495
$^{5}/_{16}$ 0.3125	0.3127	0.3128	0.3126	0.020	0.010	$^{5}/_{16}$–1$^{1}/_{2}$, 1$^{3}/_{4}$, 2–2$^{1}/_{2}$	8,000	0.3125	0.3120
$^{3}/_{8}$ 0.3750	0.3752	0.3753	0.3751	0.020	0.010	$^{3}/_{8}$–1$^{1}/_{2}$, 1$^{3}/_{4}$, 2–3	11,500	0.3750	0.3745

[a] Where specifying nominal pin size in decimals, zeros preceding decimal and in the fourth decimal place are omitted.

[b] Lengths increase in $^{1}/_{16}$-inch steps up to 1 inch; in $^{1}/_{8}$-inch steps from 1 inch to 2 inches; and then are 2$^{1}/_{4}$, 2$^{1}/_{2}$, and 3 inches.

[c] These hole sizes have been commonly used for press fitting production dowel pins into materials such as mild steels and cast iron. In soft materials such as aluminum or zinc die castings, hole size limits are usually decreased by 0.0005 inch to increase the press fit.

All dimensions are in inches.

Size: These pins have basic diameters that are 0.0002 inch over the nominal pin diameter. The diameter shall be ground, or ground and lapped, to within ±0.0001 inch of the basic diameter as specified in Table 10.

Roundness, Straightness, and Surface Roughness: These standard pins shall conform to true round within 0.0001 inch; straightness over that portion of the length not affected by the rounded ends, within an accumulative total of 0.005 inch per inch of length for nominal lengths up to 4 inches, and within 0.002 inch total for all nominal lengths over 4 inches; roughness shall not exceed 8 microinches (μin.) over the cylindrical portion of the pin, nor over 125 μin. on all other surfaces.

Designation: These pins are designated by the following data in the sequence shown: Product name (noun first), nominal pin diameter (fraction or decimal equivalent), length (fraction or decimal equivalent), material, and protective finish, if required.

Examples: Pins, Hardened Ground Production Dowel, $^{1}/_{8}$ × $^{3}/_{4}$, Steel, Phosphate Coated
Pins, Hardened Ground Production Dowel, 0.375 × 1.500, Steel

Table 12. American National Standard Chamfered and Square End Straight Pins
ANSI/ASME B18.8.2-2000 (R2010)

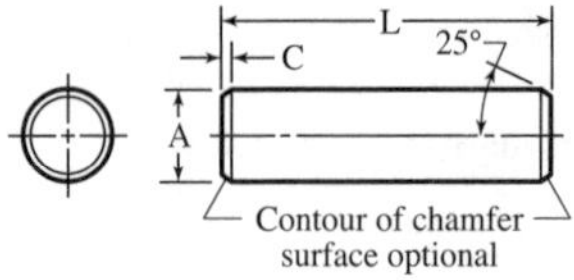

CHAMFERED STRAIGHT PIN

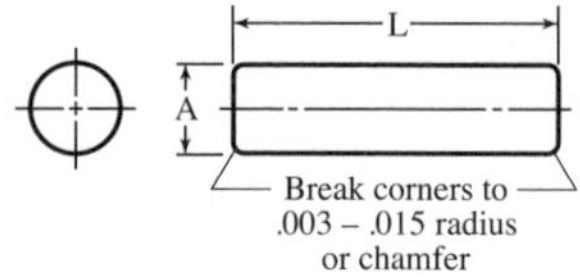

SQUARE END STRAIGHT PIN

Nominal Size[a] or Basic Pin Diameter	Pin Diameter, *A*		Chamfer Length, *C*		Nominal Size[a] or Basic Pin Diameter	Pin Diameter, *A*		Chamfer Length, *C*	
	Max	Min	Max	Min		Max	Min	Max	Min
1/16 0.062	0.0625	0.0605	0.025	0.005	5/16 0.312	0.3125	0.3105	0.040	0.020
3/32 0.094	0.0937	0.0917	0.025	0.00	3/8 0.375	0.3750	0.3730	0.040	0.020
7/64[b] 0.109	0.1094	0.1074	0.025	0.005	7/16 0.438	0.4375	0.4355	0.040	0.020
1/8 0.125	0.1250	0.1230	0.025	0.005	1/2 0.500	0.5000	0.4980	0.040	0.020
5/32 0.156	0.1562	0.1542	0.025	0.005	5/8 0.625	0.6250	0.6230	0.055	0.035
3/16 0.188	0.1875	0.1855	0.025	0.005	3/4 0.750	0.7500	0.7480	0.055	0.035
7/32 0.219	0.2187	0.2167	0.025	0.005	7/8 0.875	0.8750	0.8730	0.055	0.035
1/4 0.250	0.2500	0.2480	0.025	0.005	1 1.000	1.0000	0.9980	0.055	0.035

[a] In the Standard, zeros preceding decimal and in the fourth decimal place are omitted, when specifying nominal size in decimals. Here, they are included.

[b] Nonpreferred sizes, not recommended for use in new designs.

All dimensions are in inches.

American National Standard Straight Pins.—The diameter of both chamfered and square end straight pins is that of the commercial wire or rod from which the pins are made. The tolerances shown in Table 12 are applicable to carbon steel and some deviations in the diameter limits may be necessary for pins made from other materials.

Length Increments: Lengths are as specified by the purchaser; however, it is recommended that nominal pin lengths be limited to increments of not less than 0.062 inch.

Material: Straight pins are normally made from cold-drawn steel wire or rod having a maximum carbon content of 0.28 percent having a maximum hardness of R_c 32. Where required, pins may also be made from corrosion-resistant steel, brass, or other metals.

Designation: Straight pins are designated by the following data, in the sequence shown: Product name (noun first), nominal size (fraction or decimal equivalent), material, and protective finish, if required.

Examples: Pin, Chamfered Straight, 1/8 × 1.500, Steel
Pin, Square End Straight, 0.250 × 2.250, Steel, Zinc Plated

American National Standard Taper Pins.—Taper pins have a uniform taper over the pin length with both ends crowned. Most sizes are supplied in commercial and precision classes, the latter having generally tighter tolerances and being more closely controlled in manufacture.

Diameters: The major diameter of both commercial and precision classes of pins is the diameter of the large end and is the basis for pin size. The diameter at the small end is computed by multiplying the nominal length of the pin by the factor 0.02083 and subtracting the result from the basic pin diameter.

Taper: The taper on commercial-class pins is 0.250 ± 0.006 inch per foot and on the precision-class pins is 0.250 ± 0.004 inch per foot of length.

Materials: Unless otherwise specified, taper pins are made from AISI 1211 steel or cold-drawn AISI 1212 or 1213 steel or equivalents, and no mechanical property requirements apply.

Hole Sizes: Under most circumstances, holes for taper pins require taper reaming. Sizes and lengths of taper pins for which standard reamers are available are given in Table 13. Drilling specifications for taper pins are given below.

Table 13. American National Standard Taper Pins
ANSI/ASME B18.8.2-2000 (R2010)

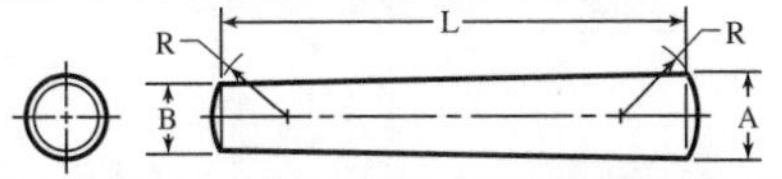

Pin Size Number and Basic Pin Diameter[a]		Major Diameter (Large End), *A*				End Crown Radius, *R*		Range of Lengths,[b] *L*	
		Commercial Class		Precision Class					
		Max	Min	Max	Min	Max	Min	Stand. Reamer Avail.[c]	Other
7/0	0.0625	0.0638	0.0618	0.0635	0.0625	0.072	0.052	…	1/4–1
6/0	0.0780	0.0793	0.0773	0.0790	0.0780	0.088	0.068	…	1/4–1/2
5/0	0.0940	0.0953	0.0933	0.0950	0.0940	0.104	0.084	1/4–1	1 1/4, 1 1/2
4/0	0.1090	0.1103	0.1083	0.1100	0.1090	0.119	0.099	1/4–1	1 1/4–2
3/0	0.1250	0.1263	0.1243	0.1260	0.1250	0.135	0.115	1/4–1	1 1/4–2
2/0	0.1410	0.1423	0.1403	0.1420	0.1410	0.151	0.131	1/2–1 1/4	1 1/2–2 1/2
0	0.1560	0.1573	0.1553	0.1570	0.1560	0.166	0.146	1/2–1 1/4	1 1/2–3
1	0.1720	0.1733	0.1713	0.1730	0.1720	0.182	0.162	3/4–1 1/4	1 1/2–3
2	0.1930	0.1943	0.1923	0.1940	0.1930	0.203	0.183	3/4–1 1/2	1 3/4–3
3	0.2190	0.2203	0.2183	0.2200	0.2190	0.229	0.209	3/4–1 3/4	2–4
4	0.2500	0.2513	0.2493	0.2510	0.2500	0.260	0.240	3/4–2	2 1/4–4
5	0.2890	0.2903	0.2883	0.2900	0.2890	0.299	0.279	1–2 1/2	2 3/4–6
6	0.3410	0.3423	0.3403	0.3420	0.3410	0.351	0.331	1 1/4–3	3 1/4–6
7	0.4090	0.4103	0.4083	0.4100	0.4090	0.419	0.399	1 1/4–3 3/4	4–8
8	0.4920	0.4933	0.4913	0.4930	0.4920	0.502	0.482	1 1/4–4 1/2	4 3/4–8
9	0.5910	0.5923	0.5903	0.5920	0.5910	0.601	0.581	1 1/4–5 1/4	5 1/2–8
10	0.7060	0.7073	0.7053	0.7070	0.7060	0.716	0.696	1 1/2–6	6 1/4–8
11	0.8600	0.8613	0.8593	…	…	0.870	0.850	…	2–8
12	1.0320	1.0333	1.0313	…	…	1.042	1.022	…	2–9
13	1.2410	1.2423	1.2403	…	…	1.251	1.231	…	3–11
14	1.5230	1.5243	1.5223	…	…	1.533	1.513	…	3–13

[a] In the Standard, zeros preceding decimal and in the fourth decimal place are omitted, when specifying nominal pin size in decimals. Here, they are included.

[b] Lengths increase in 1⁄8-inch steps up to 1 inch and in 1⁄4-inch steps above 1 inch.

[c] Standard reamers are available for pin lengths in this column. All dimensions are in inches.

All dimensions are in inches.
For nominal diameters, B, see Table 12.

Designation: Taper pins are designated by the following data in the sequence shown: Product name (noun first), class, size number (or decimal equivalent), length (fraction or three-place decimal equivalent), material, and protective finish, if required.

Examples: Pin, Taper (Commercial Class) No. 0 × 3/4, Steel
Pin, Taper (Precision Class) 0.219 × 1.750, Steel, Zinc Plated

Table 14. British Standard Parallel Steel Dowel Pins—Metric Series
BS 1804: Part 2: 1968

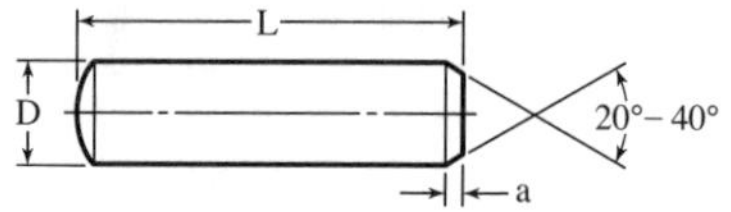

Nom. Length L, mm	Nominal Diameter D, mm													
	1	1.5	2	2.5	3	4	5	6	8	10	12	16	20	25
	Chamfer a Max., mm													
	0.3	0.3	0.3	0.4	0.45	0.6	0.75	0.9	1.2	1.5	1.8	2.5	3	4
	Standard Sizes													
4	●	●												
6	●	●	●	●										
8	●	●	●	●	●									
10		●	●	●	●	●								
12		●	●	●	●	●	●							
16			●	●	●	●	●	●	●					
20				●	●	●	●	●	●	●				
25					●	●	●	●	●	●	●			
30						●	●	●	●	●	●	●		
35							●	●	●	●	●	●		
40							●	●	●	●	●	●	●	
45								●	●	●	●	●	●	
50									●	●	●	●	●	●
60									●	●	●	●	●	●
70										●	●	●	●	●
80										●	●	●	●	●
90											●	●	●	●
100												●	●	●
110												●	●	●
120													●	●

Limits of Tolerance on Diameter							
Grade[a]		1		2		3	
Tolerance Zone		m5		h7		h11	
Nom. Dia., mm							
Over	To & Incl.	Limits of Tolerance, 0.001 mm					
	3	+7	+2	0	−12[b]	0	−60
3	6	+9	+4	0	−12	0	−75
6	10	+12	+6	0	−15	0	−90
10	14	+15	+7	0	−18	0	−110
14	18	+15	+7	0	−18	0	−110
18	24	+17	+8	0	−21	0	−130
24	30	+17	+8	0	−21	0	−130

[a] The limits of tolerance for grades 1 and 2 dowel pins have been chosen to provide satisfactory assembly when used in standard reamed holes (H7 and H8 tolerance zones). If the assembly is not satisfactory, refer to BS 1916: Part 1, Limits and Fits for Engineering, and select a different class of fit.

[b] This tolerance is larger than that given in BS 1916, and has been included because the use of a closer tolerance would involve precision grinding by the manufacturer, which is uneconomic for a grade 2 dowel pin.

The tolerance limits on the overall length of all grades of dowel pin up to and including 50 mm long are +0.5, −0.0 mm, and for pins over 50 mm long are +0.8, −0.0 mm. The Standard specifies that the roughness of the cylindrical surface of grades 1 and 2 dowel pins, when assessed in accordance with BS 1134, shall not be greater than 0.4 μm CLA (16 CLA).

American National Standard Spring Pins.—These pins are made in two types: one type has a slot throughout its length; the other is shaped into a coil.

Preferred Lengths and Sizes: The preferred lengths and sizes in which these pins are normally available are given in Table 15 and Table 16.

Materials: Spring pins are normally made from AISI 1070-1095 carbon steel, AISI 6150 H alloy steel, AISI Types 410 through 420 and 302 corrosion-resistant steels, and beryllium copper alloy, heat treated or cold worked to attain the hardness and performance characteristics set forth in ANSI/ASME B18.8.2-2000 (R2010).

Table 15. American National Standard Slotted-Type Spring Pins
ANSI/ASME 18.8.2-2000 (R2010)

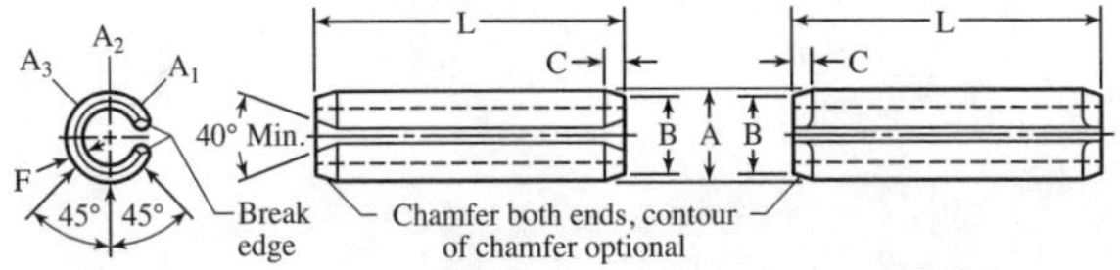

Style 1 **Style 2**

Nominal Size[a] or Basic Pin Diameter	Average Pin Diameter, *A*		Chamfer Dia., *B*	Chamfer Length, *C*		Stock Thickness, *F*	Recommended Hole Size		Material: SAE 1070, 1095, and SAE 51420	Material: SAE 30302 and 30304	Material: Beryllium Copper	Range of Practical Lengths[b]
	Max.	Min.	Max.	Max.	Min.	Basic	Max.	Min.	Double Shear Load, Min., lb			
1/16 0.062	0.069	0.066	0.059	0.028	0.007	0.012	0.065	0.062	430	250	270	3/16-1
5/64 0.078	0.086	0.083	0.075	0.032	0.008	0.018	0.081	0.078	800	460	500	3/16-1 1/2
3/32 0.094	0.103	0.099	0.091	0.038	0.008	0.022	0.097	0.094	1,150	670	710	3/16-1 1/2
1/8 0.125	0.135	0.131	0.122	0.044	0.008	0.028	0.129	0.125	1,875	1,090	1,170	5/16-2
9/64 0.141	0.149	0.145	0.137	0.044	0.008	0.028	0.144	0.140	2,175	1,260	1,350	3/8-2
5/32 0.156	0.167	0.162	0.151	0.048	0.010	0.032	0.160	0.156	2,750	1,600	1,725	7/16-2 1/2
3/16 0.188	0.199	0.194	0.182	0.055	0.011	0.040	0.192	0.187	4,150	2,425	2,600	1/2-2 1/2
7/32 0.219	0.232	0.226	0.214	0.065	0.011	0.048	0.224	0.219	5,850	3,400	3,650	1/2-3
1/4 0.250	0.264	0.258	0.245	0.065	0.012	0.048	0.256	0.250	7,050	4,100	4,400	1/2-3 1/2
5/16 0.312	0.330	0.321	0.306	0.080	0.014	0.062	0.318	0.312	10,800	6,300	6,750	3/4-4
3/8 0.375	0.395	0.385	0.368	0.095	0.016	0.077	0.382	0.375	16,300	9,500	10,200	3/4, 7/8, 1, 1 1/4, 1 1/2, 1 3/4, 2-4
7/16 0.438	0.459	0.448	0.430	0.095	0.017	0.077	0.445	0.437	19,800	11,500	12,300	1, 1 1/4, 1 1/2, 1 3/4, 2-4
1/2 0.500	0.524	0.513	0.485	0.110	0.025	0.094	0.510	0.500	27,100	15,800	17,000	1 1/4, 1 1/2, 1 3/4, 2-4
5/8 0.625	0.653	0.640	0.608	0.125	0.030	0.125	0.636	0.625	46,000	18,800	…	2-6
3/4 0.750	0.784	0.769	0.730	0.150	0.030	0.150	0.764	0.750	66,000	23,200	…	2-6

[a] Where specifying nominal size in decimals, zeros preceding decimal point are omitted.

[b] Length increments are 1/16 inch from 1/8 to 1 inch; 1/8 from 1 inch to 2 inches; and 1/4 inch from 2 inches to 6 inches.

All dimensions are in inches.

Table 16. American National Standard Coiled-Type Spring Pins *ANSI/ASME B18.8.2-2000 (R2010)*

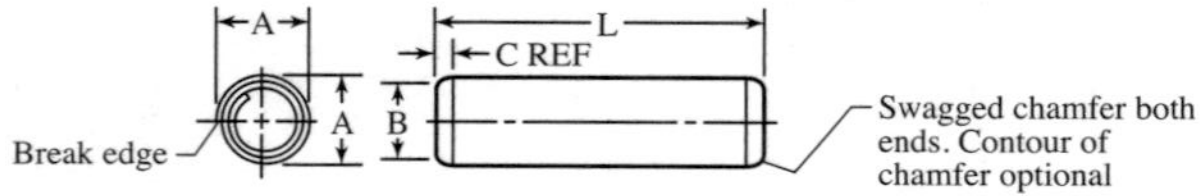

Nominal Size or Basic Pin Diameter	Pin Diameter, *A*						Chamfer		Recommended Hole Size		SAE Material Number					
	Standard Duty		Heavy Duty		Light Duty		Dia., *B*	Length, *C*			1070-1095 and 51420	30302 and 30304	1070-1095 and 51420	30302 and 30304	1070-1095 and 51420	30302 and 30304
											Double Shear Load, Min., lb					
	Max.	Min.	Max.	Min.	Max.	Min.	Max.	Ref.	Max.	Min.	Standard Duty		Heavy Duty		Light Duty	
1/32 0.031	0.035	0.033	…	…	…	…	0.029	0.024	0.032	0.031	90[a]	65	…	…	…	…
0.039	0.044	0.041	…	…	…	…	0.037	0.024	0.040	0.039	135[a]	100	…	…	…	…
3/64 0.047	0.052	0.049	…	…	…	…	0.045	0.024	0.048	0.046	190[a]	145	…	…	…	…
0.052	0.057	0.054	…	…	…	…	0.050	0.024	0.053	0.051	250[a]	190	…	…	…	…
1/16 0.062	0.072	0.067	0.070	0.066	0.073	0.067	0.059	0.028	0.065	0.061	330	265	475	360	205	160
5/64 0.078	0.088	0.083	0.086	0.082	0.089	0.083	0.075	0.032	0.081	0.077	550	425	800	575	325	250
3/32 0.094	0.105	0.099	0.103	0.098	0.106	0.099	0.091	0.038	0.097	0.093	775	600	1,150	825	475	360
7/64 0.109	0.120	0.114	0.118	0.113	0.121	0.114	0.106	0.038	0.112	0.108	1,050	825	1,500	1,150	650	500
1/8 0.125	0.138	0.131	0.136	0.130	0.139	0.131	0.121	0.044	0.129	0.124	1,400	1,100	2,000	1,700	825	650
5/32 0.156	0.171	0.163	0.168	0.161	0.172	0.163	0.152	0.048	0.160	0.155	2,200	1,700	3,100	2,400	1,300	1,000
3/16 0.188	0.205	0.196	0.202	0.194	0.207	0.196	0.182	0.055	0.192	0.185	3,150	2,400	4,500	3,500	1,900	1,450
7/32 0.219	0.238	0.228	0.235	0.226	0.240	0.228	0.214	0.065	0.224	0.217	4,200	3,300	5,900	4,600	2,600	2,000
1/4 0.250	0.271	0.260	0.268	0.258	0.273	0.260	0.243	0.065	0.256	0.247	5,500	4,300	7,800	6,200	3,300	2,600
5/16 0.312	0.337	0.324	0.334	0.322	0.339	0.324	0.304	0.080	0.319	0.308	8,700	6,700	12,000	9,300	5,200	4,000
3/8 0.375	0.403	0.388	0.400	0.386	0.405	0.388	0.366	0.095	0.383	0.370	12,600	9,600	18,000	14,000	…	…
7/16 0.438	0.469	0.452	0.466	0.450	0.471	0.452	0.427	0.095	0.446	0.431	17,000	13,300	23,500	18,000	…	…
1/2 0.500	0.535	0.516	0.532	0.514	0.537	0.516	0.488	0.110	0.510	0.493	22,500	17,500	32,000	25,000	…	…
5/8 0.625	0.661	0.642	0.658	0.640	…	…	0.613	0.125	0.635	0.618	35,000[b]	…	48,000[b]	…	…	…
3/4 0.750	0.787	0.768	0.784	0.766	…	…	0.738	0.150	0.760	0.743	50,000[b]	…	70,000[b]	…	…	…

[a] Sizes 1/32 inch through 0.052 inch are not available in SAE 1070-1095 carbon steel.

[b] Sizes 5/8 inch and larger are produced from SAE 6150H alloy steel, not SAE 1070-1095 carbon steel. Practical lengths, *L*, for sizes 1/32 through 0.052 inch are 1/8 through 5/8 inch and for the 7/64-inch size, 1/4 through 1 3/4 inches. For lengths of other sizes see Table 10.

All dimensions are in inches.

Table 17. American National Standard T-Nuts *ANSI/ASME B5.1M-1985 (R2014)*

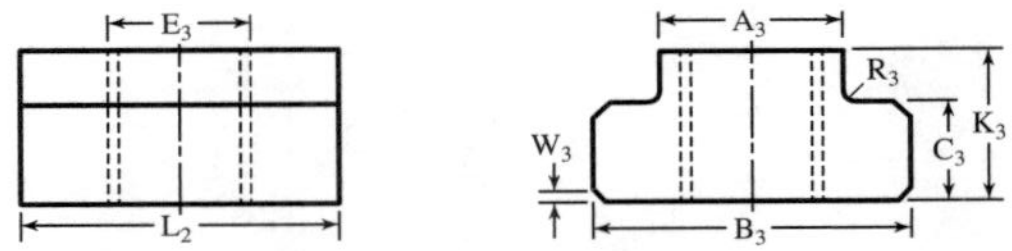

Nominal T-Bolt Size		Width of Tongue A_3				Tap for Stud E_3		Width of Nut B_3				Height of Nut C_3				Total Thickness Including Tongue[a] K_3		Length of Nut[a] L_3		Rounding of Corners			
																				R_3		W_3	
		inch		mm		inch	mm	inch		mm		inch		mm						inch	mm	inch	mm
inch	mm	max	min	max	min	UNC-3B	ISO[b]	max	min	max	min	max	min	max	min	inch	mm	inch	mm	max	max	max	max
	4		…		…		…		…	…	…			…	…		…		…		…		…
	5		…		…		…		…	…	…			…	…		…		…		…		…
0.250	6	…	…	…	…	…	…	…	…	…	…	…	…	…	…	…	…	…	…	…	…	…	…
0.312	8	0.330	0.320	8.7	8.5	0.250–20	M6	0.562	0.531	15	14	0.188	0.172	6	5.6	0.281	9	0.562	18	0.02	0.5	0.03	0.8
0.375	10	0.418	0.408	11	10.75	0.312–18	M8	0.688	0.656	18	17	0.250	0.234	7	6.6	0.375	10.5	0.688	20	0.02	0.5	0.03	0.8
0.500	12	0.543	0.533	13.5	13.25	0.375–1	6M10	0.875	0.844	22	21	0.312	0.297	8	7.6	0.531	12	0.875	23	0.02	0.5	0.06	1.5
0.625	16	0.668	0.658	17.25	17	0.500–13	M12	1.125	1.094	28	27	0.406	0.391	10	9.6	0.625	15	1.125	27	0.03	0.8	0.06	1.5
0.750	20	0.783	0.773	20.5	20.25	0.625–11	M16	1.312	1.281	34	33	0.531	0.500	14	13.2	0.781	21	1.312	35	0.03	0.8	0.06	1.5
1.000	24	1.033	1.018	26.5	26	0.750–10	M20	1.688	1.656	43	42	0.688	0.656	18	17.2	1.000	27	1.688	46	0.03	0.8	0.06	1.5
1.250	30	1.273	1.258	33	32.5	1.000–8	M24	2.062	2.031	53	52	0.938	0.906	23	22.2	1.312	34	2.062	53	0.03	0.8	0.06	1.5
1.500	36	1.523	1.508	39.25	38.75	1.250–7	M30	2.500	2.469	64	63	1.188	1.156	28	27.2	1.625	42	2.500	65	0.03	0.8	0.06	1.5
	42			46.75	46.25		M36			75	74			32	30.5		48		75		1		2
	48			52.5	51.75		M42			85	84			36	34.5		54		85		1		2

[a] No tolerances are given for "Total Thickness" or "Nut Length" as they need not be held to close limits.

[b] Metric tapped thread grade and tolerance position is 5H.

Table 18. Wrench Openings for Nuts *ANSI/ASME B18.2.2-2015, Appendix*

Max.[a] Width Across Flats of Nut	Wrench Opening[b] Min.	Wrench Opening[b] Max.	Max.[a] Width Across Flats of Nut	Wrench Opening[b] Min.	Wrench Opening[b] Max.	Max.[a] Width Across Flats of Nut	Wrench Opening[b] Min.	Wrench Opening[b] Max.
$\frac{5}{32}$	0.158	0.163	$1\frac{1}{4}$	1.257	1.267	$2\frac{15}{16}$	2.954	2.973
$\frac{3}{16}$	0.190	0.195	$1\frac{5}{16}$	1.320	1.331	*3*	*3.016*	*3.035*
$\frac{7}{32}$	0.220	0.225	*$1\frac{3}{8}$*	*1.383*	*1.394*	$3\frac{1}{8}$	3.142	3.162
$\frac{1}{4}$	0.252	0.257	$1\frac{7}{16}$	1.446	1.457	*$3\frac{3}{8}$*	*3.393*	*3.414*
$\frac{9}{32}$	0.283	0.288	$1\frac{1}{2}$	1.508	1.520	$3\frac{1}{2}$	3.518	3.540
$\frac{5}{16}$	0.316	0.322	$1\frac{5}{8}$	1.634	1.646	*$3\frac{3}{4}$*	*3.770*	*3.793*
$\frac{11}{32}$	0.347	0.353	$1\frac{11}{16}$	1.696	1.708	$3\frac{7}{8}$	3.895	3.918
$\frac{3}{8}$	0.378	0.384	$1\frac{13}{16}$	1.822	1.835	*$4\frac{1}{8}$*	*4.147*	*4.172*
$\frac{7}{16}$	0.440	0.446	$1\frac{7}{8}$	1.885	1.898	$4\frac{1}{4}$	4.272	4.297
$\frac{1}{2}$	0.504	0.510	2	2.011	2.025	$4\frac{1}{2}$	4.524	4.550
$\frac{9}{16}$	0.566	0.573	$2\frac{1}{16}$	2.074	2.088	$4\frac{5}{8}$	4.649	4.676
$\frac{5}{8}$	0.629	0.636	$2\frac{3}{16}$	2.200	2.215	*$4\frac{7}{8}$*	*4.900*	*4.928*
$\frac{11}{16}$	0.692	0.699	$2\frac{1}{4}$	2.262	2.277	5	5.026	5.055
$\frac{3}{4}$	0.755	0.763	$2\frac{3}{8}$	2.388	2.404	*$5\frac{1}{4}$*	*5.277*	*5.307*
$\frac{13}{16}$	0.818	0.826	*$2\frac{7}{16}$*	*2.450*	*2.466*	$5\frac{3}{8}$	5.403	5.434
$\frac{7}{8}$	0.880	0.888	$2\frac{9}{16}$	2.576	2.593	*$5\frac{5}{8}$*	*5.654*	*5.686*
$\frac{15}{16}$	0.944	0.953	*$2\frac{5}{8}$*	*2.639*	*2.656*	$5\frac{3}{4}$	5.780	5.813
1	1.006	1.015	$2\frac{3}{4}$	2.766	2.783	6	*6.031*	6.157
$1\frac{1}{16}$	1.068	1.077	*$2\frac{13}{16}$*	*2.827*	*2.845*	$6\frac{1}{8}$	6.065	6.192
$1\frac{1}{8}$	1.132	1.142						

[a] Wrenches are marked with the "Nominal Size of Wrench," which is equal to the basic or maximum width across flats of the corresponding nut. Minimum wrench opening is (1.005W + 0.001). Tolerance on wrench opening is (0.005W + 0.004) from minimum, where W equals nominal size of wrench.

[b] Openings for $\frac{5}{32}$ to $\frac{3}{8}$ widths from old ASA B18.2-1960 and italic values are from former ANSI B18.2.2-1972.

All dimensions given in inches.

Table 19. Clearances for Open-End Engineers Wrenches (15°)

Nominal Wrench Size		*A* Min. (in.)	*B*[a] Max. (in.)	*C* Min. (in.)	*D* Min. (in.)	*E* Min. (in.)	*F*[b] Max. (in.)	*G* Ref. (in.)	*H*[c] Max. (in.)	*J* Min.[d] in.-lbf
$\frac{5}{32}$	0.156	0.220	0.250	0.390	0.160	0.250	0.200	0.030	0.094	35
$\frac{3}{16}$	0.188	0.250	0.280	0.430	0.190	0.270	0.230	0.030	0.172	45
$\frac{1}{4}$	0.250	0.280	0.340	0.530	0.270	0.310	0.310	0.030	0.172	67
$\frac{5}{16}$	0.313	0.380	0.470	0.660	0.280	0.390	0.390	0.050	0.203	138
$\frac{11}{32}$	0.344	0.420	0.500	0.750	0.340	0.450	0.450	0.050	0.203	193
$\frac{3}{8}$	0.375	0.420	0.500	0.780	0.360	0.450	0.520	0.050	0.219	275
$\frac{7}{16}$	0.438	0.470	0.590	0.890	0.420	0.520	0.640	0.050	0.250	413
$\frac{1}{2}$	0.500	0.520	0.640	1.000	0.470	0.580	0.660	0.050	0.266	550
$\frac{9}{16}$	0.563	0.590	0.770	1.130	0.520	0.660	0.700	0.050	0.297	770
$\frac{5}{8}$	0.625	0.640	0.830	1.230	0.550	0.700	0.700	0.050	0.344	1100
$\frac{11}{16}$	0.688	0.770	0.920	1.470	0.660	0.880	0.800	0.060	0.375	1375
$\frac{3}{4}$	0.750	0.770	0.920	1.510	0.670	0.880	0.800	0.060	0.375	1650
$\frac{13}{16}$	0.813	0.910	1.120	1.660	0.720	0.970	0.860	0.060	0.406	2200
$\frac{7}{8}$	0.875	0.970	1.150	1.810	0.800	1.060	0.910	0.060	0.438	2475
$\frac{15}{16}$	0.938	0.970	1.150	1.850	0.810	1.060	0.950	0.060	0.438	3025
1	1.000	1.050	1.230	2.000	0.880	1.160	1.060	0.060	0.500	3575
$1\frac{1}{16}$	1.063	1.090	1.250	2.100	0.970	1.200	1.200	0.080	0.500	3850
$1\frac{1}{8}$	1.125	1.140	1.370	2.210	1.000	1.270	1.230	0.080	0.500	4400
$1\frac{1}{4}$	1.250	1.270	1.420	2.440	1.080	1.390	1.310	0.080	0.562	5775
$1\frac{5}{16}$	1.313	1.390	1.690	2.630	1.170	1.520	1.340	0.080	0.562	6600
$1\frac{7}{16}$	1.438	1.470	1.720	2.800	1.250	1.590	1.340	0.090	0.641	8250
$1\frac{1}{2}$	1.500	1.470	1.720	2.840	1.270	1.590	1.450	0.090	0.641	8500
$1\frac{5}{8}$	1.625	1.560	1.880	3.100	1.380	1.750	1.560	0.090	0.641	9000

[a] *B* = arc radius created by the swing of the wrench.

[b] *F* = inside arc radius of part.

[c] *H* = thickness of wrench head. (Dimension line not shown.)

[d] *J* = torque that wrench will withstand in inch-pounds. Values updated from ANSI/ASME B107.100-2010, Wrenches.

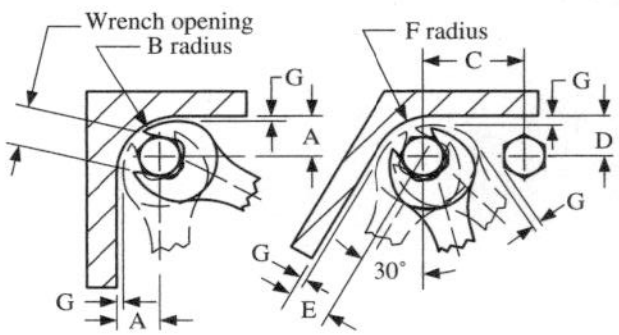

Fig. 1. Clearances for Open-End Engineers Wrench (See Table 19)

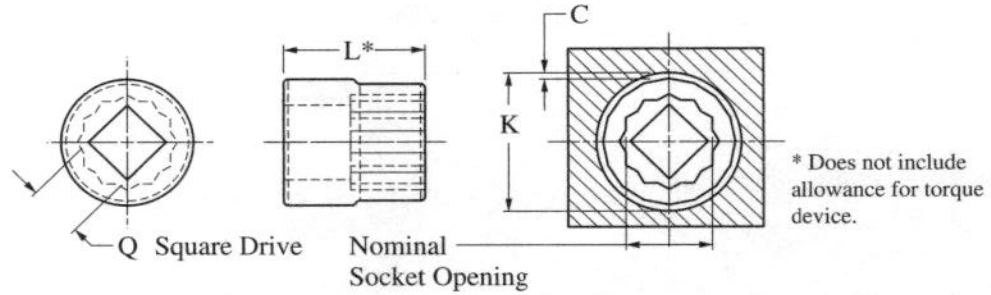

Fig. 2. Clearance Dimensions for Single and Double Hexagon Socket Wrenches (See Table 20a and Table 20b)

Table 20a. Clearances for Single and Double Hexagon Socket Wrenches, Regular Length—Inch Series

See Fig. 2, page 113 for Dimensions

Nominal Opening		Radial Clear-ance C Ref.[a]	1/4 in. Square Drive, Q					3/8 in. Square Drive, Q					1/2 in. Square Drive, Q					3/4 in. Square Drive, Q				
			Length L Max.	Nut End Dia. D1 Max.	Drive End Dia. D2 Max.	C-Bore Dia. K Min.	Proof Torque P (lbf-in.) Min.	Length L Max.	Nut End Dia. D1 Max.	Drive End Dia. D2 Max.	Counter-Bore Dia. K Min.	Proof Torque P (lbf-in.) Min.	Length L Max.	Nut End Dia. D1 Max.	Drive End Dia. D2 Max.	Counter-Bore Dia. K Min.	Proof Torque P (lbf-in.) Min.	Length L Max.	Nut End Dia. D1 Max.	Drive End Dia. D2 Max.	Counter-Bore Dia. K Min.	Proof Torque P (lbf-in.) Min.
$\frac{1}{8}$	(0.125)	0.030	1.010	0.250	0.510	0.540	35	…	…	…	…	…	…	…	…	…	…	…	…	…	…	…
$\frac{5}{32}$	(0.156)	0.030	1.010	0.281	0.510	0.540	60	…	…	…	…	…	…	…	…	…	…	…	…	…	…	…
$\frac{3}{16}$	(0.188)	0.030	1.010	0.338	0.510	0.540	95	…	…	…	…	…	…	…	…	…	…	…	…	…	…	…
$\frac{7}{32}$	(0.219)	0.030	1.010	0.382	0.510	0.540	135	…	…	…	…	…	…	…	…	…	…	…	…	…	…	…
$\frac{1}{4}$	(0.250)	0.030	1.010	0.425	0.510	0.540	190	1.260	0.472	0.690	0.720	270	…	…	…	…	…	…	…	…	…	…
$\frac{9}{32}$	(0.281)	0.030	1.010	0.457	0.510	0.540	250	1.260	0.496	0.690	0.720	350	…	…	…	…	…	…	…	…	…	…
$\frac{5}{16}$	(0.313)	0.030	1.010	0.510	0.510	0.540	320	1.260	0.521	0.690	0.720	440	…	…	…	…	…	…	…	…	…	…
$\frac{11}{32}$	(0.344)	0.030	1.010	0.547	0.547	0.577	400	1.260	0.567	0.690	0.720	550	…	…	…	…	…	…	…	…	…	…
$\frac{3}{8}$	(0.375)	0.030	1.010	0.597	0.597	0.627	500	1.260	0.613	0.690	0.720	660	1.525	0.655	0.940	0.970	1100	…	…	…	…	…
$\frac{7}{16}$	(0.438)	0.030	1.010	0.683	0.683	0.713	500	1.260	0.683	0.690	0.720	930	1.525	0.730	0.940	0.970	1500	…	…	…	…	…
$\frac{1}{2}$	(0.500)	0.030	1.010	0.697	0.697	0.727	500	1.260	0.751	0.880	0.910	1240	1.525	0.775	0.940	0.970	2000	…	…	…	…	…
$\frac{9}{16}$	(0.563)	0.030	1.010	0.778	0.778	0.808	500	1.260	0.814	0.880	0.910	1610	1.572	0.845	0.940	0.970	2600	…	…	…	…	…
$\frac{5}{8}$	(0.625)	0.030	…	…	…	…	…	1.260	0.890	0.890	0.920	200	1.572	0.942	0.970	1.000	3300	…	…	…	…	…

Table 20a. *(Continued)* **Clearances for Single and Double Hexagon Socket Wrenches, Regular Length — Inch Series**

See Fig. 2, **page** 113 **for Dimensions**

Nominal Opening	Radial Clearance C Ref.[a]	1/4 in. Square Drive, Q					3/8 in. Square Drive, Q					1/2 in. Square Drive, Q					3/4 in. Square Drive, Q				
		Length L Max.	Nut End Dia. D1 Max.	Drive End Dia. D2 Max.	C-Bore Dia. K Min.	Proof Torque P (lbf-in.) Min.	Length L Max.	Nut End Dia. D1 Max.	Drive End Dia. D2 Max.	Counter-Bore Dia. K Min.	Proof Torque P (lbf-in.) Min.	Length L Max.	Nut End Dia. D1 Max.	Drive End Dia. D2 Max.	Counter-Bore Dia. K Min.	Proof Torque P (lbf-in.) Min.	Length L Max.	Nut End Dia. D1 Max.	Drive End Dia. D2 Max.	Counter-Bore Dia. K Min.	Proof Torque P (lbf-in.) Min.
11/16 (0.688)	0.030	…	…	…	…	…	1.260	0.968	0.968	0.998	2200	1.572	1.010	1.010	1.040	4100	…	…	…	…	…
3/4 (0.750)	0.030	…	…	…	…	…	1.260	1.110	1.110	1.140	2200	1.572	1.080	1.080	1.110	5000	2.000	1.285	1.450	1.480	6000
13/16 (0.813)	0.030	…	…	…	…	…	1.406	1.141	1.141	1.171	2200	1.635	1.145	1.145	1.175	5000	2.000	1.300	1.450	1.480	6800
7/8 (0.875)	0.030	…	…	…	…	…	1.406	1.250	1.250	1.280	2200	1.760	1.218	1.218	1.248	5000	2.010	1.385	1.575	1.605	7700
15/16 (0.938)	0.030	…	…	…	…	…	1.650	1.310	1.310	1.340	2200	1.760	1.300	1.300	1.330	5000	2.010	1.450	1.575	1.605	8700
1 (1.000)	0.030	…	…	…	…	…	1.650	1.380	1.380	1.410	2200	1.760	1.375	1.375	1.405	5000	2.072	1.520	1.575	1.605	9700
1 1/16 (1.063)	0.030	…	…	…	…	…	…	…	…	…	…	1.853	1.480	1.480	1.510	5000	2.200	1.595	1.595	1.625	10,800
1 1/8 (1.125)	0.030	…	…	…	…	…	…	…	…	…	…	1.947	1.540	1.540	1.570	5000	2.322	1.600	1.680	1.710	11,900
1 3/16 (1.188)	0.030	…	…	…	…	…	…	…	…	…	…	1.947	1.675	1.675	1.705	5000	2.322	1.735	1.735	1.765	13,000
1 1/4 (1.250)	0.030	…	…	…	…	…	…	…	…	…	…	2.015	1.750	1.750	1.780	5000	2.385	1.870	1.870	1.900	14,200
1 5/16 (1.313)	0.030	…	…	…	…	…	…	…	…	…	…	2.015	1.820	1.820	1.850	5000	2.510	1.920	1.920	1.950	15,400
1 3/8 (1.375)	0.030	…	…	…	…	…	…	…	…	…	…	2.155	1.885	1.885	1.915	5000	2.635	1.980	1.980	2.010	16,700
1 7/16 (1.438)	0.030	…	…	…	…	…	…	…	…	…	…	2.295	1.955	1.955	1.985	5000	2.635	2.075	2.075	2.105	18,000
1 1/2 (1.500)	0.030	…	…	…	…	…	…	…	…	…	…	2.295	2.025	2.025	2.055	5000	2.635	2.145	2.145	2.175	18,000
1 5/8 (1.625)	0.030	…	…	…	…	…	…	…	…	…	…	…	…	…	…	…	2.760	2.260	2.260	2.290	18,000
1 3/4 (1.750)	0.030	…	…	…	…	…	…	…	…	…	…	…	…	…	…	…	2.760	2.325	2.325	2.355	18,000
1 13/16 (1.813)	0.030	…	…	…	…	…	…	…	…	…	…	…	…	…	…	…	3.135	2.400	2.400	2.430	18,000
1 7/8 (1.875)	0.030	…	…	…	…	…	…	…	…	…	…	…	…	…	…	…	3.135	2.510	2.510	2.540	18,000
2 (2.000)	0.030	…	…	…	…	…	…	…	…	…	…	…	…	…	…	…	3.260	2.575	2.575	2.605	18,000
2 1/16 (2.063)	0.030	…	…	…	…	…	…	…	…	…	…	…	…	…	…	…	3.385	2.695	2.695	2.725	18,000
2 1/8 (2.125)	0.030	…	…	…	…	…	…	…	…	…	…	…	…	…	…	…	3.510	2.885	2.885	2.915	18,000
2 3/16 (2.188)	0.030	…	…	…	…	…	…	…	…	…	…	…	…	…	…	…	3.697	3.025	3.025	3.055	18,000
2 1/4 (2.250)	0.030	…	…	…	…	…	…	…	…	…	…	…	…	…	…	…	3.697	3.075	3.075	3.105	18,000

[a] From the SAE Aeronautical Drafting Manual

All dimensions are in inches. For details not shown and additional socket sizes, see ANSI/ASME B107.1-2002, Socket Wrenches, Hand (Inch Series).

Table 20b. Clearances for Single and Double Hexagon Socket, Regular Length — Metric Series

(See Fig. 2, page 113 for Dimensions)

		6.3 mm Square Drive Q					10 mm Square Drive Q					12.5 mm Square Drive Q					20 mm Square Drive Q				
Nominal Opening	Radial Clearance C Ref[a]	Length L Max.	Nut End Dia. D1 Max.	Drive End Dia D2 Max.	C-Bore Dia. K Min.	Proof Torque P (N-m) Min.	Length L Max.	Nut End Dia. D1 Max.	Drive End Dia. D2 Max.	Counter-Bore Dia. K Min.	Proof Torque P (N-m) Min.	Length L Max.	Nut End Dia. D1 Max.	Drive End Dia. D2 Max.	Counter-Bore Dia. K Min.	Proof Torque P (N-m) Min.	Length L Max.	Nut End Dia. D1 Max.	Drive End Dia. D2 Max.	Counter-Bore Dia. K Min.	Proof Torque P (N-m) Min.
3.2	0.762	26	6.10	12.95	14.47	7	…	…	…	…	…	…	…	…	…	…	…	…	…	…	…
4	0.762	26	7.10	12.95	14.47	8	…	…	…	…	…	…	…	…	…	…	…	…	…	…	…
4.5	0.762	26	7.60	12.95	14.47	9	…	…	…	…	…	…	…	…	…	…	…	…	…	…	…
5	0.762	26	8.15	12.95	14.47	10	…	…	…	…	…	…	…	…	…	…	…	…	…	…	…
5.5	0.762	26	8.90	12.95	14.47	14	32	10.10	17.60	19.124	270	…	…	…	…	…	…	…	…	…	…
6	0.762	26	9.90	12.95	14.47	16	32	10.10	17.60	19.124	350	…	…	…	…	…	…	…	…	…	…
6.3	0.762	26	9.90	12.95	14.47	21	32	10.10	17.60	19.124	440	…	…	…	…	…	…	…	…	…	…
7	0.762	26	10.90	12.95	14.47	27	32	11.05	17.60	19.124	550	…	…	…	…	…	…	…	…	…	…
8	0.762	26	12.20	12.95	14.47	38	32	12.20	17.60	19.124	660	39	14.00	23.87	25.39	80	…	…	…	…	…
9	0.762	26	13.45	13.45	14.97	49	32	13.60	17.60	19.124	930	39	15.10	23.87	25.39	110	…	…	…	…	…
10	0.762	26	14.75	14.75	16.27	63	32	15.00	17.60	19.124	1240	39	16.80	23.87	25.39	153	…	…	…	…	…
11	0.762	26	16.00	16.00	17.52	68	32	16.75	17.60	19.124	1610	39	18.20	23.87	25.39	170	…	…	…	…	…
12	0.762	26	17.30	17.30	18.82	68	32	17.80	22.40	23.924	200	39	18.70	23.87	25.39	203	…	…	…	…	…
13	0.762	26	18.55	18.55	20.07	68	32	18.80	22.40	23.924	2200	39	20.25	23.87	25.39	249	…	…	…	…	…
14	0.762	26	19.80	19.80	21.32	68	32	20.00	22.40	23.924	2200	39	21.80	23.87	25.39	282	…	…	…	…	…
15	0.762	26	21.50	21.50	23.02	68	32	22.40	22.40	23.924	2200	40	22.40	23.87	25.39	339	…	…	…	…	…
16	0.762	26	22.00	22.00	23.52	68	32	22.50	22.50	24.024	2200	40	23.87	23.87	25.39	407	…	…	…	…	…
17	0.762	…	…	…	…	…	32	23.80	23.80	25.324	2200	40	24.75	24.75	26.27	475	…	…	…	…	…
18	0.762	…	…	…	…	…	32	24.60	24.60	26.124	2200	40	26.14	26.14	27.66	542	…	…	…	…	…
19	0.762	…	…	…	…	…	32	25.70	25.70	27.224	…	40	27.20	27.20	28.72	575	51	30.50	33.00	33.76	780
20	0.762	…	…	…	…	…	32	27.76	27.76	29.284	…	42	27.95	27.95	29.47	570	…	…	…	…	…
21	0.762	…	…	…	…	…	34	28.80	28.80	30.324	…	42	28.95	28.95	30.47	570	51	33.00	33.00	33.76	930
22	0.762	…	…	…	…	…	34	30.00	30.00	31.524	…	45	30.20	30.20	31.72	570	51	35.05	38.10	38.86	972
23	0.762	…	…	…	…	…	35	31.30	31.30	32.824	…	45	31.25	31.25	32.77	570	51	36.10	39.10	39.86	1015

Table 20b. *(Continued)* **Clearances for Single and Double Hexagon Socket, Regular Length - Metric Series**

(See Fig. 2, **page** 113 **for Dimensions)**

Nominal Opening	Radial Clearance C Ref[a]	6.3 mm Square Drive Q					10 mm Square Drive Q					12.5 mm Square Drive Q					20 mm Square Drive Q				
		Length L Max.	Nut End Dia. D1 Max.	Drive End Dia D2 Max.	C-Bore Dia. K Min.	Proof Torque P (N-m) Min.	Length L Max.	Nut End Dia. D1 Max.	Drive End Dia. D2 Max.	Counter-Bore Dia. K Min.	Proof Torque P (N-m) Min.	Length L Max.	Nut End Dia. D1 Max.	Drive End Dia. D2 Max.	Counter-Bore Dia. K Min.	Proof Torque P (N-m) Min.	Length L Max.	Nut End Dia. D1 Max.	Drive End Dia. D2 Max.	Counter-Bore Dia. K Min.	Proof Torque P (N-m) Min.
24	0.762	…	…	…	…	…	36	32.50	32.50	34.024	…	45	32.15	32.15	33.67	570	51	37.00	40.00	40.76	1085
25	0.762	…	…	…	…	…	38	33.00	33.00	34.524	…	45	33.40	33.40	34.92	570	52	37.85	40.00	40.76	1160
26	0.762	…	…	…	…	…	38	35.00	35.00	36.524	…	48	35.05	35.05	36.57	570	53	38.85	40.00	40.76	1240
27	0.762	…	…	…	…	…	…	…	…	…	…	48	36.75	36.75	38.27	570	54	41.00	41.00	41.76	1330
28	0.762	…	…	…	…	…	…	…	…	…	…	50	37.80	37.80	39.32	570	57	41.00	41.00	41.76	1420
29	0.762	…	…	…	…	…	…	…	…	…	…	50	39.50	39.50	41.02	570	59	42.10	42.10	42.86	1520
30	0.762	…	…	…	…	…	…	…	…	…	…	50	42.40	42.40	43.92	570	59	43.00	43.00	43.76	1640
31	0.762	…	…	…	…	…	…	…	…	…	…	50	43.20	43.20	44.72	570	60	45.10	45.10	45.86	1730
32	0.762	…	…	…	…	…	…	…	…	…	…	51	44.05	44.05	45.57	570	60	47.05	47.05	47.81	1820
34	0.762	…	…	…	…	…	…	…	…	…	…	…	…	…	…	…	64	49.00	49.00	49.76	2000
35	0.762	…	…	…	…	…	…	…	…	…	…	…	…	…	…	…	67	50.40	50.40	51.16	2030
36	0.762	…	…	…	…	…	…	…	…	…	…	…	…	…	…	…	67	51.80	51.80	52.56	2030
38	0.762	…	…	…	…	…	…	…	…	…	…	…	…	…	…	…	67	54.10	54.10	54.86	2030
40	0.762	…	…	…	…	…	…	…	…	…	…	…	…	…	…	…	70	57.65	57.65	58.41	2030
41	0.762	…	…	…	…	…	…	…	…	…	…	…	…	…	…	…	70	58.80	58.80	59.56	2030
42	0.762	…	…	…	…	…	…	…	…	…	…	…	…	…	…	…	70	58.80	58.80	59.56	2030
46	0.762	…	…	…	…	…	…	…	…	…	…	…	…	…	…	…	83	65.40	65.40	66.16	2030
50	0.762	…	…	…	…	…	…	…	…	…	…	…	…	…	…	…	89	72.15	72.15	72.91	2030
54	0.762	…	…	…	…	…	…	…	…	…	…	…	…	…	…	…	94	78.10	78.10	78.86	2030
55	0.762	…	…	…	…	…	…	…	…	…	…	…	…	…	…	…	95	79.10	79.10	79.86	2030
58	0.762	…	…	…	…	…	…	…	…	…	…	…	…	…	…	…	97	80.00	80.00	80.76	2030
60	0.762	…	…	…	…	…	…	…	…	…	…	…	…	…	…	…	100	84.45	84.45	85.21	2030

[a] Converted from inch dimensions given in the SAE Aeronautical Drafting Manual.

All dimensions are in mm. For details not shown and additional socket sizes, see ANSI/ASME B107.5M-2002, Socket Wrenches, Hand (Metric Series).

Table 21. Clearances for Box Wrenches — 12 Point Inch and Metric Series

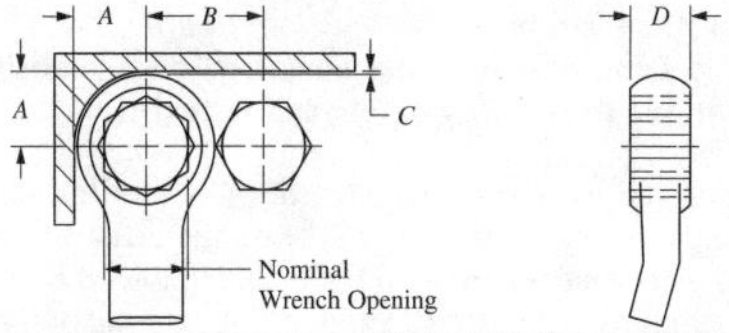

US Customary (inch)						Metric (mm)					
Nominal Wrench Opening (in.)	A Min. (in.)	B Min. (in.)	C Ref.[a] (in.)	Head Thickness D Max., (in.)	Proof Torque (lbf-in.)	Nominal Wrench Opening (mm)	A Min. (mm)	B Min. (mm)	C Ref.[b] (mm)	Head Thickness D Max., (mm)	Proof Torque (N-m)
1/8 (0.125)	0.179	0.219	0.030	0.172	60	4	4.56	6.03	0.762	4.0	12
5/32 (0.156)	0.187	0.244	0.030	0.172	90	5	5.26	7.29	0.762	4.6	17
3/16 (0.188)	0.218	0.301	0.030	0.203	150	5.5	6.66	8.97	0.762	6.0	18
7/32 (0.219)	0.233	0.325	0.030	0.234	165	6	7.11	9.69	0.762	7.4	20
1/4 (0.250)	0.269	0.378	0.030	0.295	220	7	7.91	11.05	0.762	7.7	27
9/32 (0.281)	0.280	0.407	0.030	0.280	248	8	8.26	11.98	0.762	8.2	30
5/16 (0.313)	0.316	0.461	0.030	0.330	275	9	9.46	13.76	0.762	9.0	40
11/32 (0.344)	0.336	0.499	0.030	0.335	275	10	10.16	15.04	0.762	9.0	71
3/8 (0.375)	0.362	0.543	0.030	0.344	605	11	10.71	16.15	0.762	10.0	80
7/16 (0.438)	0.395	0.612	0.030	0.391	715	12	11.46	17.47	0.762	10.0	91
1/2 (0.500)	0.442	0.694	0.030	0.394	1020	13	12.31	18.89	0.762	10.5	115
9/16 (0.563)	0.492	0.779	0.030	0.425	1500	14	12.96	20.10	0.762	11.5	158
5/8 (0.625)	0.530	0.853	0.030	0.500	2200	15	13.76	21.46	0.762	11.5	200
11/16 (0.688)	0.577	0.935	0.030	0.535	2640	16	14.26	22.53	0.762	12.1	248
3/4 (0.750)	0.618	1.012	0.030	0.594	2860	17	15.41	24.25	0.762	12.7	267
13/16 (0.813)	0.702	1.132	0.030	0.609	3300	18	15.41	24.83	0.762	12.7	304
7/8 (0.875)	0.718	1.183	0.030	0.688	3630	19	16.36	26.35	0.762	14.8	323
15/16 (0.938)	0.765	1.266	0.030	0.701	4510	20	17.21	27.77	0.762	14.8	347
1 (1.000)	0.796	1.330	0.030	0.719	5390	21	17.66	28.79	0.762	16.3	372
1 1/16 (1.063)	0.874	1.445	0.030	0.790	5940	22	18.56	30.27	0.762	16.3	408
1 1/8 (1.125)	0.892	1.498	0.030	0.860	6430	23	19.41	31.69	0.762	16.5	455
1 3/16 (1.188)	0.937	1.579	0.030	0.890	7200	24	19.81	32.65	0.762	17.8	509
1 1/4 (1.250)	0.983	1.661	0.030	0.940	7920	25	20.86	34.24	0.762	17.9	559
1 5/16 (1.313)	1.062	1.775	0.030	0.940	8400	26	12.86	26.79	0.762	18.0	608
1 3/8 (1.375)	1.087	1.836	0.030	0.940	8970	27	22.86	37.37	0.762	19.8	671
1 7/16 (1.438)	1.144	1.929	0.030	0.953	9240	28	23.41	38.49	0.762	19.8	710
1 1/2 (1.500)	1.228	2.049	0.030	1.008	10,365	29	23.41	39.06	0.762	19.8	750
1 9/16 (1.563)	1.249	2.104	0.030	1.031	11,495	30	24.51	40.73	0.762	20.0	795
1 5/8 (1.625)	1.351	2.241	0.030	1.063	12,800	31	25.06	41.85	0.762	20.5	850
1 11/16 (1.688)	1.425	2.351	0.030	1.063	13,570	32	25.66	43.03	0.762	22.0	905
1 3/4 (1.750)	1.499	2.461	0.030	1.125	14,300	33	25.91	43.84	0.762	22.3	950
1 13/16 (1.813)	1.499	2.496	0.030	1.125	15,100	34	26.76	45.26	0.762	23.2	994
1 7/8 (1.875)	1.593	2.625	0.030	1.125	15,900	36	28.81	48.47	0.762	25.1	1165
2 (2.000)	1.593	2.696	0.030	1.125	17,400	41	32.21	54.68	0.762	25.3	1579
2 1/16 (2.063)	1.687	2.825	0.030	1.234	18,200	46	34.76	60.06	0.762	25.8	2067
2 1/8 (2.125)	1.687	2.861	0.030	1.234	19,000	50	38.76	66.33	0.762	27.6	2512
2 3/16 (2.188)	1.687	2.896	0.030	1.234	19,700	…	…	…	…	…	…
2 1/4 (2.250)	1.687	2.931	0.030	1.234	20,500	…	…	…	…	…	…

[a] From SAE Aeronautical Drafting Manual

[b] Converted from SAE Aeronautical Drafting Manual. For details not shown, including material, see ANSI/ASME B107.100-2010 Wrenches

Bolts and Screws Specification

The following definitions are based on *Specification for Identification of Bolts and Screws*, ANSI/ASME B18.2.1-2012. This specification establishes a recommended procedure for determining the identity of an externally threaded fastener as a bolt or as a screw.

Bolt: A bolt is an externally threaded fastener designed for insertion through the holes in assembled parts, and is normally intended to be tightened or released by torquing a nut.

Screw: A screw is an externally threaded fastener capable of being inserted into holes in assembled parts, of mating with a preformed internal thread or forming its own thread, and of being tightened or released by torquing the head.

Primary Criteria.—1) A bolt is an externally threaded fastener that, because of head design or other feature, is prevented from being turned during assembly and can be tightened or released only by torquing a nut. *Example:* Round head bolts, track bolts, plow bolts.

2) A screw is an externally threaded fastener that has a thread form which prohibits assembly with a nut having a straight thread of multiple pitch length. *Example:* Wood screws, tapping screws.

3) A bolt is an externally threaded fastener that must be assembled with a nut to perform its intended service. *Example:* Heavy hex structural bolt.

4) A screw is an externally threaded fastener that must be torqued by its head into a tapped or preformed hole to perform its intended service. *Example:* Square head set screw.

British Unified Machine Screws and Nuts.—*Identification:* As revised by Amendment No. 1 in February 1955, this standard now requires that the above-mentioned screws and nuts that conform to this standard should have a distinguishing feature applied to identify them as Unified. All *recessed head screws* are to be identified as Unified by a groove in the form of four arcs of a circle in the upper surface of the head. All *hexagon head screws* are to be identified as Unified by: 1) a circular recess in the upper surface of the head; 2) a continuous line of circles indented on one or more of the flats of the hexagon and parallel to the screw axis; and 3) at least two contiguous circles indented on the upper surface of the head. All *machine screw* nuts of the pressed type shall be identified as Unified by means of the application of a groove indented in one face of the nut approximately midway between the major diameter of the thread and flats of the square or hexagon. *Slotted head screws* shall be identified as Unified either by a circular recess or by a circular platform or raised portion on the upper surface of the head. *Machine screw nuts* of the *precision type* shall be identified as Unified by either a groove indented on one face of the front approximately midway between the major diameter of the thread and the flats of the hexagon or a continuous line of circles indented on one or more of the flats of the hexagon and parallel to the nut axis.

Identification Markings for British Standard Unified Machine Screws

Recessed and Hexagon Head Screws

Precision Type Pressed Type

Hexagon Machine Screw Nuts

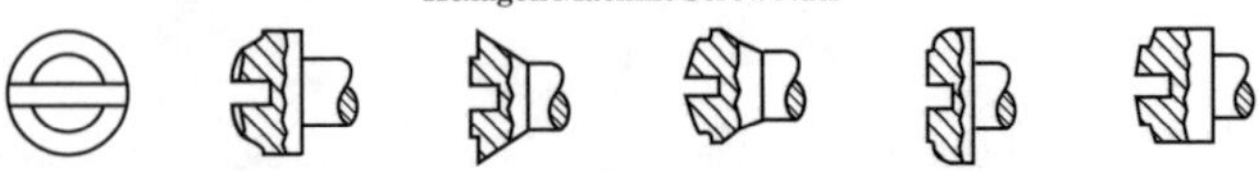

Slotted Head Screws

Table 22. British Standard Machine Screws and Nuts
BS 450:1958 (obsolescent) and BS 1981:1953

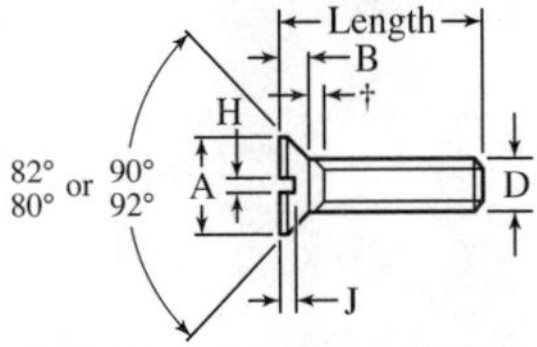

80° Countersunk head screw (Unified)
90° Countersink head screw (BSW & BSF)

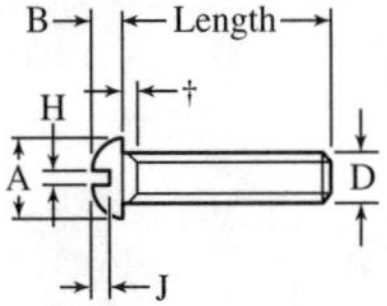

Round head screw (BSW & BSF)

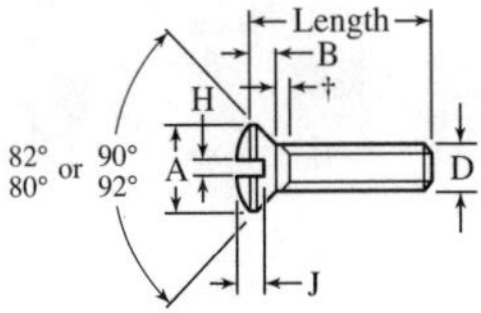

80° Raised countersunk head screw (Unified)
90° Raised countersunk head screw (BSW & BSF)

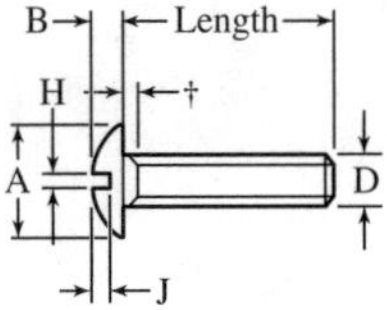

Mushroom head screw (BSW & BSF)

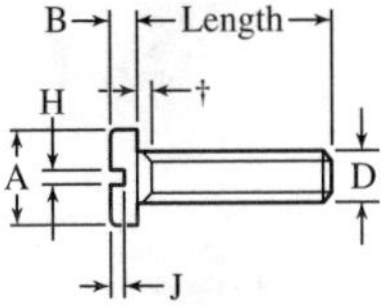

Pan head screw (Unified, BSW & BSF)

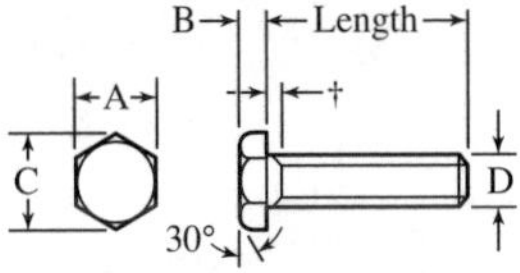

Hexagon head screw (Unified)

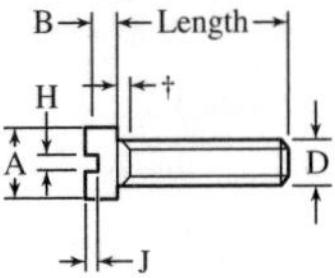

Cheese head screw (BSW & BSF)

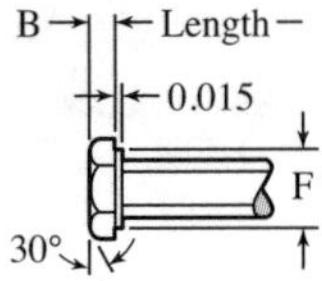

Hexagon head screw (Unified) alternate design

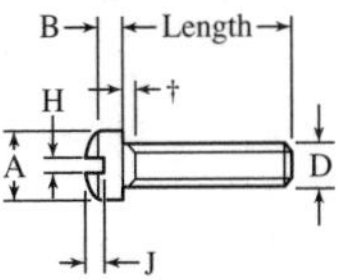

Raised cheese head screw (Unified)

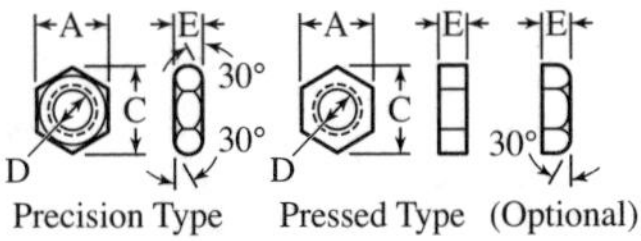

Hexagon machine screw nut (Unified)

*Countersinks to suit the screws should have a maximum angle of 80° (Unified) or 90° (BSF and BSW) with a negative tolerance.

†Unified countersunk and raised countersunk head screws 2 inches long and under are threaded right up to the head. Other Unified, BSW and BSF machine screws 2 inches long and under have an unthread shank equal to twice the pitch. All Unified, BSW and BSF machine screws longer than 2 inches have a minimum thread length of $1\frac{3}{4}$ inches.

Table 23. British Standard Unified Machine Screws and Nuts *BS 1981:1953 (R2004)*

Nom. Size of Screw	Basic Dia. *D*	Threads per Inch		Dia. of Head *A*		Depth of Head *B*		Width of Slot *H*		Depth of Slot *J*
		UNC	UNF	Max.	Min.	Max.	Min.	Max.	Min.	
80° Countersunk Head Screws[a,b]										
4	0.112	40	…	0.211	0.194	0.067	…	0.039	0.031	0.025
6	0.138	32	…	0.260	0.242	0.083	…	0.048	0.039	0.031
8	0.164	32	…	0.310	0.291	0.100	…	0.054	0.045	0.037
10	0.190	24[c]	32	0.359	0.339	0.116	…	0.060	0.050	0.044
1/4	0.250	20	28	0.473	0.450	0.153	…	0.075	0.064	0.058
5/16	0.3125	18	24	0.593	0.565	0.191	…	0.084	0.072	0.073
3/8	0.375	16	24	0.712	0.681	0.230	…	0.094	0.081	0.086
7/16	0.4375	14	20	0.753	0.719	0.223	…	0.094	0.081	0.086
1/2	0.500	13	20	0.808	0.770	0.223	…	0.106	0.091	0.086
5/8	0.625	11	18	1.041	0.996	0.298	…	0.133	0.116	0.113
3/4	0.750	10	16	1.275	1.223	0.372	…	0.149	0.131	0.141
Pan Head Screws[b]										
4	0.112	40	…	0.219	0.205	0.068	0.058	0.039	0.031	0.036
6	0.138	32	…	0.270	0.256	0.082	0.072	0.048	0.039	0.044
8	0.164	32	…	0.322	0.306	0.096	0.085	0.054	0.045	0.051
10	0.190	24[c]	32	0.373	0.357	0.110	0.099	0.060	0.050	0.059
1/4	0.250	20	28	0.492	0.473[d]	0.144	0.130	0.075	0.064	0.079
5/16	0.3125	18	24	0.615	0.594	0.178	0.162	0.084	0.072	0.101
3/8	0.375	16	24	0.740	0.716	0.212	0.195	0.094	0.081	0.122
7/16	0.4375	14	20	0.863	0.838	0.247	0.227	0.094	0.081	0.133
1/2	0.500	13	20	0.987	0.958	0.281	0.260	0.106	0.091	0.152
5/8	0.625	11	18	1.125	1.090	0.350	0.325	0.133	0.116	0.189
3/4	0.750	10	16	1.250	1.209	0.419	0.390	0.149	0.131	0.226
Raised Cheese-Head Screws[b]										
4	0.112	40	…	0.183	0.166	0.107	0.088	0.039	0.031	0.042
6	0.138	32	…	0.226	0.208	0.132	0.111	0.048	0.039	0.053
8	0.164	32	…	0.270	0.250	0.156	0.133	0.054	0.045	0.063
10	0.190	24[c]	32	0.313	0.292	0.180	0.156	0.060	0.050	0.074
1/4	0.250	20	28	0.414	0.389	0.237	0.207	0.075	0.064	0.098
5/16	0.3125	18	24	0.518	0.490	0.295	0.262	0.084	0.072	0.124
3/8	0.375	16	24	0.622	0.590	0.355	0.315	0.094	0.081	0.149
7/16	0.4375	14	20	0.625	0.589	0.368	0.321	0.094	0.081	0.153
1/2	0.500	13	20	0.750	0.710	0.412	0.362	0.106	0.091	0.171
5/8	0.625	11	18	0.875	0.827	0.521	0.461	0.133	0.116	0.217
3/4	0.750	10	16	1.000	0.945	0.612	0.542	0.149	0.131	0.254

[a] All dimensions, except *J*, given for the No. 4 to 3/8-inch sizes, incl., also apply to all the 80° Raised Countersunk Head Screws given in the Standard.
[b] Also available with recessed heads.
[c] Non-preferred.
[d] By arrangement may also be 0.468.

Nom. Size	Basic Dia. *D*	Threads per Inch		Width Across Flats *A*		Width Across Corners *C*	Head Depth *B* Nut Thick. *E*		Washer Face Dia. *F*	
		UNC	UNF	Max.	Min.	Max.	Max.	Min.	Max.	Min.
Hexagon Head Screws										
4	0.112	40	…	0.1875	0.1835	0.216	0.060	0.055	0.183	0.173
6	0.138	32	…	0.2500	0.2450	0.289	0.080	0.074	0.245	0.235
8	0.164	32	…	0.2500	0.2450	0.289	0.110	0.104	0.245	0.235
10	0.190	24[c]	32	0.3125	0.3075	0.361	0.120	0.113	0.307	0.297
Hexagon Machine Screw Nuts—Precision Type										
4	0.112	40	…	0.1875	0.1835	0.216	0.098	0.087	…	…
6	0.138	32	…	0.2500	0.2450	0.269	0.114	0.102	…	…
8	0.164	32	…	0.3125	0.3075	0.361	0.130	0.117	…	…
10	0.190	24[c]	…	0.3125	0.3075	0.361	0.130	0.117	…	…
Hexagon Machine Screw Nuts—Pressed Type										
4	0.112	40	…	0.2500	0.2410	0.289	0.087	0.077	…	…
6	0.138	32	…	0.3125	0.3020	0.361	0.114	0.102	…	…
8	0.164	32	…	0.3438	0.3320	0.397	0.130	0.117	…	…
10	0.190	24[c]	32	0.3750	0.3620	0.433	0.130	0.117	…	…
1/4	0.250	20	28	0.4375	0.4230	0.505	0.193	0.178	…	…
5/16	0.3125	18	24	0.5625	0.5450	0.649	0.225	0.208	…	…
3/8	0.375	16	24	0.6250	0.6070	0.722	0.257	0.239	…	…

All dimensions in inches. See page 119 for a pictorial representation and letter dimensions.

Table 24. British Standard Whitworth (BSW) and Fine (BSF) Machine Screws

BS 450:1958 (obsolescent)

	Nom. Size of Screw	Basic Dia. *D*	Threads per Inch BSW	Threads per Inch BSF	Dia. of Head *A* Max.	Dia. of Head *A* Min.	Depth of Head *B* Max.	Depth of Head *B* Min.	Width of Slot *H* Max.	Width of Slot *H* Min.	Depth of Slot *J*
90° Countersunk Head Screws[a,b]	1/8	0.1250	40	…	0.219	0.201	0.056	…	0.039	0.032	0.027
	3/16	0.1875	24	32[c]	0.328	0.307	0.084	…	0.050	0.042	0.041
	7/32	0.2188	…	28[c]	0.383	0.360	0.098	…	0.055	0.046	0.048
	1/4	0.2500	20	26	0.438	0.412	0.113	…	0.061	0.051	0.055
	5/16	0.3125	18	22	0.547	0.518	0.141	…	0.071	0.061	0.069
	3/8	0.3750	16	20	0.656	0.624	0.169	…	0.082	0.072	0.083
	7/16	0.4375	14	18	0.766	0.729	0.197	…	0.093	0.082	0.097
	1/2	0.5000	12	16	0.875	0.835	0.225	…	0.104	0.092	0.111
	9/16	0.5625	12[c]	16[c]	0.984	0.941	0.253	…	0.115	0.103	0.125
	5/8	0.6250	11	14	1.094	1.046	0.281	…	0.126	0.113	0.138
	3/4	0.7500	10	12	1.312	1.257	0.338	…	0.148	0.134	0.166
Round Head Screws[b]	1/8	0.1250	40	…	0.219	0.206	0.087	0.082	0.039	0.032	0.048
	3/16	0.1875	24	32[c]	0.328	0.312[d]	0.131	0.124	0.050	0.042	0.072
	7/32	0.2188	…	28[c]	0.383	0.365	0.153	0.145	0.055	0.046	0.084
	1/4	0.2500	20	26	0.438	0.417	0.175	0.165	0.061	0.051	0.096
	5/16	0.3125	18	22	0.547	0.524	0.219	0.207	0.071	0.061	0.120
	3/8	0.3750	16	20	0.656	0.629	0.262	0.249	0.082	0.072	0.144
	7/16	0.4375	14	18	0.766	0.735	0.306	0.291	0.093	0.082	0.168
	1/2	0.5000	12	16	0.875	0.840	0.350	0.333	0.104	0.092	0.192
	9/16	0.5625	12[c]	16[c]	0.984	0.946	0.394	0.375	0.115	0.103	0.217
	5/8	0.6250	11	14	1.094	1.051	0.437	0.417	0.126	0.113	0.240
	3/4	0.7500	10	12	1.312	1.262	0.525	0.500	0.148	0.134	0.288
Pan Head Screws[b]	1/8	0.1250	40	…	0.245	0.231	0.075	0.065	0.039	0.032	0.040
	3/16	0.1875	24	32[c]	0.373	0.375	0.110	0.099	0.050	0.042	0.061
	7/32	0.2188	…	28[c]	0.425	0.407	0.125	0.112	0.055	0.046	0.069
	1/4	0.2500	20	26	0.492	0.473[e]	0.144	0.130	0.061	0.051	0.078
	5/16	0.3125	18	22	0.615	0.594	0.178	0.162	0.071	0.061	0.095
	3/8	0.3750	16	20	0.740	0.716	0.212	0.195	0.082	0.072	0.112
	7/16	0.4375	14	18	0.863	0.838	0.247	0.227	0.093	0.082	0.129
	1/2	0.5000	12	16	0.987	0.958	0.281	0.260	0.104	0.092	0.145
	9/16	0.5625	12[c]	16[c]	1.031	0.999	0.315	0.293	0.115	0.103	0.162
	5/8	0.6250	11	14	1.125	1.090	0.350	0.325	0.126	0.113	0.179
	3/4	0.7500	10	12	1.250	1.209	0.419	0.390	0.148	0.134	0.213
Cheese Head Screws[b]	1/8	0.1250	40	…	0.188	0.180	0.087	0.082	0.039	0.032	0.039
	3/16	0.1875	24	32[c]	0.281	0.270	0.131	0.124	0.050	0.042	0.059
	7/32	0.2188	…	28[c]	0.328	0.315	0.153	0.145	0.055	0.046	0.069
	1/4	0.2500	20	26	0.375	0.360	0.175	0.165	0.061	0.051	0.079
	5/16	0.3125	18	22	0.469	0.450	0.219	0.207	0.071	0.061	0.098
	3/8	0.3750	16	20	0.562	0.540	0.262	0.249	0.082	0.072	0.118
	7/16	0.4375	14	18	0.656	0.630	0.306	0.291	0.093	0.082	0.138
	1/2	0.5000	12	16	0.750	0.720	0.350	0.333	0.104	0.092	0.157
	9/16	0.5625	12[c]	16[c]	0.844	0.810	0.394	0.375	0.115	0.103	0.177
	5/8	0.6250	11	14	0.938	0.900	0.437	0.417	0.126	0.113	0.197
	3/4	0.7500	10	12	1.125	1.080	0.525	0.500	0.148	0.134	0.236
Mushroom Head Screws[b]	1/8	0.1250	40	…	0.289	0.272	0.078	0.066	0.043	0.035	0.040
	3/16	0.1875	24	32[c]	0.448	0.425	0.118	0.103	0.060	0.050	0.061
	1/4	0.2500	20	26	0.573	0.546	0.150	0.133	0.075	0.064	0.079
	5/16	0.3125	18	22	0.698	0.666	0.183	0.162	0.084	0.072	0.096
	3/8	0.3750	16	20	0.823	0.787	0.215	0.191	0.094	0.081	0.112

[a] All dimensions, except *J*, given for the 1/8- through 3/8-inch sizes also apply to all the 90° Raised Countersunk Head Screw dimensions given in the Standard.

[b] These screws are also available with recessed heads; dimensions of recess are not given here but may be found in the Standard.

[c] Non-preferred size; avoid use whenever possible.

[d] By arrangement may also be 0.309.

[e] By arrangement may also be 0.468.

All dimensions in inches.

See diagram on page 119 for a pictorial representation of screws and letter dimensions.

CUTTING FLUIDS

Cutting Fluids Recommended for Machining Operations.—*Soluble Oils:* Types of oils paste compounds that form emulsions when mixed with water: Soluble oils are used extensively in machining both ferrous and nonferrous metals when the cooling quality is paramount and the chip-bearing pressure is not excessive. Care should be taken in selecting the proper soluble oil for precision grinding operations. Grinding coolants should be free from fatty materials that tend to load the wheel, thus affecting the finish on the machined part. Soluble coolants should contain rust preventive constituents to prevent corrosion.

Mineral Oils: This group includes all types of oils extracted from petroleum such as paraffin oil, mineral seal oil, and kerosene. Mineral oils are often blended with base stocks, but they are generally used in the original form for light machining operations on both free-machining steels and nonferrous metals. The coolants in this class should be of a type that has a relatively high flash point. Care should be taken to see that they are nontoxic, so that they will not be injurious to the operator. The heavier mineral oils (paraffin oils) usually have a viscosity of about 100 seconds at 100 degrees F. Mineral seal oil and kerosene have a viscosity of 35 to 60 seconds at 100 degrees F.

Cutting Fluids Recommended for Turning and Milling Operations

Material to be Cut	Turning	Milling
Aluminum[a]	Mineral Oil with 10% Fat (or) Soluble Oil	Soluble Oil (96% Water) (or) Mineral Seal Oil (or) Mineral Oil
Alloy Steels[b]	25% Sulfur-Based Oil[b] with 75% Mineral Oil	10% Lard Oil with 90% Mineral Oil
Brass	Mineral Oil with 10% Fat	Soluble Oil (96% Water)
Tool Steels and Low-Carbon Steels	25% Lard Oil with 75% Mineral Oil	Soluble Oil
Copper	Soluble Oil	Soluble Oil
Monel Metal	Soluble Oil	Soluble Oil
Cast Iron[c]	Dry	Dry
Malleable Iron	Soluble Oil	Soluble Oil
Bronze	Soluble Oil	Soluble Oil
Magnesium[d]	10% Lard Oil with 90% Mineral Oil	Mineral Seal Oil

[a] In machining aluminum, several varieties of coolants may be used. For rough machining, where the stock removal is sufficient to produce heat, water soluble mixtures can be used with good results to dissipate the heat. Other oils that may be recommended are straight mineral seal oil; a 50–50 mixture of mineral seal oil and kerosene; a mixture of 10 percent lard oil with 90 percent kerosene; and a 100-second mineral oil cut back with mineral seal oil or kerosene.

[b] The sulfur-based oil referred to contains $4\frac{1}{2}$ percent sulfur compound. Base oils are usually dark in color. As a rule, they contain sulfur compounds resulting from a thermal or catalytic refinery process. When so processed, they are more suitable for industrial coolants than when they have had such compounds as flowers of sulfur added by hand. The adding of sulfur compounds by hand to the coolant reservoir is of temporary value only, and the non-uniformity of the solution may affect the machining operation.

[c] A soluble oil or low-viscosity mineral oil may be used in machining cast iron to prevent excessive metal dust.

[d] When a cutting fluid is needed for machining magnesium, low or nonacid mineral seal or lard oils are recommended. Coolants containing water should not be used because of the fire danger when magnesium chips react with water, forming hydrogen gas.

Cutting Fluids Recommended for Drilling and Tapping Operations

Material to be Cut	Drilling	Tapping
Aluminum[a]	Soluble Oil (75 to 90% Water) (Or) 10% Lard Oil with 90% Mineral Oil	Lard Oil (or) Sperm Oil (or) Wool Grease (or) 25% Sulfur-Based Oil[b] Mixed with Mineral Oil
Alloy Steels[b]	Soluble Oil	30% Lard Oil with 70% Mineral Oil
Brass	Soluble Oil (75 to 90% Water) (Or) 30% Lard Oil with 70% Mineral Oil	10 to 20% Lard Oil with Mineral Oil
Tool Steels and Low-Carbon Steels	Soluble Oil	25 to 40% Lard Oil with Mineral Oil (or) 25% Sulfur-Based Oil[b] with 75% Mineral Oil
Copper	Soluble Oil	Soluble Oil
Monel Metal	Soluble Oil	25 to 40% Lard Oil Mixed with Mineral Oil (or) Sulfur-Based Oil[b] Mixed with Mineral Oil
Cast Iron[c]	Dry	Dry (or) 25% Lard Oil with 75% Mineral Oil
Malleable Iron	Soluble Oil	Soluble Oil
Bronze	Soluble Oil	20% Lard Oil with 80% Mineral Oil
Magnesium[d]	60-Second Mineral Oil	20% Lard Oil with 80% Mineral Oil

[a] Sulfurized oils ordinarily are not recommended for tapping aluminum; however, for some tapping operations they have proved very satisfactory, although the work should be rinsed in a solvent right after machining to prevent discoloration.

See additional notes following previous table.

Base Oils: Various types of highly sulfurized and chlorinated oils containing inorganic, animal, or fatty materials. This "base stock" usually is "cut back" or blended with a lighter oil, unless the chip-bearing pressures are high, as when cutting alloy steel. Base oils usually have a viscosity range of from 300 to 900 seconds at 100 degrees F.

Grinding: Soluble oil emulsions or emulsions made from paste compounds are used extensively in precision grinding operations. For cylindrical grinding, 1 part oil to 40 to 50 parts water is used. Solution-type fluids and translucent grinding emulsions are particularly suited for many fine–finish grinding applications. Mineral-oil-based grinding fluids are recommended for many applications where a fine surface finish is required on the ground surface. Mineral oils are used with vitrified wheels but are not recommended for wheels with rubber or shellac bonds. Under certain conditions the oil vapor mist caused by the action of the grinding wheel can be ignited by the grinding sparks and explode. To quench the grinding spark a secondary coolant line to direct a flow of grinding oil below the grinding wheel is recommended.

Broaching: For steel, a heavy mineral oil such as sulfurized oil of 300 to 500 Saybolt viscosity at 100 degrees F can be used to provide both adequate lubricating effect and a dampening of the shock loads. Soluble oil emulsions may be used for the lighter broaching operations.

DRILLING AND REAMING

Generally Used Values for Drill Points

Work Material	Point Angle	Comments
General work	118°	Lip relief angle 10°–15° Helix angle 24°–32°
High-strength (tough) steels	118°–135°	Lip relief angle[a] 7°–12° Helix angle 24°–32°
Aluminum alloys, cast iron	90°–140°	Lip relief angle[a] 10°–15°
Magnesium and copper alloys	70°–118°	Lip relief angle[a] 10°–15° Helix angle 10°–30°
Deep holes (various materials) or drilling stainless steel, titanium alloys, high-temperature alloys, nickel alloys, very high-strength materials, tool steels	118° Split point or crankshaft drill point	Lip relief angle 9° Chisel edge is entirely eliminated

[a] The lower values of these angle ranges are used for drills of larger diameter, the higher values for the smaller diameters. For drills of diameter less than 1/4-in, the lip relief angles are increased beyond the listed maximum value up to 24°. For soft and free-machining materials, 12° to 18° except for diameters less than 1/4 inch, 20° to 26°.

Note: Improperly sharpened twist drills, that is, those with unequal edge length or asymmetrical point angle, will tend to produce holes with poor diameter and directional control.

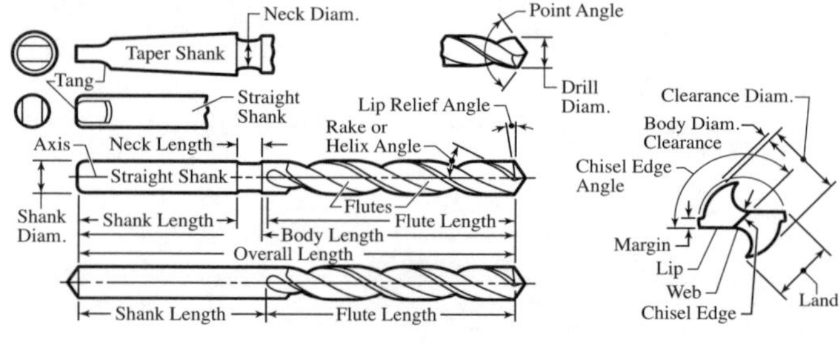

ANSI Standard Twist Drill Nomenclature

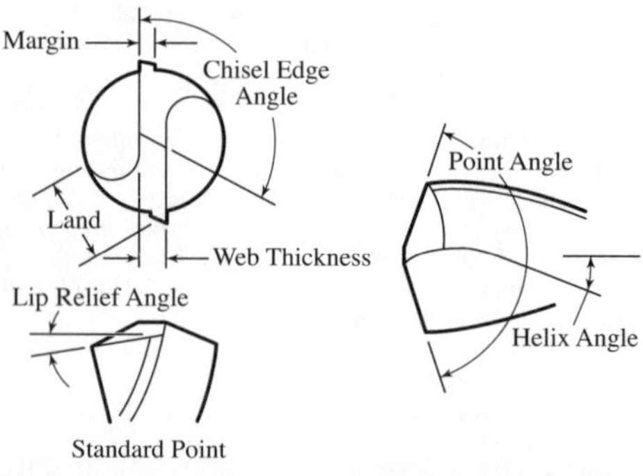

Fig. 1. Elements of Twist Drill Tool Geometry

Table 1. US and Metric Size Commercial Drills

Fraction No. or Letter	mm	Equivalent Inch	Equivalent mm	Fraction No. or Letter	mm	Equivalent Inch	Equivalent mm
80		0.0135	0.343	57		0.0430	1.092
	0.35	0.0138	0.350		1.10	0.0433	1.100
79		0.0145	0.368		1.15	0.0453	1.150
	0.38	0.0150	0.380	56		0.0465	1.181
1/64		0.0156	0.396	3/64		0.0469	1.191
	0.40	0.0157	0.400		1.20	0.0472	1.200
78		0.0160	0.406		1.25	0.0492	1.250
	0.42	0.0165	0.420		1.30	0.0512	1.300
	0.45	0.0177	0.450	55		0.0520	1.321
77		0.0180	0.457		1.35	0.0531	1.350
	0.48	0.0189	0.480	54		0.0550	1.397
	0.50	0.0197	0.500		1.40	0.0551	1.400
76		0.0200	0.508		1.45	0.0571	1.450
75		0.0210	0.533		1.50	0.0591	1.500
	0.55	0.0217	0.550	53		0.0595	1.511
74		0.0225	0.572		1.55	0.0610	1.550
	0.60	0.0236	0.600	1/16		0.0625	1.588
73		0.0240	0.610		1.60	0.0630	1.600
72		0.0250	0.635	52		0.0635	1.613
	0.65	0.0256	0.650		1.65	0.0650	1.650
71		0.0260	0.660		1.70	0.0669	1.700
	0.70	0.0276	0.700	51		0.0670	1.702
70		0.0280	0.711		1.75	0.0689	1.750
69		0.0292	0.742	50		0.0700	1.778
	0.75	0.0295	0.750		1.80	0.0709	1.800
68		0.0310	0.787		1.85	0.0728	1.850
1/32		0.0312	0.792	49		0.0730	1.854
	0.80	0.0315	0.800		1.90	0.0748	1.900
67		0.0320	0.813	48		0.0760	1.930
66		0.0330	0.838		1.95	0.0768	1.950
	0.85	0.0335	0.850	5/64		0.0781	1.984
65		0.0350	0.889	47		0.0785	1.994
	0.90	0.0354	0.899		2.00	0.0787	2.000
64		0.0360	0.914		2.05	0.0807	2.050
63		0.0370	0.940	46		0.0810	2.057
	0.95	0.0374	0.950	45		0.0820	2.083
62		0.0380	0.965		2.10	0.0827	2.100
61		0.0390	0.991		2.15	0.0846	2.150
	1.00	0.0394	1.000	44		0.0860	2.184
60		0.0400	1.016		2.20	0.0866	2.200
59		0.0410	1.041		2.25	0.0886	2.250
	1.05	0.0413	1.050	43		0.0890	2.261
58		0.0420	1.067		2.30	0.0906	2.300
					2.35	0.0925	2.350

Table 1. *(Continued)* **US and Metric Size Commercial Drills**

Fraction No. or Letter	mm	Equivalent		Fraction No. or Letter	mm	Equivalent	
		Inch	mm			Inch	mm
42		0.0935	2.375	21		0.1590	4.039
3/32		0.0938	2.383	20		0.1610	4.089
	2.40	0.0945	2.400		4.10	0.1614	4.100
41		0.0960	2.438		4.20	0.1654	4.200
	2.46	0.0965	2.450	19		0.1660	4.216
40		0.0980	2.489		4.30	0.1693	4.300
	2.50	0.0984	2.500	18		0.1695	4.305
39		0.0995	2.527	11/64		0.1719	4.366
38		0.1015	2.578	17		0.1730	4.394
	2.60	0.1024	2.600		4.40	0.1732	4.400
37		0.1040	2.642	16		0.1770	4.496
	2.70	0.1063	2.700		4.50	0.1772	4.500
36		0.1065	2.705	15		0.1800	4.572
7/64		0.1094	2.779		4.60	0.1811	4.600
35		0.1100	2.794	14		0.1820	4.623
	2.80	0.1102	2.800	13	4.70	0.1850	4.700
34		0.1110	2.819	3/16		0.1875	4.762
33		0.1130	2.870	12	4.80	0.1890	4.800
	2.90	0.1142	2.900	11		0.1910	4.851
32		0.1160	2.946		4.90	0.1929	4.900
	3.00	0.1181	3.000	10		0.1935	4.915
31		0.1200	3.048	9		0.1960	4.978
	3.10	0.1220	3.100		5.00	0.1969	5.000
1/8		0.1250	3.175	8		0.1990	5.054
	3.20	0.1260	3.200		5.10	0.2008	5.100
30		0.1285	3.264	7		0.2010	5.105
	3.30	0.1299	3.300	13/64		0.2031	5.159
	3.40	0.1339	3.400	6		0.2040	5.182
29		0.1360	3.454		5.20	0.2047	5.200
	3.50	0.1378	3.500	5		0.2055	5.220
28		0.1405	3.569		5.30	0.2087	5.300
9/64		0.1406	3.571	4		0.2090	5.309
	3.60	0.1417	3.600		5.40	0.2126	5.400
27		0.1440	3.658	3		0.2130	5.410
	3.70	0.1457	3.700		5.50	0.2165	5.500
26		0.1470	3.734	7/32		0.2188	5.558
25		0.1495	3.797		5.60	0.2205	5.600
	3.80	0.1496	3.800	2		0.2210	5.613
24		0.1520	3.861		5.70	0.2244	5.700
23		0.1540	3.912	1		0.2280	5.791
5/32		0.1562	3.967		5.80	0.2283	5.800
22		0.1570	3.988		5.90	0.2323	5.900
	4.00	0.1575	4.000	A		0.2340	5.944

Table 1. *(Continued)* **US and Metric Size Commercial Drills**

Fraction No. or Letter	mm	Equivalent		Fraction No. or Letter	mm	Equivalent	
		Inch	mm			Inch	mm
15/64		0.2344	5.954	P		0.3230	8.204
	6.00	0.2362	6.000		8.30	0.3268	8.300
B		0.2380	6.045	21/64		0.3281	8.334
	6.10	0.2402	6.100		8.40	0.3307	8.400
C		0.2420	6.147	Q		0.3320	8.433
	6.20	0.2441	6.200		8.50	0.3346	8.500
D		0.2460	6.248		8.60	0.3386	8.600
	6.30	0.2480	6.300	R		0.3390	8.611
E, 1/4		0.2500	6.350		8.70	0.3425	8.700
	6.40	0.2520	6.400	11/32		0.3438	8.733
	6.50	0.2559	6.500		8.80	0.3465	8.800
F		0.2570	6.528	S		0.3480	8.839
	6.60	0.2598	6.600		8.90	0.3504	8.900
G		0.2610	6.629		9.00	0.3543	9.000
	6.70	0.2638	6.700	T		0.3580	9.093
17/64		0.2656	6.746		9.10	0.3583	9.100
H		0.2660	6.756	23/64		0.3594	9.129
	6.80	0.2677	6.800		9.20	0.3622	9.200
	6.90	0.2717	6.900		9.30	0.3661	9.300
I		0.2720	6.909	U		0.3680	9.347
	7.00	0.2756	7.000		9.40	0.3701	9.400
J		0.2770	7.036		9.50	0.3740	9.500
	7.10	0.2795	7.100	3/8		0.3750	9.525
K		0.2810	7.137	V		0.3770	9.576
9/32		0.2812	7.142		9.60	0.3780	9.600
	7.20	0.2835	7.200		9.70	0.3819	9.700
	7.30	0.2874	7.300		9.80	0.3858	9.800
L		0.2900	7.366	W		0.3860	9.804
	7.40	0.2913	7.400		9.90	0.3898	9.900
M		0.2950	7.493	25/64		0.3906	9.921
	7.50	0.2953	7.500		10.00	0.3937	10.000
19/64		0.2969	7.541	X		0.3970	10.084
	7.60	0.2992	7.600		10.20	0.4016	10.200
N		0.3020	7.671	Y		0.4040	10.262
	7.70	0.3031	7.700	13/32		0.4062	10.317
	7.80	0.3071	7.800	Z		0.4130	10.490
	7.90	0.3110	7.900		10.50	0.4134	10.500
5/16		0.3125	7.938	27/64		0.4219	10.716
	8.00	0.3150	8.000		10.80	0.4252	10.800
O		0.3160	8.026		11.00	0.4331	11.000
	8.10	0.3189	8.100	7/16		0.4375	11.112
	8.20	0.3228	8.200		11.20	0.4409	11.200

Table 1. *(Continued)* **US and Metric Size Commercial Drills**

Fraction No. or Letter	mm	Equivalent		Fraction No. or Letter	mm	Equivalent	
		Inch	mm			Inch	mm
	11.50	0.4528	11.500	43/64		0.6719	17.066
29/64		0.4531	11.509		17.25	0.6791	17.250
	11.80	0.4646	11.800	11/16		0.6875	17.462
15/32		0.4688	11.908		17.50	0.6890	17.500
	12.00	0.4724	12.000	45/64		0.7031	17.859
	12.20	0.4803	12.200		18.00	0.7087	18.000
31/64		0.4844	12.304	23/32		0.7188	18.258
	12.50	0.4921	12.500		18.50	0.7283	18.500
1/2		0.5000	12.700	47/64		0.7344	18.654
	12.80	0.5039	12.800		19.00	0.7480	19.000
	13.00	0.5118	13.000	3/4		0.7500	19.050
33/64		0.5156	13.096	49/64		0.7656	19.446
	13.20	0.5197	13.200		19.50	0.7677	19.500
17/32		0.5312	13.492	25/32		0.7812	19.845
	13.50	0.5315	13.500		20.00	0.7879	20.000
	13.80	0.5433	13.800	51/64		0.7969	20.241
35/64		0.5469	13.891		20.50	0.8071	20.500
	14.00	0.5512	14.000	13/16		0.8125	20.638
	14.25	0.5610	14.250		21.00	0.8268	21.000
9/16		0.5625	14.288	53/64		0.8281	21.034
	14.50	0.5709	14.500	27/32		0.8438	21.433
37/64		0.5781	14.684		21.50	0.8465	21.500
	14.75	0.5807	14.750	55/64		0.8594	21.829
	15.00	0.5906	15.000		22.00	0.8661	22.000
19/32		0.5938	15.083	7/8		0.8750	22.225
	15.25	0.6004	15.250		22.50	0.8858	22.500
39/64		0.6094	15.479	57/64		0.8906	22.621
	15.50	0.6102	15.500		23.00	0.9055	23.000
	15.75	0.6201	15.750	29/32		0.9062	23.017
5/8		0.6250	15.875	59/64		0.9219	23.416
	16.00	0.6299	16.000		23.50	0.9252	23.500
	16.25	0.6398	16.250	15/16		0.9375	23.812
	16.50	0.4528	11.500		24.00	0.9449	24.000
41/64		0.6406	16.271	61/64		0.9531	24.209
	16.50	0.6496	16.500		24.50	0.9646	24.500
21/32		0.6562	16.669	31/32		0.9688	24.608
	16.75	0.6594	16.750		25.00	0.9843	25.000
	17.00	0.6693	17.000	63/64		0.9844	25.004
				1		1.0000	25.400

Table 2. Common Drilling Difficulties

Drill split at the web	Too much feed or insufficient lip clearance at center due to improper grinding.
Rapid wear of extreme outer corners of the cutting edges	Speed too high; excessive speed will draw the temper.
Chipping or breaking out at the cutting edges	Feed is too heavy or drill ground with too much lip clearance.
Checking of high-speed drill.	Cold water hitting heated drill. It is equally bad to plunge drill into cold water after the point has been heated in use.
Drill Breaks	Insufficient speed when drilling small holes with hand feed increases risk of breakage, especially at the moment the drill is breaking through the further side of the work. Small drills have heavier webs and smaller flutes in proportion to their size than do larger drills, and breakage due to clogging of chips in the flutes is more likely to occur.
Drill binds on one side and wears on one side, resulting in a hole larger than the drill.	The point is on center but the cutting edges have been ground at different angles.
Drill press spindle wobbles and weaves, resulting in a hole larger than the drill.	Angles of the chisel edge are equal but lips are of different lengths.
"Accuracy of drilled holes"	Influenced by many factors, which include: Accuracy of the drill point drill size length and shape of the chisel edge whether or not a bushing is used to guide the drill length of the drill runout of the spindle and the chuck rigidity of the machine tool, workpiece, and the setup the cutting fluid, if any work material

Note: When drilling holes deeper than three times the diameter of the drill, it is advisable to withdraw the drill at intervals to remove chips and permit coolant to reach the drill tip.

Table 3. American National Standard Combined Drills and Countersinks—Plain and Bell Types *ANSI B94.11M-1993*

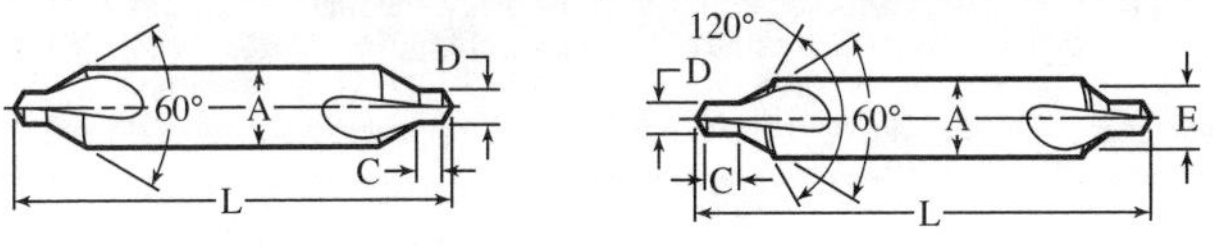

Size Designation	Plain Type: Body Diameter A, Inches	Body Diameter A, Millimeters	Drill Diameter D, Inches	Drill Diameter D, Millimeters	Drill Length C, Inches	Drill Length C, Millimeters	Overall Length L, Inches	Overall Length L, Millimeters
00	3/32	2.38	0.025	0.64	0.030	0.76	1 1/8	29
0	3/32	2.38	1/32	0.79	0.038	0.97	1 1/8	29
1	1/8	3.18	3/64	1.19	3/64	1.19	1 1/4	32
2	3/16	4.76	5/64	1.98	5/64	1.98	1 7/8	48
3	1/4	6.35	7/64	2.78	7/64	2.78	2	51
4	5/16	7.94	1/8	3.18	1/8	3.18	2 1/8	54
5	7/16	11.11	3/16	4.76	3/16	4.76	2 3/4	70
6	1/2	12.70	7/32	5.56	7/32	5.56	3	76
7	5/8	15.88	1/4	6.35	1/4	6.35	3 1/4	83
8	3/4	19.05	5/16	7.94	5/16	7.94	3 1/2	89

Table 3b. American National Standard Combined Drills and Countersinks, Bell Type *ANSI B94.11M-1993*

Bell Type										
Size Designation	Body Diameter *A*		Drill Diameter *D*		Bell Diameter *E*		Drill Length *C*		Overall Length L	
	Inches	mm	Inches	mm	Inches	mm	Inches	mm	Inches	mm
11	1/8	3.18	3/64	1.19	0.10	2.5	3/64	1.19	1 1/4	32
12	3/16	4.76	1/16	1.59	0.15	3.8	1/16	1.59	1 7/8	48
13	1/4	6.35	3/32	2.38	0.20	5.1	3/32	2.38	2	51
14	5/16	7.94	7/64	2.78	0.25	6.4	7/64	2.78	2 1/8	54
15	7/16	11.11	5/32	3.97	0.35	8.9	5/32	3.97	2 3/4	70
16	1/2	12.70	3/16	4.76	0.40	10.2	3/16	4.76	3	76
17	5/8	15.88	7/32	5.56	0.50	12.7	7/32	5.56	3 1/4	83
18	3/4	19.05	1/4	6.35	0.60	15.2	1/4	6.35	3 1/2	89

Counterboring.—Counterboring (called spot-facing if the depth is shallow) is the enlargement of a previously formed hole. Counterbores for screw holes are generally made in sets. Each set contains three counterbores: one with the body of the size of the screw head and the pilot the size of the hole to admit the body of the screw; one with the body the size of the head of the screw and the pilot the size of the tap drill; and the third with the body the size of the body of the screw and the pilot the size of the tap drill. Counterbores are usually provided with helical flutes to provide positive effective rake on the cutting edges. The four flutes are so positioned that the end teeth cut ahead of center to provide a shearing action and eliminate chatter in the cut. Three designs are most common: solid, two-piece, and three-piece. Solid designs have the body, cutter, and pilot all in one piece. Two-piece designs have an integral shank and counterbore cutter, with an interchangeable pilot, and provide true concentricity of the cutter diameter with the shank, but allowing use of various pilot diameters. Three-piece counterbores have separate holder, counterbore cutter, and pilot, so that a holder will take any size of counterbore cutter. Each counterbore cutter, in turn, can be fitted with any suitable size diameter of pilot. Counterbores for brass are fluted straight.

Small counterbores are often made with three flutes, but should then have the size plainly stamped on them before fluting, as they cannot afterwards be conveniently measured. The flutes should be deep enough to come below the surface of the pilot. The counterbore should be relieved on the end of the body only, and not on the cylindrical surface. To facilitate the relieving process, a small neck is turned between the guide and the body for clearance. The amount of clearance on the cutting edges is, for general work, from 4 to 5 degrees. The accompanying table gives dimensions for straight shank counterbores.

Table 4. Counterbores with Interchangeable Cutters and Guides

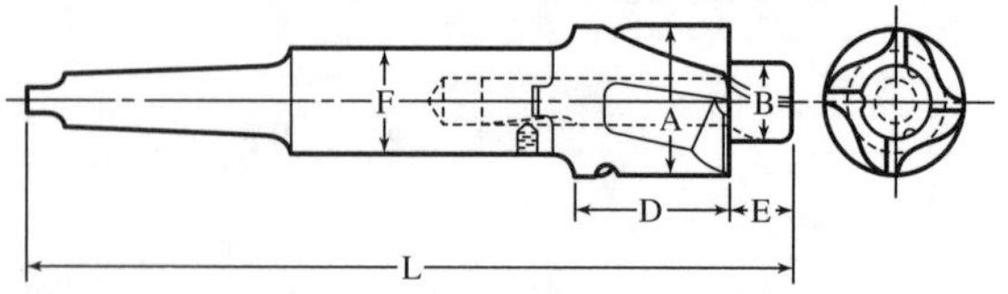

No. of Holder	No. of Morse Taper Shank	Range of Cutter Diameters, *A*	Range of Pilot Diameters, *B*	Total Length, *C*	Length of Cutter Body, *D*	Length of Pilot, *E*	Dia. of Shank, *F*
1	1 or 2	3/4-1 1/16	1/2-3/4	7 1/4	1	5/8	3/4
2	2 or 3	1 1/8-1 9/16	1 1/16-1 1/8	9 1/2	1 3/8	7/8	1 1/8
3	3 or 4	1 5/8-2 1/16	7/8-1 5/8	12 1/2	1 3/4	1 1/8	1 5/8
4	4 or 5	2 1/8-3 1/2	1-2 1/8	15	2 1/4	1 3/8	2 1/8

Table 5. Length of Point on Twist Drills and Centering Tools

Size of Drill	Decimal Equiva-lent	Length of Point when Included Angle =90°	Length of Point when Included Angle =118°	Size of Drill	Decimal Equiva-lent	Length of Point when Included Angle =90°	Length of Point when Included Angle =118°	Size or Dia. of Drill	Decimal Equivalent	Length of Point when Included Angle =90°	Length of Point when Included Angle =118°	Dia. of Drill	Decimal Equiva-lent	Length of Point when Included Angle =90°	Length of Point when Included Angle =118°
60	0.0400	0.020	0.012	37	0.1040	0.052	0.031	14	0.1820	0.091	0.055	3/8	0.3750	0.188	0.113
59	0.0410	0.021	0.012	36	0.1065	0.054	0.032	13	0.1850	0.093	0.056	25/64	0.3906	0.195	0.117
58	0.0420	0.021	0.013	35	0.1100	0.055	0.033	12	0.1890	0.095	0.057	13/32	0.4063	0.203	0.122
57	0.0430	0.022	0.013	34	0.1110	0.056	0.033	11	0.1910	0.096	0.057	27/64	0.4219	0.211	0.127
56	0.0465	0.023	0.014	33	0.1130	0.057	0.034	10	0.1935	0.097	0.058	7/16	0.4375	0.219	0.131
55	0.0520	0.026	0.016	32	0.1160	0.058	0.035	9	0.1960	0.098	0.059	29/64	0.4531	0.227	0.136
54	0.0550	0.028	0.017	31	0.1200	0.060	0.036	8	0.1990	0.100	0.060	15/32	0.4688	0.234	0.141
53	0.0595	0.030	0.018	30	0.1285	0.065	0.039	7	0.2010	0.101	0.060	31/64	0.4844	0.242	0.145
52	0.0635	0.032	0.019	29	0.1360	0.068	0.041	6	0.2040	0.102	0.061	1/2	0.5000	0.250	0.150
51	0.0670	0.034	0.020	28	0.1405	0.070	0.042	5	0.2055	0.103	0.062	33/64	0.5156	0.258	0.155
50	0.0700	0.035	0.021	27	0.1440	0.072	0.043	4	0.2090	0.105	0.063	17/32	0.5313	0.266	0.159
49	0.0730	0.037	0.022	26	0.1470	0.074	0.044	3	0.2130	0.107	0.064	35/64	0.5469	0.273	0.164
48	0.0760	0.038	0.023	25	0.1495	0.075	0.045	2	0.2210	0.111	0.067	9/16	0.5625	0.281	0.169
47	0.0785	0.040	0.024	24	0.1520	0.076	0.046	1	0.2280	0.114	0.068	37/64	0.5781	0.289	0.173
46	0.0810	0.041	0.024	23	0.1540	0.077	0.046	15/64	0.2344	0.117	0.070	19/32	0.5938	0.297	0.178
45	0.0820	0.041	0.025	22	0.1570	0.079	0.047	1/4	0.2500	0.125	0.075	39/64	0.6094	0.305	0.183
44	0.0860	0.043	0.026	21	0.1590	0.080	0.048	17/64	0.2656	0.133	0.080	5/8	0.6250	0.313	0.188
43	0.0890	0.045	0.027	20	0.1610	0.081	0.048	9/32	0.2813	0.141	0.084	41/64	0.6406	0.320	0.192
42	0.0935	0.047	0.028	19	0.1660	0.083	0.050	19/64	0.2969	0.148	0.089	21/32	0.6563	0.328	0.197
41	0.0960	0.048	0.029	18	0.1695	0.085	0.051	5/16	0.3125	0.156	0.094	43/64	0.6719	0.336	0.202
40	0.0980	0.049	0.029	17	0.1730	0.087	0.052	21/64	0.3281	0.164	0.098	11/16	0.6875	0.344	0.206
39	0.0995	0.050	0.030	16	0.1770	0.089	0.053	11/32	0.3438	0.171	0.103	23/32	0.7188	0.359	0.216
38	0.1015	0.051	0.030	15	0.1800	0.090	0.054	23/64	0.3594	0.180	0.108	3/4	0.7500	0.375	0.225

Table 6. Solid Counterbores with Integral Pilot

Counterbore Diameters	Pilot Diameters			Straight Shank Diameter	Overall Length	
	Nominal	+1/64	+1/32		Short	Long
13/32	1/4	17/64	9/32	3/8	3 1/2	5 1/2
1/2	5/16	21/64	11/32	3/8	3 1/2	5 1/2
19/32	3/8	25/64	13/32	1/2	4	6
11/16	7/16	29/64	15/32	1/2	4	6
25/32	1/2	33/64	17/32	1/2	5	7
0.110	0.060	0.076	…	7/64	2 1/2	…
0.133	0.073	0.089	…	1/8	2 1/2	…
0.155	0.086	0.102	…	5/32	2 1/2	…
0.176	0.099	0.115	…	11/64	2 1/2	…
0.198	0.112	0.128	…	3/16	2 1/2	…
0.220	0.125	0.141	…	3/16	2 1/2	…
0.241	0.138	0.154	…	7/32	2 1/2	…
0.285	0.164	0.180	…	1/4	2 1/2	…
0.327	0.190	0.206	…	9/32	2 3/4	…
0.372	0.216	0.232	…	5/16	2 3/4	…

All dimensions are in inches.

Table 7. American National Standard Solid Carbide Square Boring Tools—Style SSC for 60° Boring Bar and Style SSE for 45° Boring Bar *ANSI B212.1-2002 (R2007)*

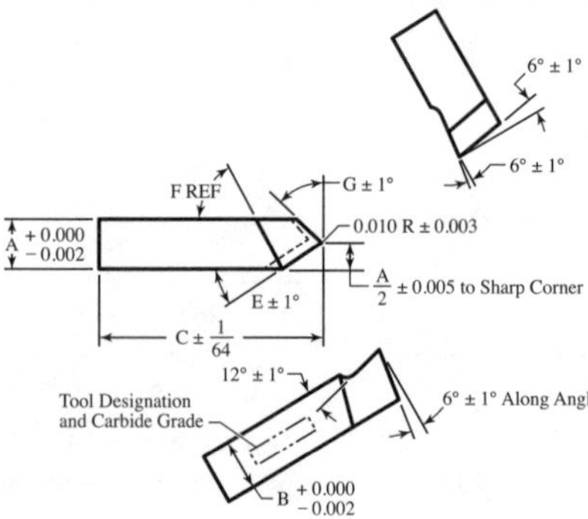

Tool Designation	Boring Bar Angle, Deg. from Axis	Shank Dimensions, Inches			Side Cutting Edge Angle *E*, Deg.	End Cutting Edge Angle *G*, Deg.	Shoulder Angle *F*, Deg.
		Width *A*	Height *B*	Length *C*			
SSC-58	60	5/32	5/32	1	30	38	60
SSE-58	45				45	53	45
SSC-610	60	3/16	3/16	1 1/4	30	38	60
SSE-610	45				45	53	45
SSC-810	60	1/4	1/4	1 1/4	30	38	60
SSE-810	45				45	53	45
SSC-1012	60	5/16	5/16	1 1/2	30	38	60
SSE-1012	45				45	53	45

Hand Reamers.—Hand reamers are made with both straight and helical flutes. Helical flutes provide a shearing cut and are especially useful in reaming holes having keyways or grooves, as these are bridged over by the helical flutes, thus preventing binding or chattering. Hand reamers are made in both solid and expansion forms. The American standard dimensions for solid forms are given in the accompanying table. The expansion type is useful whenever, in connection with repair or other work, it is necessary to enlarge a reamed hole by a few thousandths of an inch. The expansion form is split through the fluted section and a slight amount of expansion is obtained by screwing in a tapering plug. The diameter increase may vary from 0.005 to 0.008 inch for reamers up to about 1 inch diameter and from 0.010 to 0.012 inch for diameters between 1 and 2 inches. Hand reamers are tapered slightly on the end to facilitate starting them properly. The actual diameter of the shanks of commercial reamers may be from 0.002 to 0.005 inch under the reamer size. That part of the shank that is squared should be turned smaller in diameter than the shank itself, so that, when applying a wrench, no burr may be raised that may mar the reamed hole if the reamer is passed clear through it.

When fluting reamers, the cutter is so set with relation to the center of the reamer blank that the tooth gets a slight negative rake; that is, the cutter should be set *ahead* of the center, as shown in the illustration accompanying the table giving the amount to set the cutter ahead of the radial line. The amount is so selected that a tangent to the circumference of the reamer at the cutting point makes an angle of approximately 95 degrees with the front face of the cutting edge.

Illustrations of Terms Applying to Reamers

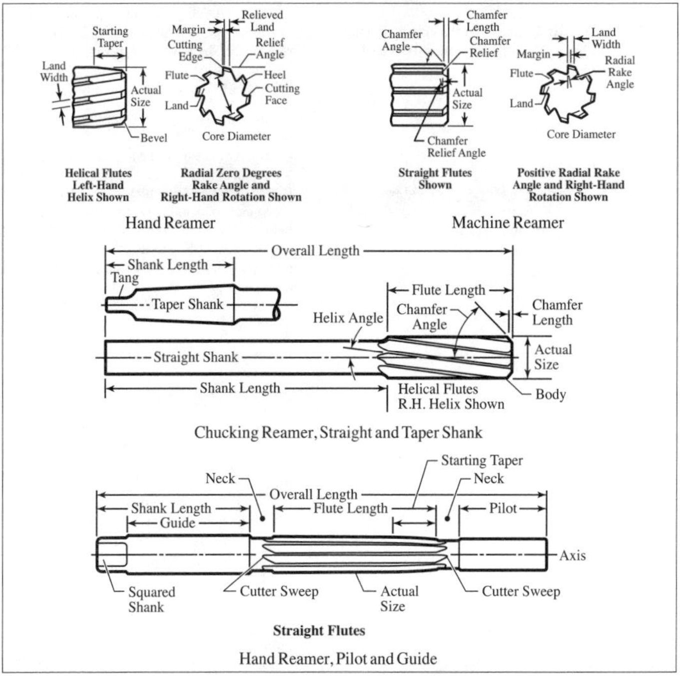

Table 8. Common Reamer Difficulties

Problem	Possible Cause	Solution
Chatter	Lack of rigidity in the machine, spindle, reamer or workpiece.	Reduce the feed. Increase the feed. Chamfer hole before reaming. Reduce clearance angle on the reamer's cutting edge. Using a reamer with a pilot and guide bushings. Note: Any amount of chatter may cause carbide-tipped reamer edges to chip, especially as the reamer initially enters the hole.
Oversize Holes	Wrong reamer for the workpiece material used. Inadequate workpiece support. Inadequate or worn guide bushings. Loose spindle bearings. Misalignment of the spindles, bushings or workpiece or runout of the spindle or reamer holder. The reamer may be defective due to chamfer runout or runout of the cutting end due to a bent or nonconcentric reamer shank. Workpiece material forming a built-up edge on reamer.	Reduce the reamer margin widths to about 0.005 to 0.010 inch. Use hard case surface treatments on high-speed reamers, either alone or in combination with black oxide treatments. Use high-grade finish on the reamer faces, margins, and chamfer relief surfaces. Check and possibly change cutting fluid or coolant.
Bell-mouth Holes	Misalignment of the cutting portion of the reamer with respect to the hole.	Provide improved guiding of the reamer by the use of accurate bushings and pilot surfaces. If the reamer is cutting in a vertical position, use a floating holder so that it has both radial and axial movement.
Bell-mouth holes in horizontal setups	Misalignment exerts a sideways force on the reamer as it is fed to depth, resulting in a tapered hole.	Shorten the bearing length of the cutting portion of the reamer. The following modifications reduce the length of the reamer tooth that caused the condition. Method 1: Reduce the reamer diameter by 0.010 to 0.030 inch, depending on size and length, behind a short full-diameter section 1/8 to inch long according to length and size, following chamfer, or Method 2: Grind a high-back taper 0.008 to 0.015 inch per inch, behind the short full-diameter section.
Poor Finish		Reduce the reamer feed per revolution. Feeds as low as 0.0002 to 0.0005 inch per tooth have been used successfully, but reamer life will be better if the maximum feasible feed is used. The minimum practical amount of stock allowance will often improve finish by reducing the volume of chips and heat generated on the cutting portion of the chamfer. Too small a stock allowance may prevent the reamer teeth from cutting freely and will deflect the work material out of the way and cause rapid reamer wear. Not enough cutting fluid or coolant being applied during reaming.
Reamer breaks	Feed too fast. Dull edges.	Slow feed. Hone edges.

TAPPING

Tap Terms and Designs

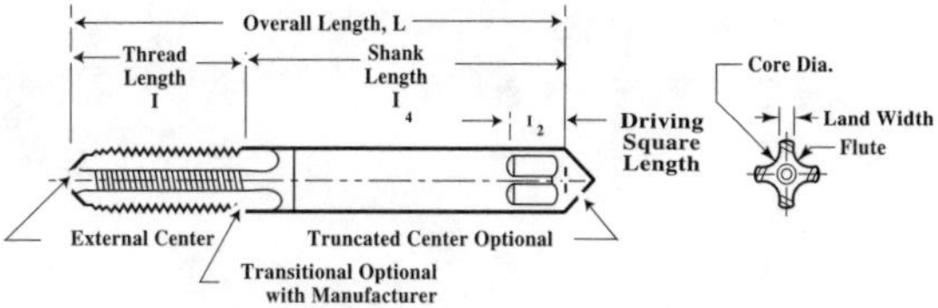

BLANK Design 1

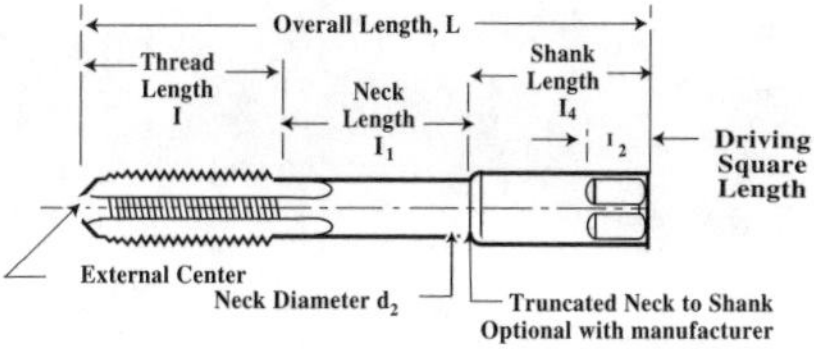

BLANK Design 2 with Optional Neck

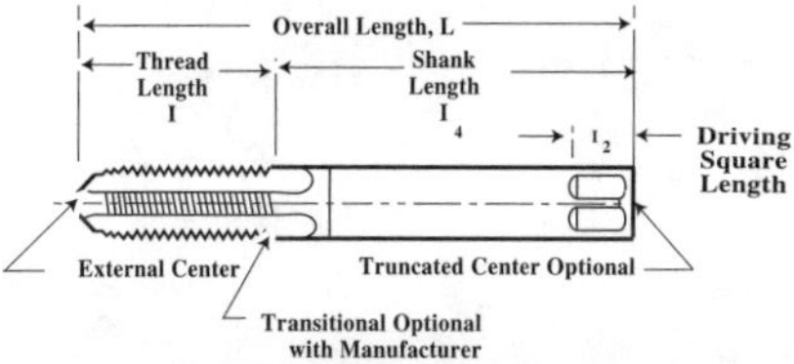

BLANK Design 2 (without optional neck)

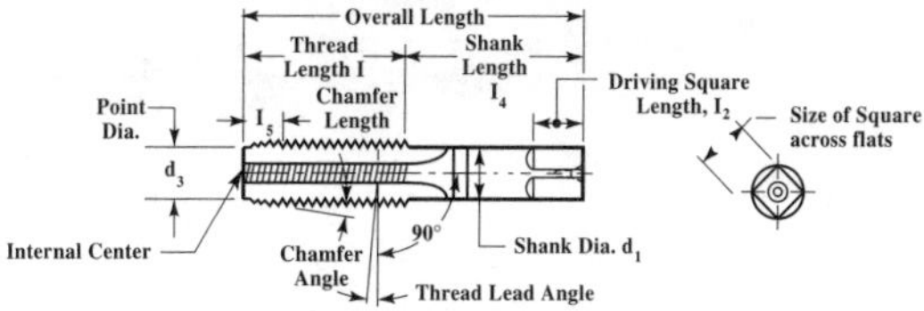

BLANK Design 3

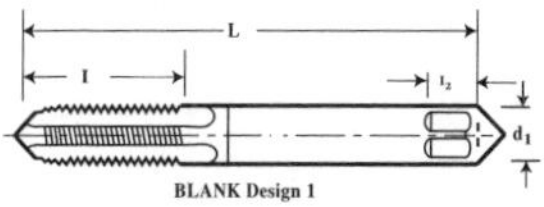

BLANK Design 1

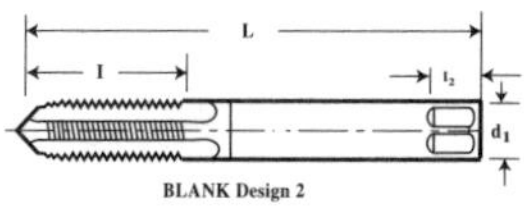

BLANK Design 2

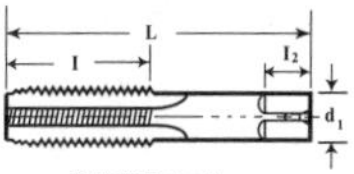

BLANK Design 3

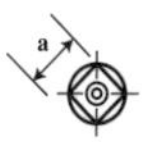

Table 1. Standard Tap Dimensions (Ground and Cut Thread) *ANSI/ASME B94.9-2008 (R2018)*

Nominal Diameter Range, inch		Nominal Diameter, inch		Nominal Metric Diameter		Blank Design No.	Tap Dimensions, inch				
Over	To	Machine Screw Size No. and Fractional Sizes	Decimal Equiv.	mm	inch		Overall Length L	Thread Length I	Square Length I_2	Shank Diameter d_1	Size of Square a
0.052	0.065	0	(0.0600)	M1.6	0.0630	1	1.63	0.31	0.19	0.141	0.110
0.065	0.078	1	(0.0730)	M1.8	0.0709	1	1.69	0.38	0.19	0.141	0.110
0.078	0.091	2	(0.0860)	M2.0	0.0787	1	1.75	0.44	0.19	0.141	0.110
				M2.2	0.0866						
0.091	0.104	3	(0.0990)	M2.5	0.0984	1	1.81	0.50	0.19	0.141	0.110
0.104	0.117	4	(0.1120)	…	…	1	1.88	0.56	0.19	0.141	0.110
0.117	0.130	5	(0.1250)	M3.0	0.1182	1	1.94	0.63	0.19	0.141	0.110
0.130	0.145	6	(0.1380)	M3.5	0.1378	1	2.00	0.69	0.19	0.141	0.110
0.145	0.171	8	(0.1640)	M4.0	0.1575	1	2.13	0.75	0.25	0.168	0.131
0.171	0.197	10	(0.1900)	M4.5	0.1772	1	2.38	0.88	0.25	0.194	0.152
				M5	0.1969						
0.197	0.223	12	(0.2160)	…	…	1	2.38	0.94	0.28	0.220	0.165
0.223	0.260	1/4	(0.2500)	M6	0.2363	2	2.50	1.00	0.31	0.255	0.191
0.260	0.323	5/16	(0.3125)	M7	0.2756	2	2.72	1.13	0.38	0.318	0.238
				M8	0.3150						
0.323	0.395	3/8	(0.3750)	M10	0.3937	2	2.94	1.25	0.44	0.381	0.286
0.395	0.448	7/16	(0.4375)	…	…	3	3.16	1.44	0.41	0.323	0.242
0.448	0.510	1/2	(0.5000)	M12	0.4724	3	3.38	1.66	0.44	0.367	0.275
0.510	0.573	9/16	(0.5625)	M14	0.5512	3	3.59	1.66	0.50	0.429	0.322
0.573	0.635	5/8	(0.6250)	M16	0.6299	3	3.81	1.81	0.56	0.480	0.360
0.635	0.709	11/16	(0.6875)	M18	0.7087	3	4.03	1.81	0.63	0.542	0.406
0.709	0.760	3/4	(0.7500)	…	…	3	4.25	2.00	0.69	0.590	0.442
0.760	0.823	13/16	(0.8125)	M20	0.7874	3	4.47	2.00	0.69	0.652	0.489
0.823	0.885	7/8	(0.8750)	M22	0.8661	3	4.69	2.22	0.75	0.697	0.523

Table 1. *(Continued)* **Standard Tap Dimensions (Ground and Cut Thread)** *ANSI/ASME B94.9-2008 (R2018)*

Nominal Diameter Range, inch		Nominal Diameter, inch		Nominal Metric Diameter			Tap Dimensions, inch				
Over	To	Machine Screw Size No. and Fractional Sizes	Decimal Equiv.	mm	inch	Blank Design No.	Overall Length L	Thread Length I	Square Length I_2	Shank Diameter d_1	Size of Square a
0.885	0.948	$^{15}/_{16}$	(0.9375)	M24	0.9449	3	4.91	2.22	0.75	0.760	0.570
0.948	1.010	1	(1.0000)	M25	0.9843	3	5.13	2.50	0.81	0.800	0.600
1.010	1.073	$1^1/_{16}$	(1.0625)	M27	1.0630	3	5.13	2.50	0.88	0.896	0.672
1.073	1.135	$1^1/_8$	(1.1250)	…	…	3	5.44	2.56	0.88	0.896	0.672
1.135	1.198	$1^3/_{16}$	(1.1875)	M30	1.1811	3	5.44	2.56	1.00	1.021	0.766
1.198	1.260	$1^1/_4$	(1.2500)	…	…	3	5.75	2.56	1.00	1.021	0.766
1.260	1.323	$1^5/_{16}$	(1.3125)	M33	1.2992	3	5.75	2.56	1.06	1.108	0.831
1.323	1.385	$1^3/_8$	(1.3750)	…	…	3	6.06	3.00	1.06	1.108	0.831
1.358	1.448	$1^7/_{16}$	(1.4375)	M36	1.4173	3	6.06	3.00	1.13	1.233	0.925
1.448	1.510	$1^1/_2$	(1.5000)	…	…	3	6.38	3.00	1.13	1.233	0.925
1.510	1.635	$1^5/_8$	(1.6250)	M39	1.5353	3	6.69	3.19	1.13	1.305	0.979
1.635	1.760	$1^3/_4$	(1.7500)	M42	1.6535	3	7.00	3.19	1.25	1.430	1.072
1.760	1.885	$1^7/_8$	(1.8750)	…	…	3	7.31	3.56	1.25	1.519	1.139
1.885	2.010	2	(2.0000)	M48	1.8898	3	7.63	3.56	1.38	1.644	1.233

Tap sizes 0.395 inch and smaller have an external center on the thread end (may be removed on bottom taps). Sizes 0.223 inch and smaller have an external center on the shank end. Sizes 0.224 inch through 0.395 inch have truncated partial cone centers on the shank end (of diameter of shank). Sizes greater than 0.395 inch have internal centers on both the thread and shank ends.

Table 2. General Threading Formulas

Tap Drill Sizes	$S = D_m - 1.0825 \times P \times \%$ (Unified Threads) $S = D_m - 1.2990 \times P \times \%$ (American Standard Threads) $S = D_m - 1.0825 \times P \times \%$ (ISO Metric Threads)	Percentage of Full Thread	$\frac{D_m - S}{1.0825 - P}$ = % of full thread (Unified Threads) $\frac{D_m - S}{1.2990 - P}$ = % of full thread (American Standard Threads) $\frac{D_m - S}{1.0825 - P}$ = % of full thread (ISO Metric Thread)

Table 3. General Threading Formulas

Determining Machine Screw Sizes	$N = \frac{D_m - 0.060}{0.013}$ $D_m = N \times 0.013 + 0.060$

All dimensions in mm

D_m = major diameter; P = pitch;% = percentage of full thread; S = size of selected tap drill; N = number of machine screw

Courtesy of the Society of Manufacturing Engineers

Table 4. Tapping Specific Materials

Material	Rake Angle, degrees	Speeds ft/min	Lubricant	Comments
Alloys, High-Temperature, Nickel or Cobalt Base Nonferrous	0–10	5–10	Sulfur-chlorinated mineral lard oil	Nitrided tap or one made from M41, M42, M43, or M44 Steel recommended. Use plug tap having 3-5 chamfered threads. To reduce rubbing of the lands, eccentric or con-eccentric relieved land should be used. To control a continuous chip use a spiral pointed tap for through holes; use low-helix angle spiral-fluted taps for blind holes. Oxide-coated tap recommended.
Aluminum	8–20 (10–15 recomended)	90–100	Heavy-duty water soluble oil or light base mineral oil	Spiral pointed tap for through holes; spiral-fluted tap for blind holes
Brass	2–7	90–100	10-20% lard oil with mineral oil	Use interrupted thread tap to reduce jamming; straight-fluted tap for machine tapping. For red brass, yellow brass, and similar alloys containing more than 35% zinc, use a fluted tap for hand tapping; spiral pointed or spiral-fluted tap for machine tapping.
Brass, Naval, Leaded and Cast	5–10		Soluble oil	Interrupted thread tap used to reduce jamming. Straight-fluted tap for machine tapping.
Bronze, Phosphor	5–12	30–60	Soluble oil	
Bronze, Manganese	5–12		Light base oil	
Bronze, Tobin	5–8		Soluble oil	
Copper	10–18		Medium heavy-duty mineral lard oil or soluble oil	For beryllium copper and silicon bronzes use plug-type taps, and keep taps as sharp as possible.
Iron, Cast (Gray)	0–3	90 for softer grades 30 for harder grades	Dry or soluble oils or chemical emulsions	Microstructure in a single casting can vary in tensile strength. Oxide-coated taps are helpful. Straight-fluted taps should be used for all applications.
Iron, Malleable	5–8	60–90 (ferritic) 40–50 (pearlitic) 30–50 (martensitic)	Soluble oil martensitic: sulfur-based oil	Microstructure tends to be uniform. Standard taps can be used.
Iron, Ductile or Nodular		15 (martensitic) 60 (ferritic)	Soluble oil	Oxide-coated tap recommended
Magnesium	10–20	20–50	20% lard oil with 80% mineral oil	Use no water due to fire or explosion hazard
Monel Metal	9–12	20–25% lard oil mixed with mineral oil, or sulfur-based oil		

Table 4. *(Continued)* **Tapping Specific Materials**

Material	Rake Angle, degrees	Speeds ft/min	Lubricant	Comments
Plastics, Thermo-Plastics	5–8 thermo-plastics 0–3 thermo-setting	50; 25 filled material. Reduce speeds for deep and blind holes, and when percentage of thread is greater than 65–75%.	Dry or forced air jet	Taps should be of M10, M7, or M1 molybdenum high-speed steel, with finish-ground and polished flutes. Two-flute taps are recommended for holes up to 0.125 inch diameter. Oversize taps may be required to make up for the elastic recovery of the plastics.
Rubber, Hard	0–3		Dry or air jet	
Steel, Free-Machining		60–80	Soluble oil	Sulfur, lead or phosphorus added to improve machinability. Usually standard tap can be used.
Steel, High-Tensile Strength (40–55 Rc)	at or near zero	do not exceed 10	Active sulfur-chlorinated oil	Taps with concentric lands; 6 to 8 chamfered threads on the end to reduce chip load per tooth. Keep chamfer relief to a minimum. Load on tap should be kept to a minimum: use largest possible tap drill size; keep hole depth to a minimum; avoid bottoming holes; and, in larger sizes, use fine instead of coarse pitches. Oxide-coated or nitrided tap to reduce tap wear.
Steel, Low-Carbon (up to 0.15% C)	5–12	40–60	Sulfur-based oil	Spiral pointed taps for through holes. Spiral-fluted tap for blind holes. Oxide-coated tap recommended.
Steel, Low Carbon (up to 0.15–0.30% C)	5–12	40–60	Sulfur-based oil	Oxide-coated tap recommended for low-carbon range.
Steel, Medium-Carbon, Annealed (0.30–0.60% C)	5–10	30–50	Sulfur-based oil	Cutting speed dependent on C content and heat treatment. Slowly tap higher C, especially if heat treatment produced pearlitic microstructure. A spheroidized microstructure will result in ease of tapping.
Steel, Heat-Treated, 225–283 Brinell (0.30 to 0.60% C)	0–8	25–35	Sulfur-based oil	
Steel, High-Carbon (More Than 0.6% C)	0–5 do not exceed	25–35	Activated sulfur-chlorinated	Use concentric tap.
Steel, High-Speed	0–5	25–35	Sulfur-based oil	
Steel, Molybdenum		10–35	Sulfur-based oil	
Steel, Stainless	8–15 10–15 Austenitic	10–35	Ferritic and martensitic: molybdenum disulfide or other sulfur-based oil. Austenitic: sulfur-chlorinated mineral lard or heavy-duty soluble oils	Ferritic and martensitic: Standard rake angle oxide-coated taps are recommended. Austenitic: Use plug tap having 3–5 chamfered threads. To reduce rub, use an eccentric or con-eccentric relieved land.
Titanium and Alloys	6–10	40–10 depends on composition of alloy	Special	Oxide-coated taps are recommended to minimize galling and welding. An eccentric or con-eccentric relief land should be used. Taps with interrupted threads are sometimes helpful. Pure Ti is comparatively easy to tap; alloys are difficult.

Table 5. American National and Unified Coarse and Fine Thread Dimensions and Tap Drill Sizes

Thread size and threads per inch	Basic major diameter inches	Pitch diameter inches	Root diameter inches	Tap drill for 75% theoretical thread	Decimal equivalent of tap drill inches
0 × 80	0.0600	0.0519	0.0438	$^{3}/_{64}''$	0.0469
1 × 64	0.0730	0.0629	0.0527	53	0.0595
1 × 72	0.0730	0.0640	0.055	53	0.0595
2 × 56	0.0860	0.0744	0.0628	50	0.0700
2 × 64	0.0860	0.0759	0.0657	50	0.0700
3 × 48	0.0990	0.0855	0.0719	47	0.0785
3 × 56	0.0990	0.0874	0.0758	46	0.0810
4 × 40	0.1120	0.0958	0.0795	43	0.0890
4 × 48	0.1120	0.0985	0.0849	42	0.0935
5 × 40	0.1250	0.1088	0.0925	38	0.1015
5 × 44	0.1250	0.1102	0.0955	37	0.1040
6 × 32	0.1380	0.1177	0.0974	36	0.1065
6 × 40	0.1380	0.1218	0.1055	33	0.1130
8 × 32	0.1640	0.1437	0.1234	29	0.1360
8 × 36	0.1640	0.1460	0.1279	29	0.1360
10 × 24	0.1900	0.1629	0.1359	26	0.1470
10 × 32	0.1900	0.1697	0.1494	21	0.1590
12 × 24	0.2160	0.1889	0.1619	16	0.1770
12 × 28	0.2160	0.1928	0.1696	15	0.1800
1/4″ × 20	0.2500	0.2175	0.185	7	0.2010
1/4″ × 28	0.2500	0.2268	0.2036	3	0.2130
5/16″ × 18	0.3125	0.2764	0.2403	F	0.2570
5/16″ × 24	0.3125	0.2854	0.2584	I	0.2720
3/8″ × 16	0.3750	0.3344	0.2938	$^{5}/_{16}''$	0.3125
3/8″ × 24	0.3750	0.3479	0.3209	Q	0.332
7/16″ × 14	0.4375	0.3911	0.3447	U	0.368
7/16″ × 20	0.4375	0.4050	0.3726	$^{25}/_{64}''$	0.3906
1/2″ × 13	0.5000	0.4500	0.4001	$^{27}/_{64}''$	0.4219
1/2″ × 20	0.5000	0.4675	0.4351	$^{29}/_{64}''$	0.4531
9/16″ × 12	0.5625	0.5084	0.4542	$^{31}/_{64}''$	0.4844
9/16″ × 18	0.5625	0.5264	0.4903	$^{33}/_{64}''$	0.5156
5/8″ × 11	0.6250	0.5660	0.5069	$^{17}/_{32}''$	0.5312
5/8″ × 18	0.6250	0.5889	0.5528	$^{37}/_{64}''$	0.5781
3/4″ × 10	0.7500	0.6850	0.6201	$^{21}/_{32}''$	0.6562
3/4″ × 16	0.7500	0.7094	0.6688	$^{11}/_{16}''$	0.6875
7/8″ × 9	0.8750	0.8028	0.7307	$^{49}/_{64}''$	0.7656
7/8″ × 14	0.8750	0.8286	0.7822	$^{13}/_{16}''$	0.8125
1″ × 8	1.0000	0.9188	0.8376	$^{7}/_{8}''$	0.8750
1″ × 12	1.0000	0.9459	0.8917	$^{59}/_{64}''$	0.9219
1-1/8″ × 7	1.1250	1.0322	0.9394	$^{63}/_{64}''$	0.9844
1-1/8″ × 12	1.1250	1.0709	1.0168	$1^{3}/_{64}''$	1.0469
1-1/4″ × 7	1.2500	1.1572	1.0644	$1^{7}/_{64}''$	1.1094
1-1/4″ × 12	1.2500	1.1959	1.1418	$1^{11}/_{16}''$	1.1719
1-3/8″ × 6	1.3750	1.2667	1.1585	$1^{7}/_{32}''$	1.2187
1-3/8″ × 12	1.3750	1.3209	1.2668	$1^{19}/_{64}''$	1.2969
1-1/2″ × 6	1.5000	1.3917	1.2835	$1^{11}/_{32}''$	1.3437
1-1/2″ × 12	1.5000	1.4459	1.3918	$1^{27/64}''$	1.4219
1-3/4″ × 5	1.7500	1.6201	1.4902	$1^{9}/_{16}''$	1.5625
2″ × 4-1/2	2.0000	1.8557	1.7113	$1^{25}/_{32}''$	1.7812

Root Diameter = Nominal Diameter − 2(0.75H) = Nominal Diameter − 2(0.75 × 0.866025 × P)

Table 6. Tap Drill Sizes for Threads of American National Form

Screw Thread		Commercial Tap Drills[a]		Screw Thread		Commercial Tap Drills[a]	
Outside Diam. Pitch	Root Diam.	Size or Number	Decimal Equiv.	Outside Diam. Pitch	Root Diam.	Size or Number	Decimal Equiv.
1/16–64	0.0422	3/64	0.0469	27	0.4519	15/32	0.4687
72	0.0445	3/64	0.0469	9/16–12	0.4542	31/64	0.4844
5/64–60	0.0563	1/16	0.0625	18	0.4903	33/64	0.5156
72	0.0601	52	0.0635	27	0.5144	17/32	0.5312
3/32–48	0.0667	49	0.0730	5/8–11	0.5069	17/32	0.5312
50	0.0678	49	0.0730	12	0.5168	35/64	0.5469
7/64–48	0.0823	43	0.0890	18	0.5528	37/64	0.5781
1/8–32	0.0844	3/32	0.0937	27	0.5769	19/32	0.5937
40	0.0925	38	0.1015	11/16–11	0.5694	19/32	0.5937
9/64–40	0.1081	32	0.1160	16	0.6063	5/8	0.6250
5/32–32	0.1157	1/8	0.1250	3/4–10	0.6201	21/32	0.6562
36	0.1202	30	0.1285	12	0.6418	43/64	0.6719
11/64–32	0.1313	9/64	0.1406	16	0.6688	11/16	0.6875
3/16–24	0.1334	26	0.1470	27	0.7019	23/32	0.7187
32	0.1469	22	0.1570	13/16–10	0.6826	23/32	0.7187
13/64–24	0.1490	20	0.1610	7/8–9	0.7307	49/64	0.7656
7/32–24	0.1646	16	0.1770	12	0.7668	51/64	0.7969
32	0.1782	12	0.1890	14	0.7822	13/16	0.8125
15/64–24	0.1806	10	0.1935	18	0.8028	53/64	0.8281
1/4–20	0.1850	7	0.2010	27	0.8269	27/32	0.8437
24	0.1959	4	0.2090	15/16–9	0.7932	53/64	0.8281
27	0.2019	3	0.2130	1–8	0.8376	7/8	0.8750
28	0.2036	3	0.2130	12	0.8918	59/64	0.9219
32	0.2094	7/32	0.2187	14	0.9072	15/16	0.9375
5/16–18	0.2403	F	0.2570	27	0.9519	31/32	0.9687
20	0.2476	17/64	0.2656	1 1/8–7	0.9394	63/64	0.9844
24	0.2584	I	0.2720	12	1.0168	1 3/64	1.0469
27	0.2644	J	0.2770	1 1/4–7	1.0644	1 7/64	1.1094
32	0.2719	9/32	0.2812	12	1.1418	1 11/64	1.1719
3/8–16	0.2938	5/16	0.3125	1 3/8–6	1.1585	1 7/32	1.2187
20	0.3100	21/64	0.3281	12	1.2668	1 19/64	1.2969
24	0.3209	Q	0.3320	1 1/2–6	1.2835	1 11/32	1.3437
27	0.3269	R	0.3390	12	1.3918	1 27/64	1.4219
7/16–14	0.3447	U	0.3680	1 5/8–5 1/2	1.3888	1 29/64	1.4531
20	0.3726	25/64	0.3906	1 3/4–5	1.4902	1 9/16	1.5625
24	0.3834	X	0.3970	1 7/8–5	1.6152	1 11/16	1.6875
27	0.3894	Y	0.4040	2–4 1/2	1.7113	1 25/32	1.7812
1/2–12	0.3918	27/64	0.4219	2 1/8–4 1/2	1.8363	1 29/32	1.9062
13	0.4001	27/64	0.4219	2 1/4–4 1/2	1.9613	2 1/32	2.0312
20	0.4351	29/64	0.4531	2 3/8–4	2.0502	2 1/8	2.1250
24	0.4459	29/64	0.4531	2 1/2–4	2.1752	2 1/4	2.2500

[a] These tap drill diameters allow approximately 75 percent of a full thread to be produced. For small thread sizes in the first column, the use of drills to produce the larger hole sizes will reduce defects caused by tap problems and breakage.

Table 7. Tap Drills and Clearance Drills for Machine Screws with American National Thread Form

Size of Screw		No. of Threads per Inch	Tap Drills		Clearance Hole Drills			
					Close Fit		Free Fit	
No. or Diam.	Decimal Equiv.		Drill Size	Decimal Equiv.	Drill Size	Decimal Equiv.	Drill Size	Decimal Equiv.
0	0.060	80	$\frac{3}{64}$	0.0469	52	0.0635	50	0.0700
1	0.073	64 72	53 53	0.0595 0.0595	48	0.0760	46	0.0810
2	0.086	56 64	50 50	0.0700 0.0700	43	0.0890	41	0.0960
3	0.099	48 56	47 45	0.0785 0.0820	37	0.1040	35	0.1100
4	0.112	36[a] 40 48	44 43 42	0.0860 0.0890 0.0935	32	0.1160	30	0.1285
5	0.125	40 44	38 37	0.1015 0.1040	30	0.1285	29	0.1360
6	0.138	32 40	36 33	0.1065 0.1130	27	0.1440	25	0.1495
8	0.164	32 36	29 29	0.1360 0.1360	18	0.1695	16	0.1770
10	0.190	24 32	25 21	0.1495 0.1590	9	0.1960	7	0.2010
12	0.216	24 28	16 14	0.1770 0.1820	2	0.2210	1	0.2280
14	0.242	20[a] 24[a]	10 7	0.1935 0.2010	D	0.2460	F	0.2570
$\frac{1}{4}$	0.250	20 28	7 3	0.2010 0.2130	F	0.2570	H	0.2660
$\frac{5}{16}$	0.3125	18 24	F I	0.2570 0.2720	P	0.3230	Q	0.3320
$\frac{3}{8}$	0.375	16 24	$\frac{5}{16}$ Q	0.3125 0.3320	W	0.3860	X	0.3970
$\frac{7}{16}$	0.4375	14 20	U $\frac{25}{64}$	0.3680 0.3906	$\frac{29}{64}$	0.4531	$\frac{15}{32}$	0.4687
$\frac{1}{2}$	0.500	13 20	$\frac{27}{64}$ $\frac{29}{64}$	0.4219 0.4531	$\frac{33}{64}$	0.5156	$\frac{17}{32}$	0.5312

[a] These screws are not in the American Standard but are from the former A.S.M.E. Standard.

Table 8. Tap Drills for Pipe Taps

Size of Tap	Drills for Briggs Pipe Taps	Drills for Whitworth Pipe Taps	Size of Tap	Drills for Briggs Pipe Taps	Drills for Whitworth Pipe Taps	Size of Tap	Drills for Briggs Pipe Taps	Drills for Whitworth Pipe Taps
$\frac{1}{8}$	$\frac{11}{32}$	$\frac{5}{16}$	$1\frac{1}{4}$	$1\frac{1}{2}$	$1\frac{15}{32}$	$3\frac{1}{4}$	…	$3\frac{1}{2}$
$\frac{1}{4}$	$\frac{7}{16}$	$\frac{27}{64}$	$1\frac{1}{2}$	$1\frac{23}{32}$	$1\frac{25}{32}$	$3\frac{1}{2}$	$3\frac{3}{4}$	$3\frac{3}{4}$
$\frac{3}{8}$	$\frac{19}{32}$	$\frac{9}{16}$	$1\frac{3}{4}$	…	$1\frac{15}{16}$	$3\frac{3}{4}$	…	4
$\frac{1}{2}$	$\frac{23}{32}$	$\frac{11}{16}$	2	$2\frac{3}{16}$	$2\frac{5}{32}$	4	$4\frac{1}{4}$	$4\frac{1}{4}$
$\frac{5}{8}$	…	$\frac{25}{32}$	$2\frac{1}{4}$	…	$2\frac{13}{32}$	$4\frac{1}{2}$	$4\frac{3}{4}$	$4\frac{3}{4}$
$\frac{3}{4}$	$\frac{15}{16}$	$\frac{29}{32}$	$2\frac{1}{2}$	$2\frac{5}{8}$	$2\frac{25}{32}$	5	$5\frac{5}{16}$	$5\frac{1}{4}$
$\frac{7}{8}$	…	$1\frac{1}{16}$	$2\frac{3}{4}$	…	$3\frac{1}{32}$	$5\frac{1}{2}$	…	$5\frac{3}{4}$
1	$1\frac{5}{32}$	$1\frac{1}{8}$	3	$3\frac{1}{4}$	$3\frac{9}{32}$	6	$6\frac{3}{8}$	$6\frac{1}{4}$

All dimensions are in inches.

To secure the best results, the hole should be reamed before tapping with a reamer having a taper of $\frac{3}{4}$ inch per foot.

Table 9. British Standard Tapping Drill Sizes for ISO Metric Coarse Pitch Series Threads *BS 1157:2004*

Nominal Size and Thread Diam.	Standard Drill Sizes[a]				Nominal. Size and Thread Diam.	Standard Drill Sizes[a]			
	Recommended		Alternative			Recommended		Alternative	
	Size	Theoretical Radial Engagement with Ext.Thread (Percent)	Size	Theoretical Radial Engagement with Ext. Thread (Percent)		Size	Theoretical Radial Engagement with Ext. Thread (Percent)	Size	Theoretical Radial Engagement with Ext. Thread (Percent)
M 1	0.75	81.5	0.78	71.7	M 12	10.20	83.7	10.40	74.5[b]
M 1.1	0.85	81.5	0.88	71.7	M 14	12.00	81.5	12.20	73.4[b]
M 1.2	0.95	81.5	0.98	71.7	M 16	14.00	81.5	14.25	71.3[c]
M 1.4	1.10	81.5	1.15	67.9	M 18	15.50	81.5	15.75	73.4[c]
M 1.6	1.25	81.5	1.30	69.9	M 20	17.50	81.5	17.75	73.4[c]
M 1.8	1.45	81.5	1.50	69.9	M 22	19.50	81.5	19.75	73.4[c]
M 2	1.60	81.5	1.65	71.3	M 24	21.00	81.5	21.25	74.7[b]
M 2.2	1.75	81.5	1.80	72.5	M 27	24.00	81.5	24.25	74.7[b]
M 2.5	2.05	81.5	2.10	72.5	M 30	26.50	81.5	26.75	75.7[b]
M 3	2.50	81.5	2.55	73.4	M 33	29.50	81.5	29.75	75.7[b]
M 3.5	2.90	81.5	2.95	74.7	M 36	32.00	81.5	…	…
M 4	3.30	81.5	3.40	69.9[b]	M 39	35.00	81.5	…	…
M 4.5	3.70	86.8	3.80	76.1	M 42	37.50	81.5	…	…
M 5	4.20	81.5	4.30	71.3[b]	M 45	40.50	81.5	…	…
M 6	5.00	81.5	5.10	73.4	M 48	43.00	81.5	…	…
M 7	6.00	81.5	6.10	73.4	M 52	47.00	81.5	…	…
M 8	6.80	78.5	6.90	71.7[b]	M 56	50.50	81.5	…	…
M 9	7.80	78.5	7.90	71.7[b]	M 60	54.50	81.5	…	…
M 10	8.50	81.5	8.60	76.1	M 64	58.00	81.5	…	…
M 11	9.50	81.5	9.60	76.1	M 68	62.00	81.5	…	…

[a] These tapping drill sizes are for fluted taps only.
[b] For tolerance class 6H and 7H threads only.
[c] For tolerance class 7H threads only.
Drill sizes are given in millimeters.

Table 10. Tap Drill or Core Hole Sizes for Cold Form Tapping ISO Metric Threads

Nominal Size of Tap	Pitch	Recommended Tap Drill Size	Nominal Size of Tap	Pitch	Recommended Tap Drill Size
1.6 mm	0.35 mm	1.45 mm	4.0 mm	0.70 mm	3.7 mm
1.8 mm	0.35 mm	1.65 mm	4.5 mm	0.75 mm	4.2 mm[a]
2.0 mm	0.40 mm	1.8 mm	5.0 mm	0.80 mm	4.6 mm
2.2 mm.	0.45 mm	2.0 mm	6.0 mm	1.00 mm	5.6 mm[a]
2.5 mm	0.45 mm	2.3 mm	7.0 mm	1.00 mm	6.5 mm
3.0 mm	0.50 mm	2.8 mm[a]	8.0 mm	1.25 mm	7.4 mm
3.5 mm	0.60 mm	3.2 mm	10.0 mm	1.50 mm	9.3 mm

[a] These diameters are the nearest stocked drill sizes and not the theoretical hole size, and may not produce 60 to 75 percent full thread.

The sizes are calculated to provide 60 to 75 percent of full thread.

SPEEDS AND FEEDS

Table 1. Recommended Cutting Speeds in Feet per Minute for Turning, Milling, Drilling and Reaming Plain Carbon and Alloy Steels

Material AISI and SAE Steels	Hardness BHN[a]	Material Condition	Cutting Speed, fpm HSS			
			Turning	Milling	Drilling	Reaming
Free-Machining Plain Carbon Steels (Resulfurized)						
1212, 1213, 1215	100–150	HR, A	150	140	120	80
	150–200	CD	160	130	125	80
1108, 1109, 1115, 1117, 1118, 1120, 1126, 1211	100–150	HR, A	130	130	110	75
	150–200	CD	120	115	120	80
1132, 1137, 1139, 1140, 1144, 1146, 1151	175–225	HR, A, N, CD	120	115	100	65
	275–325	Q and T	75	70	70	45
	325–375	Q and T	50	45	45	30
	375–425	Q and T	40	35	35	20
Free-Machining Plain Carbon Steels (Leaded)						
11L17, 11L18, 12L13, 12L14	100–150	HR, A, N, CD	140	140	130	85
	150–200	HR, A, N, CD	145	130	120	80
	200–250	N, CD	110	110	90	60
Plain Carbon Steels						
1006, 1008, 1009, 1010, 1012, 1015, 1016, 1017, 1018, 1019, 1020, 1021, 1022, 1023, 1024, 1025, 1026, 1513, 1514	100–125	HR, A, N, CD	120	110	100	65
	125–175	HR, A, N, CD	110	110	90	60
	175–225	HR, N, CD	90	90	70	45
	225–275	CD	70	65	60	40
1027, 1030, 1033, 1035, 1036, 1037, 1038, 1039, 1040, 1041, 1042, 1043, 1045, 1046, 1048, 1049, 1050, 1052, 1152, 1524, 1526, 1527, 1541	125–175	HR, A, N, CD	100	100	90	60
	175–225	HR, A, N, CD	85	85	75	50
	225–275	N, CD, Q, and T	70	70	60	40
	275–325	Q and T	60	55	50	30
	325–375	Q and T	40	35	35	20
	375–425	Q and T	30	25	25	15
1055, 1060, 1064, 1065, 1070, 1074, 1078, 1080, 1084, 1086, 1090, 1095, 1548, 1551, 1552, 1561, 1566	125–175	HR, A, N, CD	100	90	85	55
	175–225	HR, A, N, CD	80	75	70	45
	225–275	N, CD, Q, and T	65	60	50	30
	275–325	Q and T	50	45	40	25
	325–375	Q and T	35	30	30	20
	375–425	Q and T	30	15	15	10
Free-Machining Alloy Steels (Resulfurized)						
4140, 4150	175–200	HR, A, N, CD	110	100	90	60
	200–250	HR, N, CD	90	90	80	50
	250–300	Q and T	65	60	55	30
	300–375	Q and T	50	45	40	25
	375–425	Q and T	40	35	30	15
Free-Machining Alloy Steels (Leaded)						
41L30, 41L40, 41L47, 41L50, 43L47, 51L32, 52L100, 86L20, 86L40	150–200	HR, A, N, CD	120	115	100	65
	200–250	HR, N, CD	100	95	90	60
	250–300	Q and T	75	70	65	40
	300–375	Q and T	55	50	45	30
	375–425	Q and T	50	40	30	15

Table 1. *(Continued)* **Recommended Cutting Speeds in Feet per Minute for Turning, Milling, Drilling and Reaming Plain Carbon and Alloy Steels**

Material AISI and SAE Steels	Hardness BHN[a]	Material Condition	Cutting Speed, fpm HSS			
			Turning	Milling	Drilling	Reaming
Alloy Steels						
4012, 4023, 4024, 4028, 4118, 4320, 4419, 4422, 4427, 4615, 4620, 4621, 4626, 4718, 4720, 4815, 4817, 4820, 5015, 5117, 5120, 6118, 8115, 8615, 8617, 8620, 8622, 8625, 8627, 8720, 8822, 94B17	125–175	HR, A, N, CD	100	100	85	55
	175–225	HR, A, N, CD	90	90	70	45
	225–275	CD, N, Q and T	70	60	55	35
	275–325	Q and T	60	50	50	30
	325–375	Q and T	50	40	35	25
	375–425	Q and T	35	25	25	15
1330, 1335, 1340, 1345, 4032, 4037, 4042, 4047, 4130, 4135, 4137, 4140, 4142, 4145, 4147, 4150, 4161, 4337, 4340, 50B44, 50B46, 50B50, 50B60, 5130, 5132, 5140, 5145, 5147, 5150, 5160, 51B60, 6150, 81B45, 8630, 8635, 8637, 8640, 8642, 8645, 8650, 8655, 8660, 8740, 9254, 9255, 9260, 9262, 94B30	175–225	HR, A, N, CD	85	75	75	50
	225–275	N, CD, Q and T	70	60	60	40
	275–325	N, Q and T	60	50	45	30
	325–375	N, Q and T	40	35	30	15
	375–425	Q and T	30	20	20	15
E51100, E52100	175–225	HR, A, CD	70	65	60	40
	225–275	N, CD, Q and T	65	60	50	30
	275–325	N, Q and T	50	40	35	25
	325–375	N, Q and T	30	30	30	20
	375–425	Q and T	20	20	20	10
Ultra-High-Strength Steels (Not AISI)						
AMS 6421 (98B37 Mod.), AMS 6422 (98BV40), AMS 6424, AMS 6427, AMS 6428, AMS 6430, AMS 6432, AMS 6433, AMS 6434, AMS 6436, AMS 6442, 300M, D6ac	220–300	A	65	60	50	30
	300–350	N	50	45	35	20
	350–400	N	35	20	20	10
	43–48 RC	Q and T	25	…	…	…
	48–52 RC	Q and T	10	…	…	…
Maraging Steels (Not AISI)						
18% Ni Grade 200, 18% Ni Grade 250, 18% Ni Grade 300, 18% Ni Grade 350	250–325	A	60	50	50	30
	50–52 RC	Maraged	10	…	…	…
Nitriding Steels (Not AISI)						
Nitralloy 125, Nitralloy 135, Nitralloy 135 Mod., Nitralloy 225, Nitralloy 230, Nitralloy N, Nitralloy EZ, Nitrex I	200–250	A	70	60	60	40
	300–350	N, Q and T	30	25	35	20

[a] Abbreviations designate: HR, hot-rolled; CD, cold-drawn; A, annealed; N, normalized; Q and T, quenched and tempered; BHN, Brinell Hardness Number; RC, Rockwell C scale hardness number; HSS, high-speed steel.

Speeds for turning based on a feed rate of 0.012 inch per revolution and a depth of cut of 0.125 inch.

Table 2. Recommended Cutting Speeds in Feet per Minute for Turning, Milling, Drilling and Reaming Ferrous Cast Metals

Material	Hardness BHN[a]	Material Condition	Cutting Speed, fpm HSS			
			Turning	Milling	Drilling	Reaming
Gray Cast Iron						
ASTM Class 20	120–150	A	120	100	100	65
ASTM Class 25	160–200	AC	90	80	90	60
ASTM Class 30, 35, and 40	190–220	AC	80	70	80	55
ASTM Class 45 and 50	220–260	AC	60	50	60	40
ASTM Class 55 and 60	250–320	AC, HT	35	30	30	20
ASTM Type 1, 1b, 5 (Ni Resist)	100–215	AC	70	50	50	30
ASTM Type 2, 3, 6 (Ni Resist)	120–175	AC	65	40	40	25
ASTM Type 2b, 4 (Ni Resist)	150–250	AC	50	30	30	20
Malleable Iron						
(Ferritic), 32510, 35018	110–160	MHT	130	110	110	75
(Pearlitic), 40010, 43010, 45006, 45008, 48005, 50005	160–200	MHT	95	80	80	55
	200–240	MHT	75	65	70	45
(Martensitic), 53004, 60003, 60004	200–255	MHT	70	55	55	35
(Martensitic), 70002, 70003	220–260	MHT	60	50	50	30
(Martensitic), 80002	240–280	MHT	50	45	45	30
(Martensitic), 90001	250–320	MHT	30	25	25	15
Nodular (Ductile) Iron						
(Ferritic), 60-40-18, 65-45-12	140–190	A	100	75	100	65
(Ferritic-Pearlitic), 80-55-06	190–225	AC	80	60	70	45
	225–260	AC	65	50	50	30
(Pearlitic-Martensitic), 100-70-03	240–300	HT	45	40	40	25
(Martensitic), 120-90-02	270–330	HT	30	25	25	15
	330–400	HT	15	–	10	5
Cast Steels						
(Low-Carbon), 1010, 1020	100–150	AC, A, N	110	100	100	65
(Medium-Carbon), 1030, 1040, 1050	125–175	AC, A, N	100	95	90	60
	175–225	AC, A, N	90	80	70	45
	225–300	AC, HT	70	60	55	35
(Low-Carbon Alloy), 1320, 2315, 2320, 4110, 4120, 4320, 8020, 8620	150–200	AC, A, N	90	85	75	50
	200–250	AC, A, N	80	75	65	40
	250–300	AC, HT	60	50	50	30
(Medium-Carbon Alloy), 1330, 1340, 2325, 2330, 4125, 4130, 4140, 4330, 4340, 8030, 80B30, 8040, 8430, 8440, 8630, 8640, 9525, 9530, 9535	175–225	AC, A, N	80	70	70	45
	225–250	AC, A, N	70	65	60	35
	250–300	AC, HT	55	50	45	30
	300–350	AC, HT	45	30	30	20
	350–400	HT	30	…	20	10

[a] Abbreviations designate: A, annealed; AC, as cast; N, normalized; HT, heat-treated; MHT, malleabilizing heat treatment; and BHN, Brinell Hardness Number.

Speeds for turning based on a feed rate of 0.012 inch per revolution and a depth of cut of 0.125 inch.

Table 3. Recommended Cutting Speeds in Feet per Minute for Turning, Milling, Drilling and Reaming Stainless Steels

Material	Hardness BHN[a]	Material Condition	Cutting Speed, fpm HSS			
			Turning	Milling	Drilling	Reaming
Free Machining Stainless Steels (Ferritic)						
430F, 430F Se	135–185	A	110	95	90	60
(Austenitic), 203EZ, 303, 303Se, 303MA, 303Pb, 303Cu, 303 Plus X	135–185	A	100	90	85	55
	225–275	CD	80	75	70	45
(Martensitic), 416, 416Se, 416Plus X, 420F, 420FSe, 440F, 440FSe	135–185	A	110	95	90	60
	185–240	A,CD	100	80	70	45
	275–325	Q and T	60	50	40	25
	375–425	Q and T	30	20	20	10
Stainless Steels						
(Ferritic), 405, 409, 429, 430, 434, 436, 442, 446, 502	135–185	A	90	75	65	45
(Austenitic), 201, 202, 301, 302, 304, 304L, 305, 308, 321, 347, 348	135–185	A	75	60	55	35
	225–275	CD	65	50	50	30
(Austenitic), 302B, 309, 309S, 310, 310S, 314, 316, 316L, 317, 330	135–185	A	70	50	50	30
(Martensitic), 403, 410, 420, 501	135–175	A	95	75	75	50
	175–225	A	85	65	65	45
	275–325	Q and T	55	40	40	25
	375–425	Q and T	35	25	25	15
(Martensitic), 414, 431, Greek Ascoloy	225–275	A	60	55	50	30
	275–325	Q and T	50	45	40	25
	375–425	Q and T	30	25	25	15
(Martensitic), 440A, 440B, 440C	225–275	A	55	50	45	30
	275–325	Q and T	45	40	40	25
	375–425	Q and T	30	20	20	10
(Precipitation-Hardening), 15-5PH, 17-4PH, 17-7PH, AF-71, 17-14CuMo, AFC-77, AM-350, AM-355, AM-362, Custom 455, HNM, PH13-8, PH14-8Mo, PH15-7Mo, Stainless W	150–200	A	60	60	50	30
	275–325	H	50	50	45	25
	325–375	H	40	40	35	20
	375–450	H	25	25	20	10

[a] Abbreviations designate: A, annealed; CD, cold-drawn: N, normalized; H, precipitation-hardened; Q and T, quenched and tempered; and BHN, Brinell Hardness Number.

Speeds for turning based on a feed rate of 0.012 inch per revolution and a depth of cut of 0.125 inch.

Table 4. Recommended Cutting Speeds in Feet per Minute for Turning, Milling, Drilling and Reaming Tool Steels

Material Tool Steels (AISI Types)	Hardness BHN[a]	Material Condition	Cutting Speed, fpm HSS			
			Turning	Milling	Drilling	Reaming
Water-Hardening W1, W2, W5	150–200	A	100	85	85	55
Shock-Resisting S1, S2, S5, S6, S7	175–225	A	70	55	50	35
Cold-Work, Oil-Hardening O1, O2, O6, O7	175–225	A	70	50	45	30
Cold-Work, High-Carbon High-Chromium D2, D3, D4, D5, D7	200–250	A	45	40	30	20
Cold-Work, Air-Hardening A2, A3, A8, A9, A10	200–250	A	70	50	50	35
A4, A6	200–250	A	55	45	45	30
A7	225–275	A	45	40	30	20
Hot-Work, Chromium-Type H10, H11, H12, H13, H14, H19	150–200	A	80	60	60	40
	200–250	A	65	50	50	30
	325–375	Q and T	50	30	30	20
	48–50 RC	Q and T	20	…	…	…
	50–52 RC	Q and T	10	…	…	…
	52–54 RC	Q and T	…	…	…	…
	54–56 RC	Q and T	…	…	…	…
Hot-Work, Tungsten-Type H21, H22, H23, H24, H25, H26	150–200	A	60	55	55	35
	200–250	A	50	45	40	25
Hot-Work, Molybdenum-Type H41, H42, H43	150–200	A	55	55	45	30
	200–250	A	45	45	35	20
Special-Purpose, Low-Alloy L2, L3, L6	150–200	A	75	65	60	40
Mold P2, P3, P4, P5, P6	100–150	A	90	75	75	50
P20, P21	150–200	A	80	60	60	40
High-Speed Steel M1, M2, M6, M10, T1, T2, T6	200–250	A	65	50	45	30
M3-1, M4, M7, M30, M33, M34, M36, M41, M42, M43, M44, M46, M47, T5, T8	225–275	A	55	40	35	20
T15, M3-2	225–275	A	45	30	25	15

[a] Abbreviations designate: A, annealed; Q and T, quenched and tempered; BHN, Brinell Hardness Number; and RC, Rockwell C scale hardness number.

Speeds for turning based on a feed rate of 0.012 inch per revolution and a depth of cut of 0.125 inch.

Table 5. Recommended Cutting Speeds in Feet per Minute for Turning, Milling, Drilling, and Reaming Light Metals

Material Light Metals	Material Condition[a]	Cutting Speed, fpm HSS			
		Turning	Milling	Drilling	Reaming
All Wrought Aluminum Alloys	CD	600	600	400	400
	ST and A	500	500	350	350
All Aluminum Sand and Permanent Mold Casting Alloys	AC	750	750	500	500
	ST and A	600	600	350	350
All Aluminum Die Casting Alloys	AC	125	125	300	300
	ST and A	100	100	70	70
Except Alloys 390.0 and 392.0	AC	80	80	125	100
	ST and A	60	60	45	40
All Wrought Magnesium Alloys	A, CD, ST, and A	800	800	500	500
All Cast Magnesium Alloys	A, AC, ST, and A	800	800	450	450

[a] Abbreviations designate: A, annealed; AC, as cast; CD, cold drawn; ST and A, solution-treated as aged.

Table 6. Recommended Cutting Speeds in Feet per Minute for Turning and Drilling Titanium and Titanium Alloys

Material Titanium and Titanium Alloys	Material Condition[a]	Hardness, BHN[a]	Cutting Speed, fpm HSS
Commercially Pure			
99.5 Ti	A	110–150	110
99.1 Ti, 99.2 Ti	A	180–240	90
99.0 Ti	A	250–275	70
Low Alloyed			
99.5 Ti-0.15 Pd	A	110–150	100
99.2 Ti-0.15 Pd, 98.9 Ti-0.8 Ni-0.3 Mo	A	180–250	85
Alpha Alloys and Alpha-Beta Alloys			
5Al-2.5 Sn, 8Mn, 2Al-11Sn-5Zr-1Mo, 4Al-3Mo-1V, 5Al-6Sn-2Zr-1Mo, 6Al-2Sn-4Zr-2Mo, 6Al-2Sn-4Zr-6Mo, 6Al-2Sn-4Zr-2Mo-0.25Si	A	300–350	50
6Al-4V	A	310–350	40
6Al-6V-2Sn, 7Al-4Mo, 8Al-1Mo-1V	A	320–370	30
8V-5Fe-1Al	A	320–380	20
6Al-4V, 6Al-2Sn-4Zr-2Mo, 6Al-2Sn-4Zr-6Mo, 6Al-2Sn-4Zr-2Mo-0.25Si	ST and A	320–380	40
4Al-3Mo-1V, 6Al-6V-2Sn, 7Al-4Mo	ST and A	375–420	20
1Al-8V-5Fe	ST and A	375–440	20
Beta Alloys			
13V-11Cr-3Al, 8Mo-8V-2Fe-3Al, 3Al-8V-6Cr-4Mo-4Zr, 11.5Mo-6Zr-4.5Sn	A, ST	275–350	25
	ST and A	350–440	20

[a] Abbreviations designate: A, annealed; ST, solution treated; ST and A, solution-treated as aged; and BHN, Brinell Hardness Number.

Table 7. Recommended Cutting Speeds in Feet per Minute for Turning, Milling, and Drilling* Superalloys

Material Superalloys	Cutting Speed, fpm HSS		Material Superalloys	Cutting Speed, fpm HSS	
	Roughing	Finishing		Roughing	Finishing
A-286	30–35	35–40	Mar-M200, M246, M421, and M432	8–10	10–12
AF_2-1DA	8–10	10–15	Mar-M905, and M918	15–20	20–25
Air Resist 213	15–20	20–25	Mar-M302, M322, and M509	10–12	10–15
Air Resist 13, and 215	10–12	10–15	N-12M	8–12	10–15
Astroloy	5–10	5–15	N-155	15–20	15–25
B-1900	8–10	8–10	Nasa C-W-Re	10–12	10–15
CW-12M	8–12	10–15	Nimonic 75, and 80	15–20	20–25
Discalloy	15–35	35–40	Nimonic 90, and 95	10–12	12–15
FSH-H14	10–12	10–15	Refractaloy 26	15–20	20–25
GMR-235, and 235D	8–10	8–10	Rene 41	10–15	12–20
Hastelloy B, C, G, and X (wrought)	15–20	20–25	Rene 80, and 95	8–10	10–15
Hastelloy B, and C (cast)	8–12	10–15	S-590	10–20	15–30
Haynes 25, and 188	15–20	20–25	S-816	10–15	15–20
Haynes 36, and 151	10–12	10–15	T-D Nickel	70–80	80–100
HS 6, 21, 2, 31(X40), 36, and 151	10–12	10–15	Udimet 500, 700, and 710	10–15	12–20
IN 100, and 738	8–10	8–10	Udimet 630	10–20	20–25
Incoloy 800, 801, and 802	30–35	35–40	Unitemp 1753	8–10	10–15
Incoloy 804, and 825	15–20	20–25	V-36	10–15	15–20
Incoloy 901	10–20	20–35	V-57	30–35	35–40
Inconel 625, 702, 706, 718 (wrought), 721, 722, X750, 751, 901, 600, and 604	15–20	20–25	W-545	25–35	30–40
Inconel 700, and 702	10–12	12–15	WI-52	10–12	10–15
Inconel 713C, and 718 (cast)	8–10	8–10	Waspaloy	10–30	25–35
J1300	15–25	20–30	X-45	10–12	10–15
J1570	15–20	20–25	16-25-6	30–35	35–40
M252 (wrought)	15–20	20–25	19-9DL	25–35	30–40
M252 (cast)	8–10	8–10			

*For milling and drilling, use the cutting speeds recommended under roughing.

Table 8. Cutting Feeds and Speeds for Turning, Drilling, and Reaming Copper Alloys

Group 1
Architectural bronze (C38500); Extra-high-leaded brass (C35600); Forging brass (C37700); Free-cutting phosphor bronze, B2 (C54400); Free-cutting brass (C36000); Free-cutting Muntz metal (C37000); High-leaded brass (C33200; C34200); High-leaded brass tube (C35300); Leaded commercial bronze (C31400); Leaded naval brass (C48500); Medium-leaded brass (C34000)
Group 2
Aluminum brass, arsenical (C68700); Cartridge brass, 70% (C26000); High-silicon bronze, B (C65500); Admiralty brass (inhibited) (C44300, C44500); Jewelry bronze, 87.5% (C22600); Leaded Muntz metal (C36500, C36800); Leaded nickel silver (C79600); Low brass, 80% (C24000); Low-leaded brass (C33500); Low-silicon bronze, B (C65100); Manganese bronze, A (C67500); Muntz metal, 60% (C28000); Nickel silver, 55-18 (C77000); Red brass, 85% (C23000); Yellow brass (C26800)
Group 3
Aluminum bronze, D (C61400); Beryllium copper (C17000, C17200, C17500); Commercial bronze, 90% (C22000); Copper nickel, 10% (C70600); Copper nickel, 30% (C71500); Electrolytic tough pitch copper (C11000); Guilding, 95% (C21000); Nickel silver, 65-10 (C74500); Nickel silver, 65-12 (C75700); Nickel silver, 65-15 (C75400); Nickel silver, 65-18 (C75200); Oxygen-free copper (C10200); Phosphor bronze, 1.25% (C50200); Phosphor bronze, 10% D (C52400); Phosphor bronze, 5% A (C51000); Phosphor bronze, 8% C (C52100); Phosphorus deoxidized copper (C12200)

Wrought Alloys Description and UNS Alloy Numbers	Material Condition	Cutting Speed, fpm HSS		
		Turning	Drilling	Reaming
Group 1	A CD	300 350	160 175	160 175
Group 2	A CD	200 250	120 140	110 120
Group 3	A CD	100 110	60 65	50 60

Abbreviations designate: A, annealed; CD, cold-drawn.

Table 9. Cutting Speed Adjustment Factors for Turning with High-Speed Steel Tools

Feed		Feed Factor	Depth of Cut		Depth-of-Cut Factor
in.	mm	F_f	in.	mm	F_d
0.002	0.05	1.50	0.005	0.13	1.50
0.003	0.08	1.50	0.010	0.25	1.42
0.004	0.10	1.50	0.016	0.41	1.33
0.005	0.13	1.44	0.031	0.79	1.21
0.006	0.15	1.34	0.047	1.19	1.15
0.007	0.18	1.25	0.062	1.57	1.10
0.008	0.20	1.18	0.078	1.98	1.07
0.009	0.23	1.12	0.094	2.39	1.04
0.010	0.25	1.08	0.100	2.54	1.03
0.011	0.28	1.04	0.125	3.18	1.00
0.012	0.30	1.00	0.150	3.81	0.97
0.013	0.33	0.97	0.188	4.78	0.94
0.014	0.36	0.94	0.200	5.08	0.93
0.015	0.38	0.91	0.250	6.35	0.91
0.016	0.41	0.88	0.312	7.92	0.88
0.018	0.46	0.84	0.375	9.53	0.86
0.020	0.51	0.80	0.438	11.13	0.84
0.022	0.56	0.77	0.500	12.70	0.82
0.025	0.64	0.73	0.625	15.88	0.80
0.028	0.71	0.70	0.688	17.48	0.78
0.030	0.76	0.68	0.750	19.05	0.77
0.032	0.81	0.66	0.812	20.62	0.76
0.035	0.89	0.64	0.938	23.83	0.75
0.040	1.02	0.60	1.000	25.40	0.74
0.045	1.14	0.57	1.250	31.75	0.73
0.050	1.27	0.55	1.250	31.75	0.72
0.060	1.52	0.50	1.375	34.93	0.71

For use with HSS tool data only from Tables 1 through 8. Adjusted cutting speed $V = V_{HSS} \times F_f \times F_d$, where V_{HSS} is the tabular speed for turning with high-speed tools.

Table 10. Recommended Feed in Inches per Tooth (f_t) for Milling with High-Speed Steel Cutters

Material	Hardness, BHN	End Mills							Plain or Slab Mills	Form Relieved Cutters	Face Mills and Shell End Mills	Slotting and Side Mills
		Depth of Cut, .250 in			Depth of Cut, .050 in							
		Cutter Diam., in			Cutter Diam., in							
		½	¾	1 and up	¼	½	¾	1 and up				
		Feed per Tooth, inch										
Free-machining plain carbon steels	100–185	.001	.003	.004	.001	.002	.003	.004	.003–.008	.005	.004–.012	.002–.008
Plain carbon steels, AISI 1006 to 1030; 1513 to 1522	100–150	.001	.003	.003	.001	.002	.003	.004	.003–.008	.004	.004–.012	.002–.008
	150–200	.001	.002	.003	.001	.002	.002	.003	.003–.008	.004	.003–.012	.002–.008
AISI 1033 to 1095; 1524 to 1566	120–180	.001	.003	.003	.001	.002	.003	.004	.003–.008	.004	.004–.012	.002–.008
	180–220	.001	.002	.003	.001	.002	.002	.003	.003–.008	.004	.003–.012	.002–.008
	220–300	.001	.002	.002	.001	.001	.002	.003	.002–.006	.003	.002–.008	.002–.006
Alloy steels having less than 3% carbon. Typical examples: AISI 4012, 4023, 4027, 4118, 4320 4422, 4427, 4615, 4620, 4626, 4720, 4820, 5015, 5120, 6118, 8115, 8620, 8627, 8720, 8820, 8822, 9310, 93B17	125–175	.001	.003	.003	.001	.002	.003	.004	.003–.008	.004	.004–.012	.002–.008
	175–225	.001	.002	.003	.001	.002	.003	.003	.003–.008	.004	.003–.012	.002–.008
	225–275	.001	.002	.003	.001	.001	.002	.003	.002–.006	.003	.003–.008	.002–.006
	275–325	.001	.002	.002	.001	.001	.002	.002	.002–.005	.003	.002–.008	.002–.005
Alloy steels having 3% carbon or more. Typical examples: AISI 1330, 1340, 4032, 4037, 4130, 4140, 4150, 4340, 50B40, 50B60, 5130, 51B60, 6150, 81B45, 8630, 8640, 86B45, 8660, 8740, 94B30	175–225	.001	.002	.003	.001	.002	.003	.004	.003–.008	.004	.003–.012	.002–.008
	225–275	.001	.002	.003	.001	.001	.002	.003	.002–.006	.003	.003–.010	.002–.006
	275–325	.001	.002	.002	.001	.001	.002	.003	.002–.005	.003	.002–.008	.002–.005
	325–375	.001	.002	.002	.001	.001	.002	.002	.002–.004	.002	.002–.008	.002–.005
Tool steel	150–200	.001	.002	.002	.001	.002	.003	.003	.003–.008	.004	.003–.010	.002–.006
	200–250	.001	.002	.002	.001	.002	.002	.003	.002–.006	.003	.003–.008	.002–.005
Gray cast iron	120–180	.001	.003	.004	.002	.003	.004	.004	.004–.012	.005	.005–.016	.002–.010
	180–225	.001	.002	.003	.001	.002	.003	.003	.003–.010	.004	.004–.012	.002–.008
	225–300	.001	.002	.002	.001	.001	.002	.002	.002–.006	.003	.002–.008	.002–.005
Free malleable iron	110–160	.001	.003	.004	.002	.003	.004	.004	.003–.010	.005	.005–.016	.002–.010

Table 10. *(Continued)* **Recommended Feed in Inches per Tooth (f_t) for Milling with High-Speed Steel Cutters**

Material	Hardness, BHN	End Mills							Plain or Slab Mills	Form Relieved Cutters	Face Mills and Shell End Mills	Slotting and Side Mills
		Depth of Cut, .250 in			Depth of Cut, .050 in							
		Cutter Diam., in			Cutter Diam., in							
		½	¾	1 and up	¼	½	¾	1 and up				
		Feed per Tooth, inch										
Pearlitic-Martensitic malleable iron	160-200	.001	.003	.004	.001	.002	.003	.004	.003–.010	.004	.004–.012	.002–.018
	200-240	.001	.002	.003	.001	.002	.003	.003	.003–.007	.004	.003–.010	.002–.006
	240-300	.001	.002	.002	.001	.001	.002	.002	.002–.006	.003	.002–.008	.002–.005
Cast steel	100-180	.001	.003	.003	.001	.002	.003	.004	.003–.008	.004	.003–.012	.002–.008
	180-240	.001	.002	.003	.001	.002	.003	.003	.003–.008	.004	.003–.010	.002–.006
	240-300	.001	.002	.002	.005	.002	.002	.002	.002–.006	.003	.003–.008	.002–.005
Zinc alloys (die castings)	…	.002	.003	.004	.001	.003	.004	.006	.003–.010	.005	.004–.015	.002–.012
Copper alloys (brasses & bronzes)	100-150	.002	.004	.005	.002	.003	.005	.006	.003–.015	.004	.004–.020	.002–.010
	150-250	.002	.003	.004	.001	.003	.004	.005	.003–.015	.004	.003–.012	.002–.008
Free cutting brasses & bronzes	80-100	.002	.004	.005	.002	.003	.005	.006	.003–.015	.004	.004–.015	.002–.010
Cast aluminum alloys—as cast	…	.003	.004	.005	.002	.004	.005	.006	.005–.016	.006	.005–.020	.004–.012
Cast aluminum alloys—hardened	…	.003	.004	.005	.002	.003	.004	.005	.004–.012	.005	.005–.020	.004–.012
Wrought aluminum alloys— cold drawn	…	.003	.004	.005	.002	.003	.004	.005	.004–.014	.005	.005–.020	.004–.012
Wrought aluminum alloys—hardened	…	.002	.003	.004	.001	.002	.003	.004	.003–.012	.004	.005–.020	.004–.012
Magnesium alloys	…	.003	.004	.005	.003	.004	.005	.007	.005–.016	.006	.008–.020	.005–.012
Ferritic stainless steel	135-185	.001	.002	.003	.001	.002	.003	.003	.002–.006	.004	.004–.008	.002–.007
Austenitic stainless steel	135-185	.001	.002	.003	.001	.002	.003	.003	.003–.007	.004	.005–.008	.002–.007
	185-275	.001	.002	.003	.001	.002	.002	.002	.003–.006	.003	.004–.006	.002–.007
Martensitic stainless steel	135-185	.001	.002	.002	.001	.002	.003	.003	.003–.006	.004	.004–.010	.002–.007
	185-225	.001	.002	.002	.001	.002	.002	.003	.003–.006	.004	.003–.008	.002–.007
	225-300	.0005	.002	.002	.0005	.001	.002	.002	.002–.005	.003	.002–.006	.002–.005
Monel	100-160	.001	.003	.004	.001	.002	.003	.004	.002–.006	.004	.002–.008	.002–.006

Table 11. Cutting Speeds and Equivalent RPM for Drills of Number and Letter Sizes

Size No.	Cutting Speed, Feet per Minute										
	30′	40′	50′	60′	70′	80′	90′	100′	110′	130′	150′
	Revolutions per Minute for Number Sizes										
1	503	670	838	1005	1173	1340	1508	1675	1843	2179	2513
2	518	691	864	1037	1210	1382	1555	1728	1901	2247	2593
4	548	731	914	1097	1280	1462	1645	1828	2010	2376	2741
6	562	749	936	1123	1310	1498	1685	1872	2060	2434	2809
8	576	768	960	1151	1343	1535	1727	1919	2111	2495	2879
10	592	790	987	1184	1382	1579	1777	1974	2171	2566	2961
12	606	808	1010	1213	1415	1617	1819	2021	2223	2627	3032
14	630	840	1050	1259	1469	1679	1889	2099	2309	2728	3148
16	647	863	1079	1295	1511	1726	1942	2158	2374	2806	3237
18	678	904	1130	1356	1582	1808	2034	2260	2479	2930	3380
20	712	949	1186	1423	1660	1898	2135	2372	2610	3084	3559
22	730	973	1217	1460	1703	1946	2190	2433	2676	3164	3649
24	754	1005	1257	1508	1759	2010	2262	2513	2764	3267	3769
26	779	1039	1299	1559	1819	2078	2338	2598	2858	3378	3898
28	816	1088	1360	1631	1903	2175	2447	2719	2990	3534	4078
30	892	1189	1487	1784	2081	2378	2676	2973	3270	3864	4459
32	988	1317	1647	1976	2305	2634	2964	3293	3622	4281	4939
34	1032	1376	1721	2065	2409	2753	3097	3442	3785	4474	5162
36	1076	1435	1794	2152	2511	2870	3228	3587	3945	4663	5380
38	1129	1505	1882	2258	2634	3010	3387	3763	4140	4892	5645
40	1169	1559	1949	2339	2729	3118	3508	3898	4287	5067	5846
42	1226	1634	2043	2451	2860	3268	3677	4085	4494	5311	6128
44	1333	1777	2221	2665	3109	3554	3999	4442	4886	5774	6662
46	1415	1886	2358	2830	3301	3773	4244	4716	5187	6130	7074
48	1508	2010	2513	3016	3518	4021	4523	5026	5528	6534	7539
50	1637	2183	2729	3274	3820	4366	4911	5457	6002	7094	8185
52	1805	2406	3008	3609	4211	4812	5414	6015	6619	7820	9023
54	2084	2778	3473	4167	4862	5556	6251	6945	7639	9028	10417
Size	Revolutions per Minute for Letter Sizes										
A	491	654	818	982	1145	1309	1472	1636	1796	2122	2448
B	482	642	803	963	1124	1284	1445	1605	1765	2086	2407
C	473	631	789	947	1105	1262	1420	1578	1736	2052	2368
D	467	622	778	934	1089	1245	1400	1556	1708	2018	2329
E	458	611	764	917	1070	1222	1375	1528	1681	1968	2292
F	446	594	743	892	1040	1189	1337	1486	1635	1932	2229
G	440	585	732	878	1024	1170	1317	1463	1610	1903	2195
H	430	574	718	862	1005	1149	1292	1436	1580	1867	2154
I	421	562	702	842	983	1123	1264	1404	1545	1826	2106
J	414	552	690	827	965	1103	1241	1379	1517	1793	2068
K	408	544	680	815	951	1087	1223	1359	1495	1767	2039
L	395	527	659	790	922	1054	1185	1317	1449	1712	1976
M	389	518	648	777	907	1036	1166	1295	1424	1683	1942
N	380	506	633	759	886	1012	1139	1265	1391	1644	1897
O	363	484	605	725	846	967	1088	1209	1330	1571	1813
P	355	473	592	710	828	946	1065	1183	1301	1537	1774
Q	345	460	575	690	805	920	1035	1150	1266	1496	1726
R	338	451	564	676	789	902	1014	1127	1239	1465	1690
S	329	439	549	659	769	878	988	1098	1207	1427	1646
T	320	426	533	640	746	853	959	1066	1173	1387	1600
U	311	415	519	623	727	830	934	1038	1142	1349	1557
V	304	405	507	608	709	810	912	1013	1114	1317	1520
W	297	396	495	594	693	792	891	989	1088	1286	1484
X	289	385	481	576	672	769	865	962	1058	1251	1443
Y	284	378	473	567	662	756	851	945	1040	1229	1418
Z	277	370	462	555	647	740	832	925	1017	1202	1387

For fractional drill sizes, use Tables 12a and 12b.

Table 12a. Revolutions per Minute for Various Cutting Speeds and Diameters

Diameter, Inches	Cutting Speed, Feet per Minute											
	40	50	60	70	80	90	100	120	140	160	180	200
	Revolutions per Minute											
1/4	611	764	917	1070	1222	1376	1528	1834	2139	2445	2750	3056
5/16	489	611	733	856	978	1100	1222	1466	1711	1955	2200	2444
3/8	408	509	611	713	815	916	1018	1222	1425	1629	1832	2036
7/16	349	437	524	611	699	786	874	1049	1224	1398	1573	1748
1/2	306	382	459	535	611	688	764	917	1070	1222	1375	1528
9/16	272	340	407	475	543	611	679	813	951	1086	1222	1358
5/8	245	306	367	428	489	552	612	736	857	979	1102	1224
11/16	222	273	333	389	444	500	555	666	770	888	999	1101
3/4	203	254	306	357	408	458	508	610	711	813	914	1016
13/16	190	237	284	332	379	427	474	569	664	758	853	948
7/8	175	219	262	306	349	392	438	526	613	701	788	876
15/16	163	204	244	285	326	366	407	488	570	651	733	814
1	153	191	229	267	306	344	382	458	535	611	688	764
1 1/16	144	180	215	251	287	323	359	431	503	575	646	718
1 1/8	136	170	204	238	272	306	340	408	476	544	612	680
1 3/16	129	161	193	225	258	290	322	386	451	515	580	644
1 1/4	123	153	183	214	245	274	306	367	428	490	551	612
1 5/16	116	146	175	204	233	262	291	349	407	466	524	582
1 3/8	111	139	167	195	222	250	278	334	389	445	500	556
1 7/16	106	133	159	186	212	239	265	318	371	424	477	530
1 1/2	102	127	153	178	204	230	254	305	356	406	457	508
1 9/16	97.6	122	146	171	195	220	244	293	342	390	439	488
1 5/8	93.9	117	141	165	188	212	234	281	328	374	421	468
1 11/16	90.4	113	136	158	181	203	226	271	316	362	407	452
1 3/4	87.3	109	131	153	175	196	218	262	305	349	392	436
1 13/16	84.3	105	126	148	169	190	211	253	295	337	379	422
1 7/8	81.5	102	122	143	163	184	204	244	286	326	367	408
1 15/16	78.9	98	118	138	158	177	197	237	276	315	355	394
2	76.4	95.5	115	134	153	172	191	229	267	306	344	382
2 1/8	72.0	90.0	108	126	144	162	180	216	252	288	324	360
2 1/4	68.0	85.5	102	119	136	153	170	204	238	272	306	340
2 3/8	64.4	80.5	96.6	113	129	145	161	193	225	258	290	322
2 1/2	61.2	76.3	91.7	107	122	138	153	184	213	245	275	306
2 5/8	58.0	72.5	87.0	102	116	131	145	174	203	232	261	290
2 3/4	55.6	69.5	83.4	97.2	111	125	139	167	195	222	250	278
2 7/8	52.8	66.0	79.2	92.4	106	119	132	158	185	211	238	264
3	51.0	63.7	76.4	89.1	102	114	127	152	178	203	228	254
3 1/8	48.8	61.0	73.2	85.4	97.6	110	122	146	171	195	219	244
3 1/4	46.8	58.5	70.2	81.9	93.6	105	117	140	164	188	211	234
3 3/8	45.2	56.5	67.8	79.1	90.4	102	113	136	158	181	203	226
3 1/2	43.6	54.5	65.5	76.4	87.4	98.1	109	131	153	174	196	218
3 5/8	42.0	52.5	63.0	73.5	84.0	94.5	105	126	147	168	189	210
3 3/4	40.8	51.0	61.2	71.4	81.6	91.8	102	122	143	163	184	205
3 7/8	39.4	49.3	59.1	69.0	78.8	88.6	98.5	118	138	158	177	197
4	38.2	47.8	57.3	66.9	76.4	86.0	95.6	115	134	153	172	191
4 1/4	35.9	44.9	53.9	62.9	71.8	80.8	89.8	108	126	144	162	180
4 1/2	34.0	42.4	51.0	59.4	67.9	76.3	84.8	102	119	136	153	170
4 3/4	32.2	40.2	48.2	56.3	64.3	72.4	80.4	96.9	113	129	145	161
5	30.6	38.2	45.9	53.5	61.1	68.8	76.4	91.7	107	122	138	153
5 1/4	29.1	36.4	43.6	50.9	58.2	65.4	72.7	87.2	102	116	131	145
5 1/2	27.8	34.7	41.7	48.6	55.6	62.5	69.4	83.3	97.2	111	125	139
5 3/4	26.6	33.2	39.8	46.5	53.1	59.8	66.4	80.0	93.0	106	120	133
6	25.5	31.8	38.2	44.6	51.0	57.2	63.6	76.3	89.0	102	114	127
6 1/4	24.4	30.6	36.7	42.8	48.9	55.0	61.1	73.3	85.5	97.7	110	122
6 1/2	23.5	29.4	35.2	41.1	47.0	52.8	58.7	70.4	82.2	93.9	106	117
6 3/4	22.6	28.3	34.0	39.6	45.3	50.9	56.6	67.9	79.2	90.6	102	113
7	21.8	27.3	32.7	38.2	43.7	49.1	54.6	65.5	76.4	87.4	98.3	109
7 1/4	21.1	26.4	31.6	36.9	42.2	47.4	52.7	63.2	73.8	84.3	94.9	105
7 1/2	20.4	25.4	30.5	35.6	40.7	45.8	50.9	61.1	71.0	81.4	91.6	102
7 3/4	19.7	24.6	29.5	34.4	39.4	44.3	49.2	59.0	68.9	78.7	88.6	98.4
8	19.1	23.9	28.7	33.4	38.2	43.0	47.8	57.4	66.9	76.5	86.0	95.6

Table 12b. Revolutions per Minute for Various Cutting Speeds and Diameters

Diameter, Inches	Cutting Speed, Feet per Minute											
	225	250	275	300	325	350	375	400	425	450	500	550
	Revolutions per Minute											
1/4	3438	3820	4202	4584	4966	5348	5730	6112	6493	6875	7639	8403
5/16	2750	3056	3362	3667	3973	4278	4584	4889	5195	5501	6112	6723
3/8	2292	2546	2801	3056	3310	3565	3820	4074	4329	4584	5093	5602
7/16	1964	2182	2401	2619	2837	3056	3274	3492	3710	3929	4365	4802
1/2	1719	1910	2101	2292	2483	2675	2866	3057	3248	3439	3821	4203
9/16	1528	1698	1868	2037	2207	2377	2547	2717	2887	3056	3396	3736
5/8	1375	1528	1681	1834	1987	2139	2292	2445	2598	2751	3057	3362
11/16	1250	1389	1528	1667	1806	1941	2084	2223	2362	2501	2779	3056
3/4	1146	1273	1401	1528	1655	1783	1910	2038	2165	2292	2547	2802
13/16	1058	1175	1293	1410	1528	1646	1763	1881	1998	2116	2351	2586
7/8	982	1091	1200	1310	1419	1528	1637	1746	1855	1965	2183	2401
15/16	917	1019	1120	1222	1324	1426	1528	1630	1732	1834	2038	2241
1	859	955	1050	1146	1241	1337	1432	1528	1623	1719	1910	2101
1 1/16	809	899	988	1078	1168	1258	1348	1438	1528	1618	1798	1977
1 1/8	764	849	933	1018	1103	1188	1273	1358	1443	1528	1698	1867
1 3/16	724	804	884	965	1045	1126	1206	1287	1367	1448	1609	1769
1 1/4	687	764	840	917	993	1069	1146	1222	1299	1375	1528	1681
1 5/16	654	727	800	873	946	1018	1091	1164	1237	1309	1455	1601
1 3/8	625	694	764	833	903	972	1042	1111	1181	1250	1389	1528
1 7/16	598	664	730	797	863	930	996	1063	1129	1196	1329	1461
1 1/2	573	636	700	764	827	891	955	1018	1082	1146	1273	1400
1 9/16	550	611	672	733	794	855	916	978	1039	1100	1222	1344
1 5/8	528	587	646	705	764	822	881	940	999	1057	1175	1293
1 11/16	509	566	622	679	735	792	849	905	962	1018	1132	1245
1 3/4	491	545	600	654	709	764	818	873	927	982	1091	1200
1 13/16	474	527	579	632	685	737	790	843	895	948	1054	1159
1 7/8	458	509	560	611	662	713	764	815	866	917	1019	1120
1 15/16	443	493	542	591	640	690	739	788	838	887	986	1084
2	429	477	525	573	620	668	716	764	811	859	955	1050
2 1/8	404	449	494	539	584	629	674	719	764	809	899	988
2 1/4	382	424	468	509	551	594	636	679	721	764	849	933
2 3/8	362	402	442	482	522	563	603	643	683	724	804	884
2 1/2	343	382	420	458	496	534	573	611	649	687	764	840
2 5/8	327	363	400	436	472	509	545	582	618	654	727	800
2 3/4	312	347	381	416	451	486	520	555	590	625	694	763
2 7/8	299	332	365	398	431	465	498	531	564	598	664	730
3	286	318	350	381	413	445	477	509	541	572	636	700
3 1/8	274	305	336	366	397	427	458	488	519	549	611	672
3 1/4	264	293	323	352	381	411	440	470	499	528	587	646
3 3/8	254	283	311	339	367	396	424	452	481	509	566	622
3 1/2	245	272	300	327	354	381	409	436	463	490	545	600
3 5/8	237	263	289	316	342	368	395	421	447	474	527	579
3 3/4	229	254	280	305	331	356	382	407	433	458	509	560
3 7/8	221	246	271	295	320	345	369	394	419	443	493	542
4	214	238	262	286	310	334	358	382	405	429	477	525
4 1/4	202	224	247	269	292	314	337	359	383	404	449	494
4 1/2	191	212	233	254	275	297	318	339	360	382	424	466
4 3/4	180	201	221	241	261	281	301	321	341	361	402	442
5	171	191	210	229	248	267	286	305	324	343	382	420
5 1/4	163	181	199	218	236	254	272	290	308	327	363	399
5 1/2	156	173	190	208	225	242	260	277	294	312	347	381
5 3/4	149	166	182	199	215	232	249	265	282	298	332	365
6	143	159	174	190	206	222	238	254	270	286	318	349
6 1/4	137	152	168	183	198	213	229	244	259	274	305	336
6 1/2	132	146	161	176	190	205	220	234	249	264	293	322
6 3/4	127	141	155	169	183	198	212	226	240	254	283	311
7	122	136	149	163	177	190	204	218	231	245	272	299
7 1/4	118	131	144	158	171	184	197	210	223	237	263	289
7 1/2	114	127	139	152	165	178	190	203	216	229	254	279
7 3/4	111	123	135	148	160	172	185	197	209	222	246	271
8	107	119	131	143	155	167	179	191	203	215	238	262

Table 13a. Revolutions per Minute for Various Cutting Speeds and Diameters (Metric Units)

Diam., mm	Cutting Speed, Meters per Minute											
	5	6	8	10	12	16	20	25	30	35	40	45
	Revolutions per Minute											
5	318	382	509	637	764	1019	1273	1592	1910	2228	2546	2865
6	265	318	424	530	637	849	1061	1326	1592	1857	2122	2387
8	199	239	318	398	477	637	796	995	1194	1393	1592	1790
10	159	191	255	318	382	509	637	796	955	1114	1273	1432
12	133	159	212	265	318	424	531	663	796	928	1061	1194
16	99.5	119	159	199	239	318	398	497	597	696	796	895
20	79.6	95.5	127	159	191	255	318	398	477	557	637	716
25	63.7	76.4	102	127	153	204	255	318	382	446	509	573
30	53.1	63.7	84.9	106	127	170	212	265	318	371	424	477
35	45.5	54.6	72.8	90.9	109	145	182	227	273	318	364	409
40	39.8	47.7	63.7	79.6	95.5	127	159	199	239	279	318	358
45	35.4	42.4	56.6	70.7	84.9	113	141	177	212	248	283	318
50	31.8	38.2	51	63.7	76.4	102	127	159	191	223	255	286
55	28.9	34.7	46.3	57.9	69.4	92.6	116	145	174	203	231	260
60	26.6	31.8	42.4	53.1	63.7	84.9	106	133	159	186	212	239
65	24.5	29.4	39.2	49	58.8	78.4	98	122	147	171	196	220
70	22.7	27.3	36.4	45.5	54.6	72.8	90.9	114	136	159	182	205
75	21.2	25.5	34	42.4	51	68	84.9	106	127	149	170	191
80	19.9	23.9	31.8	39.8	47.7	63.7	79.6	99.5	119	139	159	179
90	17.7	21.2	28.3	35.4	42.4	56.6	70.7	88.4	106	124	141	159
100	15.9	19.1	25.5	31.8	38.2	51	63.7	79.6	95.5	111	127	143
110	14.5	17.4	23.1	28.9	34.7	46.2	57.9	72.3	86.8	101	116	130
120	13.3	15.9	21.2	26.5	31.8	42.4	53.1	66.3	79.6	92.8	106	119
130	12.2	14.7	19.6	24.5	29.4	39.2	49	61.2	73.4	85.7	97.9	110
140	11.4	13.6	18.2	22.7	27.3	36.4	45.5	56.8	68.2	79.6	90.9	102
150	10.6	12.7	17	21.2	25.5	34	42.4	53.1	63.7	74.3	84.9	95.5
160	9.9	11.9	15.9	19.9	23.9	31.8	39.8	49.7	59.7	69.6	79.6	89.5
170	9.4	11.2	15	18.7	22.5	30	37.4	46.8	56.2	65.5	74.9	84.2
180	8.8	10.6	14.1	17.7	21.2	28.3	35.4	44.2	53.1	61.9	70.7	79.6
190	8.3	10	13.4	16.8	20.1	26.8	33.5	41.9	50.3	58.6	67	75.4
200	8	9.5	12.7	15.9	19.1	25.5	31.8	39.8	47.7	55.7	63.7	71.6
220	7.2	8.7	11.6	14.5	17.4	23.1	28.9	36.2	43.4	50.6	57.9	65.1
240	6.6	8	10.6	13.3	15.9	21.2	26.5	33.2	39.8	46.4	53.1	59.7
260	6.1	7.3	9.8	12.2	14.7	19.6	24.5	30.6	36.7	42.8	49	55.1
280	5.7	6.8	9.1	11.4	13.6	18.2	22.7	28.4	34.1	39.8	45.5	51.1
300	5.3	6.4	8.5	10.6	12.7	17	21.2	26.5	31.8	37.1	42.4	47.7
350	4.5	5.4	7.3	9.1	10.9	14.6	18.2	22.7	27.3	31.8	36.4	40.9
400	4	4.8	6.4	8	9.5	12.7	15.9	19.9	23.9	27.9	31.8	35.8
450	3.5	4.2	5.7	7.1	8.5	11.3	14.1	17.7	21.2	24.8	28.3	31.8
500	3.2	3.8	5.1	6.4	7.6	10.2	12.7	15.9	19.1	22.3	25.5	28.6

Table 13b. Revolutions per Minute for Various Cutting Speeds and Diameters (Metric Units)

Diam. mm	Cutting Speed, Meters per Minute											
	50	55	60	65	70	75	80	85	90	95	100	200
	Revolutions per Minute											
5	3183	3501	3820	4138	4456	4775	5093	5411	5730	6048	6366	12,732
6	2653	2918	3183	3448	3714	3979	4244	4509	4775	5039	5305	10,610
8	1989	2188	2387	2586	2785	2984	3183	3382	3581	3780	3979	7958
10	1592	1751	1910	2069	2228	2387	2546	2706	2865	3024	3183	6366
12	1326	1459	1592	1724	1857	1989	2122	2255	2387	2520	2653	5305
16	995	1094	1194	1293	1393	1492	1591	1691	1790	1890	1989	3979
20	796	875	955	1034	1114	1194	1273	1353	1432	1512	1592	3183
25	637	700	764	828	891	955	1019	1082	1146	1210	1273	2546
30	530	584	637	690	743	796	849	902	955	1008	1061	2122
35	455	500	546	591	637	682	728	773	819	864	909	1818
40	398	438	477	517	557	597	637	676	716	756	796	1592
45	354	389	424	460	495	531	566	601	637	672	707	1415
50	318	350	382	414	446	477	509	541	573	605	637	1273
55	289	318	347	376	405	434	463	492	521	550	579	1157
60	265	292	318	345	371	398	424	451	477	504	530	1061
65	245	269	294	318	343	367	392	416	441	465	490	979
70	227	250	273	296	318	341	364	387	409	432	455	909
75	212	233	255	276	297	318	340	361	382	403	424	849
80	199	219	239	259	279	298	318	338	358	378	398	796
90	177	195	212	230	248	265	283	301	318	336	354	707
100	159	175	191	207	223	239	255	271	286	302	318	637
110	145	159	174	188	203	217	231	246	260	275	289	579
120	133	146	159	172	186	199	212	225	239	252	265	530
130	122	135	147	159	171	184	196	208	220	233	245	490
140	114	125	136	148	159	171	182	193	205	216	227	455
150	106	117	127	138	149	159	170	180	191	202	212	424
160	99.5	109	119	129	139	149	159	169	179	189	199	398
170	93.6	103	112	122	131	140	150	159	169	178	187	374
180	88.4	97.3	106	115	124	133	141	150	159	168	177	354
190	83.8	92.1	101	109	117	126	134	142	151	159	167	335
200	79.6	87.5	95.5	103	111	119	127	135	143	151	159	318
220	72.3	79.6	86.8	94	101	109	116	123	130	137	145	289
240	66.3	72.9	79.6	86.2	92.8	99.5	106	113	119	126	132	265
260	61.2	67.3	73.4	79.6	85.7	91.8	97.9	104	110	116	122	245
280	56.8	62.5	68.2	73.9	79.6	85.3	90.9	96.6	102	108	114	227
300	53.1	58.3	63.7	69	74.3	79.6	84.9	90.2	95.5	101	106	212
350	45.5	50	54.6	59.1	63.7	68.2	72.8	77.3	81.8	99.1	91	182
400	39.8	43.8	47.7	51.7	55.7	59.7	63.7	67.6	71.6	75.6	79.6	159
450	35.4	38.9	42.4	46	49.5	53.1	56.6	60.1	63.6	67.2	70.7	141
500	31.8	35	38.2	41.4	44.6	47.7	50.9	54.1	57.3	60.5	63.6	127

Speeds and Feeds in Diamond Grinding.—General recommendations are as follows:

Wheel Speeds: The generally recommended wheel speeds for diamond grinding are in the range of 5000 to 6000 surface feet per minute, with this upper limit as a maximum to avoid harmful "overspeeding." Exceptions from that general rule are diamond wheels with coarse grains and high concentration (100 percent) where the wheel wear in dry surface grinding can be reduced by lowering the speed to 2500–3000 sfpm. However, this lower speed range can cause rapid wheel breakdown in finer grit wheels or in those with reduced diamond concentration.

Work Speeds: In diamond grinding, work rotation and table traverse are usually established by experience, adjusting these values to the selected infeed so as to avoid excessive wheel wear.

Infeed per Pass: Often referred to as downfeed and usually a function of the grit size of the wheel. The following are general values which may be increased for raising the productivity, or lowered to improve finish or to reduce wheel wear.

Wheel Grit Size Range	Infeed per Pass
100 to 120	0.001 inch
150 to 220	0.0005 inch
250 and finer	0.00025 inch

Table 14. Suggested Approximate Speeds and Feeds for Drilling Various Thermoplastics[a]

Material	1/16 in. Diameter		1/4 in. Diameter		1 in. Diameter	
	Speed (RPM)	Feed (in/rev)	Speed (RPM)	Feed (in/rev)	Speed (RPM)	Feed (in/rev)
ABS	6000	0.015	2000	0.040	500	0.080
Acetal (POM)	12000	0.010	4000	0.030	1000	0.060
Acrylic (PMMA)	9000	0.010	3000	0.030	750	0.060
Polyamide, Nylon 6/6 and PA6	6000	0.010	2000	0.030	500	0.060
Polycarbonate (PC)	9000	0.010	3000	0.030	750	0.060
Polyester (PET)	9000	0.007	3000	0.020	750	0.040
Polyether ether ketone (PEEK)	12000	0.005	4000	0.015	1000	0.030
Polyethylene (PE)	12000	0.015	4000	0.040	1000	0.080
Polyphenylene sulfide (PPS)	3000	0.010	1000	0.030	250	0.060
Polypropylene (PP)	12000	0.015	4000	0.040	1000	0.080
Polystyrene (PS)	6000	0.007	2000	0.020	500	0.040
Polytetrafluoroethylene (PTFE)	9000	0.007	3000	0.020	750	0.040
Polyvinyl chloride (PVC)	12000	0.007	4000	0.020	1000	0.040
Ultem polyetherimide (PEI)	12000	0.010	4000	0.030	1000	0.040

[a] Using a two-fluted drill.

Table 15. Speeds and Numbers of Teeth for Sawing Plastics Materials with High-Carbon Steel Saw Blades

Material Thickness		Number of Teeth/inch on Blade	Peripheral Speed			
			Thermoset Cast or Molded Plastics		Thermoplastics (and Epoxy, Melamine, Phenolic and Allyl Thermosets)	
(inch)	(mm)		(ft/min)	(m/min)	(ft/min)	(m/min)
0–0.5	0–13	8–14	2000–3000	607–914	4000–5000	1219–1524
0.5–1	13–25	6–8	1800–2200	549–671	3500–4300	1067–1311
1–3	25–76	3	1500–2200	475–671	3000–3500	914–1067
>3	>76	>3	1200–1800	366–549	2500–3000	762–914

MILLING CUTTERS

Milling Cutter Terms and Parts

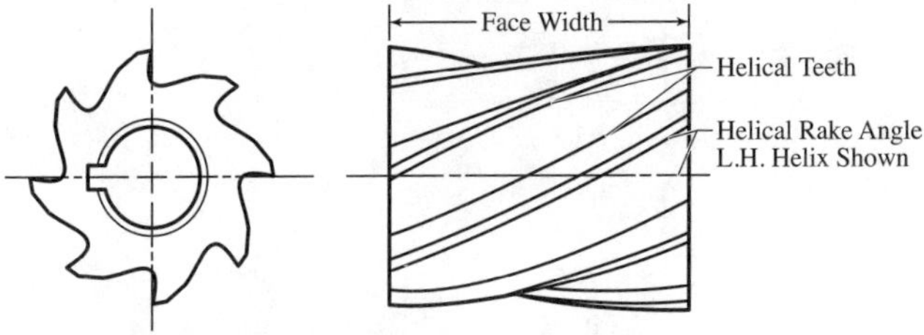

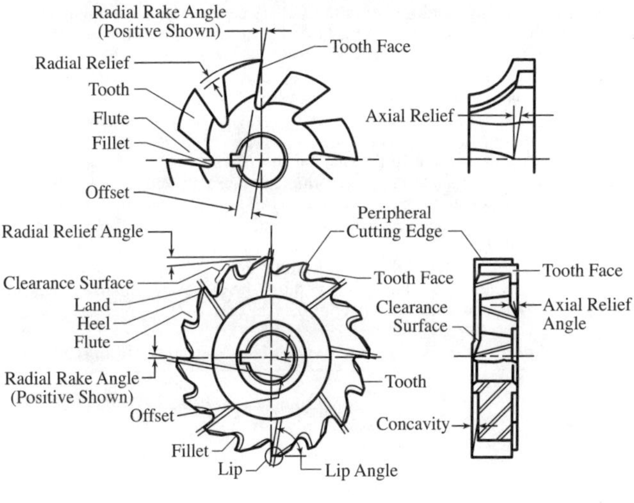

End Mill Terms and Parts

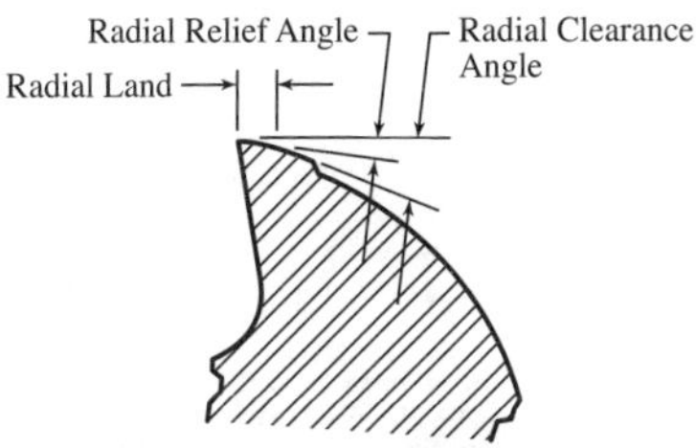

Enlarged Section of End Mill Tooth

End Mill Terms and Parts *(Continued)*

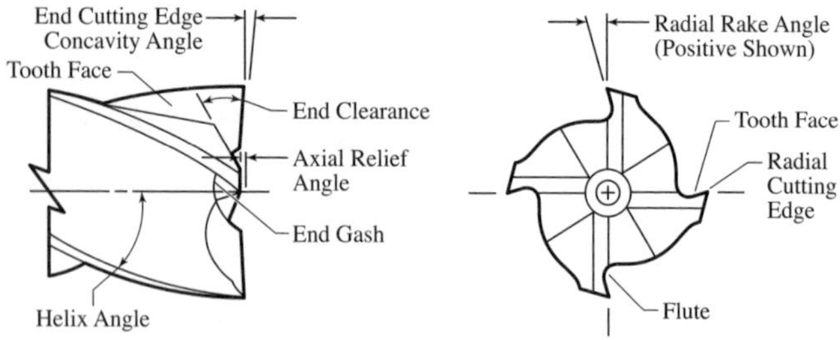

Enlarged Section of End Mill

Wheels for Sharpening Milling Cutters.—Milling cutters may be sharpened either by using the periphery of a disk wheel or the face of a cup wheel. The latter grinds the lands of the teeth flat, whereas the periphery of a disk wheel leaves the teeth slightly concave back of the cutting edges. The concavity produced by disk wheels reduces the effective clearance angle on the teeth, the effect being more pronounced for wheels of small diameter than for wheels of large diameter. For this reason, large diameter wheels are preferred when sharpening milling cutters with disk-type wheels. Irrespective of what type of wheel is used to sharpen a milling cutter, any burrs resulting from grinding should be carefully removed by a hand-stoning operation. Stoning also helps to reduce the roughness of grinding marks and improves the quality of the finish produced on the surface being machined. Unless done very carefully, hand stoning may dull the cutting edge. Stoning may be avoided and a sharper cutting edge produced if the wheel rotates toward the cutting edge, which requires that the operator maintain contact between the tool and the rest while the wheel rotation is trying to move the tool away from the rest. Though slightly more difficult, this method will eliminate the burr.

Table 1. Specifications of Grinding Wheels for Sharpening Milling Cutters

Cutter Material	Operation	Grinding Wheel			
		Abrasive Material	Grain Size	Grade	Bond
Carbon Tool Steel	Roughing	Aluminum Oxide	46–60	K	Vitrified
	Finishing	Aluminum Oxide	100	H	Vitrified
High-Speed Steel:					
18-4-1 {	Roughing	Aluminum Oxide	60	K, H	Vitrified
	Finishing	Aluminum Oxide	100	H	Vitrified
18-4-2 {	Roughing	Aluminum Oxide	80	F, G, H	Vitrified
	Finishing	Aluminum Oxide	100	H	Vitrified
Cast Nonferrous Tool Material	Roughing	Aluminum Oxide	46	H, K, L, N	Vitrified
	Finishing	Aluminum Oxide	100–120	H	Vitrified
Sintered Carbide	Roughing after Brazing	Silicon Carbide	60	G	Vitrified
	Roughing	Diamond	100	[a]	Resinoid
	Finishing	Diamond	Up to 500	[a]	Resinoid
Carbon Tool Steel and High-Speed Steel[b]	Roughing	Cubic Boron Nitride	80–100	R, P	Resinoid
	Finishing	Cubic Boron Nitride	100–120	S, T	Resinoid

[a] Not indicated in diamond wheel markings.

[b] For hardnesses above Rockwell C 56.

Wheel Speeds and Feeds for Sharpening Milling Cutters.—Relatively low cutting speeds should be used when sharpening milling cutters to avoid tempering and heat checking. Dry grinding is recommended in all cases except when diamond wheels are employed. The surface speed of grinding wheels should be in the range of 4500 to 6500 feet per minute for grinding milling cutters of high-speed steel or cast nonferrous tool material. For sintered carbide cutters, 5000 to 5500 feet per minute should be used.

The maximum stock removed per pass of the grinding wheel should not exceed about 0.0004 inch for sintered carbide cutters; 0.003 inch for large high-speed steel and cast nonferrous tool material cutters; and 0.0015 inch for narrow saws and slotting cutters of high-speed steel or cast nonferrous tool material. The stock removed per pass of the wheel may be increased for backing-off operations such as the grinding of secondary clearance behind the teeth since there is usually a sufficient body of metal to carry off the heat.

Clearance Angles for Milling Cutter Teeth.—The clearance angle provided on the cutting edges of milling cutters has an important bearing on cutter performance, cutting efficiency, and cutter life between sharpenings. It is desirable in all cases to use a clearance angle as small as possible so as to leave more metal back of the cutting edges for better heat dissipation and to provide maximum support. Excessive clearance angles not only weaken the cutting edges but also increase the likelihood of "chatter," which will result in poor finish on the machined surface and reduce the life of the cutter. According to The Cincinnati Milling Machine Co., milling cutters used for general purpose work and having diameters from $^1/_8$ to 3 inches should have clearance angles from 13 to 5 degrees, respectively, decreasing proportionately as the diameter increases. General purpose cutters over 3 inches in diameter should be provided with a clearance angle of 4 to 5 degrees. The land width is usually $^1/_{64}$, $^1/_{32}$, and $^1/_{16}$ inch, respectively, for small, medium, and large cutters.

The primary clearance or relief angle for best results varies according to the material being milled about as follows: low-carbon, high-carbon, and alloy steels, 3 to 5 degrees; cast iron and medium and hard bronze, 4 to 7 degrees; brass, soft bronze, aluminum, magnesium, plastics, etc., 10 to 12 degrees. When milling cutters are resharpened, it is customary to grind a secondary clearance angle of 3 to 5 degrees behind the primary clearance angle to reduce the land width to its original value and thus avoid interference with the surface to be milled. A general formula for plain milling cutters, face mills, and form relieved cutters which gives the clearance angle C, in degrees, necessitated by the feed per revolution F, in inches, the width of land L, in inches, the depth of cut d, in inches, the cutter diameter D, in inches, and the Brinell Hardness Number (BHN) B of the work being cut is:

$$C = \frac{45860}{DB}\ 1.5L + \frac{F}{\pi D}\sqrt{d(D-d)}$$

Rake Angles for Milling Cutters.—In peripheral milling cutters, the rake angle is generally defined as the angle in degrees that the tooth face deviates from a radial line to the cutting edge. In face milling cutters, the teeth are inclined with respect to both the radial and axial lines. These angles are called *radial* and *axial* rake, respectively. The radial and axial rake angles may be positive, zero, or negative.

Positive rake angles should be used whenever possible for all types of high-speed steel milling cutters. For sintered carbide-tipped cutters, zero and negative rake angles are frequently employed to provide more material back of the cutting edge to resist shock loads.

Rake Angles for High-Speed Steel Cutters: Positive rake angles of 10 to 15 degrees are satisfactory for milling steels of various compositions with plain milling cutters. For softer materials such as magnesium and aluminum alloys, the rake angle may be 25 degrees or more. Metal slitting saws for cutting alloy steel usually have rake angles from 5 to 10 degrees, whereas zero and sometimes negative rake angles are used for saws to cut copper and other soft nonferrous metals to reduce the tendency to "hog in." Form relieved cutters usually have rake angles of 0, 5, or 10 degrees. Commercial face milling cutters usually have 10 degrees positive radial and axial rake angles for general use in milling cast iron, forged and alloy steel, brass, and bronze; for milling castings and forgings of magnesium and free-cutting aluminum and their alloys, the rake angles may be increased to 25 degrees positive or more, depending on the operating conditions; a smaller rake angle is used for abrasive or difficult to machine aluminum alloys.

Cast Nonferrous Tool Material Milling Cutters: Positive rake angles are generally provided on milling cutters using cast nonferrous tool materials although negative rake angles may be used advantageously for some operations such as those where shock loads are encountered or where it is necessary to eliminate vibration when milling thin sections.

Sintered Carbide Milling Cutters: Peripheral milling cutters, such as slab mills, slotting cutters, saws, etc., tipped with sintered carbide, generally have negative radial rake angles of 5 degrees for soft low-carbon steel and 10 degrees or more for alloy steels. Positive axial rake angles of 5 and 10 degrees, respectively, may be provided, and for slotting saws and cutters, 0 degree axial rake may be used. On soft materials, such as free-cutting aluminum alloys, positive rake angles of 10 to so degrees are used. For milling abrasive or difficult to machine aluminum alloys, small positive or even negative rake angles are used.

Various Set-Ups Used in Grinding the Clearance Angle on Milling Cutter Teeth

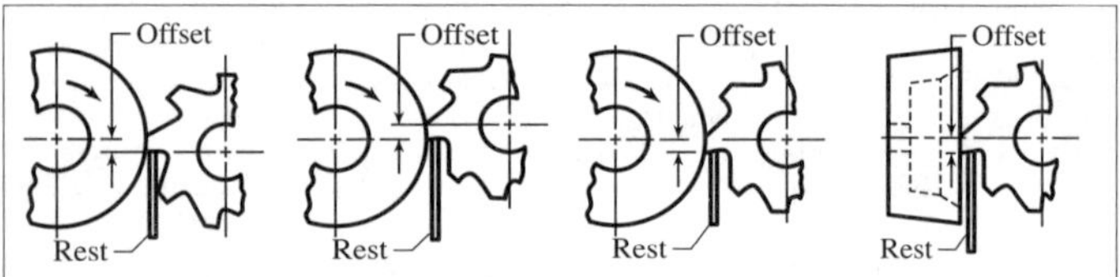

Wheel Above Center | Wheel Below Center | In-Line Centers | Cup Wheel

Table 2. Distance to Set Center of Wheel Above the Cutter Center (Disk Wheel)

Dia. of Wheel, Inches	Desired Clearance Angle, Degrees											
	1	2	3	4	5	6	7	8	9	10	11	12
	[a]Distance to Offset Wheel Center Above Cutter Center, Inches											
3	0.026	0.052	0.079	0.105	0.131	0.157	0.183	0.209	0.235	0.260	0.286	0.312
4	0.035	0.070	0.105	0.140	0.174	0.209	0.244	0.278	0.313	0.347	0.382	0.416
5	0.044	0.087	0.131	0.174	0.218	0.261	0.305	0.348	0.391	0.434	0.477	0.520
6	0.052	0.105	0.157	0.209	0.261	0.314	0.366	0.417	0.469	0.521	0.572	0.624
7	0.061	0.122	0.183	0.244	0.305	0.366	0.427	0.487	0.547	0.608	0.668	0.728
8	0.070	0.140	0.209	0.279	0.349	0.418	0.488	0.557	0.626	0.695	0.763	0.832
9	0.079	0.157	0.236	0.314	0.392	0.470	0.548	0.626	0.704	0.781	0.859	0.936
10	0.087	0.175	0.262	0.349	0.436	0.523	0.609	0.696	0.782	0.868	0.954	10.040

[a] Calculated from the formula: Offset = Cutter Diameter × $\frac{1}{2}$ × Sine of Clearance Angle.

Table 3. Distance to Set Center of Wheel Below the Cutter Center (Disk Wheel)

Dia. of Cutter, Inches	Desired Clearance Angle, Degrees											
	1	2	3	4	5	6	7	8	9	10	11	12
	[a]Distance to Offset Wheel Center Below Cutter Center, Inches											
2	0.017	0.035	0.052	0.070	0.087	0.105	0.122	0.139	0.156	0.174	0.191	0.208
3	0.026	0.052	0.079	0.105	0.131	0.157	0.183	0.209	0.235	0.260	0.286	0.312
4	0.035	0.070	0.105	0.140	0.174	0.209	0.244	0.278	0.313	0.347	0.382	0.416
5	0.044	0.087	0.131	0.174	0.218	0.261	0.305	0.348	0.391	0.434	0.477	0.520
6	0.052	0.105	0.157	0.209	0.261	0.314	0.366	0.417	0.469	0.521	0.572	0.624
7	0.061	0.122	0.183	0.244	0.305	0.366	0.427	0.487	0.547	0.608	0.668	0.728
8	0.070	0.140	0.209	0.279	0.349	0.418	0.488	0.557	0.626	0.695	0.763	0.832
9	0.079	0.157	0.236	0.314	0.392	0.470	0.548	0.626	0.704	0.781	0.859	0.936
10	0.087	0.175	0.262	0.349	0.436	0.523	0.609	0.696	0.782	0.868	0.954	10.040

Distance to Set Tooth Rest Below Center Line of Wheel and Cutter.—When the clearance angle is ground with a disk-type wheel by keeping the center-line of the wheel in line with the centerline of the cutter, the tooth rest should be lowered by an amount given by the following formula:

$$\text{Offset} = \frac{\text{Wheel Diam.} \times \text{Cutter Diam.} \times \text{Sine of One-half the Clearance Angle}}{\text{Wheel Diam.} + \text{Cutter Diam.}}$$

Distance to Set Tooth Rest Below Cutter Center When Cup Wheel is Used.—When the clearance is ground with a cup wheel, the tooth rest is set below the center of the cutter the same amount as given in Table 3.

Table 4. American National Standard Multiple- and Two-Flute Single-End Helical End Mills with Plain Straight and Weldon Shanks *ANSI/ASME B94.19-1997 (R2019)*

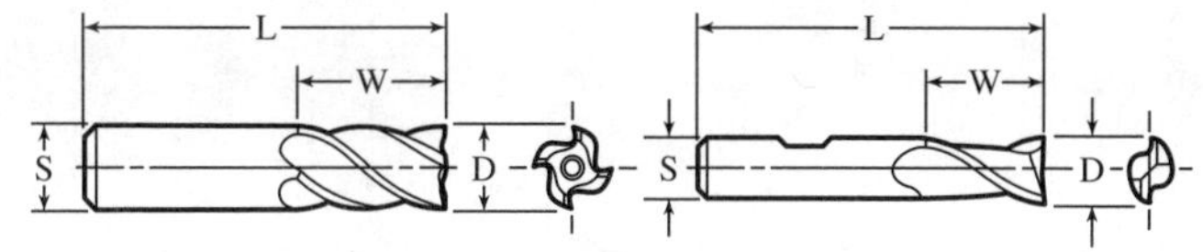

Cutter Diameter, *D*			Shank Diameter, *S*		Length of Cut, *W*	Length Overall, *L*
Nom.	Max.	Min.	Max.	Min.		
Multiple-Flute with Plain Straight Shanks						
$\frac{1}{8}$	.130	.125	.125	.1245	$\frac{5}{16}$	$1\frac{1}{4}$
$\frac{3}{16}$	.1925	.1875	.1875	.1870	$\frac{1}{2}$	$1\frac{3}{8}$
$\frac{1}{4}$	.255	.250	.250	.2495	$\frac{5}{8}$	$1\frac{11}{16}$
$\frac{3}{8}$	.380	.375	.375	.3745	$\frac{3}{4}$	$1\frac{13}{16}$
$\frac{1}{2}$	.505	.500	.500	.4995	$\frac{15}{16}$	$2\frac{1}{4}$
$\frac{3}{4}$	.755	.750	.750	.7495	$1\frac{1}{4}$	$2\frac{5}{8}$
Two-Flute for Keyway Cutting with Weldon Shanks						
$\frac{1}{8}$	.125	.1235	.375	.3745	$\frac{3}{8}$	$2\frac{5}{16}$
$\frac{3}{16}$	.1875	.1860	.375	.3745	$\frac{7}{16}$	$2\frac{5}{16}$
$\frac{1}{4}$	.250	.2485	.375	.3745	$\frac{1}{2}$	$2\frac{5}{16}$
$\frac{5}{16}$	.3125	.3110	.375	.3745	$\frac{9}{16}$	$2\frac{5}{16}$
$\frac{3}{8}$	.375	.3735	.375	.3745	$\frac{9}{16}$	$2\frac{5}{16}$
$\frac{1}{2}$	.500	.4985	.500	.4995	1	3
$\frac{5}{8}$	.625	.6235	.625	.6245	$1\frac{5}{16}$	$3\frac{7}{16}$
$\frac{3}{4}$	.750	.7485	.750	.7495	$1\frac{5}{16}$	$3\frac{9}{16}$
$\frac{7}{8}$	.875	.8735	.875	.8745	$1\frac{1}{2}$	$3\frac{3}{4}$
1	1.000	.9985	1.000	.9995	$1\frac{5}{8}$	$4\frac{1}{8}$
$1\frac{1}{4}$	1.250	1.2485	1.250	1.2495	$1\frac{5}{8}$	$4\frac{1}{8}$
$1\frac{1}{2}$	1.500	1.4985	1.250	1.2495	$1\frac{5}{8}$	$4\frac{1}{8}$

All dimensions are in inches. All cutters are high-speed steel. Right-hand cutters with right-hand helix are standard.

The helix angle is not less than 10 degrees for multiple-flute cutters with plain straight shanks; the helix angle is optional with the manufacturer for two-flute cutters with Weldon shanks.

Tolerances: On *W*, $\pm\frac{1}{32}$ inch; on *L*, $\pm\frac{1}{16}$ inch.

Table 5. ANSI Regular-, Long-, and Extra-Long-Length, Multiple-Flute Medium Helix Single-End End Mills with Weldon Shanks *ANSI/ASME B94.19-1997 (R2019)*

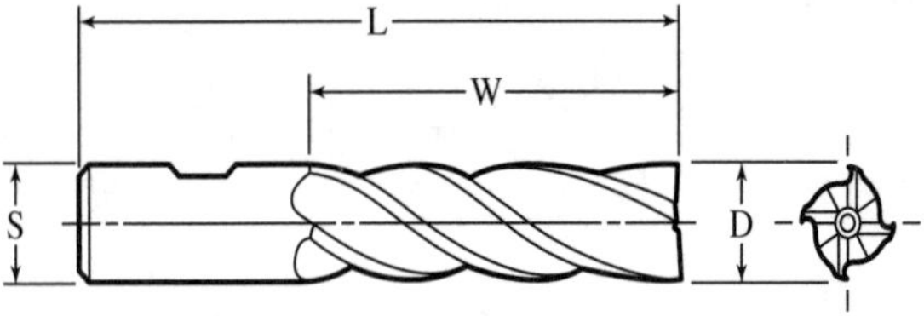

Cutter Dia., D	Regular Mills				Long Mills				Extra-Long Mills			
	S	W	L	N[a]	S	W	L	N[a]	S	W	L	N[a]
$\frac{1}{8}$ [b]	$\frac{3}{8}$	$\frac{3}{8}$	$2\frac{5}{16}$	4	…	…	…	…	…	…	…	…
$\frac{3}{16}$ [b]	$\frac{3}{8}$	$\frac{1}{2}$	$2\frac{3}{8}$	4	…	…	…	…	…	…	…	…
$\frac{1}{4}$ [b]	$\frac{3}{8}$	$\frac{5}{8}$	$2\frac{7}{16}$	4	$\frac{3}{8}$	$1\frac{1}{4}$	$3\frac{1}{16}$	4	$\frac{3}{8}$	$1\frac{3}{4}$	$3\frac{9}{16}$	4
$\frac{5}{16}$ [b]	$\frac{3}{8}$	$\frac{3}{4}$	$2\frac{1}{2}$	4	$\frac{3}{8}$	$1\frac{3}{8}$	$3\frac{1}{8}$	4	$\frac{3}{8}$	2	$3\frac{3}{4}$	4
$\frac{3}{8}$ [b]	$\frac{3}{8}$	$\frac{3}{4}$	$2\frac{1}{2}$	4	$\frac{3}{8}$	$1\frac{1}{2}$	$3\frac{1}{4}$	4	$\frac{3}{8}$	$2\frac{1}{2}$	$4\frac{1}{4}$	4
$\frac{7}{16}$	$\frac{3}{8}$	1	$2\frac{11}{16}$	4	$\frac{1}{2}$	$1\frac{3}{4}$	$3\frac{3}{4}$	4	…	…	…	…
$\frac{1}{2}$	$\frac{3}{8}$	1	$2\frac{11}{16}$	4	$\frac{1}{2}$	2	4	4	$\frac{1}{2}$	3	5	4
$\frac{1}{2}$ [b]	$\frac{1}{2}$	$1\frac{1}{4}$	$3\frac{1}{4}$	4	…	…	…	…	…	…	…	…
$\frac{9}{16}$	$\frac{1}{2}$	$1\frac{3}{8}$	$3\frac{3}{8}$	4	…	…	…	…	…	…	…	…
$\frac{5}{8}$	$\frac{1}{2}$	$1\frac{3}{8}$	$3\frac{3}{8}$	4	$\frac{5}{8}$	$2\frac{1}{2}$	$4\frac{5}{8}$	4	$\frac{5}{8}$	4	$6\frac{1}{8}$	4
$\frac{11}{16}$	$\frac{1}{2}$	$1\frac{5}{8}$	$3\frac{5}{8}$	4	…	…	…	…	…	…	…	…
$\frac{3}{4}$	$\frac{1}{2}$	$1\frac{5}{8}$	$3\frac{5}{8}$	4	$\frac{3}{4}$	3	$5\frac{1}{4}$	4	$\frac{3}{4}$	4	$6\frac{1}{4}$	4
$\frac{5}{8}$ [b]	$\frac{5}{8}$	$1\frac{5}{8}$	$3\frac{3}{4}$	4	…	…	…	…	…	…	…	…
$\frac{11}{16}$	$\frac{5}{8}$	$1\frac{5}{8}$	$3\frac{3}{4}$	4	…	…	…	…	…	…	…	…
$\frac{3}{4}$ [b]	$\frac{5}{8}$	$1\frac{5}{8}$	$3\frac{3}{4}$	4	…	…	…	…	…	…	…	…
$\frac{13}{16}$	$\frac{5}{8}$	$1\frac{7}{8}$	4	6	…	…	…	…	…	…	…	…
$\frac{7}{8}$	$\frac{5}{8}$	$1\frac{7}{8}$	4	6	$\frac{7}{8}$	$3\frac{1}{2}$	$5\frac{3}{4}$	4	$\frac{7}{8}$	5	$7\frac{1}{4}$	4
1	$\frac{5}{8}$	$1\frac{7}{8}$	4	6	1	4	$6\frac{1}{2}$	4	1	6	$8\frac{1}{2}$	4
$\frac{7}{8}$	$\frac{7}{8}$	$1\frac{7}{8}$	$4\frac{1}{8}$	4	…	…	…	…	…	…	…	…
1	$\frac{7}{8}$	$1\frac{7}{8}$	$4\frac{1}{8}$	4	…	…	…	…	…	…	…	…
$1\frac{1}{8}$	$\frac{7}{8}$	2	$4\frac{1}{4}$	6	1	4	$6\frac{1}{2}$	6	…	…	…	…
$1\frac{1}{4}$	$\frac{7}{8}$	2	$4\frac{1}{4}$	6	1	4	$6\frac{1}{2}$	6	$1\frac{1}{4}$	6	$8\frac{1}{2}$	6
1	1	2	$4\frac{1}{2}$	4	…	…	…	…	…	…	…	…
$1\frac{1}{8}$	1	2	$4\frac{1}{2}$	6	…	…	…	…	…	…	…	…
$1\frac{1}{4}$	1	2	$4\frac{1}{2}$	6	…	…	…	…	…	…	…	…
$1\frac{3}{8}$	1	2	$4\frac{1}{2}$	6	…	…	…	…	…	…	…	…
$1\frac{1}{2}$	1	2	$4\frac{1}{2}$	6	1	4	$6\frac{1}{2}$	6	…	…	…	…
$1\frac{1}{4}$	$1\frac{1}{4}$	2	$4\frac{1}{2}$	6	$1\frac{1}{4}$	4	$6\frac{1}{2}$	6	…	…	…	…
$1\frac{1}{2}$	$1\frac{1}{4}$	2	$4\frac{1}{2}$	6	$1\frac{1}{4}$	4	$6\frac{1}{2}$	6	$1\frac{1}{4}$	8	$10\frac{1}{2}$	6
$1\frac{3}{4}$	$1\frac{1}{4}$	2	$4\frac{1}{2}$	6	$1\frac{1}{4}$	4	$6\frac{1}{2}$	6	…	…	…	…
2	$1\frac{1}{4}$	2	$4\frac{1}{2}$	8	$1\frac{1}{4}$	4	$6\frac{1}{2}$	8	…	…	…	…

[a] N = Number of flutes.

[b] In this size of regular mill a left-hand cutter with left-hand helix is also standard.

All dimensions are in inches. All cutters are high-speed steel. Helix angle is greater than 19 degrees but not more than 39 degrees. Right-hand cutters with right-hand helix are standard.

As indicated in the table, shank diameter S may be larger, smaller, or the same as the cutter diameter D.

Tolerances: On D, +0.003 inch; on S, −0.0001 to −0.0005 inch; on W, $\pm\frac{1}{32}$ inch; on L, $\pm\frac{1}{16}$ inch.

Table 6. American National Standard Form Relieved Corner Rounding Cutters with Weldon Shanks *ANSI/ASME B94.19-1997 (R2019)*

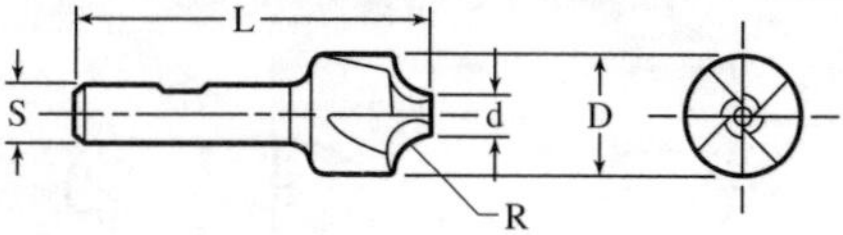

Rad., *R*	Dia., *D*	Dia., *d*	*S*	*L*	Rad., *R*	Dia., *D*	Dia., *d*	*S*	*L*
$\frac{1}{16}$	$\frac{7}{16}$	$\frac{1}{4}$	$\frac{3}{8}$	$2\frac{1}{2}$	$\frac{3}{8}$	$1\frac{1}{4}$	$\frac{3}{8}$	$\frac{1}{2}$	$3\frac{1}{2}$
$\frac{3}{32}$	$\frac{1}{2}$	$\frac{1}{4}$	$\frac{3}{8}$	$2\frac{1}{2}$	$\frac{3}{16}$	$\frac{7}{8}$	$\frac{5}{16}$	$\frac{3}{4}$	$3\frac{1}{8}$
$\frac{1}{8}$	$\frac{5}{8}$	$\frac{1}{4}$	$\frac{1}{2}$	3	$\frac{1}{4}$	1	$\frac{3}{8}$	$\frac{3}{4}$	$3\frac{1}{4}$
$\frac{5}{32}$	$\frac{3}{4}$	$\frac{5}{16}$	$\frac{1}{2}$	3	$\frac{5}{16}$	$1\frac{1}{8}$	$\frac{3}{8}$	$\frac{7}{8}$	$3\frac{1}{2}$
$\frac{3}{16}$	$\frac{7}{8}$	$\frac{5}{16}$	$\frac{1}{2}$	3	$\frac{3}{8}$	$1\frac{1}{4}$	$\frac{3}{8}$	$\frac{7}{8}$	$3\frac{3}{4}$
$\frac{1}{4}$	1	$\frac{3}{8}$	$\frac{1}{2}$	3	$\frac{7}{16}$	$1\frac{3}{8}$	$\frac{3}{8}$	1	4
$\frac{5}{16}$	$1\frac{1}{8}$	$\frac{3}{8}$	$\frac{1}{2}$	$3\frac{1}{4}$	$\frac{1}{2}$	$1\frac{1}{2}$	$\frac{3}{8}$	1	$4\frac{1}{8}$

All dimensions are in inches. All cutters are high-speed steel. Right-hand cutters are standard.

Tolerances: On D, ±0.010 inch; on diameter of circle, $2R$, ±0.001 inch for cutters up to and including $\frac{1}{8}$-inch radius, +0.002, −0.001 inch for cutters over $\frac{1}{8}$-inch radius; on S, −0.0001 to −0.0005 inch; and on L, ± $\frac{1}{16}$ inch.

Table 7. ANSI Two-Flute, High Helix, Regular-, Long-, and Extra-Long-Length, Single-End End Mills with Weldon Shanks *ANSI/ASME B94.19-1997 (R2019)*

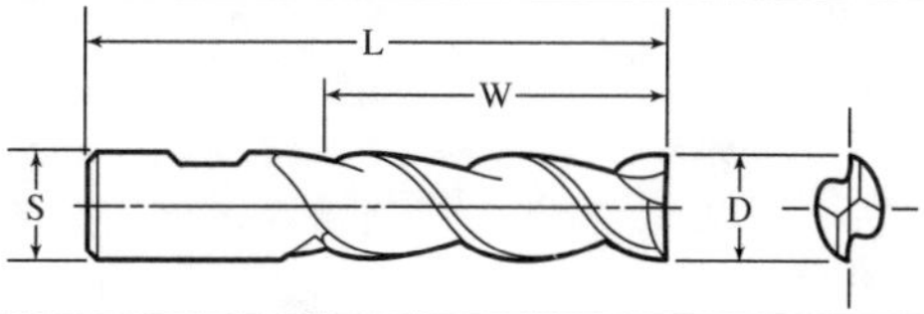

Cutter Dia., *D*	Regular Mill			Long Mill			Extra-Long Mill		
	S	*W*	*L*	*S*	*W*	*L*	*S*	*W*	*L*
$\frac{1}{4}$	$\frac{3}{8}$	$\frac{5}{8}$	$2\frac{7}{16}$	$\frac{3}{8}$	$1\frac{1}{4}$	$3\frac{1}{16}$	$\frac{3}{8}$	$1\frac{3}{4}$	$3\frac{9}{16}$
$\frac{5}{16}$	$\frac{3}{8}$	$\frac{3}{4}$	$2\frac{1}{2}$	$\frac{3}{8}$	$1\frac{3}{8}$	$3\frac{1}{8}$	$\frac{3}{8}$	2	$3\frac{3}{4}$
$\frac{3}{8}$	$\frac{3}{8}$	$\frac{3}{4}$	$2\frac{1}{2}$	$\frac{3}{8}$	$1\frac{1}{2}$	$3\frac{1}{4}$	$\frac{3}{8}$	$2\frac{1}{2}$	$4\frac{1}{4}$
$\frac{7}{16}$	$\frac{3}{8}$	1	$2\frac{11}{16}$	$\frac{1}{2}$	$1\frac{3}{4}$	$3\frac{3}{4}$	…	…	…
$\frac{1}{2}$	$\frac{1}{2}$	$1\frac{1}{4}$	$3\frac{1}{4}$	$\frac{1}{2}$	2	4	$\frac{1}{2}$	3	5
$\frac{5}{8}$	$\frac{5}{8}$	$1\frac{5}{8}$	$3\frac{3}{4}$	$\frac{5}{8}$	$2\frac{1}{2}$	$4\frac{5}{8}$	$\frac{5}{8}$	4	$6\frac{1}{8}$
$\frac{3}{4}$	$\frac{3}{4}$	$1\frac{5}{8}$	$3\frac{7}{8}$	$\frac{3}{4}$	3	$5\frac{1}{4}$	$\frac{3}{4}$	4	$6\frac{1}{4}$
$\frac{7}{8}$	$\frac{7}{8}$	$1\frac{7}{8}$	$4\frac{1}{8}$	…	…	…	…	…	…
1	1	2	$4\frac{1}{2}$	1	4	$6\frac{1}{2}$	1	6	$8\frac{1}{2}$
$1\frac{1}{4}$	$1\frac{1}{4}$	2	$4\frac{1}{2}$	$1\frac{1}{4}$	4	$6\frac{1}{2}$	$1\frac{1}{4}$	6	$8\frac{1}{2}$
$1\frac{1}{2}$	$1\frac{1}{4}$	2	$4\frac{1}{2}$	$1\frac{1}{4}$	4	$6\frac{1}{2}$	$1\frac{1}{4}$	8	$10\frac{1}{2}$
2	$1\frac{1}{4}$	2	$4\frac{1}{2}$	$1\frac{1}{4}$	4	$6\frac{1}{2}$	…	…	…

All dimensions are in inches. All cutters are high-speed steel. Right-hand cutters with right-hand helix are standard. Helix angle is greater than 39 degrees.

Tolerances: On D, +0.003 inch; on S, −0.0001 to −0.0005 inch; on W, ±$\frac{1}{32}$ inch; and on L, ±$\frac{1}{16}$ inch.

Table 8. American National Standard Stub- and Regular-Length, Two-Flute, Medium Helix, Plain- and Ball-End, Single-End End Mills with Weldon Shanks
ANSI/ASME B94.19-1997 (R2019)

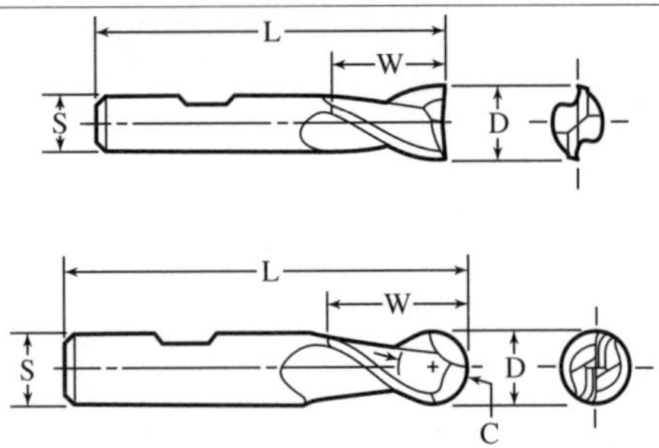

Regular Length — Plain End			
Dia., *D*	*S*	*W*	*L*
$\frac{1}{8}$	$\frac{3}{8}$	$\frac{3}{8}$	$2\frac{5}{16}$
$\frac{3}{16}$	$\frac{3}{8}$	$\frac{7}{16}$	$2\frac{5}{16}$
$\frac{1}{4}$	$\frac{3}{8}$	$\frac{1}{2}$	$2\frac{5}{16}$
$\frac{5}{16}$	$\frac{3}{8}$	$\frac{9}{16}$	$2\frac{5}{16}$
$\frac{3}{8}$	$\frac{3}{8}$	$\frac{9}{16}$	$2\frac{5}{16}$
$\frac{7}{16}$	$\frac{3}{8}$	$\frac{13}{16}$	$2\frac{1}{2}$
$\frac{1}{2}$	$\frac{3}{8}$	$\frac{13}{16}$	$2\frac{1}{2}$
$\frac{1}{2}$	$\frac{1}{2}$	1	3
$\frac{9}{16}$	$\frac{1}{2}$	$1\frac{1}{8}$	$3\frac{1}{8}$
$\frac{5}{8}$	$\frac{1}{2}$	$1\frac{1}{8}$	$3\frac{1}{8}$
$\frac{11}{16}$	$\frac{1}{2}$	$1\frac{5}{16}$	$3\frac{5}{16}$
$\frac{3}{4}$	$\frac{1}{2}$	$1\frac{5}{16}$	$3\frac{5}{16}$
$\frac{5}{8}$	$\frac{5}{8}$	$1\frac{5}{16}$	$3\frac{7}{16}$
$\frac{11}{16}$	$\frac{5}{8}$	$1\frac{5}{16}$	$3\frac{7}{16}$
$\frac{3}{4}$	$\frac{5}{8}$	$1\frac{5}{16}$	$3\frac{7}{16}$
$\frac{13}{16}$	$\frac{5}{8}$	$1\frac{1}{2}$	$3\frac{5}{8}$
$\frac{7}{8}$	$\frac{5}{8}$	$1\frac{1}{2}$	$3\frac{5}{8}$
1	$\frac{5}{8}$	$1\frac{1}{2}$	$3\frac{5}{8}$
$\frac{7}{8}$	$\frac{7}{8}$	$1\frac{1}{2}$	$3\frac{3}{4}$
1	$\frac{7}{8}$	$1\frac{1}{2}$	$3\frac{3}{4}$
$1\frac{1}{8}$	$\frac{7}{8}$	$1\frac{5}{8}$	$3\frac{7}{8}$
$1\frac{1}{4}$	$\frac{7}{8}$	$1\frac{5}{8}$	$3\frac{7}{8}$
1	1	$1\frac{5}{8}$	$4\frac{1}{8}$
$1\frac{1}{8}$	1	$1\frac{5}{8}$	$4\frac{1}{8}$
$1\frac{1}{4}$	1	$1\frac{5}{8}$	$4\frac{1}{8}$
$1\frac{3}{8}$	1	$1\frac{5}{8}$	$4\frac{1}{8}$
$1\frac{1}{2}$	1	$1\frac{5}{8}$	$4\frac{1}{8}$
$1\frac{1}{4}$	$1\frac{1}{4}$	$1\frac{5}{8}$	$4\frac{1}{8}$
$1\frac{1}{2}$	$1\frac{1}{4}$	$1\frac{5}{8}$	$4\frac{1}{8}$
$1\frac{3}{4}$	$1\frac{1}{4}$	$1\frac{5}{8}$	$4\frac{1}{8}$
2	$1\frac{1}{4}$	$1\frac{5}{8}$	$4\frac{1}{8}$

Stub Length — Plain End			
Cutter Dia., *D*	Shank Dia., *S*	Length of Cut. *W*	Length Overall. *L*
$\frac{1}{8}$	$\frac{3}{8}$	$\frac{3}{16}$	$2\frac{1}{8}$
$\frac{3}{16}$	$\frac{3}{8}$	$\frac{9}{32}$	$2\frac{3}{16}$
$\frac{1}{4}$	$\frac{3}{8}$	$\frac{3}{8}$	$2\frac{1}{4}$

Regular Length — Ball End			
Dia., *C* and *D*	Shank Dia., *S*	Length of Cut. *W*	Length Overall. *L*
$\frac{1}{8}$	$\frac{3}{8}$	$\frac{3}{8}$	$2\frac{5}{16}$
$\frac{3}{16}$	$\frac{3}{8}$	$\frac{1}{2}$	$2\frac{3}{8}$
$\frac{1}{4}$	$\frac{3}{8}$	$\frac{5}{8}$	$2\frac{7}{16}$
$\frac{5}{16}$	$\frac{3}{8}$	$\frac{3}{4}$	$2\frac{1}{2}$
$\frac{3}{8}$	$\frac{3}{8}$	$\frac{3}{4}$	$2\frac{1}{2}$
$\frac{7}{16}$	$\frac{1}{2}$	1	3
$\frac{1}{2}$	$\frac{1}{2}$	1	3
$\frac{9}{16}$	$\frac{1}{2}$	$1\frac{1}{8}$	$3\frac{1}{8}$
$\frac{5}{8}$	$\frac{1}{2}$	$1\frac{1}{8}$	$3\frac{1}{8}$
$\frac{5}{8}$	$\frac{5}{8}$	$1\frac{3}{8}$	$3\frac{1}{2}$
$\frac{3}{4}$	$\frac{1}{2}$	$1\frac{5}{16}$	$3\frac{5}{16}$
$\frac{3}{4}$	$\frac{3}{4}$	$1\frac{5}{8}$	$3\frac{7}{8}$
$\frac{7}{8}$	$\frac{7}{8}$	2	$4\frac{1}{4}$
1	1	$2\frac{1}{4}$	$4\frac{3}{4}$
$1\frac{1}{8}$	1	$2\frac{1}{4}$	$4\frac{3}{4}$
$1\frac{1}{4}$	$1\frac{1}{4}$	$2\frac{1}{2}$	5
$1\frac{1}{2}$	$1\frac{1}{4}$	$2\frac{1}{2}$	5

All dimensions are in inches. All cutters are high-speed steel. Right-hand cutters with right-hand helix are standard. Helix angle is greater than 19 degrees but not more than 39 degrees.

Tolerances: On *C* and *D*, −0.0015 inch for stub-length mills, + 0.003 inch for regular-length mills; on *S*, −0.0001 to −0.0005 inch; on *W*, $\pm\frac{1}{32}$ inch; and on *L*, $\pm\frac{1}{16}$ inch.

The following single-end end mills are available in premium high-speed steel: ball end, two flute, with *D* ranging from $\frac{1}{8}$ to $1\frac{1}{2}$ inches; ball end, multiple flute, with *D* ranging from $\frac{1}{8}$ to 1 inch; and plain end, two flute, with *D* ranging from $\frac{1}{8}$ to $1\frac{1}{2}$ inches.

Table 9. American National Standard Regular-, Long-, and Extra Long-Length, Three- and Four-Flute, Medium Helix, Center Cutting, Single-End End Mills with Weldon Shanks *ANSI/ASME B94.19-1997 (R2019)*

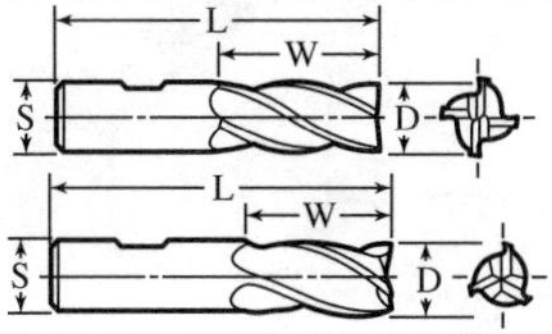

Four Flute									
Dia.,	Regular Length			Long Length			Extra Long Length		
D	*S*	*W*	*L*	*S*	*W*	*L*	*S*	*W*	*L*
$\frac{1}{8}$	$\frac{3}{8}$	$\frac{3}{8}$	$2\frac{5}{16}$	…	…	…	…	…	…
$\frac{3}{16}$	$\frac{3}{8}$	$\frac{1}{2}$	$2\frac{3}{8}$	…	…	…	…	…	…
$\frac{1}{4}$	$\frac{3}{8}$	$\frac{5}{8}$	$2\frac{7}{16}$	$\frac{3}{8}$	$1\frac{1}{4}$	$3\frac{1}{16}$	$\frac{3}{8}$	$1\frac{3}{4}$	$3\frac{9}{16}$
$\frac{5}{16}$	$\frac{3}{8}$	$\frac{3}{4}$	$2\frac{1}{2}$	$\frac{3}{8}$	$1\frac{3}{8}$	$3\frac{1}{8}$	$\frac{3}{8}$	2	$3\frac{3}{4}$
$\frac{3}{8}$	$\frac{3}{8}$	$\frac{3}{4}$	$2\frac{1}{2}$	$\frac{3}{8}$	$1\frac{1}{2}$	$3\frac{1}{4}$	$\frac{3}{8}$	$2\frac{1}{2}$	$4\frac{1}{4}$
$\frac{1}{2}$	$\frac{1}{2}$	$1\frac{1}{4}$	$3\frac{1}{4}$	$\frac{1}{2}$	2	4	$\frac{1}{2}$	3	5
$\frac{5}{8}$	$\frac{5}{8}$	$1\frac{5}{8}$	$3\frac{3}{4}$	$\frac{5}{8}$	$2\frac{1}{2}$	$4\frac{5}{8}$	$\frac{5}{8}$	4	$6\frac{1}{8}$
$\frac{11}{16}$	$\frac{5}{8}$	$1\frac{5}{8}$	$3\frac{3}{4}$	…	…	…	…	…	…
$\frac{3}{4}$	$\frac{3}{4}$	$1\frac{5}{8}$	$3\frac{7}{8}$	$\frac{3}{4}$	3	$5\frac{1}{4}$	$\frac{3}{4}$	4	$6\frac{1}{4}$
$\frac{7}{8}$	$\frac{7}{8}$	$1\frac{7}{8}$	$4\frac{1}{8}$	$\frac{7}{8}$	$3\frac{1}{2}$	$5\frac{3}{4}$	$\frac{7}{8}$	5	$7\frac{1}{4}$
1	1	2	$4\frac{1}{2}$	1	4	$6\frac{1}{2}$	1	6	$8\frac{1}{2}$
$1\frac{1}{8}$	1	2	$4\frac{1}{2}$	…	…	…	…	…	…
$1\frac{1}{4}$	$1\frac{1}{4}$	2	$4\frac{1}{2}$	$1\frac{1}{4}$	4	$6\frac{1}{2}$	$1\frac{1}{4}$	6	$8\frac{1}{2}$
$1\frac{1}{2}$	$1\frac{1}{4}$	2	$4\frac{1}{2}$	…	…	…	…	…	…

Three Flute							
Dia., *D*	*S*	*W*	*L*	Dia., *D*	*S*	*W*	*L*
Regular Length				Regular Length (*cont.*)			
$\frac{1}{8}$	$\frac{3}{8}$	$\frac{3}{8}$	$2\frac{5}{16}$	$1\frac{1}{8}$	1	2	$4\frac{1}{2}$
$\frac{3}{16}$	$\frac{3}{8}$	$\frac{1}{2}$	$2\frac{3}{8}$	$1\frac{1}{4}$	1	2	$4\frac{1}{2}$
$\frac{1}{4}$	$\frac{3}{8}$	$\frac{5}{8}$	$2\frac{7}{16}$	$1\frac{1}{2}$	1	2	$4\frac{1}{2}$
$\frac{5}{16}$	$\frac{3}{8}$	$\frac{3}{4}$	$2\frac{1}{2}$	$1\frac{1}{4}$	$1\frac{1}{4}$	2	$4\frac{1}{2}$
$\frac{3}{8}$	$\frac{3}{8}$	$\frac{3}{4}$	$2\frac{1}{2}$	$1\frac{1}{2}$	$1\frac{1}{4}$	2	$4\frac{1}{2}$
$\frac{7}{16}$	$\frac{3}{8}$	1	$2\frac{11}{16}$	$1\frac{3}{4}$	$1\frac{1}{4}$	2	$4\frac{1}{2}$
$\frac{1}{2}$	$\frac{3}{8}$	1	$2\frac{11}{16}$	2	$1\frac{1}{4}$	2	$4\frac{1}{2}$
$\frac{1}{2}$	$\frac{1}{2}$	$1\frac{1}{4}$	$3\frac{1}{4}$	Long Length			
$\frac{9}{16}$	$\frac{1}{2}$	$1\frac{3}{8}$	$3\frac{3}{8}$	$\frac{1}{4}$	$\frac{3}{8}$	$1\frac{1}{4}$	$3\frac{1}{16}$
$\frac{5}{8}$	$\frac{1}{2}$	$1\frac{3}{8}$	$3\frac{3}{8}$	$\frac{5}{16}$	$\frac{3}{8}$	$1\frac{3}{8}$	$3\frac{1}{8}$
$\frac{3}{4}$	$\frac{1}{2}$	$1\frac{5}{8}$	$3\frac{5}{8}$	$\frac{3}{8}$	$\frac{3}{8}$	$1\frac{1}{2}$	$3\frac{1}{4}$
$\frac{5}{8}$	$\frac{5}{8}$	$1\frac{5}{8}$	$3\frac{3}{4}$	$\frac{7}{16}$	$\frac{1}{2}$	$1\frac{3}{4}$	$3\frac{3}{4}$
$\frac{3}{4}$	$\frac{5}{8}$	$1\frac{5}{8}$	$3\frac{3}{4}$	$\frac{1}{2}$	$\frac{1}{2}$	2	4
$\frac{7}{8}$	$\frac{5}{8}$	$1\frac{7}{8}$	4	$\frac{5}{8}$	$\frac{5}{8}$	$2\frac{1}{2}$	$4\frac{5}{8}$
1	$\frac{5}{8}$	$1\frac{7}{8}$	4	$\frac{3}{4}$	$\frac{3}{4}$	3	$5\frac{1}{4}$
$\frac{3}{4}$	$\frac{3}{4}$	$1\frac{5}{8}$	$3\frac{7}{8}$	1	1	4	$6\frac{1}{2}$
$\frac{7}{8}$	$\frac{3}{4}$	$1\frac{7}{8}$	$4\frac{1}{8}$	$1\frac{1}{4}$	$1\frac{1}{4}$	4	$6\frac{1}{2}$
1	$\frac{3}{4}$	$1\frac{7}{8}$	$4\frac{1}{8}$	$1\frac{1}{2}$	$1\frac{1}{4}$	4	$6\frac{1}{2}$
1	$\frac{7}{8}$	$1\frac{7}{8}$	$4\frac{1}{8}$	$1\frac{3}{4}$	$1\frac{1}{4}$	4	$6\frac{1}{2}$
1	1	2	$4\frac{1}{2}$	2	$1\frac{1}{4}$	4	$6\frac{1}{2}$

All dimensions are in inches. All cutters are high-speed steel. Right-hand cutters with right-hand helix are standard. Helix angle is greater than 19 degrees but not more than 39 degrees.

Tolerances: On *D*, +0.003 inch; on *S*, −0.0001 to −0.0005 inch; on *W*, $\pm\frac{1}{32}$ inch; and on *L*, $\pm\frac{1}{16}$ inch.

The following center-cutting, single-end end mills are available in premium high-speed steel: regular length, multiple flute, with *D* ranging from $\frac{1}{8}$ to $1\frac{1}{2}$ inches; long length, multiple flute, with *D* ranging from $\frac{3}{8}$ to $1\frac{1}{4}$ inches; and extra long-length, multiple flute, with *D* ranging from $\frac{3}{8}$ to $1\frac{1}{4}$ inches.

Table 10. American National Standard 60-Degree Single-Angle Milling Cutters with Weldon Shanks *ANSI/ASME B94.19-1997 (R2019)*

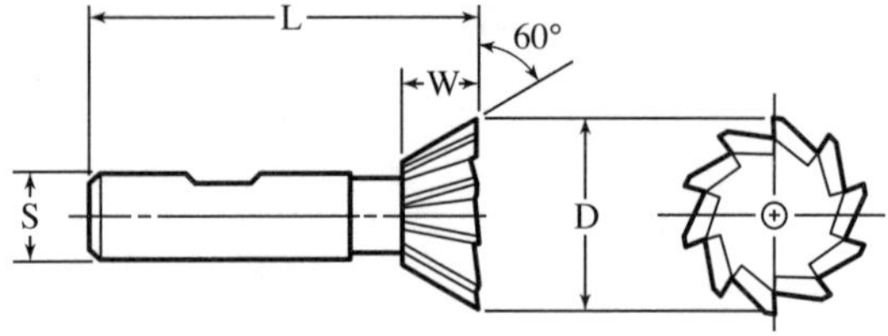

Dia., *D*	*S*	*W*	*L*	Dia., *D*	*S*	*W*	*L*
$\frac{3}{4}$	$\frac{3}{8}$	$\frac{5}{16}$	$2\frac{1}{8}$	$1\frac{7}{8}$	$\frac{7}{8}$	$\frac{13}{16}$	$3\frac{1}{4}$
$1\frac{3}{8}$	$\frac{5}{8}$	$\frac{9}{16}$	$2\frac{7}{8}$	$2\frac{1}{4}$	1	$1\frac{1}{16}$	$3\frac{3}{4}$

All dimensions are in inches. All cutters are high-speed steel. Right-hand cutters are standard. *Tolerances:* On D, ± 0.015 inch; on S, − 0.0001 to − 0.0005 inch; on W, ± 0.015 inch; and on L, $\pm\frac{1}{16}$ inch.

Table 11. Key Size Versus Shaft Diameter *ANSI B17.1-1967 (R2013)*

Nominal Shaft Diameter		Nominal Key Size			Normal Keyseat Depth	
			Height, *H*		*H*/2	
Over	To (Incl.)	Width, *W*	Square	Rectangular	Square	Rectangular
$\frac{5}{16}$	$\frac{7}{16}$	$\frac{3}{32}$	$\frac{3}{32}$	…	$\frac{3}{64}$	…
$\frac{7}{16}$	$\frac{9}{16}$	$\frac{1}{8}$	$\frac{1}{8}$	$\frac{3}{32}$	$\frac{1}{16}$	$\frac{3}{64}$
$\frac{9}{16}$	$\frac{7}{8}$	$\frac{3}{16}$	$\frac{3}{16}$	$\frac{1}{8}$	$\frac{3}{32}$	$\frac{1}{16}$
$\frac{7}{8}$	$1\frac{1}{4}$	$\frac{1}{4}$	$\frac{1}{4}$	$\frac{3}{16}$	$\frac{1}{8}$	$\frac{3}{32}$
$1\frac{1}{4}$	$1\frac{3}{8}$	$\frac{5}{16}$	$\frac{5}{16}$	$\frac{1}{4}$	$\frac{5}{32}$	$\frac{1}{8}$
$1\frac{3}{8}$	$1\frac{3}{4}$	$\frac{3}{8}$	$\frac{3}{8}$	$\frac{1}{4}$	$\frac{3}{16}$	$\frac{1}{8}$
$1\frac{3}{4}$	$2\frac{1}{4}$	$\frac{1}{2}$	$\frac{1}{2}$	$\frac{3}{8}$	$\frac{1}{4}$	$\frac{3}{16}$
$2\frac{1}{4}$	$2\frac{3}{4}$	$\frac{5}{8}$	$\frac{5}{8}$	$\frac{7}{16}$	$\frac{5}{16}$	$\frac{7}{32}$
$2\frac{3}{4}$	$3\frac{1}{4}$	$\frac{3}{4}$	$\frac{3}{4}$	$\frac{1}{2}$	$\frac{3}{8}$	$\frac{1}{4}$
$3\frac{1}{4}$	$3\frac{3}{4}$	$\frac{7}{8}$	$\frac{7}{8}$	$\frac{5}{8}$	$\frac{7}{16}$	$\frac{5}{16}$
$3\frac{3}{4}$	$4\frac{1}{2}$	1	1	$\frac{3}{4}$	$\frac{1}{2}$	$\frac{3}{8}$
$4\frac{1}{2}$	$5\frac{1}{2}$	$1\frac{1}{4}$	$1\frac{1}{4}$	$\frac{7}{8}$	$\frac{5}{8}$	$\frac{7}{16}$
$5\frac{1}{2}$	$6\frac{1}{2}$	$1\frac{1}{2}$	$1\frac{1}{2}$	1	$\frac{3}{4}$	$\frac{1}{2}$
$6\frac{1}{2}$	$7\frac{1}{2}$	$1\frac{3}{4}$	$1\frac{3}{4}$	$1\frac{1}{2}$[a]	$\frac{7}{8}$	$\frac{3}{4}$
$7\frac{1}{2}$	9	2	2	$1\frac{1}{2}$	1	$\frac{3}{4}$
9	11	$2\frac{1}{2}$	$2\frac{1}{2}$	$1\frac{3}{4}$	$1\frac{1}{4}$	$\frac{7}{8}$

[a] Some key standards show $1\frac{1}{4}$ inches; preferred height is $1\frac{1}{2}$ inches.

All dimensions are given in inches. For larger shaft sizes, see section *ANSI Standard Woodruff Keys and Keyseats* on page 177.

Square keys preferred for shaft diameters above heavy line; rectangular keys, below.

Table 12. American National Standard Keys and Keyways for Milling Cutters and Arbors *ANSI/ASME B94.19-1997 (R2019)*

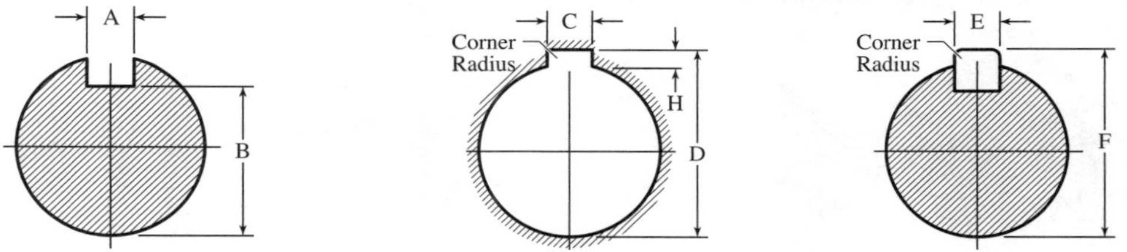

Arbor and Keyseat — Cutter Hole and Keyway — Arbor and Key

Nom. Arbor and Cutter Hole Dia.	Nom. Size Key (Square)	Arbor and Keyseat				Hole and Keyway					Arbor and Key			
		A Max.	*A* Min.	*B* Max.	*B* Min.	*C* Max.	*C* Min.	*D*[a] Min.	*H* Nom.	Corner Radius	*E* Max.	*E* Min.	*F* Max.	*F* Min.
1/2	3/32	0.0947	0.0937	0.4531	0.4481	0.106	0.099	0.5578	3/64	0.020	0.0932	0.0927	0.5468	0.5408
5/8	1/8	0.1260	0.1250	0.5625	0.5575	0.137	0.130	0.6985	1/16	1/32	0.1245	0.1240	0.6875	0.6815
3/4	1/8	0.1260	0.1250	0.6875	0.6825	0.137	0.130	0.8225	1/16	1/32	0.1245	0.1240	0.8125	0.8065
7/8	1/8	0.1260	0.1250	0.8125	0.8075	0.137	0.130	0.9475	1/16	1/32	0.1245	0.1240	0.9375	0.9315
1	1/4	0.2510	0.2500	0.8438	0.8388	0.262	0.255	1.1040	3/32	3/64	0.2495	0.2490	1.0940	1.0880
1 1/4	5/16	0.3135	0.3125	1.0630	1.0580	0.343	0.318	1.3850	1/8	1/16	0.3120	0.3115	1.3750	1.3690
1 1/2	3/8	0.3760	0.3750	1.2810	1.2760	0.410	0.385	1.6660	5/32	1/16	0.3745	0.3740	1.6560	1.6500
1 3/4	7/16	0.4385	0.4375	1.5000	1.4950	0.473	0.448	1.9480	3/16	1/16	0.4370	0.4365	1.9380	1.9320
2	1/2	0.5010	0.5000	1.6870	1.6820	0.535	0.510	2.1980	3/16	1/16	0.4995	0.4990	2.1880	2.1820
2 1/2	5/8	0.6260	0.6250	2.0940	2.0890	0.660	0.635	2.7330	7/32	1/16	0.6245	0.6240	2.7180	2.7120
3	3/4	0.7510	0.7500	2.5000	2.4950	0.785	0.760	3.2650	1/4	3/32	0.7495	0.7490	3.2500	3.2440
3 1/2	7/8	0.8760	0.8750	3.0000	2.9950	0.910	0.885	3.8900	3/8	3/32	0.8745	0.8740	3.8750	3.8690
4	1	1.0010	1.0000	3.3750	3.3700	1.035	1.010	4.3900	3/8	3/32	0.9995	0.9990	4.3750	4.3690
4 1/2	1 1/8	1.1260	1.1250	3.8130	3.8080	1.160	1.135	4.9530	7/16	1/8	1.1245	1.1240	4.9380	4.9320
5	1 1/4	1.2510	1.2500	4.2500	4.2450	1.285	1.260	5.5150	1/2	1/8	1.2495	1.2490	5.5000	5.4940

[a] *D* max. is 0.010 inch larger than *D* min.

All dimensions given in inches.

Table 13. American National Standard Woodruff Keyseat Cutters—Shank-Type Straight-Teeth and Arbor-Type Staggered-Teeth *ANSI/ASME B94.19-1997 (R2019)*

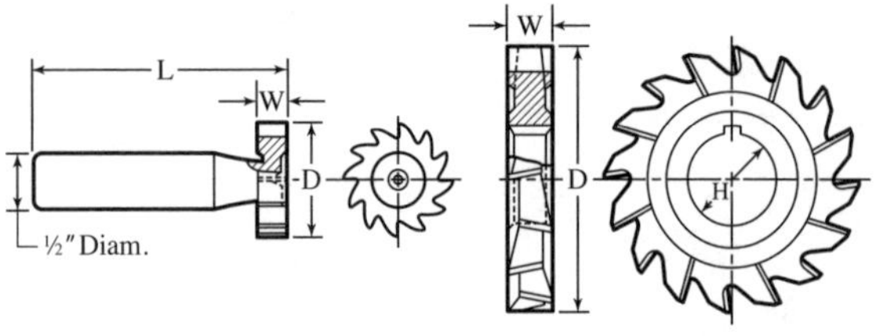

Shank-type Cutters											
Cutter Number	Nom. Dia. of Cutter, *D*	Width of Face, *W*	Length Overall,[a] *L*	Cutter Number	Nom. Dia. of Cutter, *D*	Width of Face, *W*	Length Overall,[a] *L*	Cutter Number	Nom. Dia. of Cutter, *D*	Width of Face, *W*	Length Overall,[a] *L*
202	1/4	1/16	2 1/16	506	3/4	5/32	2 5/32	809	1 1/8	1/4	2 1/4
202 1/2	5/16	1/16	2 1/16	606	3/4	3/16	2 3/16	1009	1 1/8	5/16	2 5/16
302 1/2	5/16	3/32	2 3/32	806	3/4	1/4	2 1/4	610	1 1/4	3/16	2 3/16
203	3/8	1/16	2 1/16	507	7/8	5/32	2 5/32	710	1 1/4	7/32	2 7/32
303	3/8	3/32	2 3/32	607	7/8	3/16	2 3/16	810	1 1/4	1/4	2 1/4
403	3/8	1/8	2 1/8	707	7/8	7/32	2 7/32	1010	1 1/4	5/16	2 5/16
204	1/2	1/16	2 1/16	807	7/8	1/4	2 1/4	1210	1 1/4	3/8	2 3/8
304	1/2	3/32	2 3/32	608	1	3/16	2 3/16	811	1 3/8	1/4	2 1/4
404	1/2	1/8	2 1/8	708	1	7/32	2 7/32	1011	1 3/8	5/16	2 5/16
305	5/8	3/32	2 3/32	808	1	1/4	2 1/4	1211	1 3/8	3/8	2 3/8
405	5/8	1/8	2 1/8	1008	1	5/16	2 5/16	812	1 1/2	1/4	2 1/4
505	5/8	5/32	2 5/32	1208	1	3/8	2 3/8	1012	1 1/2	5/16	2 5/16
605	5/8	3/16	2 3/16	609	1 1/8	3/16	2 3/16	1212	1 1/2	3/8	2 3/8
406	3/4	1/8	2 1/8	709	1 1/8	7/32	2 7/32	…	…	…	…
Arbor-type Cutters											
Cutter Number	Nom. Dia. of Cutter, *D*	Width of Face, *W*	Dia. of Hole, *H*	Cutter Number	Nom. Dia. of Cutter, *D*	Width of Face, *W*	Dia. of Hole, *H*	Cutter Number	Nom. Dia. of Cutter, *D*	Width of Face, *W*	Dia. of Hole, *H*
617	2 1/8	3/16	3/4	1022	2 3/4	5/16	1	1628	3 1/2	1/2	1
817	2 1/8	1/4	3/4	1222	2 3/4	3/8	1	1828	3 1/2	9/16	1
1017	2 1/8	5/16	3/4	1422	2 3/4	7/16	1	2028	3 1/2	5/8	1
1217	2 1/8	3/8	3/4	1622	2 3/4	1/2	1	2428	3 1/2	3/4	1
822	2 3/4	1/4	1	1228	3 1/2	3/8	1	…	…	…	…

All dimensions are given in inches. All cutters are high-speed steel. Cutter numbers indicate nominal key dimensions or cutter sizes.

Shank-type cutters are standard with right-hand cut and straight teeth. All sizes have 1/2-inch diameter straight shank. Arbor-type cutters have staggered teeth.

For Woodruff key and key-slot dimensions, see page 178.

Tolerances: Face with *W* for shank-type cutters: 1/16- to 5/32-inch face, + 0.0000, −0.0005; 3/16 to 7/32, − 0.0002, − 0.0007; 1/4, −0.0003, −0.0008; 5/16, −0.0004, −0.0009; 3/8, − 0.0005, −0.0010 inch. Face width *W* for arbor-type cutters; 3/16-inch face, −0.0002, −0.0007; 1/4, −0.0003, −0.0008; 5/16, −0.0004, −0.0009; 3/8 and over, −0.0005, −0.0010 inch. Hole size *H:* +0.00075, −0.0000 inch. Diameter *D* for shank-type cutters: 1/4- through 3/4-inch diameter, +0.010, +0.015, 7/8 through 1 1/8, +0.012, +0.017; 1 1/4 through 1 1/2, +0.015, +0.020 inch. These tolerances include an allowance for sharpening. For arbor-type cutters, diameter *D* is furnished 1/32 inch larger than listed and a tolerance of ±0.002 inch applies to the oversize diameter.

KEYS AND KEYSEATS

Table 1. Depth Control Values *S* and *T* for Shaft and Hub
ANSI B17.1-1967 (R2013) (See figures at end of table)

Nominal Shaft Diameter	Parallel and Taper		Parallel		Taper	
	Square *S*	Rectangular *S*	Square *T*	Rectangular *T*	Square *T*	Rectangular *T*
$\frac{1}{2}$	0.430	0.445	0.560	0.544	0.535	0.519
$\frac{9}{16}$	0.493	0.509	0.623	0.607	0.598	0.582
$\frac{5}{8}$	0.517	0.548	0.709	0.678	0.684	0.653
$\frac{11}{16}$	0.581	0.612	0.773	0.742	0.748	0.717
$\frac{3}{4}$	0.644	0.676	0.837	0.806	0.812	0.781
$\frac{13}{16}$	0.708	0.739	0.900	0.869	0.875	0.844
$\frac{7}{8}$	0.771	0.802	0.964	0.932	0.939	0.907
$\frac{15}{16}$	0.796	0.827	1.051	1.019	1.026	0.994
1	0.859	0.890	1.114	1.083	1.089	1.058
$1\frac{1}{16}$	0.923	0.954	1.178	1.146	1.153	1.121
$1\frac{1}{8}$	0.986	1.017	1.241	1.210	1.216	1.185
$1\frac{3}{16}$	1.049	1.080	1.304	1.273	1.279	1.248
$1\frac{1}{4}$	1.112	1.144	1.367	1.336	1.342	1.311
$1\frac{5}{16}$	1.137	1.169	1.455	1.424	1.430	1.399
$1\frac{3}{8}$	1.201	1.232	1.518	1.487	1.493	1.462
$1\frac{7}{16}$	1.225	1.288	1.605	1.543	1.580	1.518
$1\frac{1}{2}$	1.289	1.351	1.669	1.606	1.644	1.581
$1\frac{9}{16}$	1.352	1.415	1.732	1.670	1.707	1.645
$1\frac{5}{8}$	1.416	1.478	1.796	1.733	1.771	1.708
$1\frac{11}{16}$	1.479	1.541	1.859	1.796	1.834	1.771
$1\frac{3}{4}$	1.542	1.605	1.922	1.860	1.897	1.835
$1\frac{13}{16}$	1.527	1.590	2.032	1.970	2.007	1.945
$1\frac{7}{8}$	1.591	1.654	2.096	2.034	2.071	2.009
$1\frac{15}{16}$	1.655	1.717	2.160	2.097	2.135	2.072
2	1.718	1.781	2.223	2.161	2.198	2.136
$2\frac{1}{16}$	1.782	1.844	2.287	2.224	2.262	2.199
$2\frac{1}{8}$	1.845	1.908	2.350	2.288	2.325	2.263
$2\frac{3}{16}$	1.909	1.971	2.414	2.351	2.389	2.326
$2\frac{1}{4}$	1.972	2.034	2.477	2.414	2.452	2.389
$2\frac{5}{16}$	1.957	2.051	2.587	2.493	2.562	2.468
$2\frac{3}{8}$	2.021	2.114	2.651	2.557	2.626	2.532
$2\frac{7}{16}$	2.084	2.178	2.714	2.621	2.689	2.596
$2\frac{1}{2}$	2.148	2.242	2.778	2.684	2.753	2.659
$2\frac{9}{16}$	2.211	2.305	2.841	2.748	2.816	2.723
$2\frac{5}{8}$	2.275	2.369	2.905	2.811	2.880	2.786
$2\frac{11}{16}$	2.338	2.432	2.968	2.874	2.943	2.849
$2\frac{3}{4}$	2.402	2.495	3.032	2.938	3.007	2.913
$2\frac{13}{16}$	2.387	2.512	3.142	3.017	3.117	2.992
$2\frac{7}{8}$	2.450	2.575	3.205	3.080	3.180	3.055
$2\frac{15}{16}$	2.514	2.639	3.269	3.144	3.244	3.119
3	2.577	2.702	3.332	3.207	3.307	3.182
$3\frac{1}{16}$	2.641	2.766	3.396	3.271	3.371	3.246
$3\frac{1}{8}$	2.704	2.829	3.459	3.334	3.434	3.309
$3\frac{3}{16}$	2.768	2.893	3.523	3.398	3.498	3.373
$3\frac{1}{4}$	2.831	2.956	3.586	3.461	3.561	3.436
$3\frac{5}{16}$	2.816	2.941	3.696	3.571	3.671	3.546

Table 1. *(Continued)* **Depth Control Values *S* and *T* for Shaft and Hub**
ANSI B17.1-1967 (R2013) (See figures at end of table)

Nominal Shaft Diameter	Parallel and Taper		Parallel		Taper	
	Square	Rectangular	Square	Rectangular	Square	Rectangular
	S	*S*	*T*	*T*	*T*	*T*
$3\frac{3}{8}$	2.880	3.005	3.760	3.635	3.735	3.610
$3\frac{7}{16}$	2.943	3.068	3.823	3.698	3.798	3.673
$3\frac{1}{2}$	3.007	3.132	3.887	3.762	3.862	3.737
$3\frac{9}{16}$	3.070	3.195	3.950	3.825	3.925	3.800
$3\frac{5}{8}$	3.134	3.259	4.014	3.889	3.989	3.864
$3\frac{11}{16}$	3.197	3.322	4.077	3.952	4.052	3.927
$3\frac{3}{4}$	3.261	3.386	4.141	4.016	4.116	3.991
$3\frac{13}{16}$	3.246	3.371	4.251	4.126	4.226	4.101
$3\frac{7}{8}$	3.309	3.434	4.314	4.189	4.289	4.164
$3\frac{15}{16}$	3.373	3.498	4.378	4.253	4.353	4.228
4	3.436	3.561	4.441	4.316	4.416	4.291
$4\frac{3}{16}$	3.627	3.752	4.632	4.507	4.607	4.482
$4\frac{1}{4}$	3.690	3.815	4.695	4.570	4.670	4.545
$4\frac{3}{8}$	3.817	3.942	4.822	4.697	4.797	4.672
$4\frac{7}{16}$	3.880	4.005	4.885	4.760	4.860	4.735
$4\frac{1}{2}$	3.944	4.069	4.949	4.824	4.924	4.799
$4\frac{3}{4}$	4.041	4.229	5.296	5.109	5.271	5.084
$4\frac{7}{8}$	4.169	4.356	5.424	5.236	5.399	5.211
$4\frac{15}{16}$	4.232	4.422	5.487	5.300	5.462	5.275
5	4.296	4.483	5.551	5.363	5.526	5.338
$5\frac{3}{16}$	4.486	4.674	5.741	5.554	5.716	5.529
$5\frac{1}{4}$	4.550	4.737	5.805	5.617	5.780	5.592
$5\frac{7}{16}$	4.740	4.927	5.995	5.807	5.970	5.782
$5\frac{1}{2}$	4.803	4.991	6.058	5.871	6.033	5.846
$5\frac{3}{4}$	4.900	5.150	6.405	6.155	6.380	6.130
$5\frac{15}{16}$	5.091	5.341	6.596	6.346	6.571	6.321
6	5.155	5.405	6.660	6.410	6.635	6.385
$6\frac{1}{4}$	5.409	5.659	6.914	6.664	6.889	6.639
$6\frac{1}{2}$	5.662	5.912	7.167	6.917	7.142	6.892
$6\frac{3}{4}$	5.760	[a]5.885	7.515	[a]7.390	7.490	[a]7.365
7	6.014	[a]6.139	7.769	[a]7.644	7.744	[a]7.619
$7\frac{1}{4}$	6.268	[a]6.393	8.023	[a]7.898	7.998	[a]7.873
$7\frac{1}{2}$	6.521	[a]6.646	8.276	[a]8.151	8.251	[a]8.126
$7\frac{3}{4}$	6.619	6.869	8.624	8.374	8.599	8.349
8	6.873	7.123	8.878	8.628	8.853	8.603
9	7.887	8.137	9.892	9.642	9.867	9.617
10	8.591	8.966	11.096	10.721	11.071	10.696
11	9.606	9.981	12.111	11.736	12.086	11.711
12	10.309	10.809	13.314	12.814	13.289	12.789
13	11.325	11.825	14.330	13.830	14.305	13.805
14	12.028	12.528	15.533	15.033	15.508	15.008
15	13.043	13.543	16.548	16.048	16.523	16.023

[a] $1\frac{3}{4} \times 1\frac{1}{2}$ inch key.

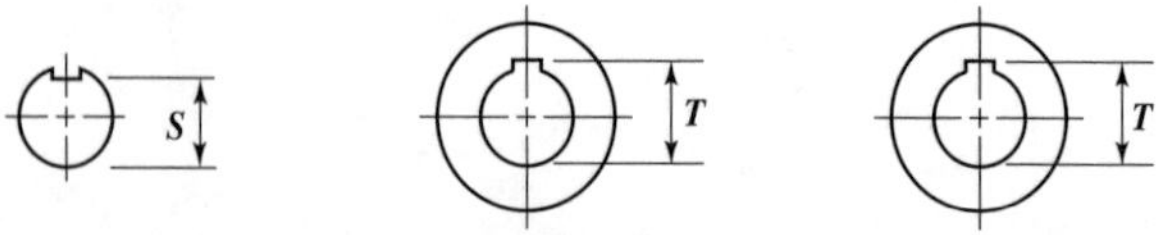

All dimensions are given in inches. See Table 2 for tolerances.

Table 2. ANSI Standard Fits for Parallel and Taper Keys *ANSI B17.1-1967 (R2013)*

Type of Key	Key Width		Side Fit			Top and Bottom Fit			
			Width Tolerance			Depth Tolerance			
	Over	To (Incl.)	Key	Keyseat	Fit Range[a]	Key	Shaft Keyseat	Hub Keyseat	Fit Range[a]
Class 1 Fit for Parallel Keys									
Square	…	1/2	+0.000	+0.002	0.004 CL	+0.000	+0.000	+0.010	0.032 CL
			−0.002	−0.000	0.000	−0.002	−0.015	−0.000	0.005 CL
	1/2	3/4	+0.000	+0.003	0.005 CL	+0.000	+0.000	+0.010	0.032 CL
			−0.002	−0.000	0.000	−0.002	−0.015	−0.000	0.005 CL
	3/4	1	+0.000	+0.003	0.006 CL	+0.000	+0.000	+0.010	0.033 CL
			−0.003	−0.000	0.000	−0.003	−0.015	−0.000	0.005 CL
	1	1 1/2	+0.000	+0.004	0.007 CL	+0.000	+0.000	+0.010	0.033 CL
			−0.003	−0.000	0.000	−0.003	−0.015	−0.000	0.005 CL
	1 1/2	2 1/2	+0.000	+0.004	0.008 CL	+0.000	+0.000	+0.010	0.034 CL
			−0.004	−0.000	0.000	−0.004	−0.015	−0.000	0.005 CL
	2 1/2	3 1/2	+0.000	+0.004	0.010 CL	+0.000	+0.000	+0.010	0.036 CL
			−0.006	−0.000	0.000	−0.006	−0.015	−0.000	0.005 CL
Rectangular	…	1/2	+0.000	+0.002	0.005 CL	+0.000	+0.000	+0.010	0.033 CL
			−0.003	−0.000	0.000	−0.003	−0.015	−0.000	0.005 CL
	1/2	3/4	+0.000	+0.003	0.006 CL	+0.000	+0.000	+0.010	0.033 CL
			−0.003	−0.000	0.000	−0.003	−0.015	−0.000	0.005 CL
	3/4	1	+0.000	+0.003	0.007 CL	+0.000	+0.000	+0.010	0.034 CL
			−0.004	−0.000	0.000	−0.004	−0.015	−0.000	0.005 CL
	1	1 1/2	+0.000	+0.004	0.008 CL	+0.000	+0.000	+0.010	0.034 CL
			−0.004	−0.000	0.000	−0.004	−0.015	−0.000	0.005 CL
	1 1/2	3	+0.000	+0.004	0.009 CL	+0.000	+0.000	+0.010	0.035 CL
			−0.005	−0.000	0.000	−0.005	−0.015	−0.000	0.005 CL
	3	4	+0.000	+0.004	0.010 CL	+0.000	+0.000	+0.010	0.036 CL
			−0.006	−0.000	0.000	−0.006	−0.015	−0.000	0.005 CL
	4	6	+0.000	+0.004	0.012 CL	+0.000	+0.000	+0.010	0.038 CL
			−0.008	−0.000	0.000	−0.008	−0.015	−0.000	0.005 CL
	6	7	+0.000	+0.004	0.017 CL	+0.000	+0.000	+0.010	0.043 CL
			−0.013	−0.000	0.000	−0.013	−0.015	−0.000	0.005 CL
Class 2 Fit for Parallel and Taper Keys									
Parallel Square	…	1 1/4	+0.001	+0.002	0.002 CL	+0.001	+0.000	+0.010	0.030 CL
			−0.000	−0.000	0.001 INT	−0.000	−0.015	−0.000	0.004 CL
	1 1/4	3	+0.002	+0.002	0.002 CL	+0.002	+0.000	+0.010	0.030 CL
			−0.000	−0.000	0.002 INT	−0.000	−0.015	−0.000	0.003 CL
	3	3 1/2	+0.003	+0.002	0.002 CL	+0.003	+0.000	+0.010	0.030 CL
			−0.000	−0.000	0.003 INT	−0.000	−0.015	−0.000	0.002 CL
Parallel Rectangular	…	1 1/4	+0.001	+0.002	0.002 CL	+0.005	+0.000	+0.010	0.035 CL
			−0.000	−0.000	0.001 INT	−0.005	−0.015	−0.000	0.000 CL
	1 1/4	3	+0.002	+0.002	0.002 CL	+0.005	+0.000	+0.010	0.035 CL
			−0.000	−0.000	0.002 INT	−0.005	−0.015	−0.000	0.000 CL
	3	7	+0.003	+0.002	0.002 CL	+0.005	+0.000	+0.010	0.035 CL
			−0.000	−0.000	0.003 INT	−0.005	−0.015	−0.000	0.000 CL
Taper	…	1 1/4	+0.001	+0.002	0.002 CL	+0.005	+0.000	+0.010	0.005 CL
			−0.000	−0.000	0.001 INT	−0.000	−0.015	−0.000	0.025 INT
	1 1/4	3	+0.002	+0.002	0.002 CL	+0.005	+0.000	+0.010	0.005 CL
			−0.000	−0.000	0.002 INT	−0.000	−0.015	−0.000	0.025 INT
	3	[b]	+0.003	+0.002	0.002 CL	+0.005	+0.000	+0.010	0.005 CL
			−0.000	−0.000	0.003 INT	−0.000	−0.015	−0.000	0.025 INT

[a] Limits of variation. CL = Clearance; INT = Interference.

[b] To (incl.) 3 1/2-inch square and 7-inch rectangular key widths.

All dimensions are given in inches. See also text on page 177.

Table 3. ANSI Standard Plain and Gib Head Keys *ANSI B17.1-1967 (R2013)*

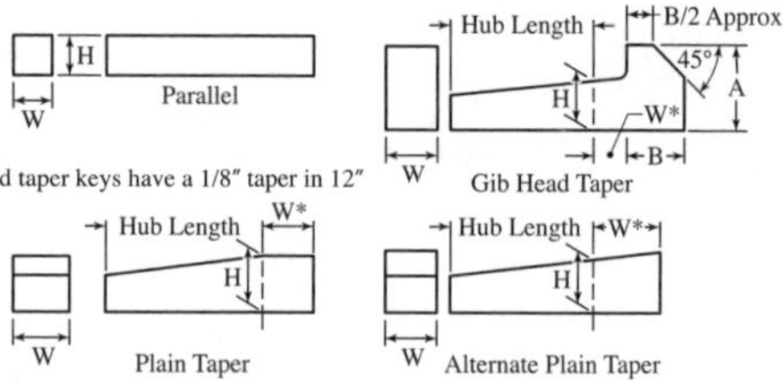

Key			Nominal Key Size Width *W* Over	To (Incl.)	Tolerance Width, *W*		Height, *H*	
Parallel	Square	Keystock	…	1 1/4	+0.001	−0.000	+0.001	−0.000
			1 1/4	3	+0.002	−0.000	+0.002	−0.000
			3	3 1/2	+0.003	−0.000	+0.003	−0.000
		Bar Stock	…	3/4	+0.000	−0.002	+0.000	−0.002
			3/4	1 1/2	+0.000	−0.003	+0.000	−0.003
			1 1/2	2 1/2	+0.000	−0.004	+0.000	−0.004
			2 1/2	3 1/2	+0.000	−0.006	+0.000	−0.006
	Rectangular	Keystock	…	1 1/4	+0.001	−0.000	+0.005	−0.005
			1 1/4	3	+0.002	−0.000	+0.005	−0.005
			3	7	+0.003	−0.000	+0.005	−0.005
		Bar Stock	…	3/4	+0.000	−0.003	+0.000	−0.003
			3/4	1 1/2	+0.000	−0.004	+0.000	−0.004
			1 1/2	3	+0.000	−0.005	+0.000	−0.005
			3	4	+0.000	−0.006	+0.000	−0.006
			4	6	+0.000	−0.008	+0.000	−0.008
			6	7	+0.000	−0.013	+0.000	−0.013
Taper	Plain or Gib Head Square or Rectangular		…	1 1/4	+0.001	−0.000	+0.005	−0.000
			1 1/4	3	+0.002	−0.000	+0.005	−0.000
			3	7	+0.003	−0.000	+0.005	−0.000

Gib Head Nominal Dimensions													
Nominal Key Size Width, *W*	Square *H*	*A*	*B*	Rectangular *H*	*A*	*B*	Nominal Key Size Width, *W*	Square *H*	*A*	*B*	Rectangular *H*	*A*	*B*
1/8	1/8	1/4	1/4	3/32	3/16	1/8	1	1	1 5/8	1 1/8	3/4	1 1/4	7/8
3/16	3/16	5/16	5/16	1/8	1/4	1/4	1 1/4	1 1/4	2	1 7/16	7/8	1 3/8	1
1/4	1/4	7/16	3/8	3/16	5/16	5/16	1 1/2	1 1/2	2 3/8	1 3/4	1	1 5/8	1 1/8
5/16	5/16	1/2	7/16	1/4	7/16	3/8	1 3/4	1 3/4	2 3/4	2	1 1/2	2 3/8	1 3/4
3/8	3/8	5/8	1/2	1/4	7/16	3/8	2	2	3 1/2	2 1/4	1 1/2	2 3/8	1 3/4
1/2	1/2	7/8	5/8	3/8	5/8	1/2	2 1/2	2 1/2	4	3	1 3/4	2 3/4	2
5/8	5/8	1	3/4	7/16	3/4	9/16	3	3	5	3 1/2	2	3 1/2	2 1/4
3/4	3/4	1 1/4	7/8	1/2	7/8	5/8	3 1/2	3 1/2	6	4	2 1/2	4	3
7/8	7/8	1 3/8	1	5/8	1	3/4	…	…	…	…	…	…	…

All dimensions are given in inches.

*For locating position of dimension *H*. Tolerance does not apply.

For larger sizes the following relationships are suggested as guides for establishing *A* and *B*: $A = 1.8H$ and $B = 1.2H$.

Table 4. Finding Depth of Keyseat and Distance from Top of Key to Bottom of Shaft

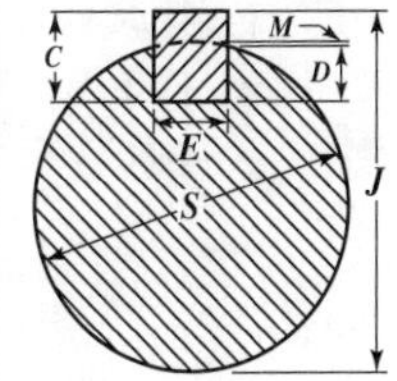

For milling keyseats, the total depth to feed cutter in from outside of shaft to bottom of keyseat is $M + D$, where D is depth of keyseat.

For checking an assembled key and shaft, caliper measurement J between top of key and bottom of shaft is used.

$$J = S - (M + D) + C$$

where C is depth of key. For Woodruff keys, dimensions C and D can be found in Table 5. Assuming shaft diameter S is normal size, the tolerance on dimension J for Woodruff keys in keyslots are +0.000, −0.010 inch.

Dia. of Shaft, S Inches	Width of Keyseat, E														
	1/16	3/32	1/8	5/32	3/16	7/32	1/4	5/16	3/8	7/16	1/2	9/16	5/8	11/16	3/4
	Dimension M, Inch														
0.3125	.0032	…	…	…	…	…	…	…	…	…	…	…	…	…	…
0.3437	.0029	.0065	…	…	…	…	…	…	…	…	…	…	…	…	…
0.3750	.0026	.0060	.0107	…	…	…	…	…	…	…	…	…	…	…	…
0.4060	.0024	.0055	.0099	…	…	…	…	…	…	…	…	…	…	…	…
0.4375	.0022	.0051	.0091	…	…	…	…	…	…	…	…	…	…	…	…
0.4687	.0021	.0047	.0085	.0134	…	…	…	…	…	…	…	…	…	…	…
0.5000	.0020	.0044	.0079	.0125	…	…	…	…	…	…	…	…	…	…	…
0.5625	…	.0039	.0070	.0111	.0161	…	…	…	…	…	…	…	…	…	…
0.6250	…	.0035	.0063	.0099	.0144	.0198	…	…	…	…	…	…	…	…	…
0.6875	…	.0032	.0057	.0090	.0130	.0179	.0235	…	…	…	…	…	…	…	…
0.7500	…	.0029	.0052	.0082	.0119	.0163	.0214	.0341	…	…	…	…	…	…	…
0.8125	…	.0027	.0048	.0076	.0110	.0150	.0197	.0312	…	…	…	…	…	…	…
0.8750	…	.0025	.0045	.0070	.0102	.0139	.0182	.0288	…	…	…	…	…	…	…
0.9375	…	…	.0042	.0066	.0095	.0129	.0170	.0263	.0391	…	…	…	…	…	…
1.0000	…	…	.0039	.0061	.0089	.0121	.0159	.0250	.0365	…	…	…	…	…	…
1.0625	…	…	.0037	.0058	.0083	.0114	.0149	.0235	.0342	…	…	…	…	…	…
1.1250	…	…	.0035	.0055	.0079	.0107	.0141	.0221	.0322	.0443	…	…	…	…	…
1.1875	…	…	.0033	.0052	.0074	.0102	.0133	.0209	.0304	.0418	…	…	…	…	…
1.2500	…	…	.0031	.0049	.0071	.0097	.0126	.0198	.0288	.0395	…	…	…	…	…
1.3750	…	…	…	.0045	.0064	.0088	.0115	.0180	.0261	.0357	.0471	…	…	…	…
1.5000	…	…	…	.0041	.0059	.0080	.0105	.0165	.0238	.0326	.0429	…	…	…	…
1.6250	…	…	…	.0038	.0054	.0074	.0097	.0152	.0219	.0300	.0394	.0502	…	…	…
1.7500	…	…	…	…	.0050	.0069	.0090	.0141	.0203	.0278	.0365	.0464	…	…	…
1.8750	…	…	…	…	.0047	.0064	.0084	.0131	.0189	.0259	.0340	.0432	.0536	…	…
2.0000	…	…	…	…	.0044	.0060	.0078	.0123	.0177	.0242	.0318	.0404	.0501	…	…
2.1250	…	…	…	…	…	.0056	.0074	.0116	.0167	.0228	.0298	.0379	.0470	.0572	.0684
2.2500	…	…	…	…	…	…	.0070	.0109	.0157	.0215	.0281	.0357	.0443	.0538	.0643
2.3750	…	…	…	…	…	…	…	.0103	.0149	.0203	.0266	.0338	.0419	.0509	.0608
2.5000	…	…	…	…	…	…	…	…	.0141	.0193	.0253	.0321	.0397	.0482	.0576
2.6250	…	…	…	…	…	…	…	…	.0135	.0184	.0240	.0305	.0377	.0457	.0547
2.7500	…	…	…	…	…	…	…	…	…	.0175	.0229	.0291	.0360	.0437	.0521
2.8750	…	…	…	…	…	…	…	…	…	.0168	.0219	.0278	.0344	.0417	.0498
3.0000	…	…	…	…	…	…	…	…	…	…	.0210	.0266	.0329	.0399	.0476

ANSI Standard Woodruff Keys and Keyseats.—American National Standard B17.2 was approved in 1967, and reaffirmed in 2013. Data from this standard are shown in Tables below. The following definitions are given in this standard:

Woodruff Key: A Remountable machinery part which, when assembled into keyseats, provides a positive means for transmitting torque between the shaft and hub.

Woodruff Key Number: An identification number by which the size of key may be readily determined.

Woodruff Keyseat—Shaft: The circular pocket in which the key is retained.

Woodruff Keyseat—Hub: An axially located rectangular groove in a hub. (This has been referred to as a keyway.)

Woodruff Keyseat Milling Cutter: An arbor-type or shank-type milling cutter normally used for milling Woodruff keyseats in shafts.

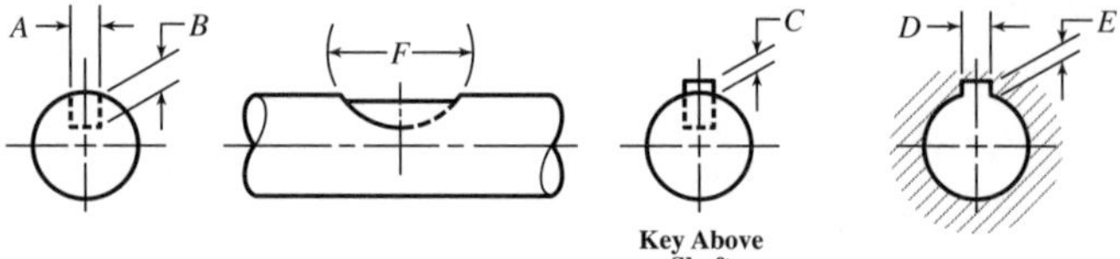

Keyseat—Shaft

Key Above Shaft

Keyseat—Hub

Table 5. ANSI Keyseat Dimensions for Woodruff Keys

ANSI B17.2-1967 (R2013)

Key No.	Nominal Size Key	Keyseat—Shaft					Key Above Shaft	Keyseat—Hub	
		Width A[a]		Depth B	Diameter F		Height C	Width D	Depth E
		Min.	Max.	+0.005 −0.000	Min.	Max.	+0.005 −0.005	+0.002 −0.000	+0.005 −0.000
202	$\frac{1}{16} \times \frac{1}{4}$	0.0615	0.0630	0.0728	0.250	0.268	0.0312	0.0635	0.0372
202.5	$\frac{1}{16} \times \frac{5}{16}$	0.0615	0.0630	0.1038	0.312	0.330	0.0312	0.0635	0.0372
302.5	$\frac{3}{32} \times \frac{5}{16}$	0.0928	0.0943	0.0882	0.312	0.330	0.0469	0.0948	0.0529
203	$\frac{1}{16} \times \frac{3}{8}$	0.0615	0.0630	0.1358	0.375	0.393	0.0312	0.0635	0.0372
303	$\frac{3}{32} \times \frac{3}{8}$	0.0928	0.0943	0.1202	0.375	0.393	0.0469	0.0948	0.0529
403	$\frac{1}{8} \times \frac{3}{8}$	0.1240	0.1255	0.1045	0.375	0.393	0.0625	0.1260	0.0685
204	$\frac{1}{16} \times \frac{1}{2}$	0.0615	0.0630	0.1668	0.500	0.518	0.0312	0.0635	0.0372
304	$\frac{3}{32} \times \frac{1}{2}$	0.0928	0.0943	0.1511	0.500	0.518	0.0469	0.0948	0.0529
404	$\frac{1}{8} \times \frac{1}{2}$	0.1240	0.1255	0.1355	0.500	0.518	0.0625	0.1260	0.0685
305	$\frac{3}{32} \times \frac{5}{8}$	0.0928	0.0943	0.1981	0.625	0.643	0.0469	0.0948	0.0529
405	$\frac{1}{8} \times \frac{5}{8}$	0.1240	0.1255	0.1825	0.625	0.643	0.0625	0.1260	0.0685
505	$\frac{5}{32} \times \frac{5}{8}$	0.1553	0.1568	0.1669	0.625	0.643	0.0781	0.1573	0.0841
605	$\frac{3}{16} \times \frac{5}{8}$	0.1863	0.1880	0.1513	0.625	0.643	0.0937	0.1885	0.0997
406	$\frac{1}{8} \times \frac{3}{4}$	0.1240	0.1255	0.2455	0.750	0.768	0.0625	0.1260	0.0685
506	$\frac{5}{32} \times \frac{3}{4}$	0.1553	0.1568	0.2299	0.750	0.768	0.0781	0.1573	0.0841
606	$\frac{3}{16} \times \frac{3}{4}$	0.1863	0.1880	0.2143	0.750	0.768	0.0937	0.1885	0.0997
806	$\frac{1}{4} \times \frac{3}{4}$	0.2487	0.2505	0.1830	0.750	0.768	0.1250	0.2510	0.1310
507	$\frac{5}{32} \times \frac{7}{8}$	0.1553	0.1568	0.2919	0.875	0.895	0.0781	0.1573	0.0841
607	$\frac{3}{16} \times \frac{7}{8}$	0.1863	0.1880	0.2763	0.875	0.895	0.0937	0.1885	0.0997
707	$\frac{7}{32} \times \frac{7}{8}$	0.2175	0.2193	0.2607	0.875	0.895	0.1093	0.2198	0.1153
807	$\frac{1}{4} \times \frac{7}{8}$	0.2487	0.2505	0.2450	0.875	0.895	0.1250	0.2510	0.1310
608	$\frac{3}{16} \times 1$	0.1863	0.1880	0.3393	1.000	1.020	0.0937	0.1885	0.0997
708	$\frac{7}{32} \times 1$	0.2175	0.2193	0.3237	1.000	1.020	0.1093	0.2198	0.1153
808	$\frac{1}{4} \times 1$	0.2487	0.2505	0.3080	1.000	1.020	0.1250	0.2510	0.1310
1008	$\frac{5}{16} \times 1$	0.3111	0.3130	0.2768	1.000	1.020	0.1562	0.3135	0.1622
1208	$\frac{3}{8} \times 1$	0.3735	0.3755	0.2455	1.000	1.020	0.1875	0.3760	0.1935
609	$\frac{3}{16} \times 1\frac{1}{8}$	0.1863	0.1880	0.3853	1.125	1.145	0.0937	0.1885	0.0997
709	$\frac{7}{32} \times 1\frac{1}{8}$	0.2175	0.2193	0.3697	1.125	1.145	0.1093	0.2198	0.1153
809	$\frac{1}{4} \times 1\frac{1}{8}$	0.2487	0.2505	0.3540	1.125	1.145	0.1250	0.2510	0.1310
1009	$\frac{5}{16} \times 1\frac{1}{8}$	0.3111	0.3130	0.3228	1.125	1.145	0.1562	0.3135	0.1622
610	$\frac{3}{16} \times 1\frac{1}{4}$	0.1863	0.1880	0.4483	1.250	1.273	0.0937	0.1885	0.0997
710	$\frac{7}{32} \times 1\frac{1}{4}$	0.2175	0.2193	0.4327	1.250	1.273	0.1093	0.2198	0.1153

Table 5. *(Continued)* **ANSI Keyseat Dimensions for Woodruff Keys**
ANSI B17.2-1967 (R2013)

Key No.	Nominal Size Key	Keyseat—Shaft					Key Above Shaft	Keyseat—Hub	
		Width A[a]		Depth B	Diameter F		Height C	Width D	Depth E
		Min.	Max.	+0.005 −0.000	Min.	Max.	+0.005 −0.005	+0.002 −0.000	+0.005 −0.000
810	$^1/_4 \times 1^1/_4$	0.2487	0.2505	0.4170	1.250	1.273	0.1250	0.2510	0.1310
1010	$^5/_{16} \times 1^1/_4$	0.3111	0.3130	0.3858	1.250	1.273	0.1562	0.3135	0.1622
1210	$^3/_8 \times 1^1/_4$	0.3735	0.3755	0.3545	1.250	1.273	0.1875	0.3760	0.1935
811	$^1/_4 \times 1^3/_8$	0.2487	0.2505	0.4640	1.375	1.398	0.1250	0.2510	0.1310
1011	$^5/_{16} \times 1^3/_8$	0.3111	0.3130	0.4328	1.375	1.398	0.1562	0.3135	0.1622
1211	$^3/_8 \times 1^3/_8$	0.3735	0.3755	0.4015	1.375	1.398	0.1875	0.3760	0.1935
812	$^1/_4 \times 1^1/_2$	0.2487	0.2505	0.5110	1.500	1.523	0.1250	0.2510	0.1310
1012	$^5/_{16} \times 1^1/_2$	0.3111	0.3130	0.4798	1.500	1.523	0.1562	0.3135	0.1622
1212	$^3/_8 \times 1^1/_2$	0.3735	0.3755	0.4485	1.500	1.523	0.1875	0.3760	0.1935
617-1	$^3/_{16} \times 2^1/_8$	0.1863	0.1880	0.3073	2.125	2.160	0.0937	0.1885	0.0997
817-1	$^1/_4 \times 2^1/_8$	0.2487	0.2505	0.2760	2.125	2.160	0.1250	0.2510	0.1310
1017-1	$^5/_{16} \times 2^1/_8$	0.3111	0.3130	0.2448	2.125	2.160	0.1562	0.3135	0.1622
1217-1	$^3/_8 \times 2^1/_8$	0.3735	0.3755	0.2135	2.125	2.160	0.1875	0.3760	0.1935
617	$^3/_{16} \times 2^1/_8$	0.1863	0.1880	0.4323	2.125	2.160	0.0937	0.1885	0.0997
817	$^1/_4 \times 2^1/_8$	0.2487	0.2505	0.4010	2.125	2.160	0.1250	0.2510	0.1310
1017	$^5/_{16} \times 2^1/_8$	0.3111	0.3130	0.3698	2.125	2.160	0.1562	0.3135	0.1622
1217	$^3/_8 \times 2^1/_8$	0.3735	0.3755	0.3385	2.125	2.160	0.1875	0.3760	0.1935
822-1	$^1/_4 \times 2^3/_4$	0.2487	0.2505	0.4640	2.750	2.785	0.1250	0.2510	0.1310
1022-1	$^5/_{16} \times 2^3/_4$	0.3111	0.3130	0.4328	2.750	2.785	0.1562	0.3135	0.1622
1222-1	$^3/_8 \times 2^3/_4$	0.3735	0.3755	0.4015	2.750	2.785	0.1875	0.3760	0.1935
1422-1	$^7/_{16} \times 2^3/_4$	0.4360	0.4380	0.3703	2.750	2.785	0.2187	0.4385	0.2247
1622-1	$^1/_2 \times 2^3/_4$	0.4985	0.5005	0.3390	2.750	2.785	0.2500	0.5010	0.2560
822	$^1/_4 \times 2^3/_4$	0.2487	0.2505	0.6200	2.750	2.785	0.1250	0.2510	0.1310
1022	$^5/_{16} \times 2^3/_4$	0.3111	0.3130	0.5888	2.750	2.785	0.1562	0.3135	0.1622
1222	$^3/_8 \times 2^3/_4$	0.3735	0.3755	0.5575	2.750	2.785	0.1875	0.3760	0.1935
1422	$^7/_{16} \times 2^3/_4$	0.4360	0.4380	0.5263	2.750	2.785	0.2187	0.4385	0.2247
1622	$^1/_2 \times 2^3/_4$	0.4985	0.5005	0.4950	2.750	2.785	0.2500	0.5010	0.2560
1228	$^3/_8 \times 3^1/_2$	0.3735	0.3755	0.7455	3.500	3.535	0.1875	0.3760	0.1935
1428	$^7/_{16} \times 3^1/_2$	0.4360	0.4380	0.7143	3.500	3.535	0.2187	0.4385	0.2247
1628	$^1/_2 \times 3^1/_2$	0.4985	0.5005	0.6830	3.500	3.535	0.2500	0.5010	0.2560
1828	$^9/_{16} \times 3^1/_2$	0.5610	0.5630	0.6518	3.500	3.535	0.2812	0.5635	0.2872
2028	$^5/_8 \times 3^1/_2$	0.6235	0.6255	0.6205	3.500	3.535	0.3125	0.6260	0.3185
2228	$^{11}/_{16} \times 3^1/_2$	0.6860	0.6880	0.5893	3.500	3.535	0.3437	0.6885	0.3497
2428	$^3/_4 \times 3^1/_2$	0.7485	0.7505	0.5580	3.500	3.535	0.3750	0.7510	0.3810

[a] These Width A values were set with the maximum keyseat (shaft) width as that figure which will receive a key with the greatest amount of looseness consistent with assuring the key's sticking in the keyseat (shaft). Minimum keyseat width is that figure permitting the largest shaft distortion acceptable when assembling maximum key in minimum keyseat. Dimensions A, B, C, D are taken at side intersection.

All dimensions are given in inches.

BROACHING

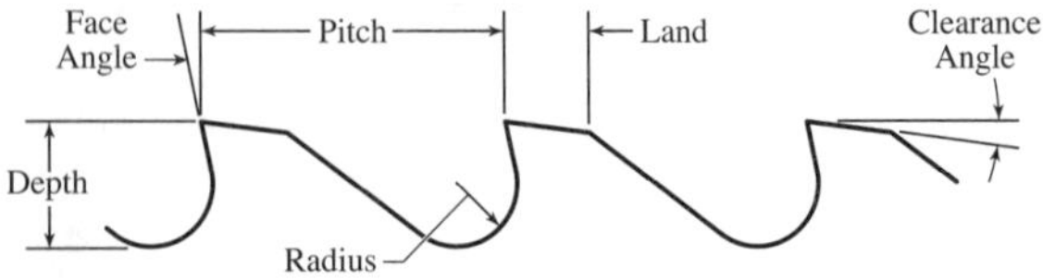

Terms Commonly Used in Broach Design

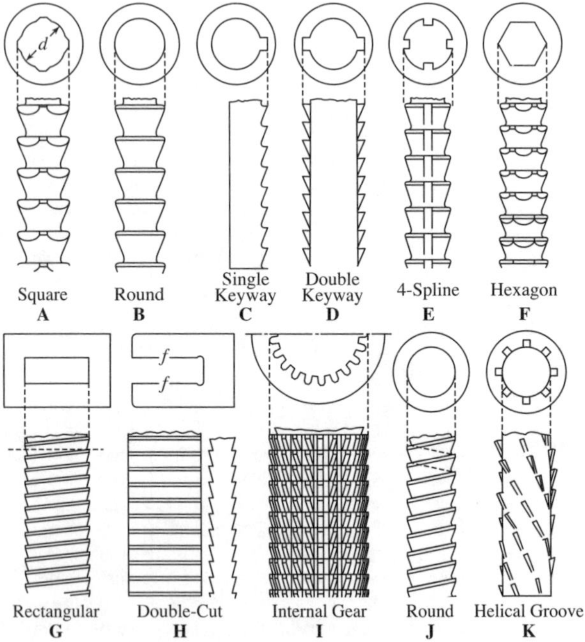

Fig. 1. Types of Broaches

Pitch of Broach Teeth.—The pitch of broach teeth depends upon the depth of cut or chip thickness, length of cut, the cutting force required, and power of the broaching machine. In the pitch formulas which follow

L = length, in inches, of layer to be removed by broaching

d = depth of cut per tooth as shown by Table 1 (For internal broaches, d = depth of cut as measured on one side of broach or one-half difference in diameters of successive teeth in case of a round broach.)

F = a factor (For brittle types of material, F = 3 or 4 for roughing teeth, and 6 for finishing teeth. For ductile types of material, F = 4 to 7 for roughing teeth and 8 for finishing teeth.)

b = width of inches, of layer to be removed by broaching

P = pressure required in tons per square inch, of an area equal to depth of cut times width of cut, in inches (Table 2)

T = usable capacity, in tons, of broaching machine = 70 percent of maximum tonnage

Table 1. Data for Designing Surface Broaches

Material to be Broached	Depth of Cut per Tooth				Face Angle or Rake, Degrees	Clearance Angle, Degrees	
	Roughing[a]		Finishing				
	inch	mm	inch	mm		Rough	Finish
Steel, High Tensile Strength	0.0015–0.002	0.04–0.05	0.0005	0.013	10–12	1.5–3	0.5–1
Steel, Med.Tensile Strength	0.0025–0.005	0.06–0.13	0.0005	0.013	14–18	1.5–3	0.5–1
Cast Steel	0.0025–0.005	0.06–0.13	0.0005	0.013	10	1.53	0.5
Malleable Iron	0.0025–0.005	0.06–0.13	0.0005	0.013	7	1.5–3	0.5
Cast Iron, Soft	0.006–0.010	0.15–0.25	0.0005	0.013	10–15	1.5–3	0.5
Cast Iron, Hard	0.003–0.005	0.08–0.13	0.0005	0.013	5	1.5–3	0.5
Zinc Die Castings	0.005–0.010	0.13–0.25	0.0010	0.025	12[b]	5	2
Cast Bronze	0.010–0.025	0.25–0.64	0.0005	0.013	8	0	0
Wrought Aluminum Alloys	0.005–0.010	0.13–0.25	0.0010	0.025	15[b]	3	1
Cast Aluminum Alloys	0.005–0.010	0.13–0.25	0.0010	0.025	12[b]	3	1
Magnesium Die Castings	0.010–0.015	0.25–0.38	0.0010	0.025	20[b]	3	1

[a] The lower depth-of-cut values for roughing are recommended when work is not very rigid, the tolerance is small, a good finish is required, or length of cut is comparatively short.

[b] In broaching these materials, smooth surfaces for tooth and chip spaces are especially recommended.

Table 2. Broaching Pressure *P* for Use in Pitch Formulas (1) and (2)

Material to be Broached	Depth d of Cut per Tooth, Inch (mm)										Side-Cutting Broaches			
	0.024 (0.60)		0.01 (0.25)		0.004 (0.10)		0.002 (0.05)		0.001 (0.025)					
	Pressure, P										Pressure, P		Cut, d	
	Ton/in²	MPa	Ton/in²	MPa	Ton/in²	MPa	Ton/in²	MPa	Ton/in²	MPa	Ton/in²	MPa	inch	mm
Steel, High Tensile Strength	…	…	…	…			250	3447	312	4302	200	2758	0.004	0.10
Steel, Med. Tensile Strength	…	…	…	…	158	2179	185	2551	243	3351	143	1972	0.006	0.15
Cast Steel	…	…	…	…	128	1765	158	2179	…	…	115	1586	0.006	0.15
Malleable Iron	…	…	…	…	108	1489	128	1765	…	…	100	1379	0.006	0.15
Cast Iron	…	…	115	1586	115	1586	143	1972	…	…	115	1586	0.020	0.51
Cast Brass	…	…	50	689	50	689	…	…	…	…	…	…	…	…
Brass, Hot-pressed	…	…	85	1172	85	1172	…	…	…	…	…	…	…	…
Zinc Die Castings	…	…	70	965	70	965	…	…	…	…	…	…	…	…
Cast Bronze	35	483	35	483	…	…	…	…	…	…	…	…	…	…
Wrought Aluminum	…	…	70	965	70	965	…	…	…	…	…	…	…	…
Cast Aluminum	…	…	85	1172	85	1172	…	…	…	…	…	…	…	…
Magnesium Alloy	35	483	35	483	…	…	…	…	…	…	…	…	…	…

The minimum pitch shown by Formula (1) is based upon the receiving capacity of the chip space. The minimum, however, should not be less than 0.2 inch unless a smaller pitch is required for exceptionally short cuts to provide at least two teeth in contact simultaneously, with the part being broached. A reduction below 0.2 inch is seldom required in surface broaching, but it may be necessary in connection with internal broaching.

$$\text{Minimum pitch} = 3\sqrt{LdF} \qquad (1)$$

Whether the minimum pitch may be used or not depends upon the power of the available machine. The factor F in the formula provides for the increase in volume as the material is broached into chips. If a broach has adjustable inserts for the finishing teeth, the pitch of the finishing teeth may be smaller than the pitch of the roughing teeth because of the smaller depth d of the cut. The higher value of F for finishing teeth prevents the pitch from becoming too small, so that the spirally curled chips will not be crowded into too small a space. The pitch of the roughing and finishing teeth should be equal for broaches without separate inserts (notwithstanding the different values of d and F) so that some of the finishing teeth may be ground into roughing teeth after wear makes this necessary.

$$\text{Allowable pitch} = \frac{dLbP}{T} \tag{2}$$

If the pitch obtained by Formula (2) is larger than the minimum obtained by Formula (1), this larger value should be used because it is based upon the usable power of the machine. As the notation indicates, 70 percent of the maximum tonnage T is taken as the usable capacity. The 30 percent reduction is to provide a margin for the increase in broaching load resulting from the gradual dulling of the cutting edges.

Table 3. Common Causes of Broaching Difficulties

Broaching Difficulty	Possible Causes
Stuck broach	Insufficient machine capacity; dulled teeth; clogged chip gullets; failure of power during cutting stroke. To remove a stuck broach, workpiece and broach are removed from the machine as a unit; never try to back out broach by reversing machine. If broach does not loosen by tapping workpiece lightly and trying to slide it off its starting end, mount workpiece and broach in a lathe and turn down workpiece to the tool surface. Workpiece may be sawed longitudinally into several sections in order to free the broach. Check broach design, perhaps tooth relief (back off) angle is too small or depth of cut per tooth is too great.
Galling and pickup	Lack of homogeneity of material being broached—uneven hardness, porosity; improper or insufficient coolant; poor broach design, mutilated broach; dull broach; improperly sharpened broach; improperly designed or outworn fixtures. Good broach design will do away with possible chip build-up on tooth faces and excessive heating. Grinding of teeth should be accurate so that the correct gullet contour is maintained. Contour should be fair and smooth.
Broach breakage	Overloading; broach dullness; improper sharpening; interrupted cutting stroke; backing up broach with workpiece in fixture; allowing broach to pass entirely through guide hole; ill-fitting and/or sharp-edged key; crooked holes; untrue locating surface; excessive hardness of workpiece; insufficient clearance angle; sharp corners on pull end of broach. When grinding bevels on pull end of broach, use wheel that is not too pointed.
Chatter	Too few teeth in cutting contact simultaneously; excessive hardness of material being broached; loose or poorly constructed tooling; surging of ram due to load variations. Chatter can be alleviated by changing the broaching speed, by using shear cutting teeth instead of right-angle teeth, and by changing the coolant and the face and relief angles of the teeth.
Drifting or misalignment of tool during cutting stroke	Lack of proper alignment when broach is sharpened in grinding machine, which may be caused by dirt in the female center of the broach; inadequate support of broach during the cutting stroke, on a horizontal machine especially; body diameter too small; cutting resistance variable around I.D. of hole due to lack of symmetry of surfaces to be cut; variations in hardness around I.D. of hole; too few teeth in cutting contact.
Streaks in broached surface	Lands too wide; presence of forging, casting or annealing scale; metal pickup; presence of grinding burrs and grinding and cleaning abrasives.
Rings in the broached hole	Due to surging resulting from uniform pitch of teeth; presence of sharpening burrs on broach; tooth clearance angle too large; locating face not smooth or square; broach not supported for all cutting teeth passing through the work. The use of differential tooth spacing or shear cutting teeth helps in preventing surging. Sharpening burrs on a broach may be removed with a wood block.

CUTTING TOOLS FOR TURNING

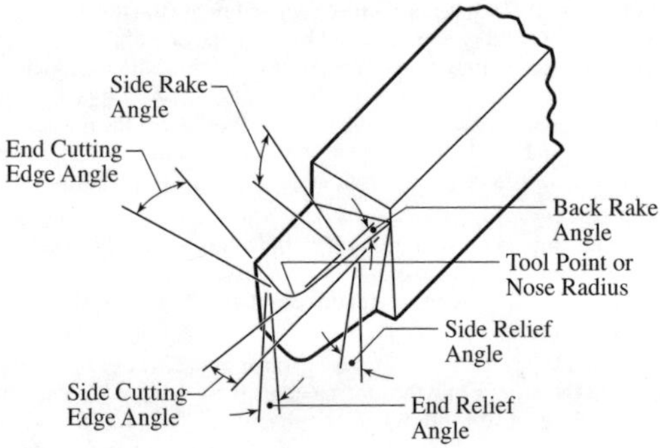

Fig. 1. Terms Applied to Single-Point Turning Tools

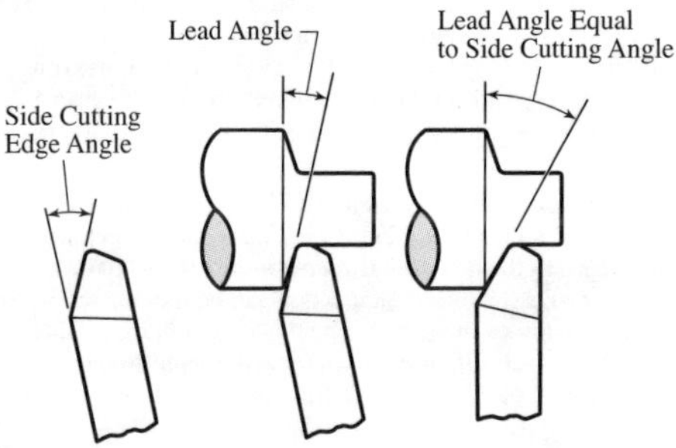

Fig. 2. Lead Angle on Single-Point Turning Tool

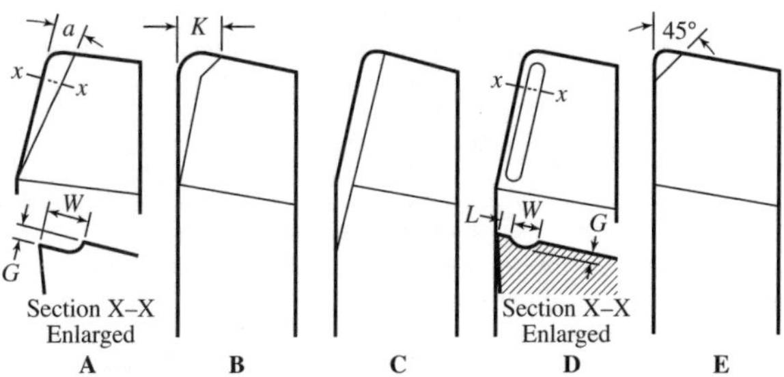

Fig. 3. Different Forms of Chipbreakers for Turning Tools

Chipbreakers.—*Angular Shoulder Type:* As shown in Fig. 3A, angle *a* between the shoulder and cutting edge may vary from 6 to 15 degrees or more, 8 degrees being a fair average. The ideal angle, width *W* and depth *G*, depends upon the speed and feed, the depth of cut, and the material. As a general rule, width *W*, at the end of the tool, varies from $\frac{3}{32}$ to $\frac{7}{32}$ inch, and the depth *G* may range from $\frac{1}{64}$ to $\frac{1}{16}$ inch. The shoulder radius equals depth *G*.

If the tool has a large nose radius, the corner of the shoulder at the nose end may be beveled off, as illustrated in Fig. 3B, to prevent it from coming into contact with the work. The width *K* for type *B* should equal approximately 1.5 times the nose radius.

Parallel Shoulder Type: Fig. 3C shows a design with a chipbreaking shoulder that is parallel with the cutting edge. With this form, the chips are likely to come off in short curled sections. The parallel form may also be applied to straight tools which do not have a side cutting edge angle. The tendency with this parallel shoulder form is to force the chips against the work and damage it.

Groove Type: This type (see Fig. 3D) has a groove in the face of the tool produced by grinding. Between the groove and the cutting edge, there is a land *L*. Under ideal conditions, this width *L*, the groove width *W*, and the groove depth *G*, would be varied to suit the feed, depth of cut and material. For average use, *L* is about $\frac{1}{32}$ inch; *G*, $\frac{1}{32}$ inch; and *W*, $\frac{1}{16}$ inch. There are differences of opinion concerning the relative merits of the groove type and the shoulder type. Both types have proved satisfactory when properly proportioned for a given class of work.

Chipbreaker for Light Cuts: Fig. 3E illustrates a form of chipbreaker that is sometimes used on tools for finishing cuts having a maximum depth of about $\frac{1}{32}$ inch. This chipbreaker is a shoulder type having an angle of 45 degrees and a maximum width of about $\frac{1}{16}$ inch. It is important in grinding all chipbreakers to give the chip-bearing surfaces a fine finish, such as would be obtained by honing. This finish greatly increases the life of the tool.

Identification System for Indexable Inserts.—The size of indexable inserts is determined by the diameter of an inscribed circle (I.C.), except for rectangular and parallelogram inserts where the length and width dimensions are used. To describe an insert in its entirety, a standard ANSI B212.4-2002 identification system is used where each position number designates a feature of the insert. The ANSI Standard includes items now commonly used and facilitates identification of items not in common use. Identification consists of up to ten positions; each position defines a characteristic of the insert as shown below:

1	2	3	4	5	6	7	8[a]	9[a]	10[a]
T	**N**	**M**	**G**	**5**	**4**	**3**			**A**

[a] Eighth, Ninth, and Tenth Positions are used only when required.

Shape: The shape of an insert is designated by a letter: **R** for round; **S,** square; **T,** triangle; **A,** 85° parallelogram; **B,** 82° parallelogram; **C,** 80° diamond; **D,** 55° diamond; **E,** 75° diamond; **H,** hexagon; **K,** 55° parallelogram; **L,** rectangle; **M,** 86° diamond; **O,** octagon; **P,** pentagon; **V,** 35° diamond; and **W,** 80° trigon.

Relief Angle (Clearances): The second position is a letter denoting the relief angles; **N** for 0°; **A**, 3°; **B**, 5°; **C**, 7°; **P**, 11°; **D**, 15°; **E**, 20°; **F**, 25°; **G**, 30°; **H**, 0° & 11°*; **J**, 0° & 14°*; **K**, 0° & 17°*; **L**, 0° & 20°*; **M**, 11° & 14°*; **R**, 11° & 17°*; **S**, 11° & 20°*. When mounted on a holder, the actual relief angle may be different from that on the insert.

Tolerances: The third position is a letter and indicates the tolerances that control the indexability of the insert. Tolerances specified do not imply the method of manufacture.

Type: The type of insert is designated by a letter. **A,** with hole; **B,** with hole and countersink; **C,** with hole and two countersinks; **F,** chip grooves both surfaces, no hole; **G,** same

*Second angle is secondary facet angle, which may vary by ± 1°.

as **F** but with hole; **H,** with hole, one countersink, and chip groove on one rake surface; **J,** with hole, two countersinks and chip grooves on two rake surfaces; **M,** with hole and chip groove on one rake surface; **N,** without hole; **Q,** with hole and two countersinks; **R,** without hole but with chip groove on one rake surface; **T,** with hole, one countersink, and chip groove on one rake face; **U,** with hole, two countersinks, and chip grooves on two rake faces; and **W,** with hole and one countersink. *Note:* a dash may be used after position 4 to separate the shape-describing portion from the following dimensional description of the insert and is not to be considered a position in the standard description.

Size: The size of the insert is designated by a one- or a two-digit number. For regular polygons and diamonds, it is the number of eighths of an inch in the nominal size of the inscribed circle, and will be a one- or two-digit number when the number of eighths is a whole number. It will be a two-digit number, including one decimal place, when it is not a whole number. Rectangular and parallelogram inserts require two digits: the first digit indicates the number of eighths of an inch width and the second digit, the number of quarters of an inch length.

Thickness: The thickness is designated by a one- or two-digit number, which indicates the number of sixteenths of an inch in the thickness of the insert. It is a one-digit number when the number of sixteenths is a whole number; it is a two-digit number carried to one decimal place when the number of sixteenths of an inch is not a whole number.

Cutting Point Configuration: The cutting point, or nose radius, is designated by a number representing $\frac{1}{64}$ths of an inch; a flat at the cutting point or nose, is designated by a letter: **0** for sharp corner; **1,** $\frac{1}{64}$-inch radius; **2,** $\frac{1}{32}$-inch radius; **3,** $\frac{3}{64}$-inch radius; **4,** $\frac{1}{16}$-inch radius; **5,** $\frac{5}{64}$-inch radius; **6,** $\frac{3}{32}$-inch radius; **7,** $\frac{7}{64}$-inch radius; **8,** $\frac{1}{8}$-inch radius; **A,** square insert with 45° chamfer; **D,** square insert with 30° chamfer; **E,** square insert with 15° chamfer; **F,** square insert with 3° chamfer; **K,** square insert with 30° double chamfer; **L,** square insert with 15° double chamfer; **M,** square insert with 3° double chamfer; **N,** truncated triangle insert; and **P,** flatted corner triangle insert.

Special Cutting Point Definition: The eighth position, if it follows a letter in the 7th position, is a number indicating the number of $\frac{1}{64}$ths of an inch measured parallel to the edge of the facet.

Hand: **R,** right; **L,** left; to be used when required in ninth position.

Other Conditions: The tenth position defines special conditions (such as edge treatment, surface finish) as follows: **A,** honed, 0.0005 inch to less than 0.003 inch; **B,** honed, 0.003 inch to less than 0.005 inch; **C,** honed, 0.005 inch to less than 0.007 inch; **J,** polished, 4 microinch arithmetic average (AA) on rake surfaces only; **T,** chamfered, manufacturer's standard negative land, rake face only.

Symbol	Tolerance (± from Nominal)		Symbol	Tolerance (± from Nominal)	
	Inscribed Circle, Inch	Thickness, Inch		Inscribed Circle, Inch	Thickness, Inch
A	0.001	0.001	**H**	0.0005	0.001
B	0.001	0.005	**J**	0.002–0.005	0.001
C	0.001	0.001	**K**	0.002–0.005	0.001
D	0.001	0.005	**L**	0.002–0.005	0.001
E	0.001	0.001	**M**	0.002–0.004[a]	0.005
F	0.0005	0.001	**U**	0.005–0.010[a]	0.005
G	0.001	0.005	**N**	0.002–0.004[a]	0.001

[a] Exact tolerance is determined by size of insert. See ANSI B94.25.

Table 1. Standard Shank Sizes for Indexable Insert Holders

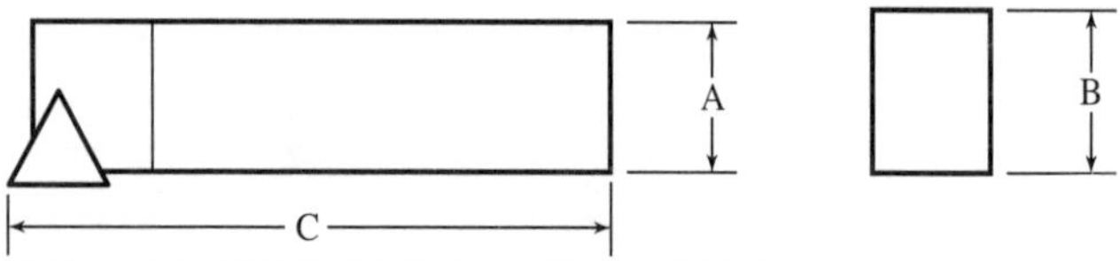

Basic Shank Size	Shank Dimensions for Indexable Insert Holders					
	A		B		C[a]	
	inch	mm	inch	mm	inch	mm
$\frac{1}{2} \times \frac{1}{2} \times 4\frac{1}{2}$	0.500	12.70	0.500	12.70	4.500	114.30
$\frac{5}{8} \times \frac{5}{8} \times 4\frac{1}{2}$	0.625	15.87	0.625	15.87	4.500	114.30
$\frac{5}{8} \times 1\frac{1}{4} \times 6$	0.625	15.87	1.250	31.75	6.000	152.40
$\frac{3}{4} \times \frac{3}{4} \times 4\frac{1}{2}$	0.750	19.05	0.750	19.05	4.500	114.30
$\frac{3}{4} \times 1 \times 6$	0.750	19.05	1.000	25.40	6.000	152.40
$\frac{3}{4} \times 1\frac{1}{4} \times 6$	0.750	19.05	1.250	31.75	6.000	152.40
$1 \times 1 \times 6$	1.000	25.40	1.000	25.40	6.000	152.40
$1 \times 1\frac{1}{4} \times 6$	1.000	25.40	1.250	31.75	6.000	152.40
$1 \times 1\frac{1}{2} \times 6$	1.000	25.40	1.500	38.10	6.000	152.40
$1\frac{1}{4} \times 1\frac{1}{4} \times 7$	1.250	31.75	1.250	31.75	7.000	177.80
$1\frac{1}{4} \times 1\frac{1}{2} \times 8$	1.250	31.75	1.500	38.10	8.000	203.20
$1\frac{3}{8} \times 2\frac{1}{16} \times 6\frac{3}{8}$	1.375	34.92	2.062	52.37	6.380	162.05
$1\frac{1}{2} \times 1\frac{1}{2} \times 7$	1.500	38.10	1.500	38.10	7.000	177.80
$1\frac{3}{4} \times 1\frac{3}{4} \times 9\frac{1}{2}$	1.750	44.45	1.750	44.45	9.500	241.30
$2 \times 2 \times 8$	2.000	50.80	2.000	50.80	8.000	203.20

[a] Holder length; may vary by manufacturer. Actual shank length depends on holder style.

Identification System for Indexable Insert Holders: The following identification system conforms to the American National Standard, ANSI B212.5-2002, Metric Holders for Indexable Inserts.

Each position in the system designates a feature of the holder in the following sequence:

1	2	3	4	5	—	6	—	7	—	8[a]	—	9	—	10[a]
C	**T**	**N**	**A**	**R**	–	**85**	–	**25**	–	**D**	–	**16**	–	**Q**

1) *Method of Holding Horizontally Mounted Insert:* The method of holding or clamping is designated by a letter: **C,** top clamping, insert without hole; **M,** top and hole clamping, insert with hole; **P,** hole clamping, insert with hole; **S,** screw clamping through hole, insert with hole; **W,** wedge clamping.

2) *Insert Shape:* The insert shape is identified by a letter: **H,** hexagonal; **O,** octagonal; **P,** pentagonal; **S,** square; **T,** triangular; **C,** rhombic, 80° included angle; **D,** rhombic, 55° included angle; **E,** rhombic, 75° included angle; **M,** rhombic, 86° included angle; **V,** rhombic, 35° included angle; **W,** hexagonal, 80° included angle; **L,** rectangular; **A,** parallelogram, 85° included angle; **B,** parallelogram, 82° included angle; **K,** parallelogram, 55° included angle; **R,** round. The included angle is always the smaller angle.

3) *Holder Style:* The holder style designates the shank style and the side cutting edge angle, or end cutting edge angle, or the purpose for which the holder is used. It is designated by a letter: **A,** for straight shank with 0° side cutting edge angle; **B,** straight shank

with 15° side cutting edge angle; **C,** straight-shank end cutting tool with 0° end cutting edge angle; **D,** straight shank with 45° side cutting edge angle; **E,** straight shank with 30° side cutting edge angle; **F,** offset shank with 0° end cutting edge angle; **G,** offset shank with 0° side cutting edge angle; **J,** offset shank with negative 3° side cutting edge angle; **K,** offset shank with 15° end cutting edge angle; **L,** offset shank with negative 5° side cutting edge angle and 5° end cutting edge angle; **M,** straight shank with 40° side cutting edge angle; **N,** straight shank with 27° side cutting edge angle; **R,** offset shank with 15° side cutting edge angle; **S,** offset shank with 45° side cutting edge angle; **T,** offset shank with 30° side cutting edge angle; **U,** offset shank with negative 3° end cutting edge angle; **V,** straight shank with 17$\frac{1}{2}$° side cutting edge angle; **W,** offset shank with 30° end cutting edge angle; **Y,** offset shank with 5° end cutting edge angle.

4) *Normal Clearances:* The normal clearances of inserts are identified by letters: **A,** 3°; **B,** 5°; **C,** 7°; **D,** 15°; **E,** 20°; **F,** 25°; **G,** 30°; **N,** 0°; **P,** 11°.

5) *Hand of Tool:* The hand of the tool is designated by a letter: **R** for right-hand; **L,** left-hand; and **N,** neutral, or either hand.

6) *Tool Height for Rectangular Shank Cross Sections:* The tool height for tool holders with a rectangular shank cross section and the height of cutting edge equal to shank height is given as a two-digit number representing this value in millimeters. For example, a height of 32 mm would be encoded as 32; 8 mm would be encoded as 08, where the one-digit value is preceded by a zero.

7) *Tool Width for Rectangular Shank Cross Sections:* The tool width for tool holders with a rectangular shank cross section is given as a two-digit number representing this value in millimeters. For example, a width of 25 mm would be encoded as 25; 8 mm would be encoded as 08, where the one-digit value is preceded by a zero.

8) *Tool Length:* The tool length is designated by a letter: **A,** 32 mm; **B,** 40 mm; **C,** 50 mm; **D,** 60 mm; **E,** 70 mm; **F,** 80 mm; **G,** 90 mm; **H,** 100 mm; **J,** 110 mm; **K,** 125 mm; **L,** 140 mm; **M,** 150 mm; **N,** 160 mm; **P,** 170 mm; **Q,** 180 mm; **R,** 200 mm; **S,** 250 mm; **T,** 300 mm; **U,** 350 mm; **V,** 400 mm; **W,** 450 mm; **X,** special length to be specified; **Y,** 500 mm.

9) *Indexable Insert Size:* The size of indexable inserts is encoded as follows: For insert shapes **C**, D, E, H, M, O, P, R, S, T, V, the side length (the diameter for R inserts) in millimeters is used as a two-digit number, with decimals being disregarded. For example, the symbol for a side length of 16.5 mm is 16. For insert shapes A, B, K, L, the length of the main cutting edge or of the longer cutting edge in millimeters is encoded as a two-digit number, disregarding decimals. If the symbol obtained has only one digit, then it should be preceded by a zero. For example, the symbol for a main cutting edge of 19.5 mm is 19; for an edge of 9.5 mm, the symbol is 09.

10) *Special Tolerances:* Special tolerances are indicated by a letter: **Q,** back and end qualified tool; **F,** front and end qualified tool; **B,** back, front, and end qualified tool. A qualified tool is one that has tolerances of ±0.08 mm for dimensions *F*, *G*, and *C*. (See Table 2.)

Table 2. Letter Symbols for Qualification of Tool Holders—Position 10
ANSI B212.5-2002

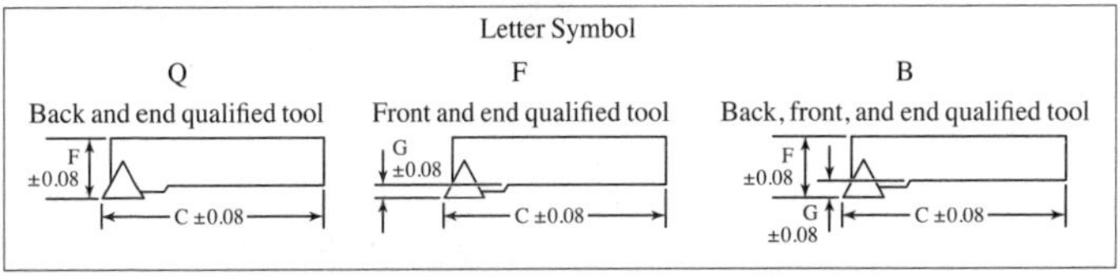

Indexable Insert Holders for Numerical Control: Indexable insert holders for numerical control lathes are usually made to more precise standards than ordinary holders. Where applicable, reference should be made to American National Standard B212.3-1986, "Precision Holders for Indexable Inserts." This standard covers the dimensional specifications, styles, and designations of precision holders for indexable inserts, which are defined as tool holders that locate the gage insert (a combination of shim and insert thicknesses) from the back or front and end surfaces to a specified dimension with a ±0.003 inch (±0.08 mm) tolerance. In NC programming, the programmed path is that followed by the center of the tool tip, which is the center of the point, or nose radius, of the insert. The surfaces produced are the result of the path of the nose and the major cutting edge, so it is necessary to compensate for the nose or point radius and the lead angle when writing the program. Table 3, from B212.3, gives the compensating dimensions for different holder styles. The reference point is determined by the intersection of extensions from the major and minor cutting edges, which would be the location of the point of a sharp pointed tool. The distances from this point to the nose radius are *L*-1 and *D*-1; *L*-2 and *D*-2 are the distances from the sharp point to the center of the nose radius. Threading tools have sharp corners and do not require a radius compensation. Other dimensions of importance in programming threading tools are also given in Table 4; the data were developed by Kennametal, Inc.

Table 3. Insert-Radius Compensation *ANSI B212.3-1986*

Square Profile

B Style[a] Also Applies to R Style

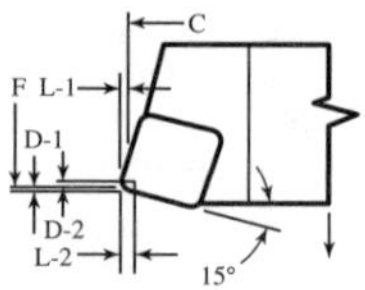

Turning 15° Lead Angle				
Rad.	*L*-1	*L*-2	*D*-1	*D*-2
1/64	0.0035	0.0191	0.0009	0.0110
1/32	0.0070	0.0383	0.0019	0.0221
3/64	0.0105	0.0574	0.0028	0.0331
1/16	0.0140	0.0765	0.0038	0.0442

D Style[a]; Also Applies to S Style

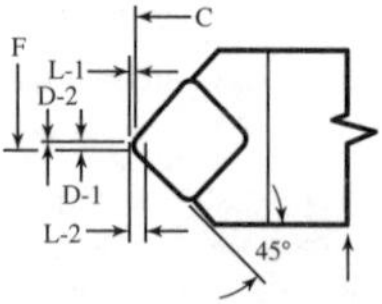

Turning 45° Lead Angle				
Rad.	*L*-1	*L*-2	*D*-1	*D*-2
1/64	0.0065	0.0221	0.0065	0
1/32	0.0129	0.0442	0.0129	0
3/64	0.0194	0.0663	0.0194	0
1/16	0.0259	0.0884	0.0259	0

K Style[a]

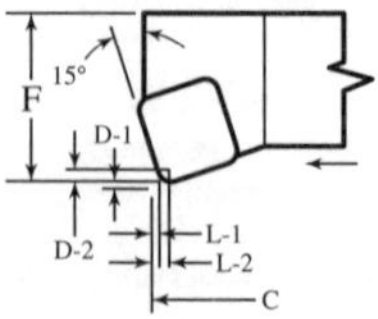

Facing 15° Lead Angle				
Rad.	*L*-1	*L*-2	*D*-1	*D*-2
1/64	0.0009	0.0110	0.0035	0.0191
1/32	0.0019	0.0221	0.0070	0.0383
3/64	0.0028	0.0331	0.0105	0.0574
1/16	0.0038	0.0442	0.0140	0.0765

Triangle Profile

G Style[a]

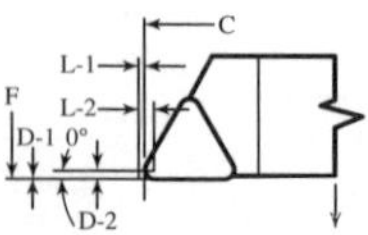

Turning 0° Lead Angle				
Rad.	*L*-1	*L*-2	*D*-1	*D*-2
1/64	0.0114	0.0271	0	0.0156
1/32	0.0229	0.0541	0	0.0312
3/64	0.0343	0.0812	0	0.0469
1/16	0.0458	0.1082	0	0.0625

Table 3. *(Continued)* **Insert-Radius Compensation** *ANSI B212.3-1986*

Style	Diagram	Rad.	*L*-1	*L*-2	*D*-1	*D*-2
Square Profile						
		Turning and Facing 15° Lead Angle				
B Style[a]; Also Applies to R Style	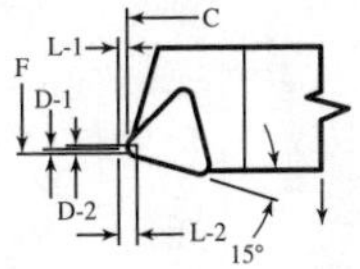	1/64	0.0146	0.0302	0.0039	0.0081
		1/32	0.0291	0.0604	0.0078	0.0162
		3/64	0.0437	0.0906	0.0117	0.0243
		1/16	0.0582	0.1207	0.0156	0.0324
		Facing 90° Lead Angle				
F Style[a]	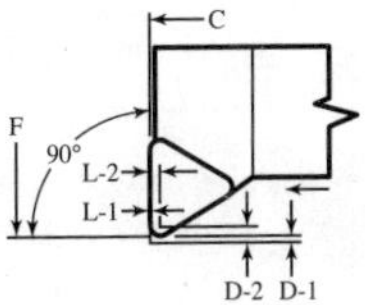	1/64	0	0.0156	0.0114	0.0271
		1/32	0	0.0312	0.0229	0.0541
		3/64	0	0.0469	0.0343	0.0812
		1/16	0	0.0625	0.0458	0.1082
		Turning & Facing 3° Lead Angle				
J Style[a]	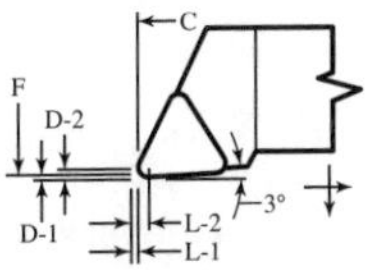	1/64	0.0106	0.0262	0.0014	0.0170
		1/32	0.0212	0.0524	0.0028	0.0340
		3/64	0.0318	0.0786	0.0042	0.0511
		1/16	0.0423	0.1048	0.0056	0.0681
80° Diamond Profile						
		Turning & Facing 0° Lead Angle				
G Style[a]	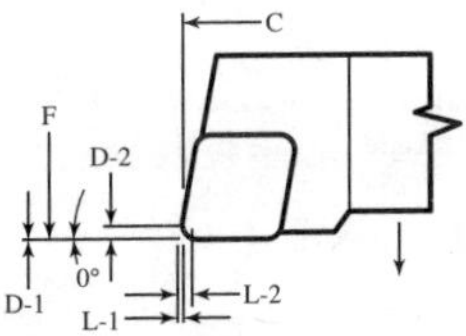	1/64	0.0030	0.0186	0	0.0156
		1/32	0.0060	0.0312	0	0.0312
		3/64	0.0090	0.0559	0	0.0469
		1/16	0.0120	0.0745	0	0.0625
		Turning & Facing 5° Reverse Lead Angle				
L Style[a]	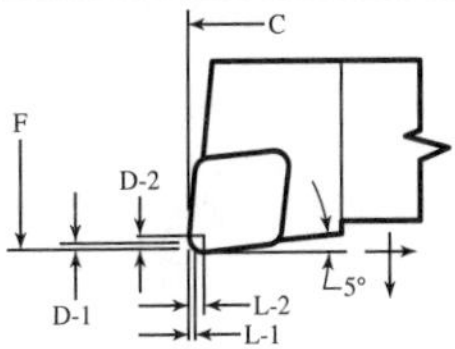	1/64	0.0016	0.0172	0.0016	0.0172
		1/32	0.0031	0.0344	0.0031	0.0344
		3/64	0.0047	0.0516	0.0047	0.0516
		1/16	0.0062	0.0688	0.0062	0.0688
		Facing 0° Lead Angle				
F Style[a]	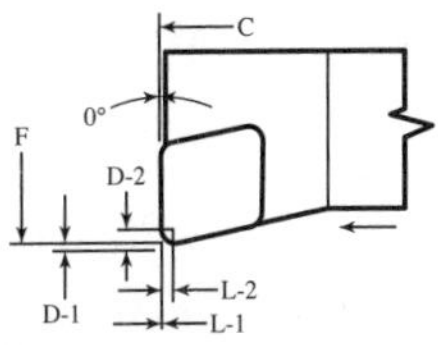	1/64	0	0.0156	0.0030	0.0186
		1/32	0	0.0312	0.0060	0.0372
		3/64	0	0.0469	0.0090	0.0559
		1/16	0	0.0625	0.0120	0.0745

Table 3. *(Continued)* **Insert-Radius Compensation** *ANSI B212.3-1986*

Style	Diagram	Rad.	*L*-1	*L*-2	*D*-1	*D*-2
Square Profile						
R Style[a]	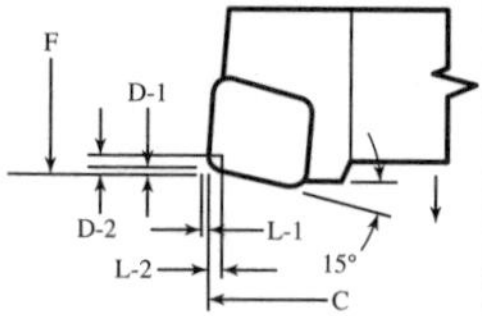	Turning 15° Lead Angle				
		1/64	0.0011	0.0167	0.0003	0.0117
		1/32	0.0022	0.0384	0.0006	0.0234
		3/64	0.0032	0.0501	0.0009	0.0351
		1/16	0.0043	0.0668	0.0012	0.0468
K Style[a]	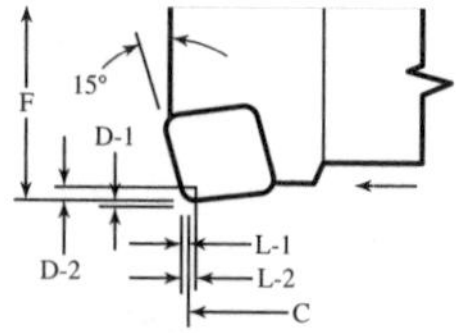	Facing 15° Lead Angle				
		1/64	0.0003	0.0117	0.0011	0.0167
		1/32	0.0006	0.0234	0.0022	0.0334
		3/64	0.0009	0.0351	0.0032	0.0501
		1/16	0.0012	0.0468	0.0043	0.0668
55° Profile						
J Style[a]	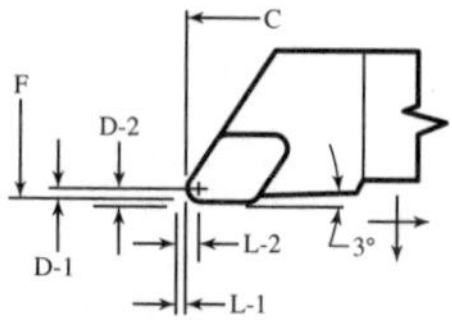	Profiling 3° Reverse Lead Angle				
		1/64	0.0135	0.0292	0.0015	0.0172
		1/32	0.0271	0.0583	0.0031	0.0343
		3/64	0.0406	0.0875	0.0046	0.0519
		1/16	0.0541	0.1166	0.0062	0.0687
35° Profile						
J Style[a]; Negative rake holders have 6°back rake and 6°side rake	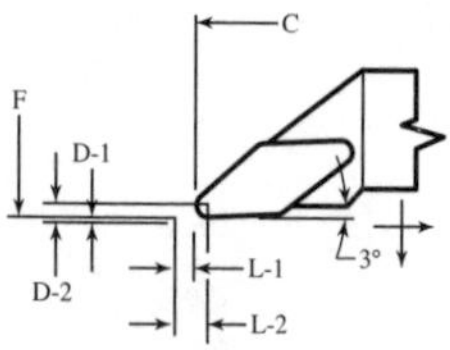	Profiling 3° Reverse Lead Angle				
		1/64	0.0330	0.0487	0.0026	0.0182
		1/32	0.0661	0.0973	0.0051	0.0364
		3/64	0.0991	0.1460	0.0077	0.0546
		1/16	0.1322	0.1947	0.0103	0.0728
L Style[a]	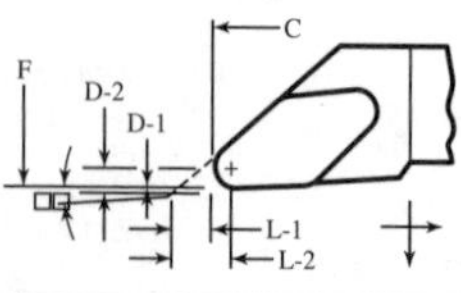	Profiling 5° Lead Angle				
		1/64	0.0324	0.0480	0.0042	0.0198
		1/32	0.0648	0.0360	0.0086	0.0398
		3/64	0.0971	0.1440	0.0128	0.0597
		1/16	0.1205	0.1920	0.0170	0.0795

[a] *L*-1 and *D*-1 over sharp point to nose radius; and *L*-2 and *D*-2 over sharp point to center of nose radius. The *D*-1 dimension for the *B, E, D, M, P, S, T,* and *V*-style tools are over the sharp point of insert to a sharp point at the intersection of a line on the lead angle on the cutting edge of the insert and the *C* dimension. The *L*-1 dimensions on *K*-style tools are over the sharp point of insert to sharp point intersection of lead angle and *F* dimensions.

All dimensions are in inches.

Table 4. Threading Tool Insert-Radius Compensation for NC Programming

Threading						
Insert Size	T	R	U	Y	X	Z
2	$^{5}/_{32}$ Wide	0.040	0.075	0.040	0.024	0.140
3	$^{3}/_{16}$ Wide	0.046	0.098	0.054	0.031	0.183
4	$^{1}/_{4}$ Wide	0.053	0.128	0.054	0.049	0.239
5	$^{3}/_{8}$ Wide	0.099	0.190	…	…	…

Buttress Threading 29° Acme 60° V-Threading

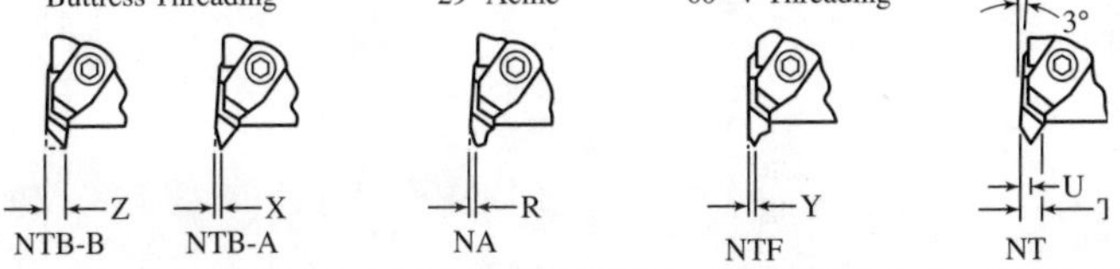

All dimensions are given in inches.
Courtesy of Kennametal, Inc.

The *C* and *F* characters are tool holder dimensions other than the shank size. In all instances, the *C* dimension is parallel to the length of the shank and the *F* dimension is parallel to the side dimension; actual dimensions must be obtained from the manufacturer. For all *K* style holders, the *C* dimension is the distance from the end of the shank to the tangent point of the nose radius and the end cutting edge of the insert. For all other holders, the *C* dimension is from the end of the shank to a tangent to the nose radius of the insert. The *F* dimension on all B, D, E, M, P, and V style holders is measured from the back side of the shank to the tangent point of the nose radius and the side cutting edge of the insert. For all A, F, G, J, K, and L style holders, the *F* dimension is the distance from the back side of the shank to the tangent of the nose radius of the insert. In all these designs, the nose radius is the standard radius corresponding to those given in the paragraph *Cutting Point Configuration* on page 185.

Table 5. Cemented Carbides

Composition	Features	Comments
Tungsten Carbide/Cobalt (WC/Co)	No porosity should be visible under the highest magnification. Has the greatest resistance to simple abrasive wear.	Hardness and abrasion resistance increases as the Co content is lowered (minimum 2–3%). Tougher and less hard grades are obtained as carbide grain or cobalt content are both increased.
Tungsten-Titanium Carbide/Cobalt (WC/TiC/Co)	Used to cut steels and other ferrous alloys. Considerably more brittle and less abrasion resistant than tungsten carbide.	Resists the high-temperature diffusive attack that causes chemical breakdown and cratering.
Tungsten-Titanium-Tantalum (-Niobium) Carbide/Cobalt	Used mainly for cutting steels. Improve on the best features of WC/TiC/Co. Can undertake very heavy cuts at high speeds on all types of steels, including austenitic stainless. Also operate well on ductile cast irons and nickel-base super alloys.	Except for coated carbides these could be the most popular class of hardmetals. Do not have the resistance to abrasive wear possessed by micrograin straight tungsten carbide grades nor the good resistance to cratering of coated grades and titanium carbide-based cermets.
Steel- and Alloy-Bonded Titanium Carbide	Used for stamping, blanking and drawing dies, machine components, and similar items where the ability to machine before hardening reduces production costs substantially.	Characterized by high binder contents (typically 50–60% by volume) and lower hardness, compared with the more usual hardmetals, and by great variation in properties obtained by heat treatment. Consists primarily of titanium carbide bonded with heat-treated steel, but some grades also contain tungsten carbide or are bonded with nickel- or copper-based alloys.

Table 6. ISO Classifications of Hardmetals (Cemented Carbides and Carbonitrides) by Application

Main Types of Chip Removal		Groups of Applications			Direction of Decrease in Characteristic	
Symbol and Color	Broad Categories of Materials to be Machined	Designation (Grade)	Specific Material to be Machined	Use and Working Conditions	of cut	of carbide
P Blue	Ferrous with long chips	P01	Steel, steel castings	Finish turning and boring; high cutting speeds, small chip sections, accurate dimensions, fine finish, vibration-free operations	↑ speed	↑ wear resistance
		P10		Turning, copying, threading, milling; high cutting speeds; small or medium chip sections		
		P20	Steel, steel castings, ductile cast iron with long chips	Turning, copying, milling; medium cutting speeds and chip sections, planing with small chip sections	↓ feed	↓ toughness
		P30		Turning, milling, planing; medium or large chip sections, unfavorable machining conditions		
		P40	Steel, steel castings with sand inclusions and cavities	Turning, planing, slotting; low cutting speeds, large chip sections, with possible large cutting angles, unfavorable cutting conditions, and work on automatic machines		
		P50	Steel, steel castings of medium or low tensile strength, with sand inclusions and cavities	Operations demanding very tough carbides; turning, planing, slotting; low cutting speeds, large chip sections, with possible large cutting angles, unfavorable conditions and work on automatic machines		
M Yellow	Ferrous metals with long or short chips, and nonferrous metals	M10	Steel, steel castings, manganese steel, gray cast iron, alloy cast iron	Turning; medium or high cutting speeds, small or medium chip sections		
		M20	Steel, steel castings, austenitic or manganese steel, gray cast iron	Turning, milling; medium cutting speeds and chip sections		
		M30	Steel, steel castings, austenitic steel, gray cast iron, high-temperature-resistant alloys	Turning, milling, planing; medium cutting speeds, medium or large chip sections		
		M40	Mild, free-cutting steel, low-tensile steel, nonferrous metals and light alloys	Turning, parting off; particularly on automatic machines		
K Red	Ferrous metals with short chips, nonferrous metals and non-metallic materials	K01	Very hard gray cast iron, chilled castings over 85 Shore, high-silicon aluminum alloys, hardened steel, highly abrasive plastics, hard cardboard, ceramics	Turning, finish turning, boring, milling, scraping		
		K10	Gray cast iron over 220 Brinell, malleable cast iron with short chips, hardened steel, silicon-aluminum and copper alloys, plastics, glass, hard rubber, hard cardboard, porcelain, stone	Turning, milling, drilling, boring, broaching, scraping		
		K20	Gray cast iron up to 220 Brinell, nonferrous metals, copper, brass, aluminum	Turning, milling, planing, boring, broaching, demanding very tough carbide		
		K30	Low-hardness gray cast iron, low-tensile steel, compressed wood	Turning, milling, planing, slotting, unfavorable conditions, and possibility of large cutting angles		
		K40	Softwood or hard wood, nonferrous metals			

MACHINING OPERATIONS

Machining Aluminum.— Some of the alloys of aluminum have been machined successfully without any lubricant or cutting compound, but some form of lubricant is desirable to obtain the best results. For many purposes, a soluble cutting oil is good.

Tools for aluminum and aluminum alloys should have larger relief and rake angles than tools for cutting steel. For high-speed steel turning tools the following angles are recommended: relief angles, 14 to 16 degrees; back rake angle, 5 to 20 degrees; side rake angle, 15 to 35 degrees. For very soft alloys even larger side rake angles are sometimes used. High-silicon aluminum alloys and some others have a very abrasive effect on the cutting tool. While these alloys can be cut successfully with high-speed steel tools, cemented carbides are recommended because of their superior abrasion resistance. The tool angles recommended for cemented carbide turning tools are: relief angles, 12 to 14 degrees; back rake angle, 0 to 15 degrees; side rake angle, 8 to 30 degrees.

Cutoff tools and necking tools for machining aluminum and its alloys should have from 12 to 20 degrees back rake angle and the end relief angle should be from 8 to 12 degrees. Excellent threads can be cut with single-point tools in even the softest aluminum. Experience seems to vary somewhat regarding the rake angle for single-point thread cutting tools. Some prefer to use a rather large back and side rake angle although this requires a modification in the included angle of the tool to produce the correct thread contour. When both rake angles are zero, the included angle of the tool is ground equal to the included angle of the thread. Excellent threads have been cut in aluminum with zero rake angle thread-cutting tools using large relief angles, which are 16 to 18 degrees opposite the front side of the thread and 12 to 14 degrees opposite the back side of the thread. In either case, the cutting edges should be ground and honed to a keen edge. It is sometimes advisable to give the face of the tool a few strokes with a hone between cuts when chasing the thread to remove any built-up edge on the cutting edge.

Fine surface finishes are often difficult to obtain on aluminum and aluminum alloys, particularly the softer metals. When a fine finish is required, the cutting tool should be honed to a keen edge and the surfaces of the face and the flank will also benefit by being honed smooth. Tool wear is inevitable, but it should not be allowed to progress too far before the tool is changed or sharpened. A sulphurized mineral oil or a heavy-duty soluble oil will sometimes be helpful in obtaining a satisfactory surface finish. For best results, however, a diamond cutting tool is recommended. Excellent surface finishes can be obtained on even the softest aluminum and aluminum alloys with these tools.

Although ordinary milling cutters can be used successfully in shops where aluminum parts are only machined occasionally, the best results are obtained with coarse-tooth, large helix-angle cutters having large rake and clearance angles. Clearance angles up to 10 to 12 degrees are recommended. When slab milling and end milling a profile, using the peripheral teeth on the end mill, climb milling (also called down milling) will generally produce a better finish on the machined surface than conventional (or up) milling. Face milling cutters should have a large axial rake angle. Standard twist drills can be used without difficulty in drilling aluminum and aluminum alloys although high helix-angle drills are preferred. The wide flutes and high helix-angle in these drills helps to clear the chips. Sometimes split-point drills are preferred. Carbide-tipped twist drills can be used for drilling aluminum and its alloys and may afford advantages in some production applications. Ordinary hand and machine taps can be used to tap aluminum and its alloys although spiral-fluted ground thread taps give superior results. Experience has shown that such taps should have a right-hand ground flute when intended to cut right-hand threads and the helix angle should be similar to that used in an ordinary twist drill.

Machining Magnesium.—Magnesium alloys are readily machined and with relatively low power consumption per cubic inch of metal removed. The usual practice is to employ high cutting speeds with relatively coarse feeds and deep cuts. Exceptionally fine finishes can be obtained so that grinding to improve the finish usually is unnecessary. The horsepower normally required in machining magnesium varies from 0.15 to 0.30 per cubic inch per minute. While this value is low, especially in comparison with power required for cast iron and steel, the total amount of power for machining magnesium usually is high because of the exceptionally rapid rate at which metal is removed.

Carbide tools are recommended for maximum efficiency, although high-speed steel frequently is employed. Tools should be designed so as to dispose of chips readily or without excessive friction, by employing polished chip-bearing surfaces, ample chip spaces, large clearances, and small contact areas. *Keen-edged tools should always be used*. In machining magnesium, it is the general but not invariable practice in the United States to use a cutting fluid. In other places, magnesium may be machined dry, except where heat generated by high cutting speeds would not be dissipated rapidly enough without a cutting fluid. This condition may exist when, for example, small tools without much heat-conducting capacity are employed on automatics. The cutting fluid for magnesium should be an anhydrous oil having, at most, a very low acid content. Various mineral-oil cutting fluids are used.

Feeds and Speeds for Magnesium: Speeds ordinarily range up to 5000 feet per minute for rough- and finish-turning, up to 3000 feet per minute for rough-milling, and up to 9000 feet per minute for finish-milling. For rough-turning, the following combinations of speed in feet per minute, feed per revolution, and depth of cut are recommended: Speed 300 to 600 feet per minute—feed 0.030 to 0.100 inch, depth of cut 0.5 inch; speed 600 to 1000—feed 0.020 to 0.080, depth of cut 0.4; speed 1000 to 1500—feed 0.010 to 0.060, depth of cut 0.3; speed 1500 to 2000—feed 0.010 to 0.040, depth of cut 0.2; speed 2000 to 5000—feed 0.010 to 0.030, depth of cut 0.15.

Lathe Tool Angles for Magnesium: The true or actual rake angle resulting from back and side rakes usually varies from 10 to 15 degrees. Back rake varies from 10 to 20, and side rake from 0 to 10 degrees. Reduced back rake may be employed to obtain better chip breakage. The back rake may also be reduced to from 2 to 8 degrees on form tools or other broad tools to prevent chatter.

Parting Tools: For parting tools, the back rake varies from 15 to 20 degrees, the front end relief 8 to 10 degrees, the side relief measured perpendicular to the top face 8 degrees, the side relief measured in the plane of the top face from 3 to 5 degrees.

Milling Magnesium: In general, the coarse-tooth type of cutter is recommended. The number of teeth or cutting blades may be one-third to one-half the number normally used; however, the two-blade fly cutter has proved to be very satisfactory. As a rule, the land relief or primary peripheral clearance is 10 degrees followed by secondary clearance of 20 degrees. The lands should be narrow, the width being about $^3/_{64}$ to $^1/_{16}$ inch. The rake, which is positive, is about 15 degrees.

For rough-milling and speeds in feet per minute up to 900—feed, inch per tooth, 0.005 to 0.025, depth of cut up to 0.5; for speeds 900 to 1500—feed 0.005 to 0.020, depth of cut up to 0.375; for speeds 1500 to 3000—feed 0.005 to 0.010, depth of cut up to 0.2.

Drilling Magnesium: If the depth of a hole is less than five times the drill diameter, an ordinary twist drill with highly polished flutes may be used. The included angle of the point may vary from 70 degrees to the usual angle of 118 degrees. The relief angle is about 12 degrees. The drill should be kept sharp and the outer corners rounded to produce a smooth finish and prevent burr formation. For deep hole drilling, use a drill having a helix angle of 40 to 45 degrees with large polished flutes of uniform cross section throughout the drill length to facilitate the flow of chips. A pyramid-shaped "spur" or "pilot point" at the tip of the drill will reduce the "spiraling or run-off."

Drilling speeds vary from 300 to 2000 feet per minute with feeds per revolution ranging from 0.015 to 0.050 inch.

Reaming Magnesium: Reamers up to 1 inch in diameter should have four flutes; larger sizes, six flutes. These flutes may be either parallel with the axis or have a negative helix angle of 10 degrees. The positive rake angle varies from 5 to 8 degrees, the relief angle from 4 to 7 degrees, and the clearance angle from 15 to 20 degrees.

Tapping Magnesium: Standard taps may be used unless Class 3B tolerances are required, in which case the tap should be designed for use in magnesium. A high-speed steel concentric type with a ground thread is recommended. The concentric form, which eliminates the radial thread relief, prevents jamming of chips while the tap is being backed out of the hole. The positive rake angle at the front may vary from 10 to 25 degrees and the "heel rake angle" at the back of the tooth from 3 to 5 degrees. The chamfer extends over two to three threads. For holes up to $\frac{1}{4}$ inch in diameter, two-fluted taps are recommended; for sizes from $\frac{1}{2}$ to $\frac{3}{4}$ inch, three flutes; and for larger holes, four flutes. Tapping speeds ordinarily range from 75 to 200 feet per minute, and mineral oil cutting fluid should be used.

Threading Dies for Magnesium: Threading dies for use on magnesium should have about the same cutting angles as taps. Narrow lands should be used to provide ample chip space. Either solid or self-opening dies may be used. The latter type is recommended when maximum smoothness is required. Threads may be cut at speeds up to 1000 feet per minute.

Grinding Magnesium: As a general rule, magnesium is ground dry. The highly inflammable dust should be formed into a sludge by means of a spray of water or low-viscosity mineral oil. Accumulations of dust or sludge should be avoided. For surface grinding, when a fine finish is desirable, a low-viscosity mineral oil may be used.

Machining Zinc Alloy Die Castings.—Machining of zinc alloy die castings is mostly done without a lubricant. For particular work, especially deep drilling and tapping, a lubricant such as lard oil and kerosene (about half and half) or a 50-50 mixture of kerosene and machine oil may be used to advantage. A mixture of turpentine and kerosene has been found effective on certain difficult jobs.

Reaming: In reaming, tools with six straight flutes are commonly used, although tools with eight flutes irregularly spaced have been found by some to yield better results. Many standard reamers have a land that is too wide for best results. A land about 0.015 inch wide is recommended, but this may often be ground down to around 0.007 or even 0.005 inch to obtain freer cutting, less tendency to loading, and reduced heating.

Turning: Tools of high-speed steel are commonly employed although the application of Stellite and carbide tools, even on short runs, is feasible. For steel or Stellite, a positive top rake of from 0 to 20 degrees and an end clearance of about 15 degrees are commonly recommended. Where side cutting is involved, a side clearance of about 4 degrees minimum is recommended. With carbide tools, the end clearance should not exceed 6 to 8 degrees, and the top rake should be from 5 to 10 degrees positive. For boring, facing, and other lathe operations, rake and clearance angles are about the same as for tools used in turning.

Machining Monel and Nickel Alloys.—These alloys are machined with high-speed steel and with cemented carbide cutting tools. High-speed steel lathe tools usually have a back rake of 6 to 8 degrees, a side rake of 10 to 15 degrees, and relief angles of 8 to 12 degrees. Broad-nose finishing tools have a back rake of 20 to 25 degrees and an end relief angle of 12 to 15 degrees. In most instances, standard commercial cemented-carbide tool holders and tool shanks can be used which provide an acceptable tool geometry. Honing the cutting edge lightly will help if chipping is encountered.

The most satisfactory tool materials for machining Monel and the softer nickel alloys, such as Nickel 200 and Nickel 230, are M2 and T5 for high-speed steel and crater-resistant grades of cemented carbides. For the harder nickel alloys such as K Monel, Permanickel, Duranickel, and Nitinol alloys, the recommended tool materials are T15, M41, M42, M43, and for high-speed steel, M42. For carbides, a grade of crater-resistant carbide is recommended when the hardness is less than 300 Bhn, and when the hardness is more than 300 Bhn, a grade of straight tungsten carbide will often work best, although some crater-resistant grades will also work well.

A sulfurized oil or a water-soluble oil is recommended for rough and finish turning. A sulfurized oil is also recommended for milling, threading, tapping, reaming, and broaching. Recommended cutting speeds for Monel and the softer nickel alloys are 70 to 100 fpm for high-speed steel tools and 200 to 300 fpm for cemented carbide tools. For the harder nickel alloys, the recommended speed for high-speed steel is 40 to 70 fpm for a hardness up to 300 Bhn and for a higher hardness, 10 to 20 fpm; for cemented carbides, 175 to 225 fpm when the hardness is less than 300 Bhn and for a higher hardness, 30 to 70 fpm.

Nickel alloys have a high tendency to work harden. To minimize work hardening caused by machining, the cutting tools should be provided with adequate relief angles and positive rake angles. Furthermore, the cutting edges should be kept sharp and replaced when dull to prevent burnishing of the work surface. The depth of cut and feed should be sufficiently large to ensure that the tool penetrates the work without rubbing.

Machining Copper Alloys.—Copper alloys can be machined by tooling and methods similar to those used for steel, but at higher surface speeds. Machinability is based on a rating of 100 percent for the free-cutting alloy C35000, which machines with small, easily broken chips. As with steels, copper alloys containing lead have the best machining properties, with alloys containing tin, and lead, having machinability ratings of 80 and 70 percent. Tellurium and sulphur are added to copper alloys to increase machinability with minimum effect on conductivity. Lead additions are made to facilitate machining, as their effect is to produce easily broken chips.

Copper alloys containing silicon, aluminum, manganese, and nickel become progressively more difficult to machine, and produce long, stringy chips, the latter alloys having only 20 percent of the machinability of the free-cutting alloys. Although copper is frequently machined dry, a cooling compound is recommended. Other lubricants that have been used include tallow for drilling, gasoline for turning, and beeswax for threading.

Machining Hard Rubber.—Tools suitable for steel may be used for hard rubber, with no top or side rake angles and 10 to 20 deg. clearance angles, of high-speed steel or tungsten carbide. Without coolant, surface speeds of about 200 ft./min. are recommended for turning, boring, and facing, and may be increased to 300 surface ft./min. with coolant.

Drilling of hard rubber requires high-speed steel drills of 35 to 40 deg. helix angle to obtain maximum cutting speeds and drill life. Feed rates for drilling range up to 0.015 in./rev. Deep-fluted taps are best for threading hard rubber, and should be 0.002 to 0.005 in. oversize if close tolerances are to be held. Machine oil is used for a lubricant. Hard rubber may be sawn with band saws having 5 to 10 points per inch, running at about 3000 ft./min. or cut with abrasive wheels. Use of coolant in grinding rubber gives a smoother finish.

Piercing and blanking of sheet rubber is best performed with the rubber or dies heated. Straightening of the often-distorted blanks may be carried out by dropping them into a pan of hot water.

Table 1a. Tool Troubleshooting and Practical Tips

Problems		Causes	Remedy
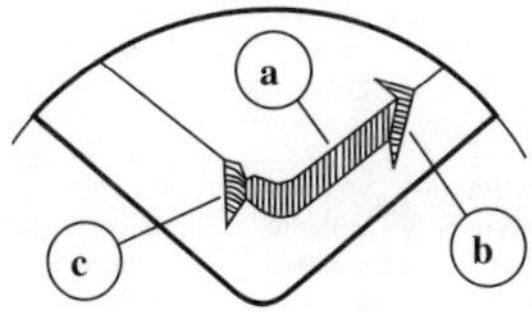 Flank and notch wear	a. Rapid flank wear causing poor surface finish or out of tolerance.	a. Cutting speed too high or insufficient wear resistance.	Reduce the cutting speed. Select a more wear-resistant grade.
	b/c. Notch wear causing poor surface finish and risk of edge breakage.	b/c. Oxidation	Select Al_2O_3 coated grade. For work hardening materials select a larger lead angle or a more wear-resistant grade.
		b/c. Attrition	Reduce the cutting speed. (When machining heat-resistant material with ceramics increase cutting speed.)
		c. Oxidation	Select a cermet grade
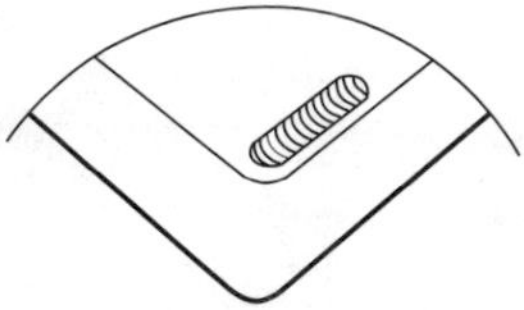 Crater wear	Excessive crater wear causing a weakened edge. Cutting-edge break-through on the training edge causes poor surface finish.	Diffusion wear due to high cutting temperatures on the rake face.	Select Al_2O_3 coated grade. Select a positive insert geometry. First reduce the speed to obtain a lower temperature, then reduce the feed.
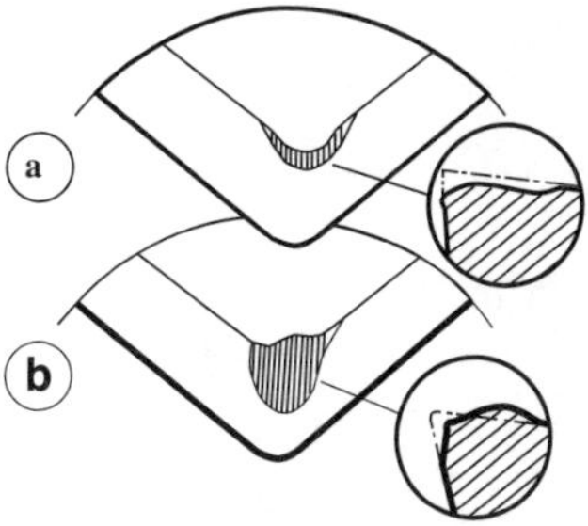 Plastic deformation	Plastic deformation a. Edge depression b. Flank impression Leading to poor chip control and poor surface finish. Risk of excessive flank wear leading to insert breakage	Cutting temperature too high combined with a high pressure.	Select a more wear-resistant grade. a. Reduce speed b. Reduce feed

Table 1a. *(Continued)* **Tool Troubleshooting and Practical Tips**

Problems		Causes	Remedy
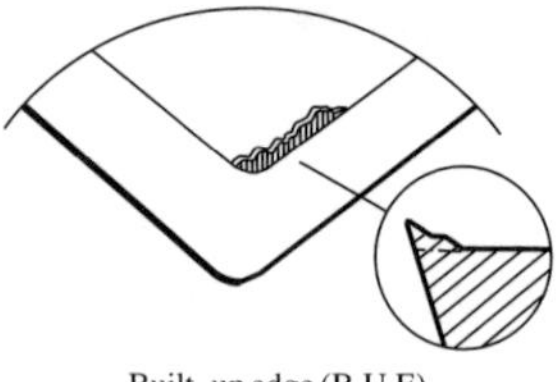 Built–up edge (B.U.E)	Built-up edge causing poor surface and cutting-edge frittering when the B.U.E. is torn away.	Workpiece material is welded to the insert due to:	
		Low cutting speed.	Increase cutting speed.
		Negative cutting geometry.	Select a positive geometry.
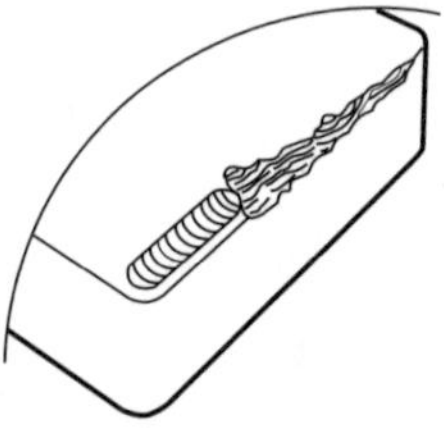 Chip hammering	The part of the cutting edge not in cut is damaged through chip hammering. Both the top side and the support for the insert can be damaged.	The chips are deflected against the cutting edge.	Change the feed. Select an alternative insert geometry.
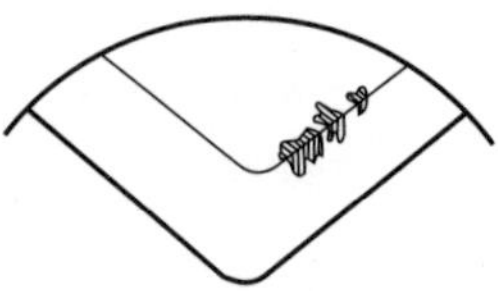 Frittering	Small cutting-edge fractures (frittering) causing poor surface finish and excessive flank wear.	Grade too brittle	Select a tougher grade.
		Insert geometry too weak.	Select an insert with a stronger geometry (bigger chamfer for ceramic inserts).
		Built-up edge	Increase cutting speed or select a positive geometry. Reduce feed at beginning of cut.
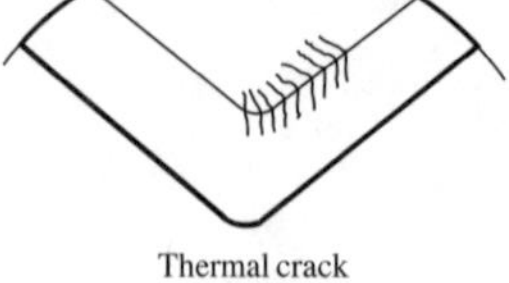 Thermal crack	Small cracks perpendicular to the cutting edge causing frittering and poor surface finish.	Thermal cracks due to temperature variations caused by:	
		–Intermittent machining	Select a tougher grade.
		–Varying coolant supply.	–Turn off coolant or –Flood coolant

Table 1a. *(Continued)* **Tool Troubleshooting and Practical Tips**

Problems		Causes	Remedy
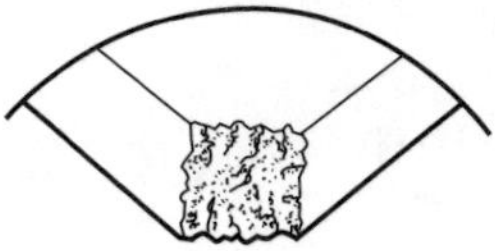 Insert breakage	Insert breakage that damages not only the insert but also the shim and workpiece.	Grade too brittle. Excessive load on the insert. Insert geometry too weak. Insert size too small.	Select a tougher grade. Reduce the feed and/or the depth of cut. Select a stronger geometry, preferably a single-sided insert. Select a thicker larger insert.
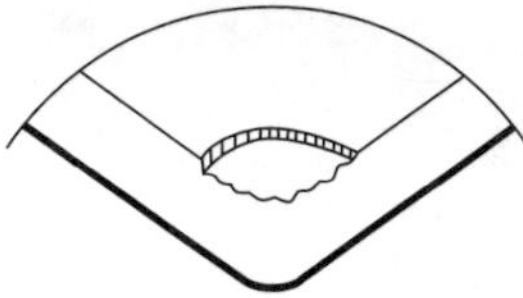 Slice fracture—Ceramics		Excessive tool presssure.	Reduce the feed. Select a tougher grade. Select an insert with smaller chamfer.

Table 1b. Tool Troubleshooting Checklist

Problem	Tool Material	Remedy
Excessive flank wear—tool life too short	Carbide	1. Change to harder, more wear-resistant grade 2. Reduce the cutting speed 3. Reduce the cutting speed and increase the feed to maintain production 4. Reduce the feed 5. For work-hardenable materials—increase the feed 6. Increase the lead angle 7. Increase the relief angles
	HSS	1. Use a coolant 2. Reduce the cutting speed 3. Reduce the cutting speed and increase the feed to maintain production 4. Reduce the feed 5. For work-hardenable materials—increase the feed 6. Increase the lead angle 7. Increase the relief angle
Excessive cratering	Carbide	1. Use a crater-resistant grade 2. Use a harder, more wear-resistant grade 3. Reduce the cutting speed 4. Reduce the feed 5. Widen the chip breaker groove
	HSS	1. Use a coolant 2. Reduce the cutting speed 3. Reduce the feed 4. Widen the chip breaker groove

Table 1b. *(Continued)* **Tool Troubleshooting Checklist**

Problem	Tool Material	Remedy
Cutting-edge chipping	Carbide	1. Increase the cutting speed
		2. Lightly hone the cutting edge
		3. Change to a tougher grade
		4. Use negative-rake tools
		5. Increase the lead angle
		6. Reduce the feed
		7. Reduce the depth of cut
		8. Reduce the relief angles
		9. If low cutting speed must be used, use a high-additive EP cutting fluid
	HSS	1. Use a high-additive EP cutting fluid
		2. Lightly hone the cutting edge before using
		3. Increase the lead angle
		4. Reduce the feed
		5. Reduce the depth of cut
		6. Use a negative rake angle
		7. Reduce the relief angles
	Carbide and HSS	1. Check the setup for cause if chatter occurs
		2. Check the grinding procedure for tool overheating
		3. Reduce the tool overhang
Cutting-edge deformation	Carbide	1. Change to a grade containing more tantalum
		2. Reduce the cutting speed
		3. Reduce the feed
Poor surface finish	Carbide	1. Increase the cutting speed
		2. If low cutting speed must be used, use a high-additive EP cutting fluid
		4. For light cuts, use straight titanium carbide grade
		5. Increase the nose radius
		6. Reduce the feed
		7. Increase the relief angles
		8. Use positive rake tools
	HSS	1. Use a high-additive EP cutting fluid
		2. Increase the nose radius
		3. Reduce the feed
		4. Increase the relief angles
		5. Increase the rake angles
	Diamond	1. Use diamond tool for soft materials
Notching at the depth of cut line	Carbide and HSS	1. Increase the lead angle
		2. Reduce the feed

Table 1c. Common Tool Faults, Failures, and Cures

Fault Description	Probable Failure	Possible Cure
Improper Tool Design		
Drastic section changes—widely different thicknesses of adjacent wall sections or protruding elements	In liquid quenching, the thin section will cool and then harden more rapidly than the adjacent thicker section, setting up stresses that may exceed the strength of the steel.	Make such parts of two pieces or use an air-hardening tool steel that avoids the harsh action of a liquid quench.
Sharp corners on shoulders or in square holes	Cracking can occur, particularly in liquid quenching, due to stress concentrations.	Apply fillets to the corners and/or use an air-hardening tool steel.
Sharp cornered keyways	Failure may arise during service, and is usually considered to be caused by fatigue.	The use of round keyways should be preferred when the general configuration of the part makes it prone to failure due to square keyways.
Abrupt section changes in battering tools	Due to impact in service, pneumatic tools are particularly sensitive to stress concentrations that lead to fatigue failures.	Use taper transitions, which are better than even generous fillets.
Functional inadequacy of tool design—e.g., insufficient guidance for a punch	Excessive wear or breakage in service may occur.	Assure solid support, avoid unnecessary play, adapt travel length to operational conditions (e.g., punch to penetrate to four-fifths of thickness in hard work material).
Improper tool clearance, such as in blanking and punching tools	Deformed and burred parts may be produced; excessive tool wear or breakage can result.	Adapt clearances to material conditions and dimensions to reduce tool load and to obtain clean sheared surfaces.
Faulty Condition or Inadequate Grade of Tool Steel		
Improper tool steel grade selection	Typical failures: Chipping—insufficient toughness. Wear—poor abrasion resistance. Softening—inadequate "red hardness."	Choose the tool steel grade by following recommendations and improve selection when needed, guided by property ratings.
Material defects—voids, streaks, tears, flakes, surface cooling cracks, etc.	When not recognized during material inspection, tools made of defective steel often prove to be useless.	Obtain tool steels from reliable sources and inspect tool material for detectable defects.
Decarburized surface layer ("bark") in rolled tool steel bars	Cracking may originate from the decarburized layer or it will not harden ("soft skin").	Provide allowance for stock to be removed from all surfaces of hot-rolled tool steel. Recommended amounts are listed in tool steel catalogs and vary according to section size, generally about 10 percent for smaller and 5 percent for larger diameters.
Brittleness caused by poor carbide distribution in high-alloy tool steels	Excessive brittleness can cause chipping or breakage during service.	Bars with large diameter (above about 4 inches) tend to be prone to nonuniform carbide distribution. Choose upset forged discs instead of large-diameter bars.

Table 1c. *(Continued)* **Common Tool Faults, Failures, and Cures**

Fault Description	Probable Failure	Possible Cure
(Continued) Faulty Condition or Inadequate Grade of Tool Steel		
Unfavorable grain flow	Improper grain flow of the steel used for milling cutters and similar tools can cause teeth to break out.	Upset forged discs made with an upset ratio of about 2 to 1 (starting to upset thickness) display radial grain flow. Highly stressed tools, such as gear-shaper cutters, may require the cross forging of blanks.
Heat Treatment Faults		
Improper preparation for heat treatment. Certain tools may require stress relieving or annealing, and often preheating, too	Tools highly stressed during machining or forming, unless stress relieved, may aggravate the thermal stresses of heat treatment, thus causing cracks. Excessive temperature gradients developed in nonpreheated tools with different section thicknesses can cause warpage.	Stress relieve, when needed, before hardening. Anneal prior to heavy machining or cold forming (e.g., hobbing). Preheat tools (a) having substantial section thickness variations or (b) requiring high quenching temperatures, as those made of high-speed tool steels.
Overheating during hardening; quenching from too high a temperature	Causes grain coarsening and a sensitivity to cracking that is more pronounced in tools with drastic section changes.	Overheated tools have a characteristic microstructure that aids recognition of the cause of failure and indicates the need for improved temperature control.
Low hardening temperature	The tool may not harden at all, or in its outer portion only, thereby setting up stresses that can lead to cracks.	Controlling both the temperature of the furnace and the time of holding the tool at quenching temperature will prevent this infrequent deficiency.
Inadequate composition or condition of the quenching media	Water-hardening tool steels are particularly sensitive to inadequate quenching media, which can cause soft spots or even violent cracking.	For water-hardening tool steels, use water free of dissolved air and contaminants, also assure sufficient quantity and proper agitation of the quench.
Improper handling during and after quenching	Cracking, particularly of tools with sharp corners, during the heat treatment can result from holding the part too long in the quench or incorrectly applied tempering.	Following the steel producer's specifications is a safe way to assure proper heat treatment handling. In general, the tool should be left in the quench until it reaches a temperature of 150 to 200°F, and should then be transferred promptly into a warm tempering furnace.
Insufficient tempering	Omission of double tempering for steel types that require it may cause early failure by heat checking in hot-work steels or make the tool abnormally sensitive to grinding checks.	Double temper highly alloyed tool steel of the high-speed, hot-work, and high-chromium categories, to remove stresses caused by martensite formed during the first tempering phase. Second temper also increases hardness of most high-speed steels.

Table 1c. *(Continued)* **Common Tool Faults, Failures, and Cures**

Fault Description	Probable Failure	Possible Cure
Decarburization and carburization	Unless hardened in a neutral atmosphere the original carbon content of the tool surface may be changed: Reduced carbon (decarburization) causes a soft layer that wears rapidly. Increased carbon (carburization) when excessive may cause brittleness.	Heating in neutral atmosphere or well-maintained salt bath and controlling the furnace temperature and the time during which the tool is subjected to heating can usually keep the carbon imbalance within acceptable limits.
Grinding Damages		
Excessive stock removal rate causing heating of the part surface beyond the applied tempering temperature	Scorched tool surface displaying temper colors varying from yellow to purple, depending on the degree of heat, causes softening of the ground surface. When coolant is used, a local rehardening can take place, often resulting in cracks.	Prevention: by reducing speed and feed, or using coarser, softer, more open structured grinding wheel, with ample coolant. Correction: eliminate the discolored layer by subsequent light stock removal. Not always a cure, because the effects of abusive grinding may not be corrected.
Improper grinding wheel specifications; grain too fine or bond too hard	Intense localized heating during grinding may set up surface stresses causing grinding cracks. These cracks are either parallel but at right angles to the direction of grinding or, when more advanced, form a network. May need cold etch or magnetic particle testing to become recognizable.	Prevention: by correcting the grinding wheel specifications. Correction: in shallow (0.002- to 0.004-inch) cracks, by removing the damaged layer, when permitted by the design of the tool, using very light grinding passes.
Incorrectly dressed or loaded grinding wheel	Heating of the work surface can cause scorching or cracking. Incorrect dressing can also cause a poor finish of the ground work surface.	Dress wheel with sharper diamond and faster diamond advance to produce coarser wheel surface. Alternate dressing methods, like crush-dressing, can improve wheel surface conditions. Dress wheel regularly to avoid loading or glazing of the wheel surface.
Inadequate coolant, with regard to composition, amount, distribution, and cleanliness	Introducing into the tool surface heat that is not adequately dissipated or absorbed by the coolant can cause softening, or even the development of cracks.	Improve coolant supply and quality, or reduce stock removal rate to reduce generation of heat in grinding.
Damage caused by abusive abrasive cutoff	The intensive heat developed during this process can cause a hardening of the steel surface, or may even result in cracks.	Reduce rate of advance; adopt wheel specifications better suited for the job. Use ample coolant or, when harmful effect not eliminated, replace abrasive cutoff by some cooler-acting stock separation method (e.g., sawing or lathe cutoff) unless damaged surface is being removed by subsequent machining.

Note: Illustrated examples of tool failures from causes such as those listed above may be found in "The Tool Steel Trouble Shooter" handbook, published by Bethlehem Steel Corporation.

COMPUTER NUMERICAL CONTROL

Format Classification.—The *format classification sheet* completely describes the format requirements of a control system and gives other important information required to program a particular control including: the type of machine, the format classification shorthand and format detail, a listing of specific letter address codes recognized by the system (for example, G-codes: G01, G02, G17, etc.) and the range of values the available codes may take (S range: 10 to 1800 rpm, for example), an explanation of any codes not specifically assigned by the Standard ANSI/EIA RS-274-D-1980, and any other unique features of the system.

The format classification shorthand is a nine- or ten-digit code that gives the type of system, the number of motion and other words available, the type and format of dimensional data required by the system, the number of motion control channels, and the number of numerically controlled axes of the system.

The *format detail* very succinctly summarizes details of the machine and control system. This NC shorthand gives the letter address words and word lengths that can be used to make up a block. The format detail defines the basic features of the control system and the type of machine tool to which it refers. For example, the format detail

N4G2X + 24Y + 24Z + 24B24I24J24F31T4M2

specifies that the NC machine is a machining center (has *X*-, *Y*-, and *Z*-axes) and a tool changer with a four-digit tool selection code (T4); the three linear axes are programmed with two digits before the decimal point and four after the decimal point (X + 24Y + 24Z + 24) and can be positive or negative; probably has a horizontal spindle and rotary table (B24 = rotary motion about the *Y*-axis); has circular interpolation (I24J24); has a feed rate range in which there are three digits before and one after the decimal point (F31); and can handle a four-digit sequence number (N4), two-digit G-words (G2), and two-digit miscellaneous words (M2). The sequence of letter addresses in the format detail is also the sequence in which words with those addresses should appear when used in a block.

The information given in the format shorthand and format detail is especially useful when programs written for one machine are to be used on different machines. Programs that use the variable block data format described in RS-274-D can be used interchangeably on systems that have the same format classification, but for complete program compatibility between machines, other features of the machine and control system must also be compatible, such as the relationships of the axes and the availability of features and control functions.

Control systems differ in the way that the numbers may be written. Most newer CNC machines accept numbers written in a decimal-point format; however, some systems require numbers to be in a fixed-length format that does not use an explicit decimal point. In the latter case, the control system evaluates a number based on the number of digits it has, including zeros.

Zero suppression in a control system is an arrangement that allows zeros before the first significant figure to be dropped (leading zero suppression) or allows zeros after the last significant figure to be dropped (trailing zero suppression). An *X*-axis movement of 05.3400, for example, could be expressed as 053400 if represented in the full field format, 53400 (leading zero suppression), or 0534 (trailing zero suppression). With decimal-point programming, the above number is expressed simply as 5.34. To ensure program compatibility between machines, all leading and trailing zeros should be included in numbers unless decimal-point programming is used.

Table 1. G-Code Addresses

Code	Description
G00[abc]	Rapid traverse, point to point (M,L)
G01[abc]	Linear interpolation (M,L)
G02[abc]	Circular interpolation—clockwise movement (M,L)
G03[abc]	Circular interpolation—counterclockwise movement (M,L)
G04[ab]	Dwell—a programmed time delay (M,L)
G05[ab]	Unassigned
G06[abc]	Parabolic interpolation (M,L)
G07[c]	Used for programming with cylindrical diameter values (L)
G08[ab]	Programmed acceleration (M,L). Also for lathe programming with cylindrical diameter values
G09[ab]	Programmed deceleration (M,L). [d]Used to stop the axis movement at a precise location (M,L)
G10–G12[ab]	Unassigned. [d]Sometimes used for machine lock and unlock devices
G13–G16[ac]	Axis selection (M,L)
G13–G16[b]	Unassigned
G13	Used for computing lines and circle intersections (M,L)
G14, G14.1[c]	Used for scaling (M,L)
G15–G16[c]	Polar coordinate programming (M)
G15, G16.1[c]	Cylindrical interpolation—C axis (L)
G16.2[c]	End face milling—C axis (L)
G17–G19[abc]	*X-Y*, *X-Z*, *Y-Z* plane selection, respectively (M,L)
G20	Unassigned
G22–G32[ab]	Unassigned
G22–G23[c]	Defines safety zones in which the machine axis may not enter (M,L)
G22.1, G233.1[c]	Defines safety zones in which the cutting tool may not exit (M,L)
G24[c]	Single-pass rough-facing cycle (L)
G27–G29	Used for automatically moving to and returning from home position (M,L)
G30	Return to an alternate home position (M,L)
G31, G31.1, G31.2, G31.3, G31.4	External skip function, moves an axis on a linear path until an external signal aborts the move (M,L)
G33[abc]	Thread cutting, constant lead (L)
G34[abc]	Thread cutting, increasing lead (L)
G35[abc]	Thread cutting, decreasing lead (L)
G36-G39[ab]	Permanently unassigned
G36[c]	Used for automatic acceleration and deceleration when the blocks are short (M,L)
G37, G37.1, G37.2, G37.3G37.4	Used for tool gaging (M,L)
G38	Used for probing to measure the diameter and center of a hole (M)
G38.1	Used with a probe to measure the parallelness of a part with respect to an axis (M)
G39, G39.1	Generates a nonprogrammed block to improve cycle time and corner cutting quality when used with cutter compensation (M)
G39	Tool tip radius compensation with linear generated block (L)
G39.1	Tool tip radius compensation used with circular generated block (L)
G40[abc]	Cancel cutter compensation/offset (M)
G41[abc]	Cutter compensation, left (M)
G42[abc]	Cutter compensation, right (M)
G43[abc]	Cutter offset, inside corner (M,L)
G44[abc]	Cutter offset, outside corner (M,L)
G45–G49[ab]	Unassigned
G50–G59[a]	Reserved for adaptive control (M,L)
G50[bb]	Unassigned
G50.1[c]	Cancel mirror image (M,L)
G51.1[c]	Program mirror image (M,L)
G52[b]	Unassigned
G52	Used to offset the axes with respect to the coordinate zero point (see G92) (M,L)
G53[bc]	Datum shift cancel
G53[c]	Call for motion in the machine coordinate system (M,L)
G54–G59[bc]	Datum shifts (M,L)
G54–G59.3[c]	Allows for presetting of work coordinate systems (M,L)
G61[c]	Modal equivalent of G09 except that rapid moves are not taken to a complete stop before the next motion block is executed (M,L)
G60–G62[abc]	Unassigned

Table 1. *(Continued)* **G-Code Addresses**

Code	Description	Code	Description
G62[c]	Automatic corner override, reduces the feed rate on an inside corner cut (M,L)	G80[abc]	Cancel fixed cycles
G63[a]	Unassigned	G81[abc]	Drill cycle, no dwell and rapid out (M,L)
G63[bc]	Tapping mode (M,L)	G82[abc]	Drill cycle, dwell and rapid out (M,L)
G64–G69[abc]	Unassigned	G83[abc]	Deep hole peck drilling cycle (M,L)
G64[c]	Cutting mode, usually set by the system installer (M,L)	G84[abc]	Right-hand tapping cycle (M,L)
G65[c]	Calls for a parametric macro (M,L)	G84.1[c]	Left-hand tapping cycle (M,L)
G66[c]	Calls for a parametric macro. Applies to motion blocks only (M,L)	G85[abc]	Boring cycle, no dwell, feed out (M,L)
G66.1[c]	Same as G66 but applies to all blocks (M,L)	G86[abc]	Boring cycle, spindle stop, rapid out (M,L)
G67[c]	Stop the modal parametric macro (see G65, G66, G66.1) (M,L)	G87[abc]	Boring cycle, manual retraction (M,L)
G68[c]	Rotates the coordinate system (i.e., the axes) (M)	G88[abc]	Boring cycle, spindle stop, manual retraction (M,L)
G69[c]	Cancel axes rotation (M)	G88.1	Pocket milling (rectangular and circular), roughing cycle (M)
G70[abc]	Inch programming (M,L)	G88.2	Pocket milling (rectangular and circular), finish cycle (M)
G71[abc]	Metric programming (M,L)	G88.3	Post milling, roughs out material around a specified area (M)
G72[ac]	Circular interpolation CW (three-dimensional) (M)	G88.4	Post milling, finish cuts material around a post (M)
G72[b]	Unassigned	G88.5	Hemisphere milling, roughing cycle (M)
G72[c]	Used to perform the finish cut on a turned part along the Z-axis after the roughing cuts initiated under G73, G74, or G75 codes (L)	G88.6	Hemisphere milling, finishing cycle (M)
		G89[abc]	Boring cycle, dwell and feed out (M,L)
G73[b]	Unassigned	G89.1	Irregular pocket milling, roughing cycle (M)
G73[c]	Deep hole peck drilling cycle (M); OD and ID roughing cycle, running parallel to the Z-axis (L)	G89.2	Irregular pocket milling, finishing cycle (M)
		G90[abc]	Absolute dimension input (M,L)
G74[ac]	Cancel multiquadrant circular interpolation (M,L)	G91[abc]	Incremental dimension input (M,L)
G74[bc]	Move to home position (M,L)	G92[abc]	Preload registers, used to shift the coordinate axes relative to the current tool position (M,L)
G74[c]	Left-hand tapping cycle (M)	G93[abc]	Inverse time feed rate (velocity/distance) (M,L)
G74	Rough facing cycle (L)	G94[c]	Feed rate in inches or millimeters per minute (ipm or mpm) (M,L)
G75[ac]	Multiquadrant circular interpolation (M,L)	G95[abc]	Feed rate given directly in inches or millimeters per revolution (ipr or mpr) (M,L)
G75[b]	Unassigned	G96[abc]	Maintains a constant surface speed, feet (meters) per minute (L)
G75	Roughing routine for castings or forgings (L)	G97[abc]	Spindle speed programmed in rpm (M,L)
G76–G79[ab]	Unassigned	G98–99[ab]	Unassigned

[a] Adheres to ANSI/EIA RS-274-D.

[b] Adheres to ISO 6983/1,2,3 Standards; where both symbols appear together, the ANSI/EIA and ISO standard codes are comparable;

[c] This code is modal. All codes that are not identified as modal are non-modal, when used according to the corresponding definition.

[d] Indicates a use of the code that does not conform with the Standard.

Symbols following a description: (M) indicates that the code applies to a mill or machining center; (L) indicates that the code applies to turning machines; (M,L) indicates that the code applies to both milling and turning machines.

Codes that appear more than once in the table are codes that are in common use, but are not defined by the Standard or are used in a manner that is different than that designated by the Standard (e.g., see G61).

Table 2. Letter Addresses Used in Numerical Control

Letter Address	Description	Refers to
A	Angular dimension about the *X*-axis, measured in decimal parts of a degree	Axis nomenclature
B	Angular dimension about the *Y*-axis, measured in decimal parts of a degree	Axis nomenclature
C	Angular dimension about the *Z*-axis, measured in decimal parts of a degree	Axis nomenclature
D	Angular dimension about a special axis, or third feed function, or tool function for selection of tool compensation	Axis nomenclature
E	Angular dimension about a special axis or second feed function	Axis nomenclature
F	Feed word (code)	Feed words
G	Preparatory word (code)	Preparatory words
H	Unassigned	
I	Interpolation parameter or thread lead parallel to the *X*-axis	Circular interpolation and threading
J	Interpolation parameter or thread lead parallel to the *Y*-axis	Circular interpolation and threading
K	Interpolation parameter or thread lead parallel to the *Z*-axis	Circular interpolation and threading
L	Unassigned	
M	Miscellaneous or auxilliary function	Miscellaneous functions
N	Sequence number	Sequence number
O	Sequence number for secondary head only	Sequence number
P	Third rapid-traverse dimension or tertiary-motion dimension parallel to *X*	Axis nomenclature
Q	Second rapid-traverse dimension or tertiary-motion dimension parallel to *Y*	Axis nomenclature
R	First rapid-traverse dimension or tertiary-motion dimension parallel to *Z* or radius for constant surface-speed calculation	Axis nomenclature
S	Spindle-speed function	Spindle speed
T	Tool function	Tool function
U	Secondary-motion dimension parallel to *X*	Axis nomenclature
V	Secondary-motion dimension parallel to *Y*	Axis nomenclature
W	Secondary-motion dimension parallel to *Z*	Axis nomenclature
X	Primary *X*-motion dimension	Axis nomenclature
Y	Primary *Y*-motion dimension	Axis nomenclature
Z	Primary *Z*-motion dimension	Axis nomenclature

Miscellaneous Functions (M–Words).—Miscellaneous functions, or M-codes, also reffered to as auxiliary functions, constitute on/off-type commands. M functions are used to control actions such as starting and stopping of motors, turning coolant on and off, changing tools, and clamping and unclamping parts. M functions are made up of the letter M followed by a two-digit code.

Table 3. Miscellaneous Function Words from *ANSI/EIA Standard RS-274-D*

Code	Description
M00	Automatically *stops* the machine. The operator must push a button to continue with the remainder of the program.
M01	An *optional stop* acted upon only when the operator has previously signaled for this command by pushing a button. The machine will automatically stop when the control system senses the M01 code.
M02	This *end-of-program* code stops the machine when all commands in the block are completed. May include rewinding of tape.
M03	Start *spindle rotation* in a *clockwise* direction—looking out from the spindle face.
M04	Start *spindle rotation* in a *counterclockwise* direction—looking out from the spindle face.
M05	*Stop* the spindle in a normal and efficient manner.
M06	Command to *change a tool* (or tools) manually or automatically. Does not cover tool selection, as is possible with the T-words.
M07 to M08	M07 (coolant 2) and M08 (coolant 1) are codes to *turn on coolant*. M07 may control *flood* coolant and M08 *mist* coolant.
M09	Shuts off the coolant.
M10 to M11	M10 applies to automatic *clamping* of the machine slides, workpiece, fixture spindle, etc. M11 is an unclamping code.
M12	An inhibiting code used to synchronize multiple sets of axes, such as a four-axis lathe having two independently operated heads (turrets).
M13	Starts *CW spindle* motion and *coolant on* in the same command.
M14	Starts *CCW spindle* motion and *coolant on* in the same command.
M15 to M16	Rapid traverse of feed motion in either the +(M15) or −(M16) direction.
M17 to M18	Unassigned.
M19	Oriented spindle stop. Causes the spindle to stop at a predetermined angular position.
M20 to M29	Permanently unassigned.
M30	An *end-of-tape* code similar to M02, but M30 will also rewind the tape; also may switch automatically to a second tape reader.
M31	A command known as *interlock bypass* for temporarily circumventing a normally provided interlock.
M32 to M35	Unassigned.
M36 to M39	Permanently unassigned.
M40 to M46	Used to signal gear changes if required at the machine; otherwise, unassigned.
M47	Continues program execution from the start of the program unless inhibited by an interlock signal.
M48 to M49	M49 deactivates a manual spindle or feed override and returns the parameter to the programmed value; M48 cancels M49.
M50 to M57	Unassigned.
M58 to M59	Holds the rpm constant at the value in use when M59 is initiated; M58 cancels M59.
M60 to M89	Unassigned.
M90 to M99	Reserved for use by the machine user.

GRINDING WHEELS

Safety in Operating Grinding Wheels.—Grinding wheels are prone to damage caused by improper handling and operation. Vitrified wheels, comprising the major part of grinding wheels used in industry, are held together by an inorganic bond which is actually a type of pottery product and therefore brittle and breakable.

It must also be understood that during the grinding process very substantial forces act on the grinding wheel, including the centrifugal force due to rotation, the grinding forces resulting from the resistance of the work material, and shocks caused by sudden contact with the work. To be able to resist these forces, the grinding wheel must have a substantial minimum strength throughout that is well beyond that needed to hold the wheel together under static conditions.

A damaged grinding wheel can disintegrate during grinding, which normally is constrained, thus presenting great hazards to both operator and equipment. Safeguards have been formulated into rules and regulations and are set forth in the ANSI B7.1-2017, entitled the "American National Standard Safety Requirements for the Use, Care, and Protection of Abrasive Wheels." All operators should be familiar with the rules.

Handling, Storage, and Inspection.—Grinding wheels should be hand carried, or transported, with proper support. A grinding wheel must not be rolled around on its periphery. The storage area, positioned near the grinding machines, should be free from excessive temperature variations and humidity. Specially built racks are recommended on which the smaller or thin wheels are stacked lying on their sides and the larger wheels in an upright position on two-point cradle supports consisting of appropriately spaced wooden bars. Partitions should separate either the individual wheels or a small group of identical wheels. Good accessibility to the stored wheels reduces the need for undesirable handling.

Inspection will primarily be directed at detecting visible damage, mostly originating from handling and shipping. Cracks that are not obvious can usually be detected by "ring testing," which consists of suspending the wheel from its hole and tapping it with a non-metallic implement. Heavy wheels may be allowed to rest vertically on a clean, hard floor while this test is performed. A clear metallic tone, a "ring," should be heard, a dead sound being indicative of a possible crack or cracks in the wheel.

Machine Conditions.—The general design of grinding machines must ensure safe operation under normal conditions. The bearings and grinding wheel spindle must be dimensioned to withstand the expected forces and ample driving power should be provided to ensure maintenance of the rated spindle speed. For the protection of the operator, stationary machines used for dry grinding should have provision made for connection to an exhaust system and, when used for offhand grinding, a work support must be available.

Wheel guards are particularly important protection elements, and their material specifications, wall thicknesses, and construction principles should agree with the Standard's specifications. The exposure of the wheel should be just enough to avoid interference with the grinding operation. The need for access of the work to the grinding wheel will define the boundary of guard opening, particularly in the direction of the operator.

Grinding Wheel Mounting.—The mass and speed of the operating grinding wheel makes it particularly sensitive to imbalance. Vibrations that result from such conditions are harmful to the machine, particularly the spindle bearings, and they also affect the ground surface, i.e., wheel imbalance causes chatter marks and interferes with size control. Grinding wheels are shipped from the manufacturer's plant in a balanced condition, but retaining the balanced state after mounting the wheel is quite uncertain. Balancing of the mounted wheel is thus required, and is particularly important for medium and large size wheels, as well as for producing accurate and smooth surfaces. The most common way of balancing mounted wheels is by using balancing flanges with adjustable weights.

The wheel and balancing flanges are mounted on a short balancing arbor, the two concentric and round stub ends of which are supported in a balancing stand.

Such stands are of two types: 1) the parallel straight-edged, which must be set up precisely level; and 2) the disk type having two pairs of ball bearing mounted overlapping disks, which form a V for containing the arbor ends without hindering the free rotation of the wheel mounted on that arbor.

The wheel will then rotate only when it is out of balance and its heavy spot is not in the lowest position. Rotating the wheel by hand to different positions will move the heavy spot, should such exist, from the bottom to a higher location where it can reveal its presence by causing the wheel to turn. Having detected the presence and location of the heavy spot, its effect can be cancelled by displacing the weights in the circular groove of the flange until a balanced condition is accomplished.

Flanges are commonly used means for holding grinding wheels on the machine spindle. For that purpose, the wheel can either be mounted directly through its hole or by means of a sleeve that slips over a tapered section of the machine spindle. Either way, the flanges must be of equal diameter, usually not less than one-third of the new wheel's diameter. The purpose is to securely hold the wheel between the flanges without interfering with the grinding operation even when the wheel becomes worn down to the point where it is ready to be discarded. Blotters or flange facings of compressible material should cover the entire contact area of the flanges.

One of the flanges is usually fixed while the other is loose and can be removed and adjusted along the machine spindle. The movable flange is held against the mounted grinding wheel by means of a nut engaging a threaded section of the machine spindle. The sense of that thread should be such that the nut will tend to tighten as the spindle revolves. In other words, to remove the nut, it must be turned in the direction that the spindle revolves when the wheel is in operation.

Safe Operating Speeds.—Safe grinding processes are predicated on the proper use of the previously discussed equipment and procedures, and are greatly dependent on the application of adequate operating speeds. The Standard establishes maximum speeds at which grinding wheels can be operated, assigning the various types of wheels to several classification groups. Different values are listed according to bond type and to wheel strength, distinguishing between low-, medium- and high-strength wheels.

For the purpose of general information, the accompanying table shows an abbreviated version of the Standard's specification. The maximum operating speeds indicated on the wheel's tag must never be exceeded. All grinding wheels of 6 inches or greater diameter must be test run in the wheel manufacturer's plant at a speed that for all wheels having operating speeds in excess of 5000 sfpm is 1.5 times the maximum speed marked on the tag of the wheel.

The table shows the permissible wheel speeds in surface feet per minute (sfpm) units, whereas the tags on the grinding wheels state, for the convenience of the user, the maximum operating speed in revolutions per minute (rpm). The sfpm unit has the advantage of remaining valid for worn wheels whose rotational speed may be increased to the applicable sfpm value. The conversion from either one to the other of these two kinds of units is a matter of simple calculation using the formulas:

$$\text{sfpm} = \text{rpm} \times \frac{D}{12} \times \pi \qquad \text{or} \qquad \text{rpm} = \frac{\text{sfpm} \times 12}{D \times \pi}$$

Where D = maximum diameter of the grinding wheel, in inches. Table 1, showing the conversion values from surface speed into rotational speed, can be used for the direct reading of the rpm values corresponding to several different wheel diameters and surface speeds.

Table 1. Revolutions per Minute for Various Grinding Speeds and Wheel Diameters *(Based on ANSI B7.1-2017)*

Wheel Diameter, Inch	Peripheral (Surface) Speed, Feet per Minute																		Wheel Diameter, Inch
	4000	4500	5000	5500	6000	6500	7000	7500	8000	8500	9000	9500	10000	12000	14200	16000	16500	17000	
	Revolutions per Minute																		
1	15279	17189	19099	21008	22918	24828	26738	28648	30558	32468	34377	36287	38197	45837	54240	61115	63025	64935	1
2	7639	8594	9549	10504	11459	12414	13369	14324	15279	16234	17189	18144	19099	22918	27120	30558	31513	32468	2
3	5093	5730	6366	7003	7639	8276	8913	9549	10186	10823	11459	12096	12732	15279	18080	20372	21008	21645	3
4	3820	4297	4775	5252	5730	6207	6685	7162	7639	8117	8594	9072	9549	11459	13560	15279	15756	16234	4
5	3056	3438	3820	4202	4584	4966	5348	5730	6112	6494	6875	7257	7639	9167	10848	12223	12605	12987	5
6	2546	2865	3183	3501	3820	4138	4456	4775	5093	5411	5730	6048	6366	7639	9040	10186	10504	10823	6
7	2183	2456	2728	3001	3274	3547	3820	4093	4365	4638	4911	5184	5457	6548	7749	8731	9004	9276	7
8	1910	2149	2387	2626	2865	3104	3342	3581	3820	4058	4297	4536	4775	5730	6780	7639	7878	8117	8
9	1698	1910	2122	2334	2546	2759	2971	3183	3395	3608	3820	4032	4244	5093	6027	6791	7003	7215	9
10	1528	1719	1910	2101	2292	2483	2674	2865	3056	3247	3438	3629	3820	4584	5424	6112	6303	6494	10
12	1273	1432	1592	1751	1910	2069	2228	2387	2546	2706	2865	3024	3183	3820	4520	5093	5252	5411	12
14	1091	1228	1364	1501	1637	1773	1910	2046	2183	2319	2456	2592	2728	3274	3874	4365	4502	4638	14
16	955	1074	1194	1313	1432	1552	1671	1790	1910	2029	2149	2268	2387	2865	3390	3820	3939	4058	16
18	849	955	1061	1167	1273	1379	1485	1592	1698	1804	1910	2016	2122	2546	3013	3395	3501	3608	18
20	764	859	955	1050	1146	1241	1337	1432	1528	1623	1719	1814	1910	2292	2712	3056	3151	3247	20
22	694	781	868	955	1042	1129	1215	1302	1389	1476	1563	1649	1736	2083	2465	2778	2865	2952	22
24	637	716	796	875	955	1035	1114	1194	1273	1353	1432	1512	1592	1910	2260	2546	2625	2706	24
26	588	661	735	808	881	955	1028	1102	1175	1249	1322	1396	1469	1763	2086	2351	2424	2498	26
28	546	614	682	750	819	887	955	1023	1091	1160	1228	1296	1364	1637	1937	2183	2251	2319	28
30	509	573	637	700	764	828	891	955	1019	1082	1146	1210	1273	1528	1808	2037	2101	2165	30
32	477	537	597	657	716	776	836	895	955	1015	1074	1134	1194	1432	1695	1910	1970	2029	32
34	449	506	562	618	674	730	786	843	899	955	1011	1067	1123	1348	1595	1798	1854	1910	34
36	424	477	531	584	637	690	743	796	849	902	955	1008	1061	1273	1507	1698	1751	1804	36
38	402	452	503	553	603	653	704	754	804	854	905	955	1005	1206	1427	1608	1659	1709	38
40	382	430	477	525	573	621	668	716	764	812	859	907	955	1146	1356	1528	1576	1623	40
42	364	409	455	500	546	591	637	682	728	773	819	864	909	1091	1291	1455	1500	1546	42
44	347	391	434	477	521	564	608	651	694	738	781	825	868	1042	1233	1389	1432	1476	44
46	332	374	415	457	498	540	581	623	664	706	747	789	830	996	1179	1329	1370	1412	46
48	318	358	398	438	477	517	557	597	637	676	716	756	796	955	1130	1273	1313	1353	48
53	288	324	360	396	432	468	504	541	577	613	649	685	721	865	1023	1153	1189	1225	53
60	255	286	318	350	382	414	446	477	509	541	573	605	637	764	904	1019	1050	1082	60
72	212	239	265	292	318	345	371	398	424	451	477	504	531	637	753	849	875	902	72

Portable Grinders.—The above discussed rules and regulations, devised primarily for stationary grinding machines, apply also to portable grinders. In addition, the details of various other regulations, specially applicable to different types of portable grinders, are discussed in the Standard, which should be consulted, particularly for safe applications of portable grinding machines.

Table 2. Maximum Peripheral Speeds for Grinding Wheels *(Based on ANSI 7.1-2017)*

Maximum Operating Speeds Depending on Strength of Bond					
Classification Number and Types of Wheels[a]		Inorganic Bonds		Organic Bonds	
		sfpm	m/min	sfpm	m/min
1	Straight wheels—Type 1, except classifications 6, 9, 11, and 12, 13, and 14 below; recessed wheels—Types 5 and 7 Type 20, 21, 22, 23, 24, 25, 26	5500–8500	1675–2590	6500–9500	1980–2895
2	Taper Side Wheels—Type 4[b]; dish wheels—Type 12; Saucer wheels—Type 13; cones and plugs—Types 16, 17, 18, 19	5000–6500	1525–1980	6500–9500	1980–2895
	Cylinder wheels—Type 2 segments	5000–6000	1525–1830	5000–7000	1525–2135
3	Cup shape tool grinding wheels—Types 6 and 11 (for fixed base machines)	4500–6000	1370–1830	6000–8500	1830–2590
4	Cup shape snagging wheels—Types 6 and 11 (for portable machines)	4500–6500	1370–1980	6000–9500	1830–2895
5	Abrasive disks and cylinder type discs solid and segmental	5500–6500	1675–1980	5500–8500	1675–2590
6	Reinforced wheels—except cutting-off wheels (depending on diameter and thickness)	…	…	9500–16000	2895–4875
7	Type 1 wheels for bench and pedestal grinders, Types 1, 5 and 7 also in certain sizes for surface grinders	5500–7550	1675–2300	6500–9500	1980–2895
8	Diamond and cubic boron nitride wheels	to 6500	to 1980	to 9500	to 2895
	Metal bond	…	to 3660	…	…
	Steel centered cutting off	to 16000	to 4875	to 16000	to 4875
9	Cutting-off wheels—maximum 4-inch diameter (including reinforced organic)	…	…	14200–20000	4330–6100
	Cutting-off wheels—maximum 16-inch diameter (including reinforced organic)	…	…	9500–16000	2895–4875
	Cutting-off wheels—larger than 16-inch diameter (including reinforced organic)	…	…	9500–14200	2895–4330
	Non-reinforced, all diameters	…	…	9500–14200	2895–4330
10	Thread and flute grinding wheels	8000–12000	2440–3660	10000–12000	3050–3660
11	Crankshaft and camshaft grinding wheels	8500–12000	2590–3660	6500–9500	1980–2895
12	Snagging wheels 16-inch or larger (including reinforced organic)—used on swing frame grinders	…	…	12500	3810
	Snagging wheels 16-inch or larger (including reinforced organic)—used on semi-automatic snagging grinders	…	…	16500	5030
13	Internal wheels—Type 1 and 4, maximum diameter 6-inch	5500–8500	1675–2590	6500–9500	1980–2895
14	Mounted wheels	10000	3050	10000	3050

[a] See Tables 7a and Tables 7b starting on page 228.

[b] Non-standard shape. For snagging wheels, 16 inches and larger—Type 1, internal wheels—Types 1 and 5, and mounted wheels, see ANSI B7.1-2017. Under no conditions should a wheel be operated faster than the maximum operating speed established by the manufacturer.

Values in this table are for general information only.

Table 3. Principal Systems of Surface Grinding

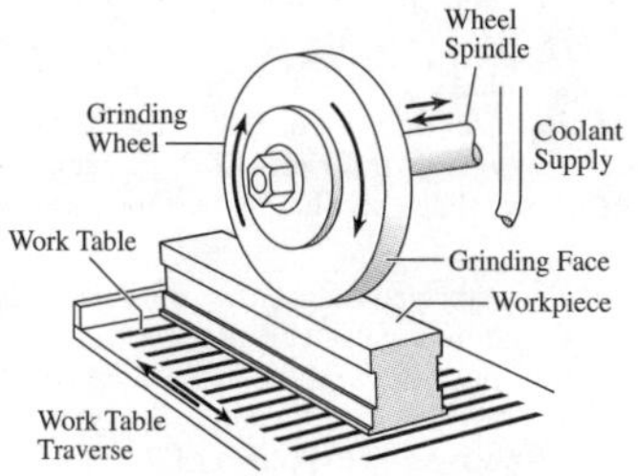

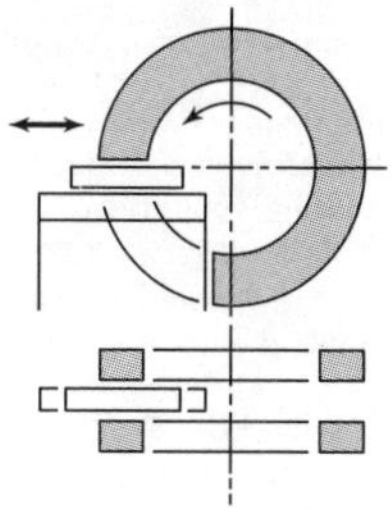

Reciprocating — Periphery of Wheel

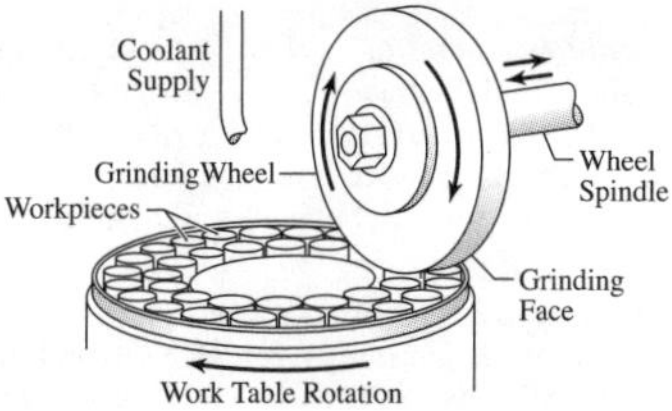

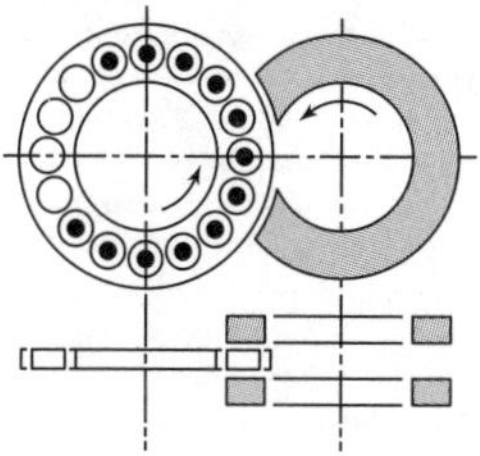

Rotary — Periphery of Wheel

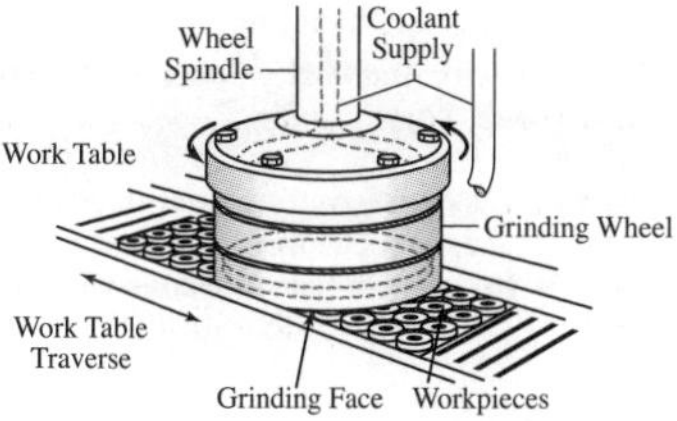

Reciprocating — Face (Side) of Wheel

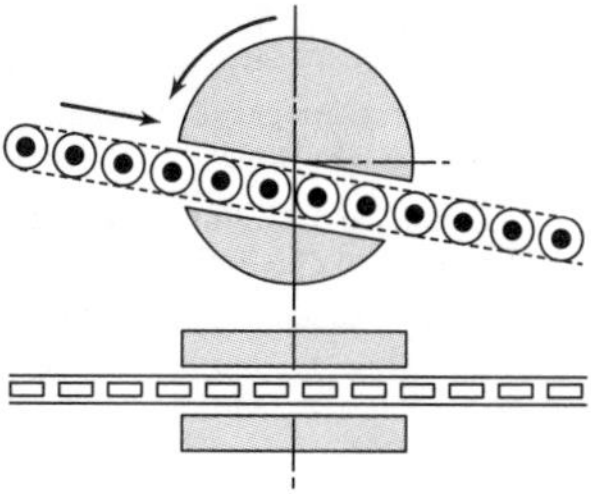

Traverse Along Straight Line or Arcuate Path — Face (Side) of Wheel

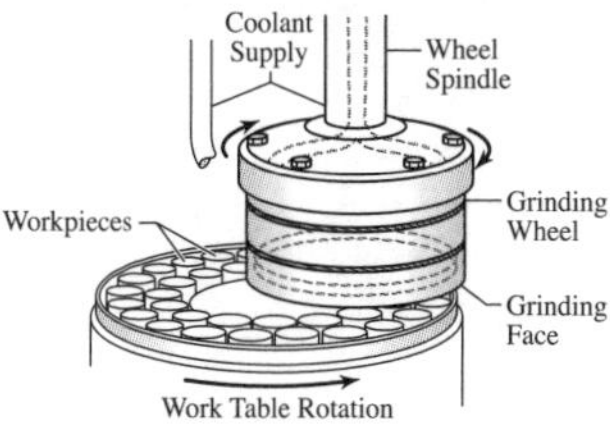

Rotary — Face (Side) of Wheel

Principles of Operations

Periphery of Wheel.—*Reciprocating:* Work is mounted on the horizontal machine table that is traversed in a reciprocating movement at a speed generally selected from a steplessly variable range. The transverse movement, called cross feed of the table or of the wheel slide, operates at the end of the reciprocating stroke and assures the gradual exposure of the entire work surface, which commonly exceeds the width of the wheel. The depth of the cut is controlled by the downfeed of the wheel, applied in increments at the reversal of the transverse movement.

Rotary: Work is mounted, usually on the full-diameter magnetic chuck of the circular machine table that rotates at a preset constant or automatically varying speed, the latter maintaining an approximately equal peripheral speed of the work surface area being ground. The wheelhead, installed on a cross slide, traverses over the table along a radial path, moving in alternating directions, toward and away from the center of the table. Infeed is by vertical movement of the saddle along the guideways of the vertical column, at the end of the radial wheelhead stroke. The saddle contains the guideways along which the wheelhead slide reciprocates.

Face (Side) of Wheel.—*Reciprocating:* Operation is similar to the reciprocating table-type peripheral surface grinder, but grinding is with the face, usually with the rim of a cup-shaped wheel, or a segmental wheel for large machines. It is capable of covering a much wider area of the work surface than the peripheral grinder, thus frequently there is no need for cross feed. It also provides efficient stock removal, but is less adaptable than the reciprocating table-type peripheral grinder.

Rotary: The grinding wheel, usually of segmental type, is set in a position to cover either an annular area near the periphery of the table or, more commonly, to reach beyond the table center. A large circular magnetic chuck generally covers the entire table surface and facilitates the mounting of workpieces, even of fixtures, when needed. The uninterrupted passage of the work in contact with the large wheel face permits a very high rate of stock removal, and the machine, with single or double wheelhead, can be adapted also to automatic operation with continuous part feed by mechanized work handling.

Traverse Along Straight or Arcuate Path: Operates with practically the entire face of the wheel, which is designated as an abrasive disc (hence "disc grinding") because of its narrow width in relation to the large diameter. Built either for one or, more frequently, for two discs operating with opposed faces for the simultaneous grinding of both sides of the workpiece. The parts pass between the operating faces of the wheel (a) pushed-in and retracted by the drawerlike movement of a feed slide; (b) in an arcuate movement carried in the nests of a rotating feed wheel; (c) nearly diagonally advancing along a rail. Very well adapted to fully mechanized work handling.

Table 3a. Grinding Wheel Recommendations for Surface Grinding Using Type 2 Cylinder Wheels, Type 6 Cup Wheels, and Wheel Segments

Material	Type 2 Cylinder Wheels	Type 6 Cup Wheels	Wheel Segments
High-tensile cast iron and nonferrous metals	37C24-HKV	37C24-HVK	37C24-HVK
Soft steel, malleable cast iron, steel castings, boiler plate	23A24-I8VBE or 23A30-G12VBEP	23A24-I8VBE	23A24-I8VSM or 23A30-H12VSM
Hardened steel—broad contact	32A46-G8VBE or 32A36-E12VBEP	32A46-G8VBE or 32A60-E12VBEP	32A36-G8VBE or 32A46-E12VBEP
Hardened steel—narrow contact or interrupt cut	32A46-H8VBE	32A60-H8VBE	32A46-G8VBE or 32A60-G12VBEP
General-purpose use	23A30-H8VBE or 23A30-E12VBEP	…	23A30-H8VSM or 23A30-G12VSM

The wheel markings in the tables are those used by the Norton Co., complementing the basic standard markings with Norton symbols. The complementary symbols used in these tables, that is, those preceding the letter designating A (aluminum oxide) or C (silicon carbide), indicate the special type of basic abrasive that has the friability best suited for particular work materials. Those preceding A (aluminum oxide) are

57—a versatile abrasive suitable for grinding steel in either a hard or soft state.
38—the most friable abrasive.
32—the abrasive suited for tool steel grinding.
23—an abrasive with intermediate grinding action, and
19—the abrasive produced for less heat-sensitive steels.

Those preceding C (silicon carbide) are
37—a general application abrasive, and
39—an abrasive for grinding hard cemented carbide.

Table 4. Basic Process Data for Peripheral Surface Grinding on Reciprocating Table Surface Grinders

Work Material	Hardness	Material Condition	Wheel Speed, fpm	Table Speed, fpm	Downfeed, in. per pass		Crossfeed per pass, fraction of wheel width
					Rough	Finish	
Plain carbon steel	52 RC max.	Annealed, Cold drawn	5500–6500	50–100	0.003	0.0005 max.	1/4
	52 to 65 RC	Carburized and/or quenched and tempered	5500–6500	50–100	0.003	0.0005 max.	1/10
Alloy steels	52 RC max.	Annealed or quenched and tempered	5500–6500	50–100	0.003	0.001 max.	1/4
	52 to 65 RC	Carburized and/or quenched and tempered	5500–6500	50–100	0.003	0.0005 max.	1/10
Tool steels	150 to 275 BHN	Annealed	5500–6500	50–100	0.002	0.0005 max.	1/5
	56 to 65 RC	Quenched and tempered	5500–6500	50–100	0.002	0.0005 max.	1/10
Nitriding steels	200 to 350 BHN	Normalized, annealed	5500–6500	50–100	0.003	0.001 max.	1/4
	60 to 65 RC	Nitrided	5500–6500	50–100	0.003	0.0005 max.	1/10
Cast steels	52 RC max.	Normalized, annealed	5500–6500	50–100	0.003	0.001 max.	1/4
	Over 52 RC	Carburized and/or quenched and tempered	5500–6500	50–100	0.003	0.0005 max.	1/10
Gray irons	52 RC max.	As cast, annealed, and/or quenched and tempered	5000–6500	50–100	0.003	0.001 max.	1/3
Ductile irons	52 RC max.	As cast, annealed or quenched and tempered	5500–6500	50–100	0.003	0.001 max.	1/5
Stainless steels, martensitic	135 to 235 BHN	Annealed or cold drawn	5500–6500	50–100	0.002	0.0005 max.	1/4
	Over 275 BHN	Quenched and tempered	5500–6500	50–100	0.001	0.0005 max.	1/8
Aluminum alloys	30 to 150 BHN	As cast, cold drawn or treated	5500–6500	50–100	0.003	0.001 max.	1/3

Table 5. Common Faults and Possible Causes in Surface Grinding

Causes	Faults	Work Dimension			Metallurgical Defects		Surface Quality				Wheel Condition			Work Retainment	
		Work not flat	Work not parallel	Poor size holding	Burnishing of work	Burning or checking	Feed lines	Chatter marks	Scratches on surface	Poor finish	Wheel loading	Wheel glazing	Rapid wheel wear	Not firmly seated	Work sliding on chuck
Work Condition	Heat-treat stresses	✓	…	…	…	…	…	…	…	…	…	…	…	…	…
	Work too thin	✓	✓	…	…	…	…	…	…	…	…	…	…	…	…
	Work warped	✓	…	…	…	…	…	…	…	…	…	…	…	✓	…
	Abrupt section changes	✓	✓	…	…	…	…	…	…	…	…	…	…	…	…
Grinding Wheel	Grit too fine	…	…	…	✓	✓	…	…	…	…	✓	✓	…	…	…
	Grit too coarse	…	…	…	…	…	…	…	…	✓	…	…	…	…	…
	Grade too hard	✓	…	…	✓	✓	…	✓	…	…	✓	✓	…	…	…
	Grade too soft	…	…	✓	…	…	…	✓	✓	…	…	…	✓	…	…
	Wheel not balanced	…	…	…	…	…	…	✓	…	…	…	…	…	…	…
	Dense structure	…	…	…	…	…	…	…	…	…	✓	✓	…	…	…
Tooling And Coolant	Improper coolant	…	…	…	…	…	…	…	…	…	✓	…	…	…	…
	Insufficient coolant	✓	✓	…	✓	✓	…	…	…	…	…	✓	…	…	…
	Dirty coolant	…	…	…	…	…	…	…	✓	…	✓	…	…	…	…
	Diamond loose or chipped	✓	✓	…	…	…	…	…	✓	…	…	…	…	…	…
	Diamond dull	…	…	✓	…	…	…	…	…	✓	✓	✓	…	…	…
	No or poor magnetic force	…	…	✓	…	…	…	…	✓	…	…	…	…	✓	✓
	Chuck surface worn or burred	✓	✓	…	…	…	…	…	✓	…	…	…	…	✓	…
Machine And Setup	Chuck not aligned	✓	✓	…	…	…	…	…	…	…	…	…	…	…	…
	Vibrations in machine	…	…	…	…	…	…	✓	…	…	…	…	…	…	…
	Plane of movement out of parallel	✓	✓	…	…	…	…	…	…	…	…	…	…	…	…
Operational Conditions	Too low work speed	…	…	…	…	…	…	…	…	…	✓	…	…	…	…
	Too light feed	…	…	…	…	…	…	…	…	…	…	✓	…	…	…
	Too heavy cut	✓	…	…	✓	…	…	…	…	✓	…	…	…	…	…
	Chuck retained swarf	✓	✓	…	…	…	…	…	…	…	…	…	…	…	✓
	Chuck loading improper	✓	✓	…	…	…	…	…	…	…	…	…	…	…	✓
	Insufficient blocking of parts	…	…	…	…	…	…	…	✓	…	…	…	…	…	…
	Wheel runs off the work	…	✓	✓	…	✓	…	…	…	…	…	…	✓	…	…
	Wheel dressing too fine	✓	…	…	…	…	…	…	…	…	…	…	…	…	…
	Wheel edge not chamfered	…	…	…	…	…	✓	…	…	…	…	…	…	…	…
	Loose dirt under guard	…	…	…	…	…	…	…	✓	…	…	…	…	…	…

American National Standard Grinding Wheel Markings.—ANSI Standard B74.13-2016 "Markings for Identifying Grinding Wheels and Other Bonded Abrasives," applies to grinding wheels and other bonded abrasives, segments, bricks, sticks, hones, rubs, and other shapes that are used to remove material or produce a desired surface or dimension. It does not apply to specialities such as sharpening stones and provides only a standard system of markings. Wheels having the same standard markings but made by different wheel manufacturers may not—and probably will not—produce exactly the same grinding action. This desirable result cannot be obtained because of the impossibility of closely correlating any measurable physical properties of bonded abrasive products in terms of their grinding action.

Sequence of Markings.—The accompanying illustration taken from ANSI B74.13-2016 shows the makeup of a typical wheel or bonded abrasive marking.

Prefix	1 Abrasive Type	2 Grain Size	3 Grade	4 Structure	5 Bond Type	6 Manufacturer's Record
51 –	**A –**	**36 –**	**L –**	**5 –**	**V –**	**23**

The meaning of each letter and number in this or other markings is indicated by the following complete list.

1) *Abrasive Letters:* The letter (A) is used for aluminum oxide, (C) for silicon carbide, and (Z) for aluminum zirconium. The manufacturer may designate some particular type in any one of these broad classes by using its own symbol as a prefix (example, 51).

2) *Grain Size:* The grain sizes commonly used and varying from coarse to very fine are indicated by the following numbers: 8, 10, 12, 14, 16, 20, 24, 30, 36, 46, 54, 60, 70, 80, 90, 100, 120, 150, 180, and 220. The following additional sizes are used occasionally: 240, 280, 320, 400, 500, and 600. The wheel manufacturer may add to the regular grain number an additional symbol to indicate a special grain combination.

3) *Grade:* Grades are indicated by letters of the alphabet from A to Z in all bonds or processes. Wheel grades from A to Z range from soft to hard.

4) *Structure:* The use of a structure symbol is optional. The structure is indicated by numbers 1 to 16 (or higher, if necessary) with progressively higher numbers indicating a progressively wider grain spacing (more open structure).

5) *Bond or Process:* Bonds are indicated by the following letters: V, vitrified; S, silicate; E, shellac or elastic; R, rubber; RF, rubber reinforced; B, resinoid (synthetic resins); BF, resinoid reinforced; O, oxychloride.

6) *Manufacturer's Record:* The sixth position may be used for manufacturer's private factory records; this is optional.

Composition of Diamond and Cubic Boron Nitride Wheels.—According to American National Standard ANSI B74.13-2016, a series of symbols is used to designate the composition of these wheels. An example is shown below.

Prefix	Abrasive	Grain Size	Grade	Concentration	Bond Type	Bond Modification	Depth of Abrasive	Manufacturer's Identification Symbol
M	D	120	R	100	B	56	⅛	*

Designation Symbols for Composition of Diamond and Cubic Boron Nitride Wheels

The meaning of each symbol is indicated by the following list:

1) *Prefix:* The prefix is a manufacturer's symbol indicating the exact kind of abrasive. Its use is optional.

2) *Abrasive Type:* The letter (B) is used for cubic boron nitride and (D) for diamond.

3) *Grain Size:* The grain sizes commonly used and varying from coarse to very fine are indicated by the following numbers: 8, 10, 12, 14, 16, 20, 24, 30, 36, 46, 54, 60, 70, 80, 90, 100, 120, 150, 180, and 220. The following additional sizes are used occasionally: 240, 280, 320, 400, 500, and 600. The wheel manufacturer may add to the regular grain number an additional symbol to indicate a special grain combination.

4) *Grade:* Grades are indicated by letters of the alphabet from A to Z in all bonds or processes. Wheel grades from A to Z range from soft to hard.

5) *Concentration:* The concentration symbol is a manufacturer's designation. It may be a number or a symbol.

6) *Bond:* Bonds are indicated by the following letters: B, resinoid; V, vitrified; M, metal.

7) *Bond Modification:* Within each bond type a manufacturer may have modifications to tailor the bond to a specific application. These modifications may be identified by either letters or numbers.

8) *Abrasive Depth:* Abrasive section depth, in inches or millimeters (inches illustrated), is indicated by a number or letter which is the amount of total dimensional wear a user may expect from the abrasive portion of the product. Most diamond and CBN wheels are made with a depth of coating on the order of $^1/_{16}$ in., $^1/_8$ in., or more as specified. In some cases the diamond is applied in thinner layers, as thin as one thickness of diamond grains. The L is included in the marking system to identify a layered-type product.

9) *Manufacturer's Identification Symbol:* The use of this symbol is optional.

Table 6. Conventional Abrasives—Grinding Wheel Recommendations

Characteristics	Recommendations
Alnico	
Offhand	23AC36–N5B5
Cylindrical	3SGP60–JVS or 53A60–J8V127
Surfacing (Straight Wheel)	3SGP60–IVS or 86A60–H10VH
Surfacing (Segments)	86A46–D12VBEP
Centerless	57A60–K8VCN or 53A60–K8VCN
Internal	32A60–J6VBE
Aluminum	
Cylindrical	86A54–J8V127, 53A54–J8VBE or 37C54–KVK
Centerless (Hard)	32A46–L7VBE or 86A46–LV127
Centerless (Soft)	37C46–LVK or 23AC46–LB24
Bars	32AC54–QB
Surfacing (Straight Wheel)	37C36–J8V
Surfacing (Segments)	5SG46–E12VSP, 86A46–D12VBEP, or Pacesetter 30E
Internal	37C36–K5V
Mounted Wheels	WNA25
Floor Stands	AC202–Q5B38S
Portable Grinders	AC24–P
Aluminum Alloys	
Cylindrical	37C54–JVK #12 Treat
Bolts (Screws and Studs)	
Cylindrical	64A60–M8V127
Centerless (Shoulder Grinder)	57A60–M8VCN
Brass and Soft Bronze	
Centerless	37C36–LVK
Cylindrical	37C36–KVK
Internal	37C36–K8VK or 37C46–J5V
Surfacing[a] (Straight Wheels)	37C36–J8V
Surfacing[a] (Cylinder, Cups)	37C24–H8V
Surfacing[a] (Segments)	Pacesetter 30G
Snagging (Floor Stands) up to 12,500 SFPM	AC202–Q5B38S
Broaches	
Sharpening	5SG60–LVS or 5SG60–JVSP
Backing Off	5SG46–KVS

Table 6. *(Continued)* **Conventional Abrasives—Grinding Wheel Recommendations**

Characteristics	Recommendations
Bronze (Hard)	
Centerless	57A46–L8VCN, 64A46–MCVE, or AC46–PB24X813
Cylinder, Cups	53A30–G12VBEP
Segments[a]	Pacesetter 30G
Cylindrical	64A46–K8V127 or 57A46–L8VBE
Internal[a]	57A60–LVFL
Portable	AC24–P
Snagging (Floor Stands) up to 12,500 SFPM	AC202–Q5B38S
Cutting off (Dry)	4NZ24–VB65B or 4NZ24–ZBNC
Surfacing (Straight Wheels)	53A36–K8VBE
Bushings (Hardened Steel)	
Hardened Steel	
Centerless	57A60–L8VCN or 86A60–L8VCN
Cylindrical	3SGP60–LVS, 86A60–KV8127, or 23A60–L5VBE
Internal	53A60–K6VBE
Bronze, Centerless	37C46–OVK
Cast Iron	
Cylindrical	3A46–J8BVE, 86A46–I8V127, or 37C46–KVK
Internal	37C46–J5V or 32A60–K6VBE
Toolroom	32A60–H8VBE or 39C60–I8VK
Surface (Dry)	38A46–H8VBE
Cast Iron	
Cam Grinding	
Roughing	3SGP60–L10VH or 57A54 –L8V127
Finishing	57A80–L8V127
Dual Cycle	57A60–M8V128
Regrinding	57A54–L8V127
Crankshaft Grinding	
Pins	86A60–NVS
Bearings	86A60–MVS
Center Thrust Bearing	86A60–MVS
Centerless	3SG46–T23B80, 37C46–LVK, 64A60–LVCE, 57A54–K8VCN, or 32AC54–QB
Cylindrical	37C46–JVK, 86A46–I8V127, or 32A46–J8VBE
Internal[a]	37C46–J5V or 53A60–JVFL
Offhand (Rough Blending), Mounted Wheels	A36–SB or 3NZG36–WB25
Surfacing	
Cylinders, Cups (Ductile, Gray)	53A30–G12VBEP
Cylinders, Cups (Chilled)	37C24–H8V
Cylinders, Cups (Ni Hard)	53A30–G12VBEP
Segments[a], Ductile, Gray, Ni Hard	Pacesetter 30G
Snagging	
Floor Stands, up to 12,500 SFPM	
Light Pressure	4ZF1634–Q5B38S
Heavy Pressure	4ZF1434–R5B38S
Swing Frame, up to 12,500 SFPM	
Light Pressure	4ZF1634–R5B38S
Heavy Pressure	4ZF1234–R5B38S
Portable Grinder	
Type 01, up to 9500 SFPM	4NZ1634–R5BSLX348
Type 06 & 11	4NZ1634–R5BX348
Chrome Plating	
Internal (Small Parts)	37C80–KVK, 5SGG80–KVS, or 32A100–JVFL
Internal (Large Parts)	3SG80–KVS, 32A80–112VBEP, or 53A80–K6VBE
Surfacing (Straight Wheels)	32A80–I8VBE, 5SG80–IVS, or 3SG80–GVSP
Cylindrical (Commercial Finish)	3SGP80–JVS or 53A80–J8V127
Cylindrical (Good Commercial Finish)	AI50–K5E
Cylindrical (High Finish Reflective)	37C500–I984

Table 6. *(Continued)* **Conventional Abrasives—Grinding Wheel Recommendations**

Characteristics	Recommendations
Copper	
Cylindrical	37C60–KVK
Cylindrical (Cups and Cylinders)	37C16–JVK
Copper Alloys	
Cylindrical	37C46–KVK
Surfacing (Horizontal Spindle)	39C36–I8V #12 Treat
Surfacing (Vertical Spindle)	
Roughing	57AC46–JB24
Finishing	57AC60–JB24
Knives, Carbon Steel and Stainless Steel	
Butcher, Hemming and Klotz Machines	53A120–OP1
Kitchen, Hemming Machines	53A801–UP1
Hollow Grinding	A60–F2RR
Cylinders (Air Craft) Internal[a]	
Molybdenum Steel	
Roughing	53A80–JVFL or 5TG120–JVFL
Finishing	32A100–JVFL or 53A100–JVFL
Regrinding	5TG120–JVFL
Nitrided	
Before Nitriding	37C80–I5V
After Nitriding	32A80–JVFL
Regrinding	37C80–J5V
Dies (Blanking and Drawing) Internal[a]	
Carbon Steel	5TG120–KVFL or 53A80–KVFL
High-Carbon, High-Chrome	3SG80–KVS or 53A80–K6VBE
Die Forging	
Offhand–Portable Grinding Mounted Points and Wheels	
Coarse	5SG60–PVS or 38A80–PVME
Medium	5SG90–QVS or 38A90–QVME
Fine	5SG120–SVS or 38A120–QVM
Straight Wheels, Roughing	
5000 – 6500 SFPM	23A46–QVBE
7000 – 9500 SFPM	A36–Q2BH
Dies (Steel) Drawing	
Surfacing–(Hardened)	
Straight Wheels (Dry)	5SG60–GVSP, 5SG60–IVS, 32A60–F12VBEP, 32AA60–HVTRP, or 32A46–H8VBE
Straight Wheels (Fast, Traverse, Wet)	5SG60–IVS, 32A60–I8VBE, or 32AA60–IVTR
Cup Wheels (Wet)	38A46–G8VBE
Segments[a]	5SG46–DVSP
Surfacing–(Annealed)	
Straight Wheels (Dry)	5SG60–JVS, 5SG60–HVSP, or 32AA60–IVTRP
Cup Wheels (Wet)	32A24–H8VBE
Segments[a]	86A30–F12VBEP or 5SG30–FVSP
Drills (Manufacturing)	
Cylindrical	57A60–L8V127
Centerless (Soft)	57A60–M8VCN
Centerless (Hard)	53A60–L8VCN or 57A60–L8VCN
Fluting	57A1001–UB467
Pointing	57A1003–T9BX340
Grinding Relief	57A100–R4R30
Drills (Resharpening)	
$^1/_4''$ and smaller	
Machine	5SG100–IVS
Offhand	57A80–L5VBE
$^1/_4''$ to 1″	
Machine	5SG54–LVS

Table 6. *(Continued)* **Conventional Abrasives—Grinding Wheel Recommendations**

Characteristics	Recommendations
Drills (Resharpening) (Continued)	
Offhand	5SG60–LVS
1″ and larger—Machine	5SG46–HVSP
Winslowmatic Machine	
Web Thinning	23A60–L7B5
Pointing	23A70–M7B5 or 5SGP80–KVSB
Hi-Production	
5 hp Machine	57A1003–R9BX340
30 hp Machine	57A1003–T9BX340
Ductile Iron	
General Reinforced Cutoff	U57A244–VB65B or U57A244 –TBNC
Surfacing (Segments)	See *Cast Iron*
Fasteners (Steel)	
Centerless	57A80–M8VCN
Forgings	
Centerless	57A60–M8VCN
Cylindrical	64A54–L8V127 or 57A54–M8VBE
Gages	
Plug	
Cylindrical	64A80–J8V127 or 57A80–K8VBE
Cylindrical, High-Finish	37C500–J9E
Thread	
Threads, 12-pitch and coarser	32A100–K8VBE or 32A100–KBVH
Threads, 13–20-pitch	32A120–K8VBE or 32A120–L8VH
Threads, 24-pitch and finer	32A180–N9VG or 32A180–N10VH
Ring	
Internal (Roughing)	5TG120–KVFL or 63A80–LVFL
Internal (Finishing)	32A120–JVFL
Internal (Fine Finishing)	37C320–J9E
Gears	
Case Hardened, Precut	
18–20 DP	A120–K8BL or 32A120–K9VG
5–18 DP	A80–I8BL or 32A60–J8VG
2–5 DP	32A60–J8VG
Case Hardened from a Solid, 18 DP or finer	A120–K8BL
Cast Iron, cleaning between teeth (offhand)	37C24–T6R30
Hardened Steel	
Internal[a]	3SG60–KVS or 53A60–K6VBE
Surfacing (Cups and Cylinders)	32A36–I8VBE
Surfacing (Segments)[a]	86A36–E12VBEP
Surfacing (Straight Wheels)	5SG60–JVS, 3SGP60–JVS, 5SG60–H12VSP, 3SG60–H12VSP, or 32A46–J8VBE
Hastelloy	
Surfacing	
Straight Wheel	86A46–G10VH or 32A60 –E25VCP
Straight Wheel (Creep Feed)	38A80–E19VCF2 or 38A80–F16VCF2
Segments[a]	5SG46–EVSP
Internal[a]	5TG120–KVFL or 32A80–KVFL
Cylindrical	5SGP80–JVS or 86A80–J8V127
Centerless	53A60–J8VCN or 57A54–K8VCN
Inconel or Inconel X (with heavy-duty soluble or straight oil)	
Surfacing	
Straight Wheel	3SGP60–H10VH, 32A60–F19VCP, 32AA60–IVTR, or 86A60 –H10VH
Straight Wheel (Creep Feed)	38A60–E25VCF2 or 38A602–F25VCF2
Segments[a]	5SG46–EVSP
Form Grinding	3SGP60–J8VH or 53A60–J8VJN
Internal[a]	5TG120–KVFL or 32A80–JVFL

Table 6. *(Continued)* **Conventional Abrasives—Grinding Wheel Recommendations**

Characteristics	Recommendations
Inconel or Inconel X (with heavy-duty soluble or straight oil) (Continued)	
Cylindrical	3SGP60–I10VH or 86A80–J8V127
Centerless	5SG60–LVS or 57A60–K8VCN
Offhand (Blending Mounted Wheels)	5SG90–QVS or 5SG90–RVH
Thread	38A180–N10VH or 38A180–N9VG
Cutting Off (Dry)	4NZ30–TB65W
Cutting Off (Wet)	A461–P4R55
Jet Blades	
Form Grinding	38A602–F16VCF2
Aerospace Alloys, Cutting Off Investment Casting Gates & Risers	
Chop Stroke	90A244–VB97B
Locked Head–Push Thru	90A244–VB97N
General Industrial	4NZ30–TB65N
Lapping (General Purpose)	
Aluminum	39C280–JVX142C
Brass	37C180–J9V
Cast Iron	37C180–J9V
Copper	39C320–JVX142C
Stainless	39C280–JVX142C
Steel	39C220–I9V
Lawn Mowers	
Resharpening	53A60–M8VBE
Lucite	
Centerless	37C60–MVK
Magnesium Alloys	
Cylindrical	37C60–KVK
Malleable Castings	
Portable Cutoff	U57A244–TB25N
General Reinforced Cutoff	U57A244–TB25N or U57A244 –TBNC
Floor Stands and Swing Frames Up to 12,500 SFPM	
Light Pressure	4ZF1434–Q5B38S
Heavy Pressure	4ZF1434–R5838S
Portable Grinders Type 01, up to 9500 SFPM	4NZ1634–R5BSLX348
Portable Grinders Types 06 & 11	4NZ1634–R5SBX348
Minerals—Lapidary (Offhand)	
Moh's Hardness 7 or less	
Roughing	37C100–NVK
Finishing	37C220–LVK
Moh's Hardness over 7	
Roughing	37C100–MVK
Finishing	37C220–K8V
Cutting off (Wet)	
Molybdenum[a]	
Cylindrical	57A60–K8V127
Surfacing	5SG60–IVS
Surfacing (Segments)	5SG46–DVSP
Monel Metal	
Portable Cutoff	U57A244–TB25N
General Reinforced Cutoff	U57A244–VB65B
Internal	37C60–K6V
Cylindrical	37C60–JVK
Nickel-Based Superalloys	
Surfacing	32A60–E25VCP
Cutting Off (Dry) Chop Stroke	90A244–VB97B
Cutting Off (Dry) Lcked Hd. Push Thru	90A244–VB97N

Table 6. *(Continued)* **Conventional Abrasives—Grinding Wheel Recommendations**

Characteristics	Recommendations
Nickel Plate	
Surfacing (Straight Wheel)	37C60–H8V
Surfacing (Creep Feed)	32A60–D28VCF2
Nickel Rods and Bars	
General Reinforced Cutoff	90A244–VB97B
Ni Hard	
Centerless	53A80–K8VCN
Cylindrical	3SGP80–JVS or 86A80–J8V127
Internal	53A80–K6VBE
Surfacing Wheels	32A46–I8VBE
Surfacing Segments	Pacesetter 30G
Cutting Off (General Reinforced)	U57A244–TBNC or 90A304–RB97B
Nitralloy (Cylindrical)	
Before Nitriding	86A60–K8V127
After Nitriding Commercial Finish	35GP80–JVS or 86A80–J8V127
After Nitriding High-Finish	37C100–IVK
After Nitriding Reflective Finish	37C500–19E
Pipe	
Cast Iron	
Cleaning Inside	4ZF1434–R5B38SL
General Reinforced Cutoff	3NZF244–ZB65N
Steel, Finish Unimportant	
Cutting Off (General Reinforced)	90A244–VB97N
Pipe Balls	
Centerless	57A30–T5VBE
Regrind	57A24–Q5VBE
Pistons	
Aluminum	
Cylindrical	86A46–H8V127 or 53A46–I8V127
Centerless	37C46–KVK
Regrinding	86A46–H8V127
Cast Iron	
Cylindrical	39C46–J8VK or 37C36–KVK
Centerless	37C46–KVK
Regrinding	23A46–I8VBE or 53A46–18V127
Piston Pins	
Centerless Machine Roughing	5SG60–JVS, 32A54–QB, or 57A60–M8VCN
Centerless Machine Semi-finishing	57A70–RB24X813 or 57A80–M8VCN
Piston Rings	
Cast Iron	
Surfacing Rough (Cylinders)	32A30–H8VBE
Surfacing (Straight Wheels)	32A80–K8VBE or 5SGG80–KVS
Internal (Snagging)	5SGG46–KVS
Plasma Spray—Carbides, Chrome	
Centerless	
Roughing and Finishing	39C80–H8VK
Finishing	37C80–PB24
Plastics	
Cylindrical (Thermoplastics)	
Wet	37C46–JVK or 32A46–I12VBEP
Dry	37C36–I5B
Thermosetting	37C30–I5B
Surfacing (Straight Wheel), Thermoplastic	37C46–JVK
Nylon	
Centerless	37C46–KVK or 37C46–LVK
Surfacing	23A36–L8VBE
Plexiglass	
Cutoff (Wet)	37C60–M4R55

Table 6. *(Continued)* **Conventional Abrasives—Grinding Wheel Recommendations**

Characteristics	Recommendations
Plastics (Continued)	
Surfacing	38A46–H12VBEP
Polystyrene, Centerless	37C46–KVK
Propeller Hubs (Cone Seals) Internal	
Rough and Finish	38A60–K6VBE
Fine Finish	A120–M2R30
Pulleys (Cast Iron)	
Cylindrical	37C36–JVK
Rails	
Surfacing, Welds up to 9500 SFPM	
Cup Wheels	4NZ1634–R5BX348
Straight Wheels	4NZ1634–R5BSX348
Removing Corrugations	4NZ1634–R5BSX348
Reamers	
Backing Off	32A46–K5VBE or 5SG46–K6VH
Cylindrical	57A60–L8VBE
Rene	
Surfacing (Form Grinding)	5SG60–JVS, 3SG60–J10VH, or 53A60–J8VJN
Straight Wheel (Creep Feed)	38A80–F19VCF2
Cutting Off	90A244–RB97B
Rods (Centerless)	
Miscellaneous Steel	57A60–M8VCN or 32A54–QB
300 Series Stainless	37C54–NVK, 86A60–L8V127, 53A60–L8VCN, or 32AC54–QB
Nitralloy (Before Nitriding)	57A60–L8VCN
Silichrome Steel	57A60–M8VCN or 32AC54–SB
Brass and Bronze	37C60–KVK
Hard Rubber	37C30–KVR
Carbon	37C36–NVK
Plastic	32A80–N7VBE
Roller Bearing Cups	
Centerless O.D.	57A60–M8VCN or 64A60–NVCE
Internal	5TG120–KVFL or 53A80–LVFL
Rollers for Bearings	
Rollers (Cylinders)	
Small	57A100–RB24
Large	57A80–NB24
Fine Finish	A100–R2R30
Rollers (Needle)	
Up to 1/8″ diameter	57AC120–TB24
Over 1/8 to 3/8″ diameter	5780–QB17X344
Rotors (Laminated) (Cylindrical)	
Roughing	86A100–H8V127 or 57A100–I8VBE
Finishing	37C500–G9E
Rubber (Soft)	
Cylindrical (Dry)	23A20–K5B7 or 32A46–G12VBEP
Rubber (Hard)	
Cylindrical	37C36–J5V
Scissors and Shears	
Cast Iron, Surfacing Sides of Blades	37C100–S8V
Steel, Resharpening, Small Wheels	32A120–M7VBE
Steel, Resharpening, Large Wheels	57A901–MV5
Shafts (Centerless)	
Pinion	57A60–L8VCN or 32A54–QB
Spline	57A60–M8VCN
Shear Blades (Power Metal Shears)	
Sharpening (Segments)	23A30–H8VBE
Sharpening (Cylinders)	23A30–G8VBE

Table 6. *(Continued)* **Conventional Abrasives—Grinding Wheel Recommendations**

Characteristics	Recommendations
Spline Shafts	
Centerless	57A60–M8VCN or 64A60–NVCE
Cylindrical	86A60–M8V 127
Grinding Splines	23A60–L5VBE
Steel Castings (Low Carbon)	
Cutting Off (Reinforced)	90A244–TB97B or U57A244–XBNC
Floorstands up to 12,500 SFPM Light Pressure	4ZF1434–Q5B38S
Floorstands up to 12,500 SFPM Heavy Pressure	4ZF1434–R5B38S
Portable Grinders Type 01	4NZ1634–R5BSX348
Portable Grinders Types 06 & 11	4NZ1634–R5BX348
Steel Castings (Manganese)	
Floorstands up to 12,500 SFPM Light Pressure	4ZF1634–Q5838S
Floorstands up to 12,500 SFPM Heavy Pressure	4ZF1434–R5B38S
Portable Grinders Type 01	4NZ1634–R5BSX348
Portable Grinders Types 06 & 11	4NZ1634–R5BX348
Portable Internal—Rough Grinding up to 9500 SFPM	Gemini
General Reinforced Cutoff	90A244–TB97B or U57A244–XBNC
Steel Forgings (Disc)	
Small—Light Work	23A16–JB14
Large—Heavy Work	23A30–QB14
Steel (Hard) (Rc 45 and harder)[b]	
Centerless (Fine Finish)	A120–P4R30
Centerless (Commercial Finish)	53A60–K8VCN
Centerless (Feed Wheel)	A8O–RR51
Cylindrical Parts smaller than 1″ diameter	3SGP80–JVS or 86A80–J8V127
Cylindrical Parts 1″ diameter & larger	35GP60–JVS or 86A60–J8V127
Internal	5TG120–KVFL or 53A80–KVFL
Surfacing (Straight Wheels)	5SG60–GVSP, 5SG46–IVS, 35GP60–JVS, 32A46–IVTR, 32A46–I8VBE, 86A60–F25VCP, or 32AA46–HVTRP
Surfacing (Segments)[a] broad area of contact	5SG30–EVSP, 86A30–EL2VBEP, or Pacesetter 30G
Surfacing (Segments)[a] medium area of contact	5SG30–FVSP, 86A30–F12VBEP or Pacesetter 30F
Surfacing (Segments)[a] narrow area of contact	5SG30–GVSP, 86A30–G12VBEP or Pacesetter 30G
Surfacing (Cylinders)	38A46–G8VBE
Steel (Soft) (Up to Rc 45)	
Portable Cutoff	U57A244–TB25N
General Reinforced Cutoff	90A244–TB97B
Cylindrical 1″ diameter and less	57A60–L8V127
Cylindrical Over 1" diameter	57A54–K8V127
Internal	32A60–KVBE or 53A80–KVFL
Surfacing Straight Wheel	53A36–K8VBE
Surfacing Segments	86A30–F12VBEP, Pacesetter 30G, or 5SG30–GVSP
Steel (High Speed)[c]	
Centerless Commercial Finish	57A60–K8VCN
Centerless Fine Finish	A120–P4R30
Feed Wheel	A8O–RR51 or A80–SR51
Cylindrical 14" and smaller	3SGP60–LVS, 53A60–L5VBE, or 32A46–HI2VBEP
Cylindrical 16" and larger	3SGP60–MVS or 86A60–L8V127
Internal	5TG120–KVFL, 3SG60–KVS, 3SG60–FVSP, or 53A80–JVFL
Surfacing (Straight Wheels)	5SG60–GVSP, 32AA60–HVTRP, 32A60–G25VCF2
Surfacing (Cylinders)	38A46–G8VBE
Surfacing (Segments)	86A46–DI2VBEP or 55G46–EVSP

Table 6. *(Continued)* **Conventional Abrasives—Grinding Wheel Recommendations**

Characteristics	Recommendations
Steel (Stainless 17–4 PH)[d]	
Surfacing Straight Wheel	3SGP6O–IVS, 32A60–I8VBE, 32A60–F25VCP, or 37C60–JVK
Internal	23A60–K6VBE or 37C60–K6V
Cylindrical	86A60–J8V127, 37C54–KVK, or 57A60–K5VBE
Centerless	57A60–K8VCN
Steel (Stainless — 300 Series)	
Centerless	53A54–K8VCN or 64A60–KVCE
Centerless (Feed Wheel)	A80–RR51
Cylindrical	37C54–JVK or 86A54–18V127
Internal	37C46–JVK
Offhand (Rough Blending) Mounted Wheels	4NZ36–WB25 or 3NZG36–WB25
Surfacing (Straight Wheels)	5SG60–IVSP, 3SGP60–IVSP, 32A46–J8VBE, or 32AA46–JVTR
Surfacing (Creep Feed)	32A80–E19VCF2 or 39C80–F24VCC
Surfacing (Cups)	38A46–I8VBE
Surfacing (Cylinders)	32A46–G8VBE
Surfacing (Segments)	86A46–D12VBEP, 5SG46–DVSP, or 57AC46–FB17
Surfacing (Form Grinding)	53A60–J8VJN
Steel (Stainless steal — 400 Series Hardened)[d]	
Centerless Commercial Finish	57A60–K8VCN or 64A60–LVCE
Centerless Fine Finish	A120–P4R30
Centerless Feed Wheel	A80–RR51 or ASO–SR51
Cylindrical Small Wheel	53A60–K8VBE
Cylindrical Large Wheel	86A60–J8V127
Internal	5TG120–KVFL or 53A80–KVFL
Offhand (Rough Blending) Mounted Wheels	4NZ36–UB25 or 3NZG36–UB25
Surfacing (Straight Wheels)	5SG60–IVS, 32A46–IVS, or 32AA46–IVTR
High-Speed	5SG60–IVS or 32A46–H8VBE
Surfacing (Creep Feed)	38A60–F19VCF2 or 86A60–F25VCP
Surfacing (Cylinders)	32A36–G8VBE
Surfacing (Segments)	5SG30–FVSP, 86A30–E12VBEP, or Pacesetter 305
Surfacing (Form Grinding)	3SG60–I10VH, 53A60–I8VJN, or 53A60–J8VJN
Stellite (also Rexalloy, Tantung)	
Cylindrical	3SGP80–JVS, 38A80–J8V127, or 86A80–J8V127
Cutter Grinding	5SG46–JVS, 32A46–J8VBE, or 32AA46–JVTR
Internal	3SG60–JVS or 53A60–J6VBE
Surfacing (Cups and Cylinders)	32A46–G8VBE or 5SG46–IVS
Surfacing (Straight Wheels)	5SG60–IVS, 5SG60–GVSP, 32A46–H8VBE, or 32AA46–IVTR
Tools Offhand	57A46–NSVBE
Tools Machine	5SG46–LVS, 32A46–L8VBE, or 32A46–LVTR
Tantalum	
Cylindrical	86A60–J8V127
Surfacing	23A46–J8VBE
Tappets (Centerless)	
Steel Roughing	57A60–M8VCN
Steel Finishing	57A80–M8VCN
Cast Iron Roughing	37C46–NVK
Cast Iron Finishing	37C80–MVK
Taps	
Fluting (Taps)	57A1003–UB354
Grinding Relief	5SG60–KVS or 32A60–K8VBE
Squaring Ends	32A801–Q8B5 or 5SG80–JVS
Shanks (Cylindrical)	5SG80–LVS or 57A80–L8V127
Titanium Precision Grinding	
Surfacing, Straight Wheel, Rust Inhibitor Coolant	
2000 SFPM	32A60–L8VBE or 5SGG60–LVS
5500 SFPM	5SG60–JVS, 39C60–J8VK, or 5SGG60–JVS

Table 6. *(Continued)* **Conventional Abrasives—Grinding Wheel Recommendations**

Characteristics	Recommendations
Titanium Precision Grinding (Continued)	
Vertical Spindle	39C80–I8VK
Offhand (Blending) Mounted Wheels	5SG60–QVS or TG60–QVH
Cylindrical	37C60–JVK
Centerless	37C54–LVK or 37C54–PB24
Creep Feed	39C46–G24VX530
Tools — Single-Point Carbon and High-Speed Steel	
Offhand Grinding	
Bench and Pedestal Grinders	
Coarse	57A36–O5VBE or General Purpose Coarse
Fine	57A60–M5VBE or General Purpose Fine
Combination (Roughing and Finishing)	57A46–N5VBE or General Purpose Medium
Wet Tool Grinders	
12" to 24" diameter wheels	57A36–O5VBE
Over 24" diameter wheels	57A24–M5VBE
Machine Grinding	
Straight Wheels	
15" diameter wheels	23A46–L5VBE
24" diameter wheels	23A24–M5VBE
Cup or Cylinder Wheels	38A46–K5VBE
Tungsten	
Cylindrical, Rolled Tungsten	86A54–K8V127
Cylindrical, Sintered Tungsten	37C60–JVK
Centerless, Rolled Tungsten	32A46–N5VBE
Centerless, Sintered Tungsten	37C601–KVK
Internal	5SG60–IVS
Surfacing 2000 SFPM	23A46–J8VBE
Surfacing 5000 SFPM	37C46–J8V
Udimet	
Surfacing (Form Grinding)	3SG60–J8VH, 5SG60–JVS, or 53A60–J8VJN
Valves (Automotive)	
Refacing	37C80–NVK or 57A80–J5VBE
Stems Centerless	3TG120/3– P8VH, 3SGP70–OVH, or 57A60–M8VCN
Waspalloy (with Straight Oil)	
Surfacing (Form Grinding)	53A60–J8VJN, 5SG60–JVS or 3SG60–K8VH
Surfacing (Straight Wheels)	5SG60–JVS, 3SGP60–JVS, 32A60–F19VCP, 32A46–IBVBE, or 32AA46–IVTR
Vertical Spindle	32A36–E19VBEP or 32A46–E19VCP
Internal	3SG60–JVS or 32A60–J6VBE
Cylindrical	3SGP60–JVS, 86A60–J8VBE, or 53A60–J8VBE
Centerless	53A60–J8VCN or 86A60–JV127
Cutting Off (General Reinforced)	90A244–TB97B
Welds (Carbon Alloy Steels)	
Portable Grinders Type 01 (to 9,500 SFPM)	4NZ1634–Q5BSX348
Portable Grinders Type 27 (to 16,000 SFPM)	NORZON
Stainless Steel Offhand (Rough Blending) Mounted Wheels	4NZ30–WB25 or 3NZG30–WB25
Portable Grinders Type 01 (9,500 SFPM)	4NZ1634–Q5BSX348
Portable Grinders Type 27 (to 16,000 SFPM)	NORZON

[a] Use of wax stick recommended

[b] CBN wheels are recommended for hard steel where tolerance, productivity, and/or problems exist with conventional abrasives

[c] CBN wheels have successfully ground HSS under certain conditions

[d] CBN wheels successfully grind stainless under certain conditions

In addition to the abrasive specifications in this table, so-called super abrasives are available for special applications involving difficult to grind materials and where specific surface finish requirements must be met. Consult grinding wheel manufacturers for specific recommendations concerning super abrasives.

American National Standard Shapes and Sizes of Grinding Wheels.—ANSI Standard B74.2-2003 includes shapes and sizes of grinding wheels and gives a wide variety of grinding wheel shape and size combinations suitable for the majority of applications. Although grinding wheels can be manufactured to shapes and dimensions different from those listed, it is advisable, for reasons of cost and inventory control, to avoid using special shapes and sizes, unless technically warranted.

Standard shapes and size ranges as given in this Standard together with typical applications are shown in Table 7a for inch dimensions and in Table 7b for metric dimensions.

Table 7a. Standard Shapes and Inch Size Ranges of Grinding Wheels *ANSI B74.2-2003*

Applications	Size Ranges of Principal Dimensions, Inches		
	D = Dia.	*T* = Thick.	*H* = Hole
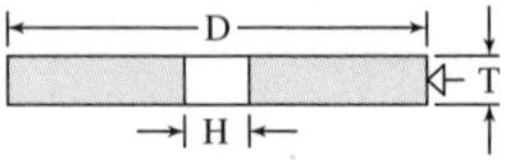	**Type 1. Straight Wheel** For peripheral grinding.		
Cylindrical Grinding			
Between centers	12 to 48	$\frac{1}{2}$ to 6	5 to 20
Centerless grinding wheels	14 to 30	1 to 20	5 or 12
Centerless regulating wheels	8 to 14	1 to 12	3 to 6
Offhand Grinding, Grinding on the Periphery			
General purpose	6 to 36	$\frac{1}{2}$ to 4	$\frac{1}{2}$ to 3
For wet tool grinding only	30 or 36	3 or 4	20
Snagging			
Floor stand machines	12 to 24	1 to 3	$1\frac{1}{4}$ to $2\frac{1}{2}$
Floor stand machines (organic bond, wheel speed over 6500 sfpm)	20 to 36	2 to 4	6 or 12
Mechanical grinders (organic bond, wheel speed up to 16,500 sfpm)	24	2 to 3	12
Portable machines	3 to 8	$\frac{1}{4}$ to 1	$\frac{3}{8}$ to $\frac{5}{8}$
Portable machines (reinforced organic bond, 17,000 sfpm)	6 or 8	$\frac{3}{4}$ or 1	1
Swing frame machines	12 to 24	2 to 3	$3\frac{1}{2}$ to 12
Other			
Cutting off, organic bonds only	1 to 48	$\frac{1}{64}$ to $\frac{3}{8}$	$\frac{1}{16}$ to 6
Internal grinding	$\frac{1}{4}$ to 4	$\frac{1}{4}$ to 2	$\frac{3}{32}$ to $\frac{7}{8}$
Saw gumming, F-type face	6 to 12	$\frac{1}{4}$ to $1\frac{1}{2}$	$\frac{1}{2}$ to $1\frac{1}{4}$
Surface grinding, horizontal spindle machines	6 to 24	$\frac{1}{2}$ to 6	$1\frac{1}{4}$ to 12
Tool grinding, broaches, cutters, mills, reamers, taps, etc.	6 to 10	$\frac{1}{4}$ to $\frac{1}{2}$	$\frac{5}{8}$ to 5
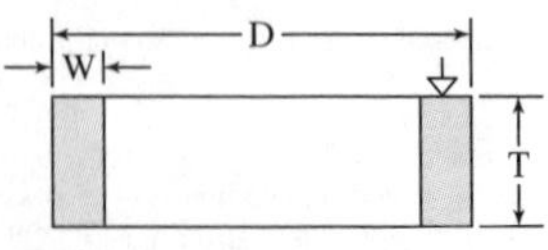	**Type 2. Cylindrical Wheel** Side grinding wheel—mounted on the diameter; may also be mounted in a chuck or on a plate.		
			W = Wall
Surface Grinding, vertical spindle machines	8 to 20	4 or 5	1 to 4

Table 7a. *(Continued)* **Standard Shapes and Inch Size Ranges of Grinding Wheels** *ANSI B74.2-2003*

Applications	Size Ranges of Principal Dimensions, Inches		
	D = Dia.	*T* = Thick.	*H* = Hole
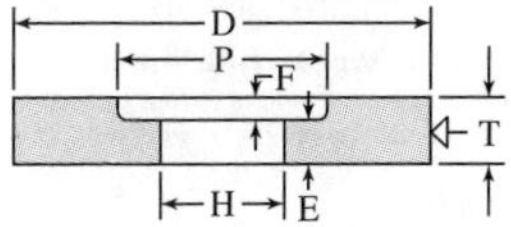	**Type 5. Wheel, Recessed One Side** For peripheral grinding. Allows wider faced wheels than the available mounting thickness, also grinding clearance for the nut and flange.		
Cylindrical grinding, between centers	12 to 36	$1\frac{1}{2}$ to 4	5 or 12
Cylindrical grinding, centerless regulating wheel	8 to 14	3 to 6	3 or 5
Internal grinding	$\frac{3}{8}$ to 4	$\frac{3}{8}$ to 2	$\frac{1}{8}$ to $\frac{7}{8}$
Surface grinding, horizontal spindle machines	7 to 24	$\frac{3}{4}$ to 6	$1\frac{1}{4}$ to 12
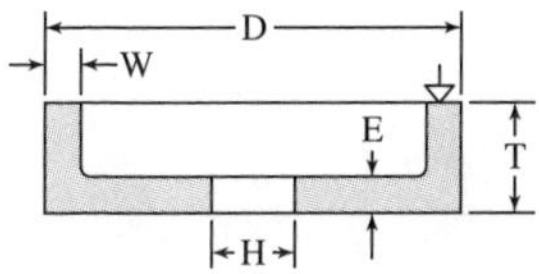	**Type 6. Straight-Cup Wheel** Side grinding wheel, in whose dimensioning the wall thickness (*W*) takes precedence over the diameter of the recess. Hole is $\frac{5}{8}$-11UNC-2B threaded for the snagging wheels and $\frac{1}{2}$ or $1\frac{1}{4}''$ for the tool grinding wheels.		
			W = Wall
Snagging, portable machines, organic bond only	4 to 6	2	$\frac{3}{4}$ to $1\frac{1}{2}$
Tool grinding, broaches, cutters, mills, reamers, taps, etc.	2 to 6	$1\frac{1}{4}$ to 2	$\frac{5}{16}$ or $\frac{3}{8}$
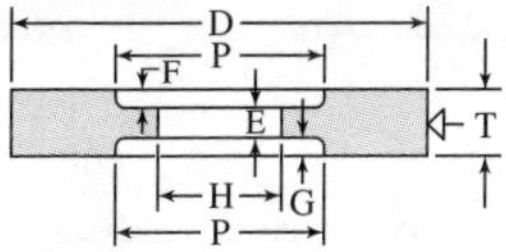	**Type 7. Wheel, Recessed Two Sides** Peripheral grinding. Recesses allow grinding clearance for both flanges and narrower mounting thickness than overall thickness.		
Cylindrical grinding, between centers	12 to 36	$1\frac{1}{2}$ to 4	5 or 12
Cylindrical grinding, centerless regulating wheel	8 to 14	4 to 20	3 to 6
Surface grinding, horizontal spindle machines	12 to 24	2 to 6	5 to 12
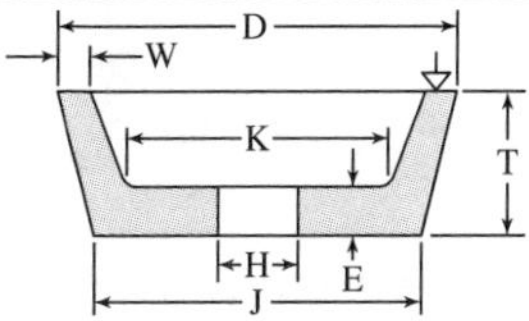	**Type 11. Flaring-Cup Wheel** Side grinding wheel with wall tapered outward from the back; wall generally thicker in the back.		
Snagging, portable machines, organic bonds only, threaded hole	4 to 6	2	$\frac{5}{8}$-11 UNC-2B
Tool grinding, broaches, cutters, mills, reamers, taps, etc.	2 to 5	$1\frac{1}{4}$ to 2	$\frac{1}{2}$ to $1\frac{1}{4}$

Table 7a. *(Continued)* **Standard Shapes and Inch Size Ranges of Grinding Wheels** *ANSI B74.2-2003*

Applications	Size Ranges of Principal Dimensions, Inches		
	D = Dia.	T = Thick.	H = Hole
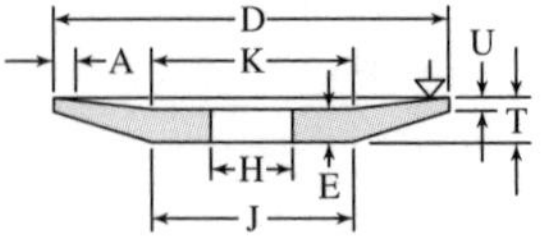	**Type 12. Dish Wheel** Grinding on the side or on the U-face of the wheel, the U-face being always present in this type.		
Tool grinding, broaches, cutters, mills, reamers, taps, etc.	3 to 8	$\frac{1}{2}$ or $\frac{3}{4}$	$\frac{1}{2}$ to 1 $\frac{1}{4}$
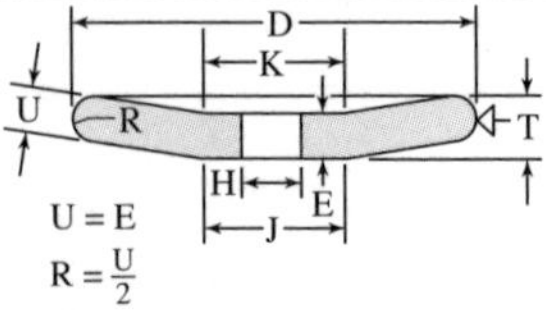	**Type 13. Saucer Wheel** Peripheral grinding wheel, resembling the shape of a saucer, with cross section equal throughout.		
Saw gumming, saw-tooth shaping and sharpening	8 to 12	$\frac{1}{2}$ to 1 $\frac{3}{4}$ *U* & *E* $\frac{1}{4}$ to 1$\frac{1}{2}$	$\frac{3}{4}$ to 1 $\frac{1}{4}$
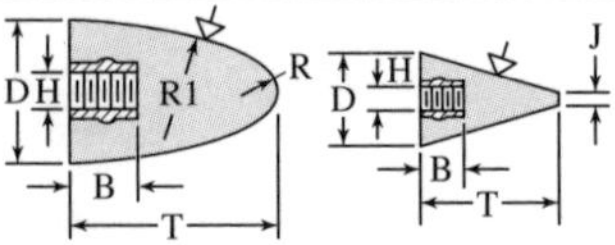	**Type 16. Cone, Curved Side** **Type 17. Cone, Straight Side, Square Tip** **Type 17R. Cone, Straight Side, Round Tip** (Tip Radius $R = J/2$)		
Snagging, portable machine, threaded holes	1$\frac{1}{4}$ to 3	2 to 3$\frac{1}{2}$	$\frac{3}{8}$-24UNF-2B to $\frac{5}{8}$-11UNC-2B
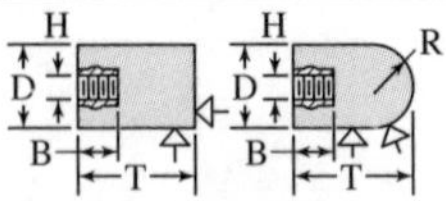	**Type 18. Plug, Square End** **Type 18R. Plug, Round End** $R = D/2$		
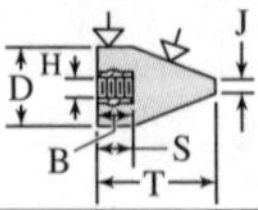	**Type 19. Plugs, Conical End, Square Tip** **Type 19R. Plugs, Conical End, Round Tip** (Tip Radius $R = J/2$)		
Snagging, portable machine, threaded holes	1$\frac{1}{4}$ to 3	2 to 3$\frac{1}{2}$	$\frac{3}{8}$-24UNF-2B to $\frac{5}{8}$-11UNC-2B
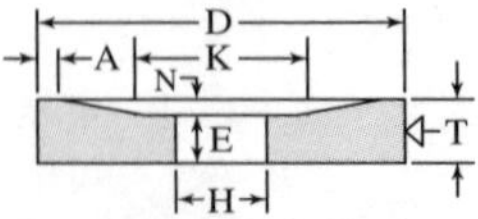	**Type 20. Wheel, Relieved One Side** Peripheral grinding wheel, one side flat, the other side relieved to a flat.		
Cylindrical grinding, between centers	12 to 36	$\frac{3}{4}$ to 4	5 to 20

Table 7a. *(Continued)* **Standard Shapes and Inch Size Ranges of Grinding Wheels** *ANSI B74.2-2003*

Applications	Size Ranges of Principal Dimensions, Inches		
	D = Dia.	T = Thick.	H = Hole
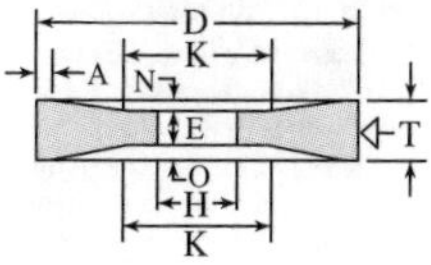	**Type 21. Wheel, Relieved Two Sides** Both sides relieved to a flat.		
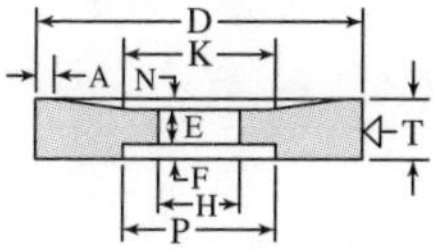	**Type 22. Wheel, Relieved One Side, Recessed Other Side** One side relieved to a flat.		
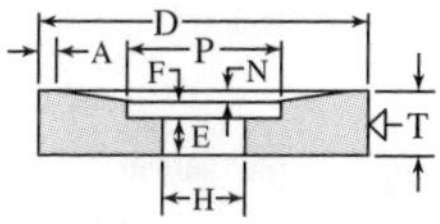	**Type 23. Wheel, Relieved and Recessed Same Side** The other side is straight.		
Cylindrical grinding, between centers, with wheel periphery	20 to 36	2 to 4	12 or 20
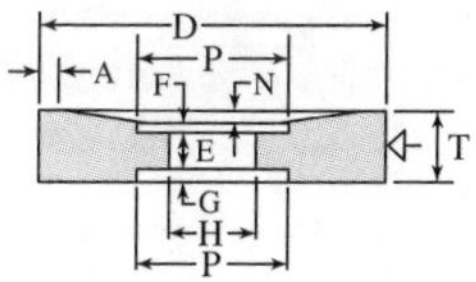	**Type 24. Wheel, Relieved and Recessed One Side, Recessed Other Side** One side recessed, the other side is relieved to a recess.		
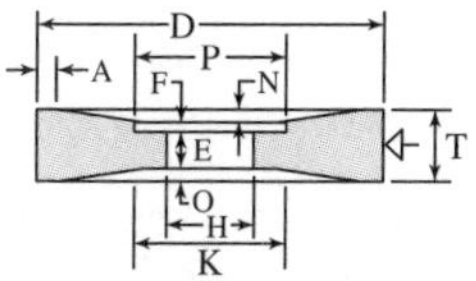	**Type 25. Wheel, Relieved and Recessed One Side, Relieved Other Side** One side relieved to a flat, the other side relieved to a recess.		
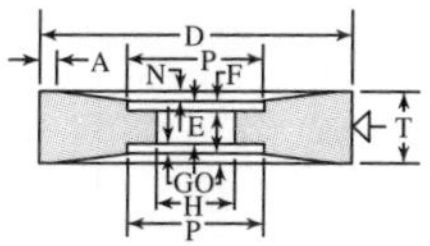	**Type 26. Wheel, Relieved and Recessed Both Sides**		
Cylindrical grinding, between centers, with the periphery of the wheel	20 to 36	2 to 4	12 or 20

Table 7a. *(Continued)* **Standard Shapes and Inch Size Ranges of Grinding Wheels** *ANSI B74.2-2003*

Applications	Size Ranges of Principal Dimensions, Inches		
	D = Dia.	T = Thick.	H = Hole
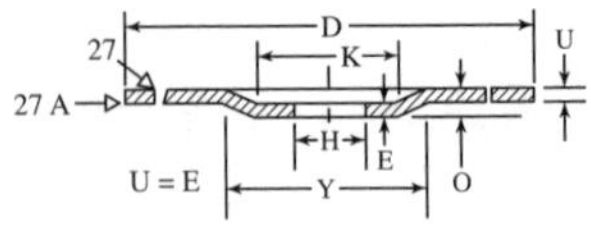	**Types 27 & 27A. Wheel, Depressed Center** 27. *Portable grinding:* Grinding normally done by contact with work at approx. a 15° angle with face of the wheel. 27A. *Cutting off:* Using the periphery as grinding face.		
Cutting off, reinforced organic bonds only	16 to 30	$U = E = \frac{5}{32}$ to $\frac{1}{4}$	1 or 1 $\frac{1}{2}$
Snagging, portable machine	3 to 9	U = Uniform thick. $\frac{1}{8}$ to $\frac{3}{8}$	$\frac{3}{8}$ or $\frac{7}{8}$
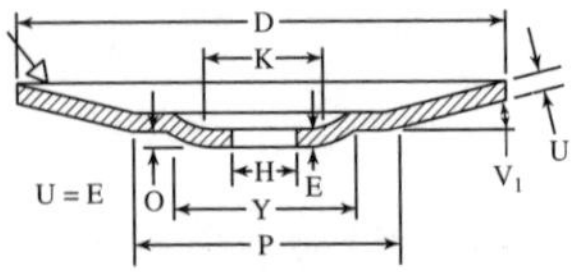	**Type 28. Wheel, Depressed Center (Saucer Shaped Grinding Face)** Grinding at approx. 15° angle with wheel face.		
Snagging portable machine	7 or 9	U = Uniform thickness $\frac{1}{4}$	$\frac{7}{8}$

Throughout Table 7, large open-head arrows indicate grinding surfaces.

Table 7b. Standard Shapes and Metric Size Ranges of Grinding Wheels *ANSI B74.2-2003*

Applications	Size Ranges of Principal Dimensions, Millimeters		
	D = Diam.	T = Thick.	H = Hole
Type 1. Straight Wheel			
Cutting off (nonreinforced and reinforced organic bonds only)	150 to 1250	0.8 to 10	16 to 152.4
Cylindrical grinding, between centers	300 to 1250	20 to 160	127 to 508
Cylindrical grinding, centerless grinding wheels	350 to 750	25 to 500	127 or 304.8
Cylindrical grinding, centerless regulating wheels	200 to 350	25 to 315	76.2 to 152.4
Internal grinding	6 to 100	6 to 50	2.5 to 25
Offhand grinding on the periphery			
— General purpose	150 to 900	13 to 100	20 to 76.2
— For wet tool grinding only	750 or 900	80 or 100	508
Saw gumming (F-type face)	150 to 300	6 to 40	32
Snagging, floor stand machines	300 to 600	25 to 80	32 to 76.2
Snagging, floor stand machines (organic bond, wheel speed over 33 meters per second)	500 to 900	50 to 100	152.4 or 304.8

Table 7b. *(Continued)* **Standard Shapes and Metric Size Ranges of Grinding Wheels** *ANSI B74.2-2003*

Applications	Size Ranges of Principal Dimensions, Millimeters		
	D = Diam.	T = Thick.	H = Hole
Snagging, mechanical grinders (organic bond, wheel speed up to 84 meters per second)	600	50 to 80	304.8
Snagging, portable machines	80 to 200	6 to 25	10 to 16
Snagging, swing frame machines (organic bond)	300 to 600	50 to 80	88.9 to 304.8
Surface grinding, horizontal spindle machines	150 to 600	13 to 160	32 to 304.8
Tool Grinding, Broaches, cutters, mills, reamers, taps, etc.	150 to 250	6 to 20	32 to 127
Type 2. Cylindrical Wheel			W = Wall
Surface grinding, vertical spindle machines	200 to 500	100 or 125	25 to 100
Type 5. Wheel, recessed one side			
Cylindrical grinding, between centers	300 to 900	40 to 100	127 or 304.8
Cylindrical grinding, centerless regulating wheels	200 to 350	80 to 160	76.2 or 127
Internal grinding	10 to 100	10 to 50	3.18 to 25
Type 6. Straight-Cup Wheel			W = Wall
Snagging, portable machines, organic bond only (hole is ⅝-11 UNC-2B)	100 to 150	50	20 to 40
Tool grinding, broaches, cutters, mills, reamers, taps, etc. (Hole is 13 to 32 mm)	50 to 150	32 to 50	8 or 10
Type 7. Wheel, recessed two sides			
Cylindrical grinding, between centers	300 to 900	40 to 100	127 or 304.8
Cylindrical grinding, centerless regulating wheels	200 to 350	100 to 500	76.2 to 152.4
Type 11. Flaring-Cup Wheel			
Snagging, portable machines, organic bonds only, threaded hole	100 to 150	50	⅝-11 UNC-2B
Tool grinding, broaches, cutters, mills, reamers, taps, etc.	50 to 125	32 to 50	13 to 32
Type 12. Dish Wheel			
Tool grinding, broaches, cutters, mills, reamers, taps, etc.	80 to 200	13 or 20	13 to 32
Type 27 and 27A. Wheel, depressed center			
Cutting off, reinforced organic bonds only	400 to 750	$U = E = 6$	25.4 or 38.1
Snagging, portable machines	80 to 230	$U = E = 3.2$ to 10	9.53 or 22.23

All dimensions in millimeters.

See Table 7a for diagrams and descriptions of each wheel type.

The operating surface of the grinding wheel is often referred to as the wheel face. In the majority of cases it is the periphery of the grinding wheel which, when not specified otherwise, has a straight profile. However, other face shapes can also be supplied by the grinding wheel manufacturers, and also reproduced during usage by appropriate truing. ANSI B74.2-2003 standard offers 15 different shapes for grinding wheel faces, which are shown in Table 8.

Table 8. Standard Shapes of Grinding Wheel Faces *ANSI B74.2-2003*

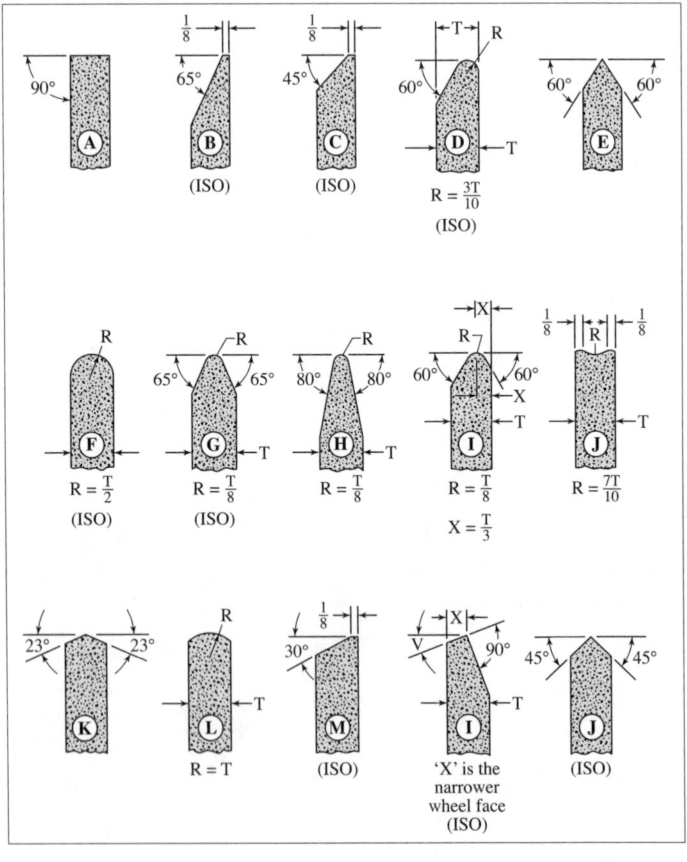

Table 9. Diamond Wheel Core Shapes and Designations *ANSI B74.3-2003 (R2014)*

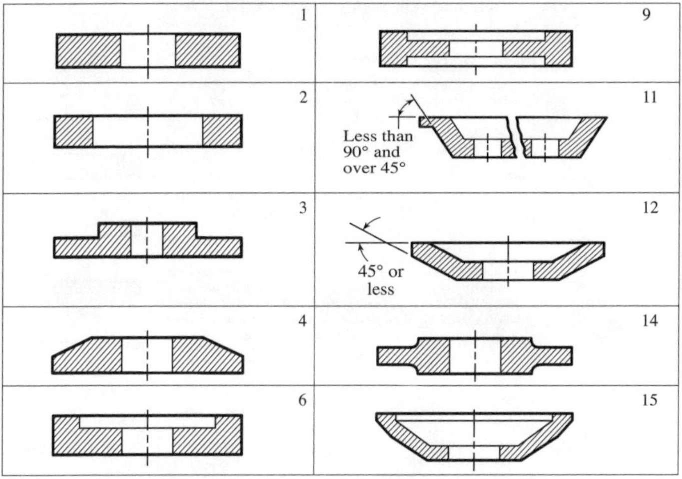

Table 10. Diamond Cross Sections and Designations
ANSI B74.3-2003 (R2014)

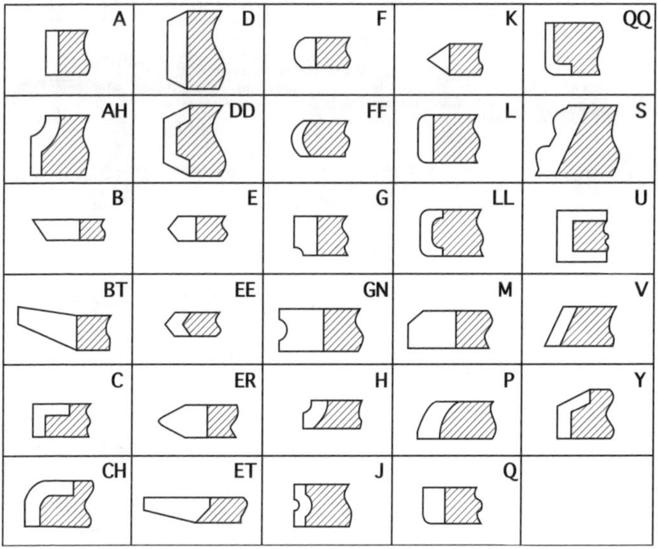

Table 11. Designations for Location of Diamond Section on Diamond Wheel *ANSI B74.3-2003 (R2014)*

Designation No. and Location	Description	Illustration
1 — Periphery	The diamond section shall be placed on the periphery of the core and shall extend the full thickness of the wheel. The axial length of this section may be greater than, equal to, or less than the depth of diamond, measured radially. A hub or hubs shall not be considered as part of the wheel thickness for this definition.	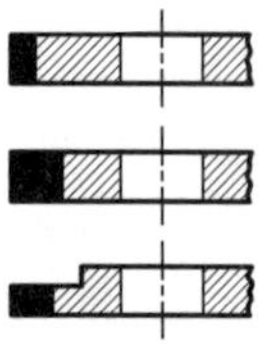
2 — Side	The diamond section shall be placed on the side of the wheel and the length of the diamond section shall extend from the periphery toward the center. It may or may not include the entire side and shall be greater than the diamond depth measured axially. It shall be on the side of the wheel that is commonly used for grinding purposes.	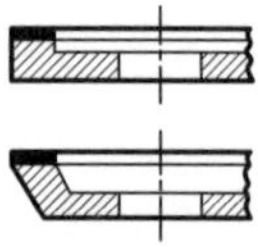
3 — Both Sides	The diamond sections shall be placed on both sides of the wheel and shall extend from the periphery toward the center. They may or may not include the entire sides, and the radial length of the diamond section shall exceed the axial diamond depth.	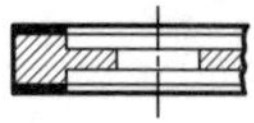
4 — Inside Bevel or Arc	This designation shall apply to the general wheel types 2, 6, 11, 12, and 15 and shall locate the diamond section on the side wall. This wall shall have an angle or arc extending from a higher point at the wheel periphery to a lower point toward the wheel center	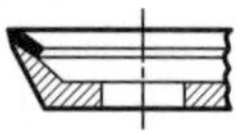
5 — Outside Bevel or Arc	This designation shall apply to the general wheel types, 2, 6, 11, and 15 and shall locate the diamond section on the side wall. This wall shall have an angle or arc extending from a lower point at the wheel periphery to a higher point toward the wheel center.	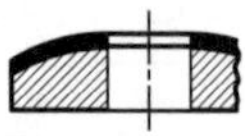
6 — Part of Periphery	The diamond section shall be placed on the periphery of the core but shall not extend the full thickness of the wheel and shall not reach to either side	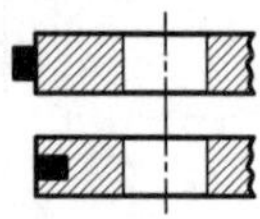
7 — Part of Side	The diamond section shall be placed on the side of the core and shall not extend to the wheel periphery. It may or may not extend to the center.	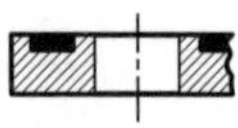
8 — Throughout	Designates wheels of solid diamond abrasive section without cores.	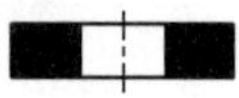
9 — Corner	Designates a location that would commonly be considered to be on the periphery except that thediamond section shall be on the corner but shall not extend to the other corner.	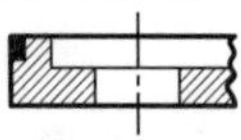
10 — Annular	Designates a location of the diamond abrasive section on the inner annular surface of the wheel.	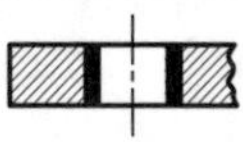

Table 12. Designation Letters for Modifications of Diamond Wheels

ANSI B74.3-2003 (R2014)

Designation Letter	Description	Illustration
B — Drilled and Counterbored	Holes drilled and counterbored in core.	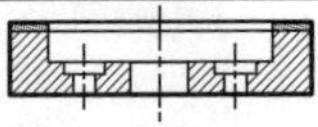 6A2B
C — Drilled and Countersunk	Holes drilled and countersunk in core.	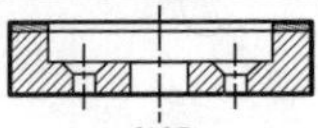 6A2C
F — Hub	Hub on basic wheel cover	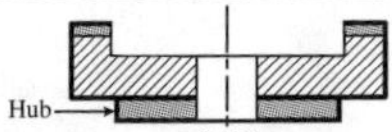 6A2F
H — Plain Hole	Straight hole drilled in core.	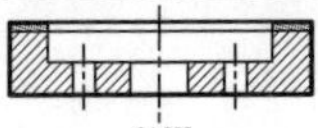 6A2H
J — Recess One Side		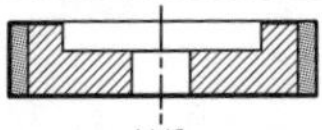 1A1J
JJ — Recess Two Sides		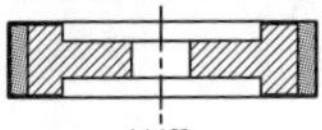 1A1JJ
K — Keyway	Arbor hole with keyway	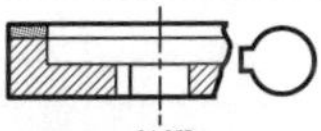 6A2K
M — Holes Plain and Threaded	Mixed holes, some plain, some threaded, are in core.	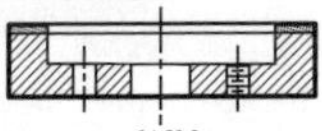 6A2M
N — Nonsteel Core	Nonsteel core for 1A1R and related shapes	
P — Relieved One Side	Core relieved on one side of wheel. Thickness of core is less than wheel thickness.	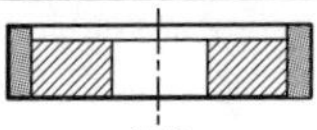 1A1P
R — Relieved Two Sides	Core relieved on both sides of wheel. Thickness of core is less than wheel thickness.	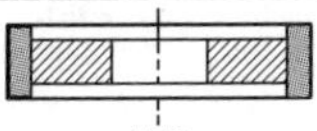 1A1R
S — Segmented-Diamond Section	Wheel has segmental diamond section mounted on core. (Clearance between segments has no bearing on definition.)	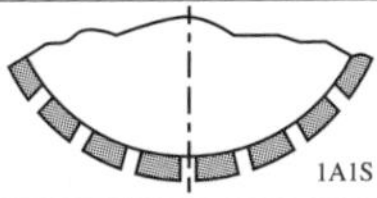 1A1S
SS — Segmental and Slotted	Wheel has separated segments mounted on a slotted core.	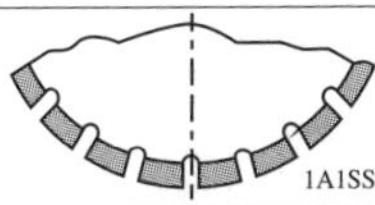 1A1SS

Table 12. *(Continued)* **Designation Letters for Modifications of Diamond Wheels** *ANSI B74.3-2003 (R2014)*

Designation Letter	Description	Illustration
T — Threaded Holes	Threaded holes are in core.	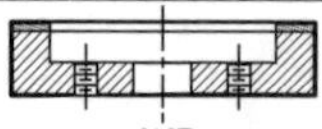 6A2T
Q — Diamond Inserted	Three surfaces of the diamond section are partially or completely enclosed by the core.	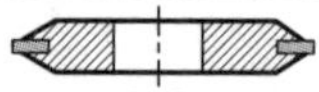 1A6Q
V — Diamond Inverted	Any diamond cross section that is mounted on the core so that the interior point of any angle, or the concave side of any arc, is exposed shall be considered inverted. *Exception:* Diamond cross section AH shall be placed on the core with the concave side of the arc exposed.	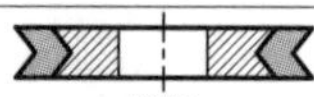 1EE1V 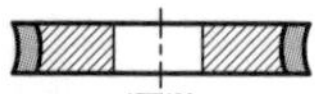 1FF1V
W — Wheels Affixed on a Mandrel		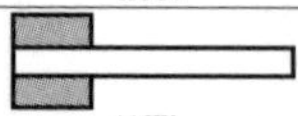 1A8W 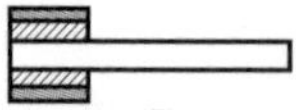 1A1W
Y — Diamond Inserted and Inverted	See definitions for Q and V.	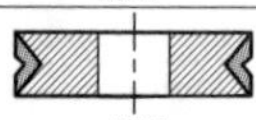 1EE6Y 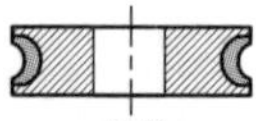 1FF6Y

Table 13. General Diamond Wheel Recommendations for Wheel Type and Abrasive Specification

Typical Applications or Operation	Basic Wheel Type	Abrasive Specification	
Single-Point Tools (offhand grinding)	D6A2C	*Rough:*	MD100-N100-B1/8
		Finish:	MD220-P75-B1/8
Single-Point Tools (machine ground)	D6A2H	*Rough:*	MD180-J100-B1/8
		Finish:	MD320-L75-B1/8
Chip Breakers	D1A1		MD150-R100-B1/8
Multitooth Tools and Cutters (face mills, end mills, reamers, broaches, etc.) Sharpening and Backing Off	D11V9	*Rough:*	MD100-R100-B1/8
		Combination:	MD150-R100-B1/8
		Finish:	MD220-R100-B1/8
Multitooth Tools and Cutters (face mills, end mills, reamers, broaches, etc.) Fluting	D12A2		MD180-N100-B1/8
Saw Sharpening	D12A2		MD180-R100-B1/8
Surface Grinding (horizontal spindle)	D1A1	Rough:	MD120-N100-B1/8
		Finish:	MD240-P100-B1/8
Surface Grinding (vertical spindle)	D2A2T		MD80-R75-B1/8
Cylindrical or Centertype Grinding	D1A1		MD120-P100-B1/8
Internal Grinding	D1A1		MD150-N100-B1/8
Slotting and Cutoff	D1A1R		MD150-R100-B1/4
Lapping	Disc		MD400-L50-B1/16
Hand Honing	DH1, DH2	*Rough:*	MD220-B1/16
		Finish:	MD320-B1/8

Table 14a. Standard Shapes and Inch Sizes of Mounted Wheels and Points

ANSI B74.2-2003

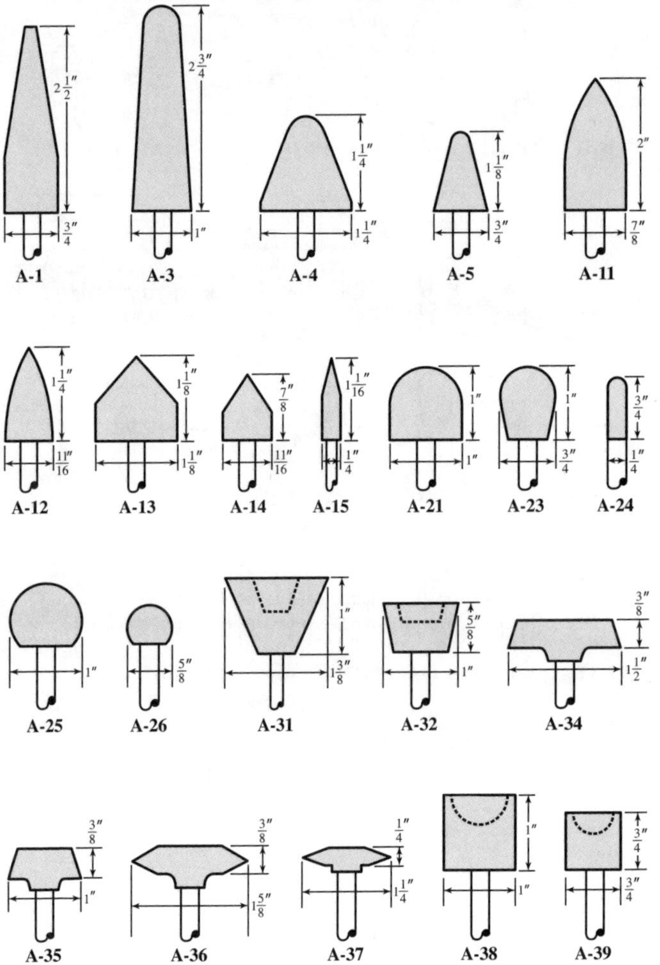

The maximum speeds of mounted vitrified wheels and points of average grade range from about 38,000 to 152,000 rpm for diameters of 1 inch down to $\frac{1}{4}$ inch. However, the safe operating speed usually is limited by the critical speed (speed at which vibration or whip tends to become excessive) which varies according to wheel or point dimensions, spindle diameter, and overhang.

Table 14b. Standard Shapes and Inch Sizes of Mounted Wheels and Points
ANSI B74.2-2003

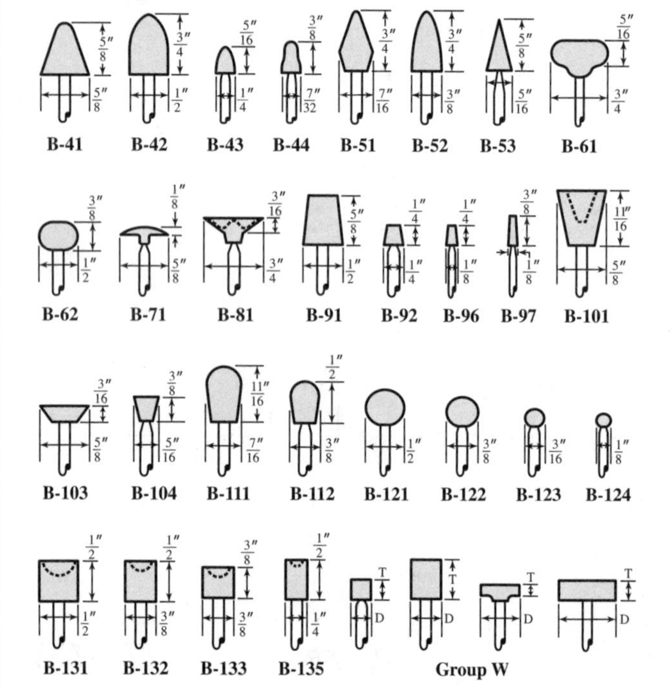

Abrasive Shape No.	Abrasive Shape Size		Abrasive Shape No.	Abrasive Shape Size		Abrasive Shape No.	Abrasive Shape Size	
	D	*T*		*D*	*T*		*D*	*T*
W 144	1/8	1/4	W 182	1/2	1/8	W 208	3/4	2
W 145	1/8	3/8	W 183	1/2	1/4	W 215	1	1/8
W 146	1/8	1/2	W 184	1/2	3/8	W 216	1	1/4
W 152	3/16	1/4	W 185	1/2	1/2	W 217	1	3/8
W 153	3/16	3/8	W 186	1/2	3/4	W 218	1	1/2
W 154	3/16	1/2	W 187	1/2	1	W 220	1	1
W 158	1/4	1/8	W 188	1/2	1 1/2	W 221	1	1 1/2
W 160	1/4	1/4	W 189	1/2	2	W 222	1	2
W 162	1/4	3/8	W 195	5/8	3/4	W 225	1 1/4	1/4
W 163	1/4	1/2	W 196	5/8	1	W 226	1 1/4	3/8
W 164	1/4	3/4	W 197	5/8	2	W 228	1 1/4	3/4
W 174	3/8	1/4	W 200	3/4	1/8	W 230	1 1/4	1 1/4
W 175	3/8	3/8	W 201	3/4	1/4	W 232	1 1/4	2
W 176	3/8	1/2	W 202	3/4	3/8	W 235	1 1/2	1/4
W 177	3/8	3/4	W 203	3/4	1/2	W 236	1 1/2	1/2
W 178	3/8	1	W 204	3/4	3/4	W 237	1 1/2	1
W 179	3/8	1 1/4	W 205	3/4	1	W 238	1 1/2	1 1/2
W 181	1/2	1/16	W 207	3/4	1 1/2	W 242	2	1

Table 15. Standard Shapes and Metric Sizes of Mounted Wheels and Points *ANSI B74.2-2003*

Abrasive Shape No.[a]	Abrasive Shape Size		Abrasive Shape No.[a]	Abrasive Shape Size	
	Diameter	Thickness		Diameter	Thickness
A 1	20	65	A 24	6	20
A 3	22	70	A 25	25	…
A 4	30	30	A 26	16	…
A 5	20	28	A 31	35	26
A 11	21	45	A 32	25	20
A 12	18	30	A 34	38	10
A 13	25	25	A 35	25	10
A 14	18	22	A 36	40	10
A 15	6	25	A 37	30	6
A 21	25	25	A 38	25	25
A 23	20	25	A 39	20	20
B 41	16	16	B 97	3	10
B 42	13	20	B 101	16	18
B 43	6	8	B 103	16	5
B 44	5.6	10	B 104	8	10
B 51	11	20	B 111	11	18
B 52	10	20	B 112	10	13
B 53	8	16	B 121	13	…
B 61	20	8	B 122	10	…
B 62	13	10	B 123	5	…
B 71	16	3	B 124	3	…
B 81	20	5	B 131	13	13
B 91	13	16	B 132	10	13
B 92	6	6	B 133	10	10
B 96	3	6	B 135	6	13
W 144	3	6	W 196	16	26
W 145	3	10	W 197	16	50
W 146	3	13	W 200	20	3
W 152	5	6	W 201	20	6
W 153	5	10	W 202	20	10
W 154	5	13	W 203	20	13
W 158	6	3	W 204	20	20
W 160	6	6	W 205	20	25
W 162	6	10	W 207	20	40
W 163	6	13	W 208	20	50
W 164	6	20	W 215	25	3
W 174	10	6	W 216	25	6
W 175	10	10	W 217	25	10
W 176	10	13	W 218	25	13
W 177	10	20	W 220	25	25
W 178	10	25	W 221	25	40
W 179	10	30	W 222	25	50
W 181	13	1.5	W 225	30	6
W 182	13	3	W 226	30	10
W 183	13	6	W 228	30	20
W 184	13	10	W 230	30	30
W 185	13	13	W 232	30	50
W 186	13	20	W 235	40	6
W 187	13	25	W 236	40	13
W 188	13	40	W 237	40	25
W 189	13	50	W 238	40	40
W 195	16	20	W 242	50	25

[a] See shape diagrams on pages 239 and 240.
All dimensions are in millimeters.

Table 16. Lapping Lubricants

Lubricant	Use
Lard Oil and Machine Oil	Machine and lard oil are the best lubricants for use with copper and steel laps, but the least effective with a cast-iron lap. Lard oil gives the higher rate of cutting. In general, the initial rate of cutting is higher with machine oil, but falls off more rapidly as work continues. With lard oil, the highest results are obtained with a carborundum-charged steel lap. Lowest results were obtained with machine oil when using an emery-charged, cast-iron lap.
Gasoline and Kerosene	Gasoline and kerosene are the best lubricants for use with cast- iron laps, and the poorest on steel. Gasoline is superior to any lubricant tested on cast-iron laps. Kerosene is used with rotary diamond lap for finishing very small holes. Values obtained with carborundum were higher than those obtained with emery, except when used on a copper lap.
Turpentine	Turpentine was found to work well with carborundum on any lap and works fairly well with emery on copper laps, but it was inferior with emery on cast-iron and steel laps.
Soda Water	Medium results with any combination of abrasives; best on copper and poorest on steel. Better than machine or lard oil on cast iron, but not as effective as gasoline or kerosene. Highest result when used with aluminum on copper lap.

Notes: The initial rate of cutting does not greatly differ for different abrasives. There is no advantage in using an abrasive coarser than No. 150. The rate of cutting is practically proportional to the pressure.

Sharpening Carbide Tools.—Cemented carbide indexable inserts are usually not resharpened, but sometimes they require a special grind in order to form a contour on the cutting edge to suit a special purpose. Brazed-type carbide cutting tools are resharpened after the cutting edge has become worn. On brazed carbide tools the cutting-edge wear should not be allowed to become excessive before the tool is resharpened. One method of determining when brazed carbide tools need resharpening is by periodic inspection of the flank wear and the condition of the face. Another method is to determine the amount of production normally obtained before excessive wear has taken place, or to determine the equivalent period of time. One disadvantage of this method is that slight variations in the work material will often cause the wear rate not to be uniform and the number of parts machined before regrinding to be different each time. Usually, sharpening should not require the removal of more than 0.005 to 0.010 inch of carbide.

General Procedure in Carbide Tool Grinding: The general procedure depends upon the kind of grinding operation required. If the operation is to resharpen a dull tool, a diamond wheel of 100- to 120-grain size is recommended although a finer wheel—up to 150-grain size—is sometimes used to obtain a better finish. If the tool is new or is a "standard" design and changes in shape are necessary, a 100-grit diamond wheel is recommended for roughing and a finer grit diamond wheel can be used for finishing. Some shops prefer to rough grind the carbide with a vitrified silicon carbide wheel, the finish grinding being done with a diamond wheel. A final operation commonly designated as lapping may or may not be employed for obtaining an extra-fine finish.

Wheel Speeds: The speed of silicon carbide wheels usually is about 5000 feet per minute. The speeds of diamond wheels generally range from 5000 to 6000 feet per minute; yet lower speeds (550 to 3000 fpm) can be effective.

Offhand Grinding: In grinding single-point tools (excepting chip breakers) the common practice is to hold the tool by hand, press it against the wheel face and traverse it continuously across the wheel face while the tool is supported on the machine rest or table, which is adjusted to the required angle. This is known as "offhand grinding" to distinguish it from the machine grinding of cutters as in regular cutter grinding practice. The selection of wheels adapted to carbide tool grinding is very important.

Silicon Carbide Wheels.—The green colored silicon carbide wheels generally are preferred to the dark gray or gray-black variety, although the latter are sometimes used.

Grain or Grit Sizes: For roughing, a grain size of 60 is very generally used. For finish grinding with silicon carbide wheels, a finer grain size of 100 or 120 is common. A silicon carbide wheel such as C60-I-7V may be used for grinding both the steel shank and carbide tip. However, for under-cutting steel shanks up to the carbide tip, it may be advantageous to use an aluminum oxide wheel suitable for grinding softer, carbon steel.

Grade: According to the standard system of marking, different grades from soft to hard are indicated by letters from A to Z. For carbide tool grinding fairly soft grades such as G, H, I, and J are used. The usual grades for roughing are I or J and for finishing H, I, and J. The grade should be such that a sharp free-cutting wheel will be maintained without excessive grinding pressure. Harder grades than those indicated tend to overheat and crack the carbide.

Structure: The common structure numbers for carbide tool grinding are 7 and 8. The larger cup-wheels (10 to 14 inches) may be of the porous type and be designated as 12P. The standard structure numbers range from 1 to 15 with progressively higher numbers indicating less density and more open wheel structure.

Diamond Wheels.—Wheels with diamond-impregnated grinding faces are fast and cool cutting and have a very low rate of wear. They are used extensively both for resharpening and for finish grinding of carbide tools when preliminary roughing is required. Diamond wheels are also adapted for sharpening multi-tooth cutters such as milling cutters and reamers, which are ground in a cutter grinding machine.

Resinoid bonded wheels are commonly used for grinding chip breakers, milling cutters, reamers, or other multi-tooth cutters. They are also applicable to precision grinding of carbide dies, gages, and various external, internal and surface grinding operations. Fast, cool cutting action is characteristic of these wheels.

Metal bonded wheels are often used for offhand grinding of single-point tools, especially when durability or long life and resistance to grooving of the cutting face are considered more important than the rate of cutting. *Vitrified bonded* wheels are used both for roughing of chipped or very dull tools and for ordinary resharpening and finishing. They provide rigidity for precision grinding, a porous structure for fast cool cutting, sharp cutting action and durability.

Diamond Wheel Grit Sizes.—For roughing with diamond wheels a grit size of 100 is the most common both for offhand and machine grinding.

Grit sizes of 120 and 150 are frequently used in offhand grinding of single-point tools 1) for resharpening; 2) for a combination roughing and finishing wheel; and 3) for chipbreaker grinding.

Grit sizes of 220 or 240 are used for ordinary finish grinding all types of tools (offhand and machine) and also for cylindrical, internal, and surface finish grinding. Grits of 320 and 400 are used for "lapping" to obtain very fine finishes, and for hand hones. A grit of 500 is for lapping to a mirror finish on such work as carbide gages and boring or other tools for exceptionally fine finishes.

Diamond Wheel Grades.—Diamond wheels are made in several different grades to better adapt them to different classes of work. The grades vary for different types and shapes of wheels. Standard Norton grades are H, J, and L, for resinoid bonded wheels; grade N for metal bonded wheels; and grades J, L, N, and P, for vitrified wheels. Harder and softer grades than standard may at times be used to advantage.

Diamond Concentration.—The relative amount (by carat weight) of diamond in the diamond section of the wheel is known as the "diamond concentration." Concentrations of 100 (high), 50 (medium) and 25 (low) ordinarily are supplied. A concentration of 50 represents one-half the diamond content of 100 (if the depth of the diamond is the same in each), and 25 equals one-fourth the content of 100 or one-half the content of 50 concentration.

100 Concentration: Generally interpreted to mean 72 carats of diamond/in.^{3} of abrasive section. (A 75 concentration indicates 54 carats/in.^{3}.) Recommended (especially in grit sizes up to about 220) for general machine grinding of carbides, and for grinding cutters and chip breakers. Vitrified and metal bonded wheels usually have 100 concentration.

50 Concentration: In the finer grit sizes of 220, 240, 320, 400, and 500, a 50 concentration is recommended for offhand grinding with resinoid bonded cup-wheels.

25 Concentration: A low concentration of 25 is recommended for offhand grinding with resinoid bonded cup-wheels with grit sizes of 100, 120 and 150.

Depth of Diamond Section: The radial depth of the diamond section usually varies from $^1/_{16}$ to $^1/_4$ inch. The depth varies somewhat according to the wheel size and type of bond.

Dry versus Wet Grinding of Carbide Tools.—In using silicon carbide wheels, grinding should be done either absolutely dry or with enough coolant to flood the wheel and tool. Satisfactory results may be obtained either by the wet or dry method. However, dry grinding is the most prevalent usually because, in wet grinding, operators tend to use an inadequate supply of coolant to obtain better visibility of the grinding operation and avoid getting wet; hence checking or cracking is more likely to occur in wet grinding than in dry grinding.

Wet Grinding with Silicon Carbide Wheels: One advantage commonly cited in connection with wet grinding is that an ample supply of coolant permits using wheels about one grade harder than in dry grinding thus increasing the wheel life. Plenty of coolant also prevents thermal stresses and the resulting cracks, and there is less tendency for the wheel to load. A dust exhaust system also is unnecessary.

Wet Grinding with Diamond Wheels: In grinding with diamond wheels the general practice is to use a coolant to keep the wheel face clean and promote free cutting. The amount of coolant may vary from a small stream to a coating applied to the wheel face by a felt pad.

Coolants for Carbide Tool Grinding.—In grinding either with silicon carbide or diamond wheels a coolant that is used extensively consists of water plus a small amount either of soluble oil, sal soda, or soda ash to prevent corrosion. One prominent manufacturer recommends for silicon carbide wheels about 1 ounce of soda ash per gallon of water and for diamond wheels, kerosene. The use of kerosene is quite general for diamond wheels and is usually applied to the wheel face by a felt pad. Another coolant recommended for diamond wheels consists of 80 percent water and 20 percent soluble oil.

Peripheral versus Flat Side Grinding.—In grinding single-point carbide tools with silicon carbide wheels, the roughing preparatory to finishing with diamond wheels may be done either by using the flat face of a cup-shaped wheel (side grinding) or the periphery of a "straight" or disk-shaped wheel. Even where side grinding is preferred, the periphery of a straight wheel may be used for heavy roughing as in grinding back chipped or broken tools (see left-hand diagram on page 245). Reasons for preferring peripheral grinding include faster cutting with less danger of localized heating and checking especially in grinding broad surfaces. The advantages usually claimed for side grinding are that proper rake or relief angles are easier to obtain and the relief or land is ground flat. The diamond wheels used for tool sharpening are designed for side grinding. (See right-hand diagram on page 245.)

Lapping Carbide Tools.—Carbide tools may be finished by lapping, especially if an exceptionally fine finish is required on the work as, for example, tools used for precision boring or turning nonferrous metals. If the finishing is done by using a diamond wheel of very fine grit (such as 240, 320, or 400), the operation is often called "lapping." A second lapping method is by means of a power-driven lapping disk charged with diamond dust,

Norbide powder, or silicon carbide finishing compound. A third method is by using a hand lap or hone usually of 320 or 400 grit. In many plants the finishes obtained with carbide tools meet requirements without a special lapping operation. Any feather edge which may be left on tools should always be removed, and it is good practice to bevel the edges of roughing tools at 45 degrees to leave a chamfer 0.005 to 0.010 inch wide. This is done by hand honing and the object is to prevent crumbling or flaking off at the edges when hard scale or heavy chip pressure is encountered.

Hand Honing: The cutting edge of carbide tools, and tools made from other tool materials, is sometimes hand honed before it is used in order to strengthen the cutting edge. When interrupted cuts or heavy roughing cuts are to be taken, or when the grade of carbide is slightly too hard, hand honing is beneficial because it will prevent chipping, or even possibly, breakage of the cutting edge. Whenever chipping is encountered, hand honing the cutting edge before use will be helpful. It is important, however, to hone the edge lightly and only when necessary. Heavy honing will always cause a reduction in tool life. Normally, removing 0.002 to 0.004 inch from the cutting edge is sufficient. When indexable inserts are used, the use of pre-honed inserts is preferred to hand honing although sometimes an additional amount of honing is required. Hand honing of carbide tools in between cuts is sometimes done to defer grinding or to increase the life of a cutting edge on an indexable insert. If correctly done, so as not to change the relief angle, this procedure is sometimes helpful. If improperly done, it can result in a reduction in tool life.

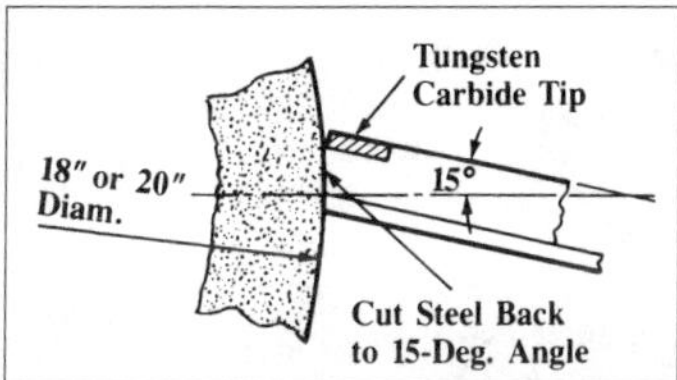

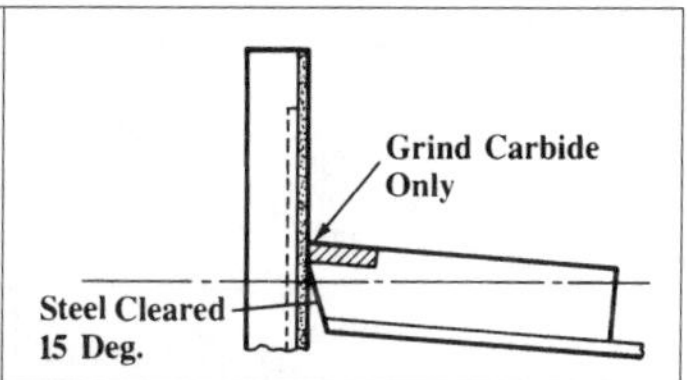

Chipbreaker Grinding.—For this operation a straight diamond wheel is used on a universal tool and cutter grinder, a small surface grinder, or a special chipbreaker grinder. A resinoid bonded wheel of the grade J or N commonly is used, and the tool is held rigidly in an adjustable holder or vise. The width of the diamond wheel usually varies from $\frac{1}{8}$ to $\frac{1}{4}$ inch. A vitrified bond may be used for wheels as thick as $\frac{1}{4}$ inch, and a resinoid bond for relatively narrow wheels.

Summary of Miscellaneous Points.—In grinding a single-point carbide tool, traverse it across the wheel face continuously to avoid localized heating. This traverse movement should be quite rapid in using silicon carbide wheels and comparatively slow with diamond wheels. A hand traversing and feeding movement, whenever practicable, is generally recommended because of greater sensitivity. In grinding, maintain a constant, moderate pressure. Manipulating the tool so as to keep the contact area with the wheel as small as possible will reduce heating and increase the rate of stock removal. Never cool a hot tool by dipping it in a liquid, as this may crack the tip. Wheel rotation should preferably be *against* the cutting edge or from the front face toward the back. If the grinder is driven by a reversing motor, opposite sides of a cup wheel can be used for grinding right-and left-hand tools and with rotation against the cutting edge. If it is necessary to grind the top face of a single-point tool, this should precede the grinding of the side and front relief, and top-face grinding should be minimized to maintain the tip thickness. In machine grinding with a diamond wheel, limit the feed per traverse to 0.001 inch for 100 to 120 grit; 0.0005 inch for 150 to 240 grit; and 0.0002 inch for 320 grit and finer.

GEARING

Nomenclature and Comparative Sizes of Gear Teeth

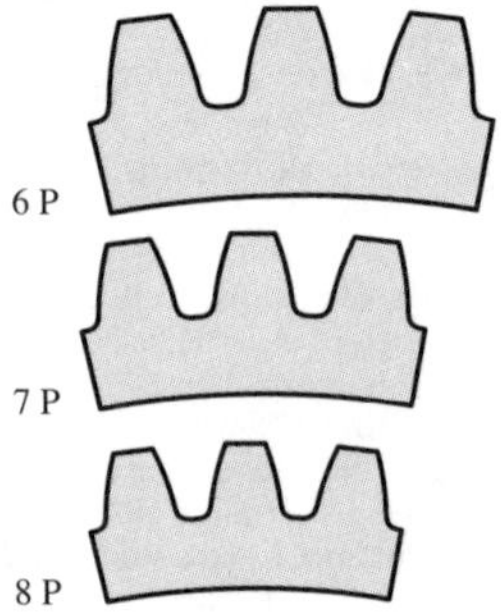

Gear Teeth of Different Diametral Pitch, Full Size

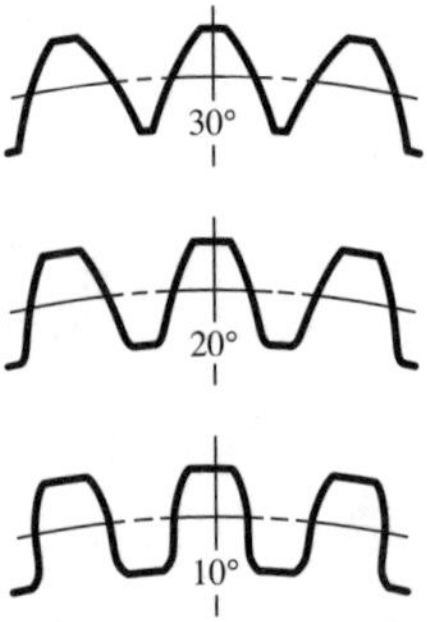

Shapes of Gear Teeth of Different Pressure Angles

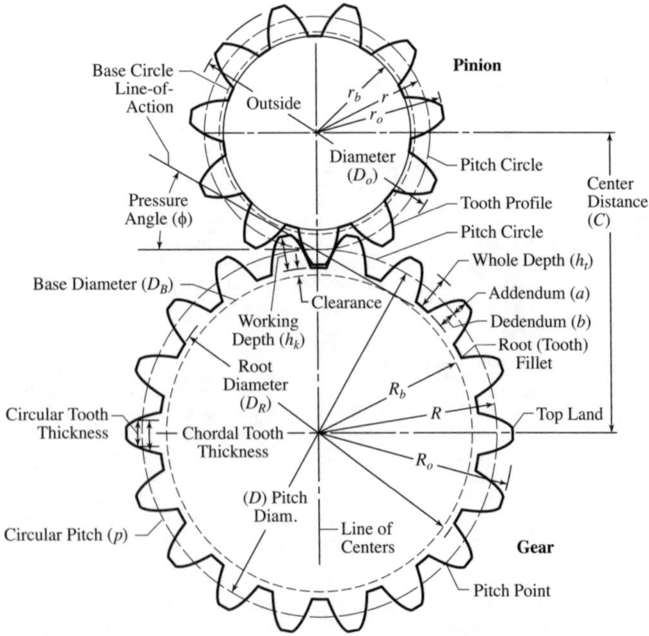

Terms Used in Gear Geometry from Table 1 on page 248

American National Standard and Former American Standard Gear Tooth Forms

ANSI B6.1-1968 (R1974) and ASA B6.1-1932

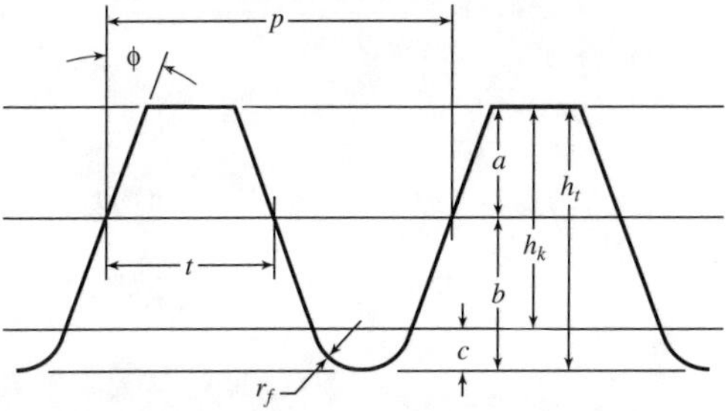

a = addendum h_k = working depth r_f = fillet radius of basic rack
b = dedendum h_t = whole depth t = circular tooth thickness—basic
c = clearance p = circular pitch ϕ = pressure angle

Basic Rack of the 20-Degree and 25-Degree Full-Depth Involute Systems

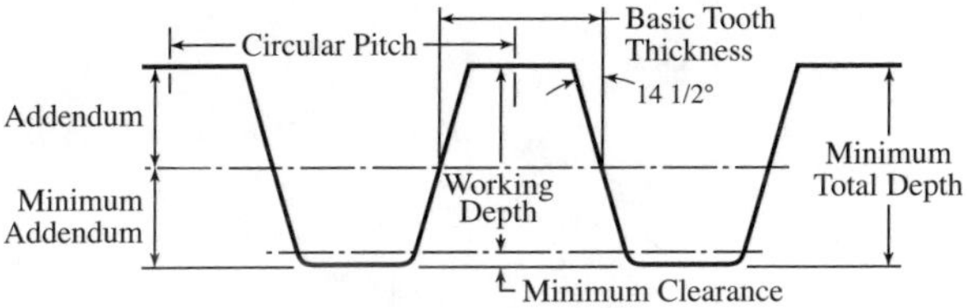

Basic Rack of the $14\frac{1}{2}$-Degree Full-Depth Involute System

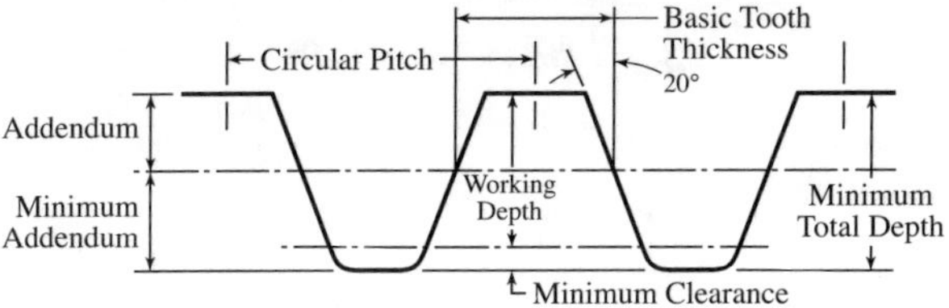

Basic Rack of the 20-Degree Stub Involute System

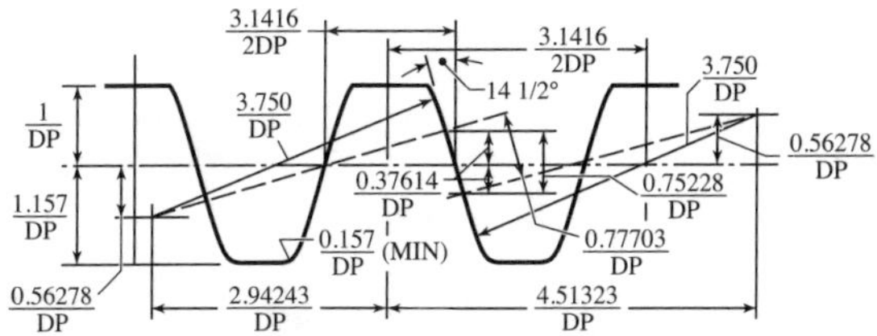

Approximation of Basic Rack for the $14\frac{1}{2}$-Degree Composite System

Table 1. Formulas for Dimensions of Standard Spur Gears

No.	To Find	Formula	No.	To Find	Formula
General Formulas					
1	Base Circle Diameter	$D_B = D\cos\phi$	6a	Number of Teeth	$N = P \times D$
2a	Circular Pitch	$p = \frac{3.1416D}{N}$	6b	Number of Teeth	$N = \frac{3.1416D}{p}$
2b	Circular Pitch	$p = \frac{3.1416}{P}$	7a	Outside Diameter (Full-depth Teeth)	$D_O = \frac{N+2}{P}$
3a	Center Distance	$C = \frac{N_P(m_G+1)}{2P}$	7b	Outside Diameter (Full-depth Teeth)	$D_O = \frac{(N+2)p}{3.1416}$
3b	Center Distance	$C = \frac{D_P+D_G}{2}$	8a	Outside Diameter (Amer. Stnd. Stub Teeth)	$D_O = \frac{N+1.6}{P}$
3c	Center Distance	$C = \frac{N_G+N_P}{2P}$	8b	Outside Diameter (Amer. Stnd. Stub Teeth)	$D_O = \frac{(N+1.6)p}{3.1416}$
3d	Center Distance	$C = \frac{(N_G+N_P)p}{6.2832}$	9	Outside Diameter	$D_O = D+2a$
4a	Diametral Pitch	$P = \frac{3.1416}{p}$	10a	Pitch Diameter	$D = \frac{N}{P}$
4b	Diametral Pitch	$P = \frac{N}{D}$	10b	Pitch Diameter	$D = \frac{Np}{3.1416}$
4c	Diametral Pitch	$P = \frac{N_P(m_G+1)}{2C}$	11	Root Diameter	$D_R = D-2b$
5	Gear Ratio	$m_G = \frac{N_G}{N_P}$	12	Whole Depth	$a+b$
			13	Working Depth	a_G+a_P

Notation	
ϕ = Pressure Angle	D_O = Outside Diameter
a = Addendum = $1/P$	D_R = Root Diameter
a_G = Addendum of Gear	F = Face Width
a_P = Addendum of Pinion	h_k = Working Depth of Tooth
b = Dedendum	h_t = Whole Depth of Tooth
c = Clearance	m_G = Gear Ratio
C = Center Distance	N = Number of Teeth
D = Pitch Diameter	N_G = Number of Teeth in Gear
D_G = Pitch Diameter of Gear	N_P = Number of Teeth in Pinion
D_P = Pitch Diameter of Pinion	p = Circular Pitch
D_B = Base Circle Diameter	P = Diametral Pitch

Table 2. Circular Pitch in Gears— Pitch Diameters, Outside Diameters, and Root Diameters

For any particular circular pitch and number of teeth, use the table as shown in the example to find the pitch diameter, outside diameter, and root diameter. *Example:* Pitch diameter for 57 teeth of 6-inch circular pitch = 10× pitch diameter given under factor for 5 teeth plus pitch diameter given under factor for 7 teeth. (10 × 9.5493) + 13.3690 = 108.862 inches.

Outside diameter of gear equals pitch diameter plus outside diameter factor from next-to-last column in table = 108.862 + 3.8197 = 112.682 inches.

Root diameter of gear equals pitch diameter minus root diameter factor from last column in table = 108.862 − 4.4194 = 104.443 inches.

Circular Pitch in Inches	Factor for Number of Teeth									Outside Dia. Factor	Root Diameter Factor
	1	2	3	4	5	6	7	8	9		
	Pitch Diameter Corresponding to Factor for Number of Teeth										
6	1.9099	3.8197	5.7296	7.6394	9.5493	11.4591	13.3690	15.2788	17.1887	3.8197	4.4194
$5\frac{1}{2}$	1.7507	3.5014	5.2521	7.0028	8.7535	10.5042	12.2549	14.0056	15.7563	3.5014	4.0511
5	1.5915	3.1831	4.7746	6.3662	7.9577	9.5493	11.1408	12.7324	14.3239	3.1831	3.6828
$4\frac{1}{2}$	1.4324	2.8648	4.2972	5.7296	7.1620	8.5943	10.0267	11.4591	12.8915	2.8648	3.3146
4	1.2732	2.5465	3.8197	5.0929	6.3662	7.6394	8.9127	10.1859	11.4591	2.5465	2.9463
$3\frac{1}{2}$	1.1141	2.2282	3.3422	4.4563	5.5704	6.6845	7.7986	8.9127	10.0267	2.2282	2.5780
3	0.9549	1.9099	2.8648	3.8197	4.7746	5.7296	6.6845	7.6394	8.5943	1.9099	2.2097
$2\frac{1}{2}$	0.7958	1.5915	2.3873	3.1831	3.9789	4.7746	5.5704	6.3662	7.1620	1.5915	1.8414
2	0.6366	1.2732	1.9099	2.5465	3.1831	3.8197	4.4563	5.0929	5.7296	1.2732	1.4731
$1\frac{7}{8}$	0.5968	1.1937	1.7905	2.3873	2.9841	3.5810	4.1778	4.7746	5.3715	1.1937	1.3811
$1\frac{3}{4}$	0.5570	1.1141	1.6711	2.2282	2.7852	3.3422	3.8993	4.4563	5.0134	1.1141	1.2890
$1\frac{5}{8}$	0.5173	1.0345	1.5518	2.0690	2.5863	3.1035	3.6208	4.1380	4.6553	1.0345	1.1969
$1\frac{1}{2}$	0.4775	0.9549	1.4324	1.9099	2.3873	2.8648	3.3422	3.8197	4.2972	0.9549	1.1049
$1\frac{7}{16}$	0.4576	0.9151	1.3727	1.8303	2.2878	2.7454	3.2030	3.6606	4.1181	0.9151	1.0588
$1\frac{3}{8}$	0.4377	0.8754	1.3130	1.7507	2.1884	2.6261	3.0637	3.5014	3.9391	0.8754	1.0128
$1\frac{5}{16}$	0.4178	0.8356	1.2533	1.6711	2.0889	2.5067	2.9245	3.3422	3.7600	0.8356	0.9667
$1\frac{1}{4}$	0.3979	0.7958	1.1937	1.5915	1.9894	2.3873	2.7852	3.1831	3.5810	0.7958	0.9207
$1\frac{3}{16}$	0.3780	0.7560	1.1340	1.5120	1.8900	2.2680	2.6459	3.0239	3.4019	0.7560	0.8747
$1\frac{1}{8}$	0.3581	0.7162	1.0743	1.4324	1.7905	2.1486	2.5067	2.8648	3.2229	0.7162	0.8286
$1\frac{1}{16}$	0.3382	0.6764	1.0146	1.3528	1.6910	2.0292	2.3674	2.7056	3.0438	0.6764	0.7826
1	0.3183	0.6366	0.9549	1.2732	1.5915	1.9099	2.2282	2.5465	2.8648	0.6366	0.7366
$\frac{15}{16}$	0.2984	0.5968	0.8952	1.1937	1.4921	1.7905	2.0889	2.3873	2.6857	0.5968	0.6905
$\frac{7}{8}$	0.2785	0.5570	0.8356	1.1141	1.3926	1.6711	1.9496	2.2282	2.5067	0.5570	0.6445
$\frac{13}{16}$	0.2586	0.5173	0.7759	1.0345	1.2931	1.5518	1.8104	2.0690	2.3276	0.5173	0.5985
$\frac{3}{4}$	0.2387	0.4475	0.7162	0.9549	1.1937	1.4324	1.6711	1.9099	2.1486	0.4775	0.5524
$\frac{11}{16}$	0.2188	0.4377	0.6565	0.8754	1.0942	1.3130	1.5319	1.7507	1.9695	0.4377	0.5064
$\frac{2}{3}$	0.2122	0.4244	0.6366	0.8488	1.0610	1.2732	1.4854	1.6977	1.9099	0.4244	0.4910
$\frac{5}{8}$	0.1989	0.3979	0.5968	0.7958	0.9947	1.1937	1.3926	1.5915	1.7905	0.3979	0.4604
$\frac{9}{16}$	0.1790	0.3581	0.5371	0.7162	0.8952	1.0743	1.2533	1.4324	1.6114	0.3581	0.4143
$\frac{1}{2}$	0.1592	0.3183	0.4775	0.6366	0.7958	0.9549	1.1141	1.2732	1.4324	0.3183	0.3683
$\frac{7}{16}$	0.1393	0.2785	0.4178	0.5570	0.6963	0.8356	0.9748	1.1141	1.2533	0.2785	0.3222
$\frac{3}{8}$	0.1194	0.2387	0.3581	0.4775	0.5968	0.7162	0.8356	0.9549	1.0743	0.2387	0.2762
$\frac{1}{3}$	0.1061	0.2122	0.3183	0.4244	0.5305	0.6366	0.7427	0.8488	0.9549	0.2122	0.2455
$\frac{5}{16}$	0.0995	0.1989	0.2984	0.3979	0.4974	0.5968	0.6963	0.7958	0.8952	0.1989	0.2302
$\frac{1}{4}$	0.0796	0.1592	0.2387	0.3183	0.3979	0.4775	0.5570	0.6366	0.7162	0.1592	0.1841
$\frac{3}{16}$	0.0597	0.1194	0.1790	0.2387	0.2984	0.3581	0.4178	0.4775	0.5371	0.1194	0.1381
$\frac{1}{8}$	0.0398	0.0796	0.1194	0.1592	0.1989	0.2387	0.2785	0.3183	0.3581	0.0796	0.0921
$\frac{1}{16}$	0.0199	0.0398	0.0597	0.0796	0.0995	0.1194	0.1393	0.1592	0.1790	0.0398	0.0460

Table 3a. Chordal Thicknesses and Chordal Addenda of Milled, Full-Depth Gear Teeth and of Gear Milling Cutters

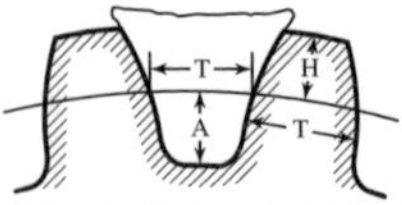

T = chordal thickness of gear tooth and cutter tooth at pitch line;
H = chordal addendum for full-depth gear tooth;
A = chordal addendum of cutter = (2.157 ÷ diametral pitch) − H
= (0.6866 × circular pitch) − H.

Diametral Pitch	Dimension	Number of Gear Cutter and Corresponding Number of Teeth							
		No. 1 135 Teeth	No. 2 55 Teeth	No. 3 35 Teeth	No. 4 26 Teeth	No. 5 21 Teeth	No. 6 17 Teeth	No. 7 14 Teeth	No. 8 12 Teeth
1	T	1.5707	1.5706	1.5702	1.5698	1.5694	1.5686	1.5675	1.5663
	H	1.0047	1.0112	1.0176	1.0237	1.0294	1.0362	1.0440	1.0514
$1\frac{1}{2}$	T	1.0471	1.0470	1.0468	1.0465	1.0462	1.0457	1.0450	1.0442
	H	0.6698	0.6741	0.6784	0.6824	0.6862	0.6908	0.6960	0.7009
2	T	0.7853	0.7853	0.7851	0.7849	0.7847	0.7843	0.7837	0.7831
	H	0.5023	0.5056	0.5088	0.5118	0.5147	0.5181	0.5220	0.5257
$2\frac{1}{2}$	T	0.6283	0.6282	0.6281	0.6279	0.6277	0.6274	0.6270	0.6265
	H	0.4018	0.4044	0.4070	0.4094	0.4117	0.4144	0.4176	0.4205
3	T	0.5235	0.5235	0.5234	0.5232	0.5231	0.5228	0.5225	0.5221
	H	0.3349	0.3370	0.3392	0.3412	0.3431	0.3454	0.3480	0.3504
$3\frac{1}{2}$	T	0.4487	0.4487	0.4486	0.4485	0.4484	0.4481	0.4478	0.4475
	H	0.2870	0.2889	0.2907	0.2919	0.2935	0.2954	0.2977	0.3004
4	T	0.3926	0.3926	0.3926	0.3924	0.3923	0.3921	0.3919	0.3915
	H	0.2511	0.2528	0.2544	0.2559	0.2573	0.2590	0.2610	0.2628
5	T	0.3141	0.3141	0.3140	0.3139	0.3138	0.3137	0.3135	0.3132
	H	0.2009	0.2022	0.2035	0.2047	0.2058	0.2072	0.2088	0.2102
6	T	0.2618	0.2617	0.2617	0.2616	0.2615	0.2614	0.2612	0.2610
	H	0.1674	0.1685	0.1696	0.1706	0.1715	0.1727	0.1740	0.1752
7	T	0.2244	0.2243	0.2243	0.2242	0.2242	0.2240	0.2239	0.2237
	H	0.1435	0.1444	0.1453	0.1462	0.1470	0.1480	0.1491	0.1502
8	T	0.1963	0.1963	0.1962	0.1962	0.1961	0.1960	0.1959	0.1958
	H	0.1255	0.1264	0.1272	0.1279	0.1286	0.1295	0.1305	0.1314
9	T	0.1745	0.1745	0.1744	0.1744	0.1743	0.1743	0.1741	0.1740
	H	0.1116	0.1123	0.1130	0.1137	0.1143	0.1151	0.1160	0.1168
10	T	0.1570	0.1570	0.1570	0.1569	0.1569	0.1568	0.1567	0.1566
	H	0.1004	0.1011	0.1017	0.1023	0.1029	0.1036	0.1044	0.1051
11	T	0.1428	0.1428	0.1427	0.1427	0.1426	0.1426	0.1425	0.1424
	H	0.0913	0.0919	0.0925	0.0930	0.0935	0.0942	0.0949	0.0955
12	T	0.1309	0.1309	0.1308	0.1308	0.1308	0.1307	0.1306	0.1305
	H	0.0837	0.0842	0.0848	0.0853	0.0857	0.0863	0.0870	0.0876
14	T	0.1122	0.1122	0.1121	0.1121	0.1121	0.1120	0.1119	0.1118
	H	0.0717	0.0722	0.0726	0.0731	0.0735	0.0740	0.0745	0.0751
16	T	0.0981	0.0981	0.0981	0.0981	0.0980	0.0980	0.0979	0.0979
	H	0.0628	0.0632	0.0636	0.0639	0.0643	0.0647	0.0652	0.0657
18	T	0.0872	0.0872	0.0872	0.0872	0.0872	0.0871	0.0870	0.0870
	H	0.0558	0.0561	0.0565	0.0568	0.0571	0.0575	0.0580	0.0584
20	T	0.0785	0.0785	0.0785	0.0785	0.0784	0.0784	0.0783	0.0783
	H	0.0502	0.0505	0.0508	0.0511	0.0514	0.0518	0.0522	0.0525

Table 3b. Chordal Thicknesses and Chordal Addenda of Milled, Full-Depth Gear Teeth and of Gear Milling Cutters

Circular Pitch	Dimension	Number of Gear Cutter and Corresponding Number of Teeth							
		No. 1 135 Teeth	No. 2 55 Teeth	No. 3 35 Teeth	No. 4 26 Teeth	No. 5 21 Teeth	No. 6 17 Teeth	No. 7 14 Teeth	No. 8 12 Teeth
$^1/_4$	T	0.1250	0.1250	0.1249	0.1249	0.1249	0.1248	0.1247	0.1246
	H	0.0799	0.0804	0.0809	0.0814	0.0819	0.0824	0.0830	0.0836
$^5/_{16}$	T	0.1562	0.1562	0.1562	0.1561	0.1561	0.1560	0.1559	0.1558
	H	0.0999	0.1006	0.1012	0.1018	0.1023	0.1030	0.1038	0.1045
$^3/_8$	T	0.1875	0.1875	0.1874	0.1873	0.1873	0.1872	0.1871	0.1870
	H	0.1199	0.1207	0.1214	0.1221	0.1228	0.1236	0.1245	0.1254
$^7/_{16}$	T	0.2187	0.2187	0.2186	0.2186	0.2185	0.2184	0.2183	0.2181
	H	0.1399	0.1408	0.1416	0.1425	0.1433	0.1443	0.1453	0.1464
$^1/_2$	T	0.2500	0.2500	0.2499	0.2498	0.2498	0.2496	0.2495	0.2493
	H	0.1599	0.1609	0.1619	0.1629	0.1638	0.1649	0.1661	0.1673
$^9/_{16}$	T	0.2812	0.2812	0.2811	0.2810	0.2810	0.2808	0.2806	0.2804
	H	0.1799	0.1810	0.1821	0.1832	0.1842	0.1855	0.1868	0.1882
$^5/_8$	T	0.3125	0.3125	0.3123	0.3123	0.3122	0.3120	0.3118	0.3116
	H	0.1998	0.2012	0.2023	0.2036	0.2047	0.2061	0.2076	0.2091
$^{11}/_{16}$	T	0.3437	0.3437	0.3436	0.3435	0.3434	0.3432	0.3430	0.3427
	H	0.2198	0.2213	0.2226	0.2239	0.2252	0.2267	0.2283	0.2300
$^3/_4$	T	0.3750	0.3750	0.3748	0.3747	0.3747	0.3744	0.3742	0.3740
	H	0.2398	0.2414	0.2428	0.2443	0.2457	0.2473	0.2491	0.2509
$^{13}/_{16}$	T	0.4062	0.4062	0.4060	0.4059	0.4059	0.4056	0.4054	0.4050
	H	0.2598	0.2615	0.2631	0.2647	0.2661	0.2679	0.2699	0.2718
$^7/_8$	T	0.4375	0.4375	0.4373	0.4372	0.4371	0.4368	0.4366	0.4362
	H	0.2798	0.2816	0.2833	0.2850	0.2866	0.2885	0.2906	0.2927
$^{15}/_{16}$	T	0.4687	0.4687	0.4685	0.4684	0.4683	0.4680	0.4678	0.4674
	H	0.2998	0.3018	0.3035	0.3054	0.3071	0.3092	0.3114	0.3137
1	T	0.5000	0.5000	0.4998	0.4997	0.4996	0.4993	0.4990	0.4986
	H	0.3198	0.3219	0.3238	0.3258	0.3276	0.3298	0.3322	0.3346
$1^1/_8$	T	0.5625	0.5625	0.5623	0.5621	0.5620	0.5617	0.5613	0.5610
	H	0.3597	0.3621	0.3642	0.3665	0.3685	0.3710	0.3737	0.3764
$1^1/_4$	T	0.6250	0.6250	0.6247	0.6246	0.6245	0.6241	0.6237	0.6232
	H	0.3997	0.4023	0.4047	0.4072	0.4095	0.4122	0.4152	0.4182
$1^3/_8$	T	0.6875	0.6875	0.6872	0.6870	0.6869	0.6865	0.6861	0.6856
	H	0.4397	0.4426	0.4452	0.4479	0.4504	0.4534	0.4567	0.4600
$1^1/_2$	T	0.7500	0.7500	0.7497	0.7495	0.7494	0.7489	0.7485	0.7480
	H	0.4797	0.4828	0.4857	0.4887	0.4914	0.4947	0.4983	0.5019
$1^3/_4$	T	0.8750	0.8750	0.8746	0.8744	0.8743	0.8737	0.8732	0.8726
	H	0.5596	0.5633	0.5666	0.5701	0.5733	0.5771	0.5813	0.5855
2	T	1.0000	1.0000	0.9996	0.9994	0.9992	0.9986	0.9980	0.9972
	H	0.6396	0.6438	0.6476	0.6516	0.6552	0.6596	0.6644	0.6692
$2^1/_4$	T	1.1250	1.1250	1.1246	1.1242	1.1240	1.1234	1.1226	1.1220
	H	0.7195	0.7242	0.7285	0.7330	0.7371	0.7420	0.7474	0.7528
$2^1/_2$	T	1.2500	1.2500	1.2494	1.2492	1.2490	1.2482	1.2474	1.2464
	H	0.7995	0.8047	0.8095	0.8145	0.8190	0.8245	0.8305	0.8365
3	T	1.5000	1.5000	1.4994	1.4990	1.4990	1.4978	1.4970	1.4960
	H	0.9594	0.9657	0.9714	0.9774	0.9828	0.9894	0.9966	1.0038

Table 4. Series of Involute, Finishing Gear Milling Cutters for Each Pitch

Number of Cutter	Will cut Gears from	Number of Cutter	Will cut Gears from
1	135 teeth to a rack	5	21 to 25 teeth
2	55 to 134 teeth	6	17 to 20 teeth
3	35 to 54 teeth	7	14 to 16 teeth
4	26 to 34 teeth	8	12 to 13 teeth
The regular cutters listed above are used ordinarily. The cutters listed below (an intermediate series having half numbers) may be used when greater accuracy of tooth shape is essential in cases where the number of teeth is between the numbers for which the regular cutters are intended.			
Number of Cutter	**Will cut Gears from**	**Number of Cutter**	**Will cut Gears from**
$1^1/_2$	80 to 134 teeth	$5^1/_2$	19 to 20 teeth
$2^1/_2$	42 to 54 teeth	$6^1/_2$	15 to 16 teeth
$3^1/_2$	30 to 34 teeth	$7^1/_2$	13 teeth
$4^1/_2$	23 to 25 teeth	…	…

Roughing cutters are made with No. 1 form only.

Gear Design Based upon Module System.—The *module* of a gear is equal to the pitch diameter divided by the number of teeth, whereas *diametral pitch* is equal to the number of teeth divided by the pitch diameter. The module system (see accompanying table and diagram) is in general use in countries that have adopted the metric system; hence, the term "module" is usually understood to mean the pitch diameter *in millimeters* divided by the number of teeth. The module system, however, may also be based on inch measurements, and then it is known as the English module to avoid confusion with the metric module. Module is an actual dimension, whereas diametral pitch is only a ratio. Thus, if the pitch diameter of a gear is 50 millimeters and the number of teeth 25, the module is 2, which means that there are 2 millimeters of pitch diameter for each tooth. The table *Tooth Dimensions Based Upon Module System* shows the relation among module, diametral pitch, and circular pitch.

Table 5. German Standard Tooth Form for Spur and Bevel Gears *DIN 867*

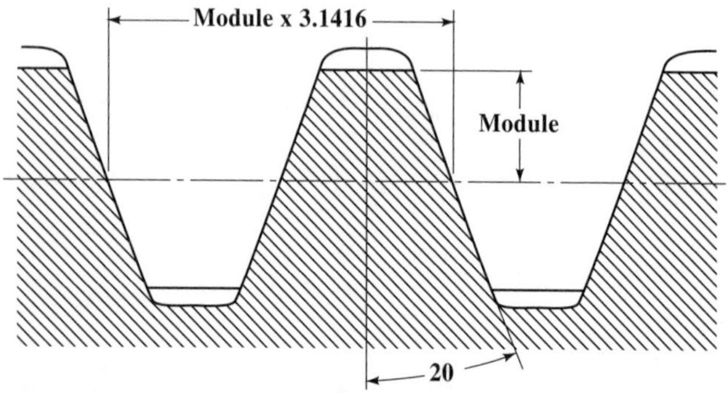

The flanks or sides are straight (involute system) and the pressure angle is 20 degrees. The shape of the root clearance space and the amount of clearance depend upon the method of cutting and special requirements. The amount of clearance may vary from 0.1 × module to 0.3 × module.

To Find	Module Known	Circular Pitch Known
Addendum	Equals module	0.31823 × Circular pitch
Dedendum	1.157 × module* 1.167 × module**	0.3683 × Circular pitch* 0.3714 × Circular pitch**
Working Depth	2 × module	0.6366 × Circular pitch
Total Depth	2.157 × module* 2.167 × module**	0.6866 × Circular pitch* 0.6898 × Circulate pitch**
Tooth Thickness on Pitch Line	1.5708 × module	0.5 × Circular pitch

Formulas for dedendum and total depth, marked (*) are used when clearance equals 0.157 × module. Formulas marked (**) are used when clearance equals one-sixth module. It is common practice among American cutter manufacturers to make the clearance of metric or module cutters equal to 0.157 × module.

Table 6. Tooth Dimensions Based Upon Module System

Module, *DIN* Standard Series	Equivalent Diametral Pitch	Circular Pitch		Addendum, Millimeters[a]	Dedendum, Millimeters[a]	Whole Depth,[a] Millimeters	Whole Depth,[b] Millimeters
		Millimeters	Inches				
0.3	84.667	0.943	0.0371	0.30	0.35	0.650	0.647
0.4	63.500	1.257	0.0495	0.40	0.467	0.867	0.863
0.5	50.800	1.571	0.0618	0.50	0.583	1.083	1.079
0.6	42.333	1.885	0.0742	0.60	0.700	1.300	1.294
0.7	36.286	2.199	0.0865	0.70	0.817	1.517	1.510
0.8	31.750	2.513	0.0989	0.80	0.933	1.733	1.726
0.9	28.222	2.827	0.1113	0.90	1.050	1.950	1.941
1	25.400	3.142	0.1237	1.00	1.167	2.167	2.157
1.25	20.320	3.927	0.1546	1.25	1.458	2.708	2.697
1.5	16.933	4.712	0.1855	1.50	1.750	3.250	3.236
1.75	14.514	5.498	0.2164	1.75	2.042	3.792	3.774
2	12.700	6.283	0.2474	2.00	2.333	4.333	4.314
2.25	11.289	7.069	0.2783	2.25	2.625	4.875	4.853
2.5	10.160	7.854	0.3092	2.50	2.917	5.417	5.392
2.75	9.236	8.639	0.3401	2.75	3.208	5.958	5.932
3	8.466	9.425	0.3711	3.00	3.500	6.500	6.471
3.25	7.815	10.210	0.4020	3.25	3.791	7.041	7.010
3.5	7.257	10.996	0.4329	3.50	4.083	7.583	7.550
3.75	6.773	11.781	0.4638	3.75	4.375	8.125	8.089
4	6.350	12.566	0.4947	4.00	4.666	8.666	8.628
4.5	5.644	14.137	0.5566	4.50	5.25	9.750	9.707
5	5.080	15.708	0.6184	5.00	5.833	10.833	10.785
5.5	4.618	17.279	0.6803	5.50	6.416	11.916	11.864
6	4.233	18.850	0.7421	6.00	7.000	13.000	12.942
6.5	3.908	20.420	0.8035	6.50	7.583	14.083	14.021
7	3.628	21.991	0.8658	7	8.166	15.166	15.099
8	3.175	25.132	0.9895	8	9.333	17.333	17.256
9	2.822	28.274	1.1132	9	10.499	19.499	19.413
10	2.540	31.416	1.2368	10	11.666	21.666	21.571
11	2.309	34.558	1.3606	11	12.833	23.833	23.728
12	2.117	37.699	1.4843	12	14.000	26.000	25.884
13	1.954	40.841	1.6079	13	15.166	28.166	28.041
14	1.814	43.982	1.7317	14	16.332	30.332	30.198
15	1.693	47.124	1.8541	15	17.499	32.499	32.355
16	1.587	50.266	1.9790	16	18.666	34.666	34.512
18	1.411	56.549	2.2263	18	21.000	39.000	38.826
20	1.270	62.832	2.4737	20	23.332	43.332	43.142
22	1.155	69.115	2.7210	22	25.665	47.665	47.454
24	1.058	75.398	2.9685	24	28.000	52.000	51.768
27	0.941	84.823	3.339	27	31.498	58.498	58.239
30	0.847	94.248	3.711	30	35.000	65.000	64.713
33	0.770	103.673	4.082	33	38.498	71.498	71.181
36	0.706	113.097	4.453	36	41.998	77.998	77.652
39	0.651	122.522	4.824	39	45.497	84.497	84.123
42	0.605	131.947	5.195	42	48.997	90.997	90.594
45	0.564	141.372	5.566	45	52.497	97.497	97.065
50	0.508	157.080	6.184	50	58.330	108.330	107.855
55	0.462	172.788	6.803	55	64.163	119.163	118.635
60	0.423	188.496	7.421	60	69.996	129.996	129.426
65	0.391	204.204	8.040	65	75.829	140.829	140.205
70	0.363	219.911	8.658	70	81.662	151.662	150.997
75	0.339	235.619	9.276	75	87.495	162.495	161.775

[a] Dedendum and total depth when clearance = 0.1666 × module, or one-sixth module.

[b] Total depth equivalent to American standard full-depth teeth. (Clearance = 0.157 × module.)

Table 7. Rules for Module System of Gearing

To Find	Rule
Metric Module	*Rule 1:* To find the metric module, divide the pitch diameter in millimeters by the number of teeth. *Example 1:* The pitch diameter of a gear is 200 millimeters and the number of teeth, 40; then $\text{Module} = \frac{200}{40} = 5$ *Rule 2:* Multiply circular pitch in millimeters by 0.3183. *Example 2:* (Same as Example 1. Circular pitch of this gear equals 15.708 millimeters.) $\text{Module} = 15.708 \times 0.3183 = 5$ *Rule 3:* Divide outside diameter in millimeters by the number of teeth plus 2.
English Module	*Note:* The module system is usually applied when gear dimensions are expressed in millimeters, but module may also be based on inch measurements. *Rule:* To find the English module, divide pitch diameter in inches by the number of teeth. *Example:* A gear has 48 teeth and a pitch diameter of 12 inches. $\text{Module} = \frac{12}{48} = \frac{1}{4}$ module or 4 diametral pitch
Metric Module Equivalent to Diametral Pitch	*Rule:* To find the metric module equivalent to a given diametral pitch, divide 25.4 by the diametral pitch. *Example:* Determine metric module equivalent to 10 diameteral pitch. $\text{Equivalent module} = \frac{25.4}{10} = 2.54$ *Note:* The nearest standard module is 2.5.
Diametral Pitch Equivalent to Metric Module	*Rule:* To find the diametral pitch equivalent to a given module, divide 25.4 by the module. (25.4 = number of millimeters per inch.) *Example:* The module is 12; determine equivalent diametral pitch. $\text{Equivalent diametral pitch} = \frac{25.4}{12} = 2.117$ *Note:* A diametral pitch of 2 is the nearest *standard* equivalent.
Pitch Diameter	*Rule:* Multiply number of teeth by module. *Example:* The metric module is 8 and the gear has 40 teeth; then $D = 40 \times 8 = 320 \text{ millimeters} = 12.598 \text{ inches}$
Outside Diameter	*Rule:* Add 2 to the number of teeth and multiply sum by the module. *Example:* A gear has 40 teeth and module is 6. Find outside or blank diameter. $\text{Outside diameter} = (40 + 2) \times 6 = 252 \text{ millimeters}$

For tooth dimensions, see table *Tooth Dimensions Based Upon Module System*; also see formulas in *German Standard Tooth Form for Spur and Bevel Gears DIN 867.*

Table 8. Equivalent Diametral Pitches, Circular Pitches, and Metric Modules Commonly Used Pitches and Modules in Bold Type

Diametral Pitch	Circular Pitch, Inches	Module Millimeters	Diametral Pitch	Circular Pitch, Inches	Module Millimeters	Diametral Pitch	Circular Pitch, Inches	Module Millimeters
$\frac{1}{2}$	6.2832	50.8000	2.2848	**$1\frac{3}{8}$**	11.1170	10.0531	$\frac{5}{16}$	2.5266
0.5080	6.1842	**50**	2.3091	1.3605	**11**	10.1600	0.3092	**$2\frac{1}{2}$**
0.5236	**6**	48.5104	**$2\frac{1}{2}$**	1.2566	10.1600	**11**	0.2856	2.3091
0.5644	5.5658	**45**	2.5133	**$1\frac{1}{4}$**	10.1063	**12**	0.2618	2.1167
0.5712	**$5\frac{1}{2}$**	44.4679	2.5400	1.2368	**10**	12.5664	$\frac{1}{4}$	2.0213
0.6283	**5**	40.4253	**$2\frac{3}{4}$**	1.1424	9.2364	12.7000	0.2474	**2**
0.6350	4.9474	**40**	2.7925	**$1\frac{1}{8}$**	9.0957	**13**	0.2417	1.9538
0.6981	**$4\frac{1}{2}$**	36.3828	2.8222	1.1132	**9**	**14**	0.2244	1.8143
0.7257	4.3290	**35**	**3**	1.0472	8.4667	**15**	0.2094	1.6933
$\frac{3}{4}$	4.1888	33.8667	3.1416	**1**	8.0851	**16**	0.1963	1.5875
0.7854	**4**	32.3403	3.1750	0.9895	**8**	16.7552	$\frac{3}{16}$	1.5160
0.8378	**$3\frac{3}{4}$**	30.3190	3.3510	$\frac{15}{16}$	7.5797	16.9333	0.1855	**$1\frac{1}{2}$**
0.8467	3.7105	**30**	**$3\frac{1}{2}$**	0.8976	7.2571	**17**	0.1848	1.4941
0.8976	**$3\frac{1}{2}$**	28.2977	3.5904	$\frac{7}{8}$	7.0744	**18**	0.1745	1.4111
0.9666	**$3\frac{1}{4}$**	26.2765	3.6286	0.8658	**7**	**19**	0.1653	1.3368
1	3.1416	25.4000	3.8666	$\frac{13}{16}$	6.5691	**20**	0.1571	1.2700
1.0160	3.0921	**25**	3.9078	0.8040	**$6\frac{1}{2}$**	**22**	0.1428	1.1545
1.0472	**3**	24.2552	**4**	0.7854	6.3500	**24**	0.1309	1.0583
1.1424	**$2\frac{3}{4}$**	22.2339	4.1888	$\frac{3}{4}$	6.0638	**25**	0.1257	1.0160
$1\frac{1}{4}$	2.5133	20.3200	4.2333	0.7421	**6**	25.1328	$\frac{1}{8}$	1.0106
1.2566	**$2\frac{1}{2}$**	20.2127	4.5696	$\frac{11}{16}$	5.5585	25.4000	0.1237	**1**
1.2700	2.4737	**20**	4.6182	0.6803	**$5\frac{1}{2}$**	**26**	0.1208	0.9769
1.3963	**$2\frac{1}{4}$**	18.1914	**5**	0.6283	5.0800	**28**	0.1122	0.9071
1.4111	2.2263	**18**	5.0265	$\frac{5}{8}$	5.0532	**30**	0.1047	0.8467
$1\frac{1}{2}$	2.0944	16.9333	5.0800	0.6184	**5**	**32**	0.0982	0.7937
1.5708	**2**	16.1701	5.5851	$\frac{9}{16}$	4.5478	**34**	0.0924	0.7470
1.5875	1.9790	**16**	5.6443	0.5566	**$4\frac{1}{2}$**	**36**	0.0873	0.7056
1.6755	**$1\frac{7}{8}$**	15.1595	**6**	0.5236	4.2333	**38**	0.0827	0.6684
1.6933	1.8553	**15**	6.2832	$\frac{1}{2}$	4.0425	**40**	0.0785	0.6350
$1\frac{3}{4}$	1.7952	14.5143	6.3500	0.4947	**4**	**42**	0.0748	0.6048
1.7952	**$1\frac{3}{4}$**	14.1489	**7**	0.4488	3.6286	**44**	0.0714	0.5773
1.8143	1.7316	**14**	7.1808	$\frac{7}{16}$	3.5372	**46**	0.0683	0.5522
1.9333	**$1\frac{5}{8}$**	13.1382	7.2571	0.4329	**$3\frac{1}{2}$**	**48**	0.0654	0.5292
1.9538	1.6079	**13**	**8**	0.3927	3.1750	**50**	0.0628	0.5080
2	1.5708	12.7000	8.3776	$\frac{3}{8}$	3.0319	50.2656	$\frac{1}{16}$	0.5053
2.0944	**$1\frac{1}{2}$**	12.1276	8.4667	0.3711	**3**	50.8000	0.0618	$\frac{1}{2}$
2.1167	1.4842	**12**	**9**	0.3491	2.8222	**56**	0.0561	0.4536
$2\frac{1}{4}$	1.3963	11.2889	**10**	0.3142	2.5400	**60**	0.0524	0.4233

The module of a gear is the pitch diameter divided by the number of teeth. The module may be expressed in any units; but when no units are stated, it is understood to be in millimeters. The metric module, therefore, equals the pitch diameter in millimeters divided by the number of teeth. To find the metric module equivalent to a given diametral pitch, divide 25.4 by the diametral pitch. To find the diametral pitch equivalent to a given module, divide 25.4 by the module. (25.4 = number of millimeters per inch.)

Caliper Measurement of Gear Tooth.—In cutting gear teeth, the general practice is to adjust the cutter or hob until it grazes the outside diameter of the blank; the cutter is then sunk to the total depth of the tooth space plus whatever slight additional amount may be required to provide the necessary play or backlash between the teeth. If the outside diameter of the gear blank is correct, the tooth thickness should also be correct after the cutter has been sunk to the depth required for a given pitch and backlash. However, it is advisable to check the tooth thickness by measuring it, and the vernier gear-tooth caliper (see illustration) is commonly used in measuring the thickness.

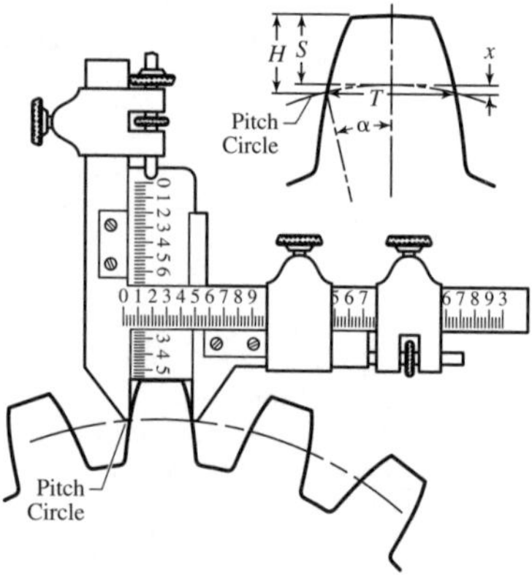

Method of Setting a Gear Tooth Caliper

The vertical scale of this caliper is set so that when it rests upon the top of the tooth as shown, the lower ends of the caliper jaws will be at the height of the pitch circle; the horizontal scale then shows the chordal thickness of the tooth at this point. If the gear is being cut on a milling machine or with the type of gear-cutting machine employing a formed milling cutter, the tooth thickness is checked by first taking a trial cut for a short distance at one side of the blank; then the gear blank is indexed for the next space and another cut is taken far enough to mill the full outline of the tooth. The tooth thickness is then measured.

Before the gear-tooth caliper can be used, it is necessary to determine both the correct chordal thickness and the chordal addendum (or "corrected addendum," as it is sometimes called). The vertical scale is set to the chordal addendum, thus locating the ends of the jaws at the height of the pitch circle. The rules or formulas to use in determining the chordal thickness and chordal addendum will depend upon the outside diameter of the gear; for example, if the outside diameter of a small pinion is enlarged to avoid undercut and improve the tooth action, this must be taken into account in figuring the chordal thickness and chordal addendum as shown by the accompanying rules. The detail of a gear tooth, included with the gear-tooth caliper illustration, represents the chordal thickness T, the addendum S, and the chordal addendum H.

Checking Spur Gear Size by Chordal Measurement Over Two or More Teeth.— Another method of checking gear sizes that is generally available is illustrated by the diagram accompanying Table 9. A vernier caliper is used to measure the distance M over two or more teeth. The diagram illustrates the measurement over two teeth (or with one intervening tooth space), but three or more teeth might be included, depending upon the pitch. The jaws of the caliper are merely held in contact with the sides or profiles of the teeth and perpendicular to the axis of the gear. Measurement M for involute teeth of the correct size is determined as follows.

Table 9. Chordal Dimensions over Spur Gear Teeth of 1 Diametral Pitch

Find value of *M* under pressure angle and opposite number of teeth; divide *M* by diametral pitch of gear to be measured and then subtract one-half total backlash to obtain a measurement *M* equivalent to given pitch and backlash. The number of teeth to gage or measure over is shown by the next Table 10

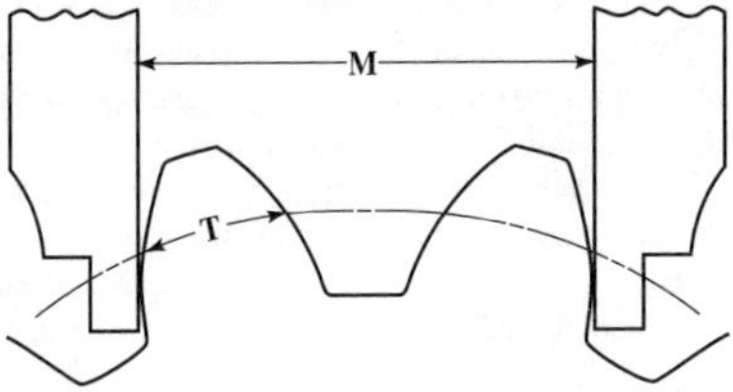

Number of Gear Teeth	*M* in Inches for 1 D.P.	Number of Gear Teeth	*M* in Inches for 1 D.P.	Number of Gear Teeth	*M* in Inches for 1 D.P.	Number of Gear Teeth	*M* in Inches for 1 D.P.
Pressure Angle, 14½ Degrees							
12	4.6267	37	7.8024	62	14.0197	87	20.2370
13	4.6321	38	10.8493	63	17.0666	88	23.2838
14	4.6374	39	10.8547	64	17.0720	89	23.2892
15	4.6428	40	10.8601	65	17.0773	90	23.2946
16	4.6482	41	10.8654	66	17.0827	91	23.2999
17	4.6536	42	10.8708	67	17.0881	92	23.3053
18	4.6589	43	10.8762	68	17.0934	93	23.3107
19	7.7058	44	10.8815	69	17.0988	94	23.3160
20	7.7112	45	10.8869	70	17.1042	95	23.3214
21	7.7166	46	10.8923	71	17.1095	96	23.3268
22	7.7219	47	10.8976	72	17.1149	97	23.3322
23	7.7273	48	10.9030	73	17.1203	98	23.3375
24	7.7326	49	10.9084	74	17.1256	99	23.3429
25	7.7380	50	10.9137	75	17.1310	100	23.3483
26	7.7434	51	13.9606	76	20.1779	101	26.3952
27	7.7488	52	13.9660	77	20.1833	102	26.4005
28	7.7541	53	13.9714	78	20.1886	103	26.4059
29	7.7595	54	13.9767	79	20.1940	104	26.4113
30	7.7649	55	13.9821	80	20.1994	105	26.4166
31	7.7702	56	13.9875	81	20.2047	106	26.4220
32	7.7756	57	13.9929	82	20.2101	107	26.4274
33	7.7810	58	13.9982	83	20.2155	108	26.4327
34	7.7683	59	14.0036	84	20.2208	109	26.4381
35	7.7917	60	14.0090	85	20.2262	110	26.4435
36	7.7971	61	14.0143	86	20.2316	…	…
Pressure Angle, 20 Degrees							
12	4.5963	30	10.7526	48	16.9090	66	23.0653
13	4.6103	31	10.7666	49	16.9230	67	23.0793
14	4.6243	32	10.7806	50	16.9370	68	23.0933
15	4.6383	33	10.7946	51	16.9510	69	23.1073
16	4.6523	34	10.8086	52	16.9650	70	23.1214
17	4.6663	35	10.8226	53	16.9790	71	23.1354
18	4.6803	36	10.8366	54	16.9930	72	23.1494
19	7.6464	37	13.8028	55	19.9591	73	26.1155
20	7.6604	38	13.8168	56	19.9731	74	26.1295
21	7.6744	39	13.8307	57	19.9872	75	26.1435
22	7.6884	40	13.8447	58	20.0012	76	26.1575
23	7.7024	41	13.8587	59	20.0152	77	26.1715
24	7.7165	42	13.8727	60	20.0292	78	26.1855
25	7.7305	43	13.8867	61	20.0432	79	26.1995
26	7.7445	44	13.9007	62	20.0572	80	26.2135
27	7.7585	45	13.9147	63	20.0712	81	26.2275
28	10.7246	46	16.8810	64	23.0373	…	…
29	10.7386	47	16.8950	65	23.0513	…	…

Table for Determining the Chordal Dimension: Table 9 above gives the chordal dimensions for one diametral pitch when measuring over the number of teeth indicated in the next Table. To obtain any chordal dimension, it is simply necessary to divide chord *M* in the table (opposite the given number of teeth) by the diametral pitch of the gear to be measured and then subtract from the quotient one-half the total backlash between the mating pair of gears. Where a small pinion is used with a large gear and all of the backlash is to be obtained by reducing the gear teeth, the total amount of backlash is subtracted from the chordal dimension of the gear and nothing from the chordal dimension of the pinion. The application of the tables will be illustrated by an example.

Table 10. Number of Teeth Included in Chordal Measurement

Tooth Range for $14\frac{1}{2}°$ Pressure Angle	Tooth Range for 20° Pressure Angle	Number of Teeth to Gage Over	Tooth Range for $14\frac{1}{2}°$ Pressure Angle	Tooth Range for 20° Pressure Angle	Number of Teeth to Gage Over
12 to 18	12 to 18	2	63 to 75	46 to 54	6
19 to 37	19 to 27	3	76 to 87	55 to 63	7
38 to 50	28 to 36	4	88 to 100	64 to 72	8
51 to 62	37 to 45	5	101 to 110	73 to 81	9

This table shows the number of teeth to be included between the jaws of the vernier caliper in measuring dimension *M* as explained in connection with Table 9.

Example: Determine the chordal dimension for checking the size of a gear having 30 teeth of 5 diametral pitch and a pressure angle of 20 degrees. A total backlash of 0.008 inch is to be obtained by reducing equally the teeth of both mating gears.

Table 9 shows that the chordal distance for 30 teeth of one diametral pitch and a pressure angle of 20 degrees is 10.7526 inches; one-half of the backlash equals 0.004 inch; hence,

$$\text{Chordal dimension} = \frac{10.7526}{5} - 0.004 = 2.1465 \text{ inches}$$

Table 10 shows that this is the chordal dimension when the vernier caliper spans four teeth, this being the number of teeth to gage over whenever gears of 20-degree pressure angle have any number of teeth from 28 to 36, inclusive. If it is considered necessary to leave enough stock on the gear teeth for a shaving or finishing cut, this allowance is simply added to the chordal dimension of the finished teeth to obtain the required measurement over the teeth for the roughing operation. It may be advisable to place this chordal dimension for rough machining on the detail drawing.

Formulas for Chordal Dimension *M*.—The required measurement *M* over spur gear teeth may be obtained by the following formula in which R = pitch radius of gear, A = pressure angle, T = tooth thickness along pitch circle, N = number of gear teeth, S = number of tooth *spaces* between caliper jaws, F = a factor depending on the pressure angle = 0.01109 for $14\frac{1}{2}°$; = 0.01973 for $17\frac{1}{2}°$; = 0.0298 for 20°; = 0.04303 for $22\frac{1}{2}°$; = 0.05995 for 25°. This factor *F* equals twice the involute function of the pressure angle.

$$M = R \times \cos A \times \frac{T}{R} + \frac{6.2832 \times S}{N} + F$$

Example: A spur gear has 30 teeth of 6 diametral pitch and a pressure angle of $14\frac{1}{2}$ degrees. Determine measurement *M* over three teeth, there being two intervening tooth spaces.

The pitch radius = $2\frac{1}{2}$ inches, the arc tooth thickness equivalent to 6 diametral pitch is 0.2618 inch (if no allowance is made for backlash) and factor *F* for $14\frac{1}{2}$ degrees = 0.01109 inch.

$$M = 2.5 \times 0.96815 \times \frac{0.2618}{2.5} + \frac{6.2832 \times 2}{30} + 0.01109 = 1.2941 \text{ inches}$$

PROPERTIES OF MATERIALS

Table 1. Standard Steel Classification

Main Group	Content	Comments
Carbon Steels	When maximum content of the main elements does not exceed the following: Mn ≤ 1.65% Si ≤ 0.60% C ≤ 0.60%	May be used with or without final heat treatment. May be annealed, normalized, case hardened, or quenched and tempered. May be killed[a], semikilled, capped, or rimmed, and, when necessary, the method of deoxidation may be specified.
Alloy Steels	The maximum range of elements exceeds the above amounts. Steels containing up to 3.99% Cr, and smaller amounts (generally 1–4%) of other alloying elements.	Alloy steels are always killed, but special deoxidation or melting practices, including vacuum, may be specified for special critical applications.
Stainless Steels	Generally contains at least 10% Cr, with or without other elements. Few contain more than 30% Cr or less than 50% Fe. In the US the stainless steel classification includes those steels containing 4% Cr.	In the broadest sense, this category can be divided into three groups based on structure: austenitic-(400 Series) nonmagnetic in the annealed condition. Nonhardenable; can be hardened by cold working. The general purpose grade is widely known as 18-8 (Cr-Ni). Ferritic-(400 Series) always magnetic and contain Cr but no Ni. Basic grade contains 17% Cr. This group also contains a 12% Cr steel with other elements, such as Al or Ti, added to prevent hardening. Martensitic-(300 Series) Magnetic and can be hardened by quenching and tempering. Basic grade contains 12% Cr. This series contains more than 10 standard compositions that include small amounts of Ni and other elements.

[a] Killed (defined)–Deoxidized with a strong deoxidizing agent such as silicon, aluminum, or manganese in order to reduce the oxygen content to such a level that no reaction occurs between carbon and oxygen during solidification.

Cr-chromium; Fe-iron; Si-silicon; C-copper; Mn-manganese; Ni-nickel; Ti-titanium

Table 2. Classification of Tool Steels

Category Designation	Letter Symbol	Group Designation	Application Type
High-speed tool steels	M	Molybdenum types	Cutting tools
	M50–M52	Intermediate types	
	T	Tungsten types	
Hot-work tool steels	H1–H19	Chromium types	Hot-work
	H20–H39	Tungsten types	
	H40–H59	Molybdenum types	
Cold-work tool steels	D	High-carbon, high-chromium types	Cold-work
	A	Medium-alloy, air-hardening types	
	O	Oil-hardening types	
Shock-resistant tool steels	S	—	Cold-work; hot-work (some grades)
Low-carbon tool and mold steels	P	—	Cold-work
Special-purpose tool steels	L	Low-alloy types	Cold-work
	F	Carbon-tungsten types	
Water-hardening tool steels	W	—	Cold-work

The following detailed discussion of tool steels will be in agreement with these categories, showing for each type the percentages of the major alloying elements. However, these values are for identification only; elements in tool steels of different producers in the mean analysis of the individual types may deviate from the listed percentages.

Table 3. AISI–SAE System of Designating Carbon and Alloy Steels

AISI-SAE Designation[a]		Type of Steel and Nominal Alloy Content (%)
		Carbon Steels
10xx		Plain Carbon (Mn 1.00% max.)
11xx		Resulfurized
12xx		Resulfurized and Rephosphorized
15xx		Plain Carbon (Max. Mn range 1.00 to 1.65%)
		Manganese(Mn) Steels
13xx		Mn 1.75
		Nickel(Ni) Steels
23xx		Ni 3.50
25xx		Ni 5.00
		Nickel(Ni)–Chromium(Cr) Steels
31xx		Ni 1.25; Cr 0.65 and 0.80
32xx		Ni 1.75; Cr 1.07
33xx		Ni 3.50; Cr 1.50 and 1.57
34xx		Ni 3.00; Cr 0.77
		Molybdenum (Mo) Steels
40xx		Mo 0.20 and 0.25
44xx		Mo 0.40 and 0.52
		Chromium(Cr)–Molybdenum(Mo) Steels
41xx		Cr 0.50, 0.80, and 0.95; Mo 0.12, 0.20, 0.25, and 0.30
		Nickel(Ni)–Chromium(Cr)–Molybdenum(Mo) Steels
43xx		Ni 1.82; Cr 0.50 and 0.80; Mo 0.25
43BVxx		Ni 1.82; Cr 0.50; Mo 0.12 and 0.35; V 0.03 min.
47xx		Ni 1.05; Cr 0.45; Mo 0.20 and 0.35
81xx		Ni 0.30; Cr 0.40; Mo 0.12
86xx		Ni 0.55; Cr 0.50; Mo 0.20
87xx		Ni 0.55; Cr 0.50; Mo 0.25
88xx		Ni 0.55; Cr 0.50; Mo 0.35
93xx		Ni 3.25; Cr 1.20; Mo 0.12
94xx		Ni 0.45; Cr 0.40; Mo 0.12
97xx		Ni 0.55; Cr 0.20; Mo 0.20
98xx		Ni 1.00; Cr 0.80; Mo 0.25
		Nickel(Ni)–Molybdenum(Mo) Steels
46xx		Ni 0.85 and 1.82; Mo 0.20 and 0.25
48xx		Ni 3.50; Mo 0.25
		Chromium(Cr) Steels
50xx		Cr 0.27, 0.40, 0.50, and 0.65
51xx		Cr 0.80, 0.87, 0.92, 0.95, 1.00, and 1.05
50xxx		Cr 0.50; C 1.00 min.
51xxx		Cr 1.02; C 1.00 min.
52xxx		Cr 1.45; C 1.00 min.
		Chromium(Cr)–Vanadium(V) Steels
61xx		Cr 0.60, 0.80, and 0.95; V 0.10 and 0.15 min
		Tungsten(W)–Chromium(V) Steels
72xx		W 1.75; Cr 0.75
		Silicon(Si)–Manganese(Mn) Steels
92xx		Si 1.40 and 2.00; Mn 0.65, 0.82, and 0.85; Cr 0.00 and 0.65
		High-Strength Low-Alloy Steels
9xx		Various SAE grades
xxBxx		B denotes boron steels
xxLxx		L denotes leaded steels
AISI	SAE	Stainless Steels
2xx	302xx	Chromium(Cr)–Manganese(Mn)–Nickel(Ni) Steels
3xx	303xx	Chromium(Cr)–Nickel(Ni) Steels
4xx	514xx	Chromium(Cr) Steels
5xx	515xx	Chromium(Cr) Steels

[a] xx in the last two digits of the carbon and low-alloy designations (but not the stainless steels) indicates that the carbon content (in hundredths of a percent) is to be inserted.

Table 4. Classification, Approximate Compositions, and Properties Affecting Selection of Tool and Die Steels
(From SAE Recommended Practice)

Type of Tool Steel	Chemical Composition[a]								Non-warping Prop.	Safety in Hardening	Tough-ness	Depth of Hardening	Wear Resis-tance
	C	Mn	Si	Cr	V	W	Mo	Co					
Water-Hardening													
0.80 Carbon	70–0.85	[b]	[b]	[b]	…	…	…	…	Poor	Fair	Good[c]	Shallow	Fair
0.90 Carbon	0.85–0.95	[b]	[b]	[b]	…	…	…	…	Poor	Fair	Good[c]	Shallow	Fair
1.00 Carbon	0.95–1.10	[b]	[b]	[b]	…	…	…	…	Poor	Fair	Good[c]	Shallow	Good
1.20 Carbon	1.10–1.30	[b]	[b]	[b]	…	…	…	…	Poor	Fair	Good[c]	Shallow	Good
0.90 Carbon–V	0.85–0.95	[b]	[b]	[b]	0.15–0.35	…	…	…	Poor	Fair	Good	Shallow	Fair
1.00 Carbon–V	0.95–1.10	[b]	[b]	[b]	0.15–0.35	…	…	…	Poor	Fair	Good	Shallow	Good
1.00 Carbon–VV	0.90–1.10	[b]	[b]	[b]	0.35–0.50	…	…	…	Poor	Fair	Good	Shallow	Good
Oil-Hardening													
Low-Manganese	0.90	1.20	0.25	0.50	0.20[d]	0.50	…	…	Good	Good	Fair	Deep	Good
High-Manganese	0.90	1.60	0.25	0.35[d]	0.20[d]	…	0.30[d]	…	Good	Good	Fair	Deep	Good
High-Carbon, High-Chromium[e]	2.15	0.35	0.35	12.00	0.80[d]	0.75[d]	0.80[d]	…	Good	Good	Poor	Through	Best
Chromium	1.00	0.35	0.25	1.40	…	…	0.40	…	Fair	Good	Fair	Deep	Good
Molybdenum Graphitic	1.45	0.75	1.00	…	…	…	0.25	…	Fair	Good	Fair	Deep	Good
Nickel–Chromium[f]	0.75	0.70	0.25	0.85	0.25[d]	…	0.50[d]	…	Fair	Good	Fair	Deep	Fair
Air Hardening													
High-Carbon, High-Chromium	1.50	0.40	0.40	12.00	0.80[d]	…	0.90	0.60[d]	Best	Best	Fair	Through	Best
5 Percent Chromium	1.00	0.60	0.25	5.25	0.40[d]	…	1.10	…	Best	Best	Fair	Through	Good
High-Carbon, High-Chromium–Cobalt	1.50	0.40	0.40	12.00	0.80[d]	…	0.90	3.10	Best	Best	Fair	Through	Best
Shock-Resisting													
Chromium–Tungsten	0.50	0.25	0.35	1.40	0.20	2.25	0.40[d]	…	Fair	Good	Good	Deep	Fair
Silicon–Molybdenum	0.50	0.40	1.00	…	0.25[d]	…	0.50	…	Poor[g]	Poor[h]	Best	Deep	Fair
Silicon–Manganese	0.55	0.80	2.00	0.30[d]	0.25[d]	…	0.40[d]	…	Poor[g]	Poor[h]	Best	Deep	Fair
Hot-Work													
Chromium–Molybdenum–Tungsten	0.35	0.30	1.00	5.00	0.25[d]	1.25	1.50	…	Good	Good	Good	Through	Fair
Chromium–Molybdenum–V	0.35	0.30	1.00	5.00	0.40	…	1.50	…	Good	Good	Good	Through	Fair
Chromium–Molybdenum–VV	0.35	0.30	1.00	5.00	0.90	…	1.50	…	Good	Good	Good	Through	Fair

Table 4. *(Continued)* **Classification, Approximate Compositions, and Properties Affecting Selection of Tool and Die Steels**
(From SAE Recommended Practice)

Type of Tool Steel	Chemical Composition[a]								Non-warping Prop.	Safety in Hardening	Tough-ness	Depth of Hardening	Wear Resis-tance
	C	Mn	Si	Cr	V	W	Mo	Co					
Tungsten	0.32	0.30	0.20	3.25	0.40	9.00	…	…	Good	Good	Good	Through	Fair
High-Speed													
Tungsten, 18–4–1	0.70	0.30	0.30	4.10	1.10	18.00	…	…	Good	Good	Poor	Through	Good
Tungsten, 18–4–2	0.80	0.30	0.30	4.10	2.10	18.50	0.80	…	Good	Good	Poor	Through	Good
Tungsten, 18–4–3	1.05	0.30	0.30	4.10	3.25	18.50	0.70	…	Good	Good	Poor	Through	Best
Cobalt–Tungsten, 14–4–2–5	0.80	0.30	0.30	4.10	2.00	14.00	0.80	5.00	Good	Fair	Poor	Through	Good
Cobalt–Tungsten, 18–4–1–5	0.75	0.30	0.30	4.10	1.00	18.00	0.80	5.00	Good	Fair	Poor	Through	Good
Cobalt–Tungsten, 18–4–2–8	0.80	0.30	0.30	4.10	1.75	18.50	0.80	8.00	Good	Fair	Poor	Through	Good
Cobalt–Tungsten, 18–4–2–12	0.80	0.30	0.30	4.10	1.75	20.00	0.80	12.00	Good	Fair	Poor	Through	Good
Molybdenum, 8–2–1	0.80	0.30	0.30	4.00	1.15	1.50	8.50	…	Good	Fair	Poor	Through	Good
Molybdenum–Tungsten, 6–6–2	0.83	0.30	0.30	4.10	1.90	6.25	5.00	…	Good	Fair	Poor	Through	Good
Molybdenum–Tungsten, 6–6–3	1.15	0.30	0.30	4.10	3.25	5.75	5.25	…	Good	Fair	Poor	Through	Best
Molybdenum–Tungsten, 6–6–4	1.30	0.30	0.30	4.25	4.25	5.75	5.25	…	Good	Fair	Poor	Through	Best
Cobalt–Molybdenum–Tungsten, 6–6–2–8	0.85	0.30	0.30	4.10	2.00	6.00	5.00	8.00	Good	Fair	Poor	Through	Good

[a] C = carbon; Mn = manganese; Si = silicon; Cr = chromium; V = vanadium; W = tungsten; Mo = molybdenum; Co = cobalt.

[b] Carbon tool steels are usually available in four grades or qualities: *Special (Grade 1)*—The highest quality water-hardening carbon tool steel, controlled for hardenability, chemistry held to closest limits, and subject to rigid tests to ensure maximum uniformity in performance; *Extra (Grade 2)*—A high-quality water-hardening carbon tool steel, controlled for hardenability, subject to tests to ensure good service; *Standard (Grade 3)*—A good-quality water-hardening carbon tool steel, not controlled for hardenability, recommended for application where some latitude with respect to uniformity is permissible; *Commercial (Grade 4)*—A commercial-quality water-hardening carbon tool steel, not controlled for hardenability, not subject to special tests. On *special* and *extra* grades, limits on manganese, silicon, and chromium are not generally required if Shepherd hardenability limits are specified. For *standard* and *commercial* grades, limits are 0.35 max. each for Mn and Si; 0.15 max. Cr for standard; 0.20 max. Cr for commercial.

[c] Toughness decreases somewhat when depth of hardening is increased.

[d] Optional element. Steels have found satisfactory application either with or without the element present. In silicon–manganese steel listed under Shock-Resisting Steels, if chromium, vanadium, and molybdenum are not present, then hardenability will be affected.

[e] This steel may have 0.50 percent nickel as an optional element. The steel has been found to give satisfactory application either with or without the element present.

[f] Approximate nickel content of this steel is 1.50 percent.

[g] Poor when water quenched, fair when oil quenched.

[h] Poor when water quenched, good when oil quenched.

Table 5. Quick Reference Guide for Tool Steel Selection

Application Areas	Tool Steel Categories and AISI Letter Symbol						
	High-Speed Tool Steels, M and T	Hot-Work Tool Steels, H	Cold-Work Tool Steels, D, A, and O	Shock-Resisting Tool Steels, S	Mold Steels, P	Special-Purpose Tool Steels, L and F	Water-Hardening Tool Steels, W
	Examples of Typical Applications						
Cutting Tools Single-point types (lathe, planer, boring) Milling cutters Drills Reamers Taps Threading dies Form cutters	General purpose production tools: M2, T1 For increased abrasion resistance M3,M4, and M10 Heavy-duty work calling for high hot T5, T15 Heavy-duty work calling for high abrasion resistance: M42, M44		Tools with keen edges (knives, razors).Tools for operations where no high-speed is involved, yet stability in heat treatment and substantial abrasion resistance are needed	Pipe cutter wheels			Uses that do not require hot hardness or high abrasion resistance. Examples with carbon content of applicable group: Taps (1.05⁄1.10% C) Reamers (1.10⁄1.15% C) Twist drills (1.20⁄1.25% C) Files (1.35⁄1.40% C)
Hot Forging Tools and Dies Dies and inserts Forging machine plungers and pierces	To combine hot hardness with high abrasion resistance: M2, T1	Dies for presses and hammers: H20, H21 For severe conditions over extended service periods: H22 to H26, also H43	Hot trimming dies: D2	Hot trimming dies Blacksmith tools Hot swaging dies			Smith's tools (1.65⁄0.70% C) Hot chisels (0.70⁄0.75% C) Drop forging dies (0.90⁄1.00% C) Applications limited to short run production
Hot Extrusion Tools and Dies Extrusion dies and mandrels Dummy blocks Valve extrusion tools	Brass extrusion dies: T1	Extrusion dies and dummy blocks: H20 to H26 For tools that are exposed to less heat: H10 to H19		Compression molding: S1			

Table 5. *(Continued)* **Quick Reference Guide for Tool Steel Selection**

Application Areas	Tool Steel Categories and AISI Letter Symbol						
	High-Speed Tool Steels, M and T	Hot-Work Tool Steels, H	Cold-Work Tool Steels, D, A, and O	Shock-Resisting Tool Steels, S	Mold Steels, P	Special-Purpose Tool Steels, L and F	Water-Hardening Tool Steels, W
	Examples of Typical Applications						
Cold-Forming Dies Bending, forming, drawing, and deep drawing dies and punches	Burnishing tools: M1, T1	Cold heading: die-casting dies: H13	Drawing dies: 01 Coining tools: O1, D2 Forming and bending dies: A2 Thread rolling dies: D2	Hobbing and short-run applications:S1, S7 Rivet sets and rivet busters		Blanking, forming, and trimmer dies when toughness has precedence over abrasion resistance: L6	Cold-heading dies: W1 or W2 (C ≅ 1.00%) Bending dies: W1 (C ≅ 1.00%)
Shearing Tools Dies for piercing, punching, and trimming Shear blades	Special dies for cold- and hot-work: T1 For work requiring high abrasion resistance: M2, M3	For shearing knives: H11, H12 For severe hot shearing applications: M21, M25	Dies for medium runs: A2, A6 also O1 and O4 Dies for long runs: D2, D3 Trimming dies (also for hot trimming): A2	Cold and hot shear blades Hot punching and piercing tools Boilermaker's tools		Knives for work requiring high toughness: L6	Trimming dies (0.90/0.95% C) Cold blanking and punching dies (1.00% C)
Die-Casting Dies and Plastics Molds		For aluminum and lead: H11and H13 For brass: H21	A2 and A6 O1		Plastics molds: P2 to P4, and P20		
Structural Parts for Severe Service Conditions	Roller bearings for high-temperature environment: T1 Lathe centers: M2 and T1	For aircraft components (landing gear, arrester hooks, rocket cases): H11	Lathe centers: D2, D3 Arbors: O1 Bushings: A4 Gages: D2	Pawls Clutch parts		Spindles, clutch parts (where high toughness is needed): L6	Spring steel (1.10/1.15% C)
Battering Tools for Hand and Power Tool Use				Pneumatic chisels for cold-work: S5 For higher performance: S7			For intermittent use: W1 (0.80% C)

Table 6. Molybdenum High-Speed Steels

Identifying Chemical Composition and Typical Heat-Treatment Data																				
	AISI Type		M1	M2	M3 Cl. 1	M3 Cl. 2	M4	M6	M7	M10	M30	M33	M34	M36	M41	M42	M43	M44	M46	M47
Identifying Chemical Elements in Percent	C		0.80	0.85 1.00	1.05	1.20	1.30	0.80	1.00	0.85 1.00	0.80	0.90	0.90	0.80	1.10	1.10	1.20	1.15	1.25	1.10
	W		1.50	6.00	6.00	6.00	5.50	4.00	1.75	…	2.00	1.50	2.00	6.00	6.75	1.50	2.75	5.25	2.00	1.50
	Mo		8.00	5.00	5.00	5.00	4.50	5.00	8.75	8.00	8.00	9.50	8.00	5.00	3.75	9.50	8.00	6.25	8.25	9.50
	Cr		4.00	4.00	4.00	4.00	4.00	4.00	4.00	4.00	4.00	4.00	4.00	4.00	4.25	3.75	3.75	4.25	4.00	3.75
	V		1.00	2.00	2.40	3.00	4.00	1.50	2.00	2.00	1.25	1.15	2.00	2.00	2.00	1.15	1.60	2.25	3.20	1.25
	Co		…	…	…	…	…	12.00	…	…	5.00	8.00	8.00	8.00	5.00	8.00	8.25	12.00	8.25	5.00
Heat-Treatment Data	Hardening Temperature Range	°F	2150–2225	2175–2225	2200–2250	2200–2250	2200–2250	2150–2200	2150–2225	2150–2225	2200–2250	2200–2250	2200–2250	2225–2275	2175–2220	2175–2210	2175–2220	2190–2240	2175–2225	2150–2200
		°C	1177–1218	1191–1218	1204–1232	1204–1232	1204–1232	1177–1204	1177–1218	1177–1218	1204–1232	1204–1232	1204–1232	1218–1246	1191–1216	1191–1210	1191–1216	1199–1227	1191–1218	1177–1204
	Tempering Temperature Range	°F	1000–1100	1000–1160	1000–1100	1000–1100	1000–1100	1000–1100	1000–1100	1000–1100	1000–1100	1000–1100	1000–1100	1000–1100	1000–1100	950–1100	950–1100	1000–1160	975–1050	975–1100
		°C	538–593	538–627	538–593	538–593	538–593	538–593	538–593	538–593	538–593	538–593	538–593	538–593	538–593	510–593	510–593	538–627	524–566	524–594
	Approx. Tempered Hardness, RC (Rockwell C scale)		65–60	65–60	66–61	66–61	66–61	66–61	66–61	65–60	65–60	65–60	65–60	65–60	70–65	70–65	70–65	70–62	69–67	70–65
Relative Ratings of Properties (A = greatest to E = least)																				
Characteristics in Heat Treatment	Safety in Hardening		D	D	D	D	D	D	D	D	D	D	D	D	D	D	D	D	D	D
	Depth of Hardening		A	A	A	A	A	A	A	A	A	A	A	A	A	A	A	A	A	A
	Resistance to Decarburization		C	B	B	B	B	C	C	C	C	C	C	C	C	C	C	C	C	C
	Stability of Shape in Heat Treatment — Quenching Medium	Air or Salt	C	C	C	C	C	C	C	C	C	C	C	C	C	C	C	C	C	C
		Oil	D	D	D	D	D	D	D	D	D	D	D	D	D	D	D	D	D	D
Service Properties	Machinability		D	D	D	D/E	D	D	D	D	D	D	D	D	D	D	D	D	D	D
	Hot Hardness		B	B	B	B	B	A	B	B	A	A	A	A	A	A	A	A	A	A
	Wear Resistance		B	B	B	B	A	B	B	B	B	B	B	B	B	B	B	B	B	B
	Toughness		E	E	E	E	E	E	E	E	E	E	E	E	E	E	E	E	E	E

Table 7. Hot-Work Tool Steels

Identifying Chemical Composition and Typical Heat-Treatment Data																	
AISI	Group		Chromium Types						Tungsten Types						Molybdenum Types		
	Type		H10	H11	H12	H13	H14	H19	H21	H22	H23	H24	H25	H26	H41	H42	H43
Identifying Chemical Elements in Percent	C		0.40	0.35	0.35	0.35	0.40	0.40	0.35	0.35	0.35	0.45	0.25	0.50	0.65	0.60	0.55
	W		…	…	1.50	…	5.00	4.25	9.00	11.00	12.00	15.00	15.00	18.00	1.50	6.00	…
	Mo		2.50	1.50	1.50	1.50	…	…	…	…	…	…	…	…	8.00	5.00	8.00
	Cr		3.25	5.00	5.00	5.00	5.00	4.25	3.50	2.00	12.00	3.00	4.00	4.00	4.00	4.00	4.00
	V		0.40	0.40	0.40	1.00	…	2.00	…	…	…	…	…	1.00	1.00	2.00	2.00
	Co		…	…	…	…	…	4.25	…	…	…	…	…	…	…	…	…
Heat-Treatment Data	Hardening Temperature Range	°F	1850–1900	1825–1875	1825–1875	1825–1900	1850–1950	2000–2200	2000–2200	2000–2200	2000–2300	2000–2250	2100–2300	2150–2300	2000–2175	2050–2225	2000–2175
		°C	1010–1038	996–1024	996–1024	996–1038	1010–1066	1093–1204	1093–1204	1093–1204	1093–1260	1093–1232	1149–1260	1177–1260	1093–1191	1121–1218	1093–1191
	Tempering Temperature Range	°F	1000–1200	1000–1200	1000–1200	1000–1200	1100–1200	1000–1300	1100–1250	1100–1250	1200–1500	1050–1200	1050–1250	1050–1250	1050–1200	1050–1200	1050–1200
		°C	538–649	538–649	538–649	538–649	593–649	538–704	593–677	593–677	649–816	566–649	566–677	566–677	566–649	566–649	566–649
	Approx. Tempered Hardness, RC (Rockwell C scale)		56–39	54–38	55–38	53–38	47–40	59–40	54–36	52–39	47–30	55–45	44–35	58–43	60–50	60–50	58–45
Relative Ratings of Properties (A = greatest to D = least)																	
Characteristics in Heat Treatment	Safety in Hardening		A	A	A	A	A	B	B	B	B	B	B	B	C	C	C
	Depth of Hardening		A	A	A	A	A	A	A	A	A	A	A	A	A	A	A
	Resistance to Decarburization		B	B	B	B	B	B	B	B	B	B	B	B	C	B	C
	Stability of Shape in Heat Treatment — Quenching Medium	Air or Salt	B	B	B	B	C	C	C	C	…	C	C	C	C	C	C
		Oil	…	…	…	…	…	D	D	D	D	D	D	D	D	D	D
Service Properties	Machinability		C/D	C/D	C/D	C/D	D	D	D	D	D	D	D	D	D	D	D
	Hot Hardness		C	C	C	C	C	C	C	C	B	B	B	B	B	B	B
	Wear Resistance		D	D	D	D	D	C/D	C/D	C/D	C/D	C	D	C	C	C	C
	Toughness		C	B	B	B	C	C	C	C	D	D	C	D	D	D	D

Table 8. Tungsten High-Speed Tool Steels—Identifying Chemical Composition and Typical Heat-Treatment Data

AISI Type			T1	T2	T4	T5	T6	T8	T15
Identifying Chemical Elements in Percent									
C			0.75	0.80	0.75	0.80	0.80	0.75	1.50
W			18.00	18.00	18.00	18.00	20.00	14.00	12.00
Cr			4.00	4.00	4.00	4.00	4.50	4.00	4.00
V			1.00	2.00	1.00	2.00	1.50	2.00	5.00
Co			…	…	5.00	…	…	5.00	5.00
Heat-Treatment Data									
Hardening Temperature Range		°F	2300–2375	2300–2375	2300–2375	2325–2375	2325–2375	2300–2375	2200–2300
		°C	1260–1302	1260–1302	1260–1302	1274–1302	1274–1302	1260–1302	1204–1260
Tempering Temperature Range		°F	1000–1100	1000–1100	1000–1100	1000–1100	1000–1100	1000–1100	1000–1200
		°C	538–593	538–593	538–593	538–593	538–593	538–593	538–649
Approx. Tempered Hardness, RC (Rockwell C scale)			65–60	66–61	66–62	65–60	65–60	65–60	68–63
Characteristics in Heat Treatment[a]									
Safety in Hardening			C	C	D	D	D	D	D
Depth of Hardening			A	A	A	A	A	A	A
Resistance to Decarburization			A	A	B	C	C	B	B
Stability of Shape in Heat Treatment	Quenching Medium	Air or Salt	C	C	C	C	C	C	C
		Oil	D	D	D	D	D	D	D
Service Properties									
Machinability			D	D	D	D	D/E	D	D/E
Hot Hardness			B	B	A	A	A	A	A
Wear Resistance			B	B	B	B	B	B	A
Toughness			E	E	E	E	E	E	E

[a] Relative Ratings of Properties (A = greatest to E = least)

Table 9. Cold-Work Tool Steels

Identifying Chemical Composition and Typical Heat-Treatment Data																			
AISI	Group		High-Carbon, High-Chromium Types					Medium-Alloy, Air-Hardening Types								Oil-Hardening Types			
	Types		D2	D3	D4	D5	D7	A2	A3	A4	A6	A7	A8	A9	A10	O1	O2	O6	O7
Identifying Chemical Elements in Percent	C		1.50	2.25	2.25	1.50	2.35	1.00	1.25	1.00	0.70	2.25	0.55	0.50	1.35	0.90	0.90	1.45	1.20
	Mn		…	…	…	…	…	…	…	2.00	2.00	…	…	…	1.80	1.00	1.60	…	…
	Si		…	…	…	…	…	…	…	…	…	…	…	…	1.25	…	…	1.00	…
	W		…	…	…	…	…	…	…	…	…	1.00	1.25	…	…	0.50	…	…	1.75
	Mo		1.00	…	1.00	1.00	1.00	1.00	1.00	1.00	1.25	1.00	1.25	1.40	1.50	…	…	0.25	…
	Cr		12.00	12.00	12.00	12.00	12.00	5.00	5.00	1.00	1.00	5.25	5.00	5.00	…	0.50	…	…	0.75
	V		1.00	…	…	…	4.00	…	1.00	…	…	4.75	…	1.00	…	…	…	…	…
	Co		…	…	…	3.00	…	…	…	…	…	…	…	…	…	…	…	…	…
	Ni		…	…	…	…	…	…	…	…	…	…	…	1.50	1.80	…	…	…	…
Heat-Treatment Data	Hardening Temperature Range	°F	1800–1875	1700–1800	1775–1850	1800–1875	1850–1950	1700–1800	1750–1850	1500–1600	1525–1600	1750–1800	1800–1850	1800–1875	1450–1500	1450–1500	1400–1475	1450–1500	1550–1525
		°C	982–1024	927–982	968–1010	982–1024	1010–1066	927–982	954–1010	816–871	829–871	954–982	982–1010	982–1024	788–816	788–816	760–802	788–816	843–829
	Quenching Medium		Air	Oil	Air	Air	Air	Air	Air	Air	Air	Air	Air	Air	Air	Oil	Oil	Oil	Oil
	Tempering Temperature Range	°F	400–1000	400–1000	400–1000	400–1000	300–1000	350–1000	350–1000	350–800	300–800	300–1000	350–1100	950–1150	350–800	350–500	350–500	350–600	350–550
		°C	204–538	204–538	204–538	204–538	149–538	177–538	177–538	177–427	149–427	149–538	177–593	510–621	177–427	177–260	177–260	177–316	177–288
	Approx. Tempered Hardness, RC (Rockwell C scale)		61–54	61–54	61–54	61–54	65–58	62–57	65–57	62–54	60–54	67–57	60–50	56–35	62–55	62–57	62–57	63–58	64–58
Relative Ratings of Properties (A = greatest to E = least)																			
Characteristics in Heat Treatment	Safety in Hardening		A	C	A	A	A	A	A	A	A	A	A	A	A	B	B	B	B
	Depth of Hardening		A	A	A	A	A	A	A	A	A	A	A	A	A	B	B	B	B
	Resistance to Decarburization		B	B	B	B	B	B	B	A/B	A/B	B	B	B	A/B	A	A	A	A
	Stability of Shape in Heat Treatment		A	B	A	A	A	A	A	A	A	A	A	A	A	B	B	B	B
Service Properties	Machinability		E	E	E	E	E	D	D	D/E	D/E	E	D	D	C/D	C	C	B	C
	Hot Hardness		C	C	C	C	C	C	C	D	D	C	C	C	D	E	E	E	E
	Wear Resistance		B/C	B	B	B/C	A	C	B	C/D	C/D	A	C/D	C/D	C	D	D	D	D
	Toughness		E	E	E	E	E	D	D	D	D	E	C	C	D	D	D	D	C

Table 10. Shock-Resisting, Mold, and Special-Purpose Tool Steels

Identifying Chemical Composition and Typical Heat-Treatment Data																		
AISI	Category		Shock-Resisting Tool Steels				Mold Steels							Special-Purpose Tool Steels				
	Types		S1	S2	S5	S7	P2	P3	P4	P5	P6	P20	P21[a]	L2[b]	L3[b]	L6	F1	F2
Identifying Elements in Percent	C		0.50	0.50	0.55	0.50	0.07	0.10	0.07	0.10	0.10	0.35	0.20	0.50/1.10	1.00	0.70	1.00	1.25
	Mn		...	...	0.80	...	...	...	...	...	...	...	...	...	...	...	...	...
	Si		...	1.00	2.00	...	...	...	...	...	...	...	...	...	...	...	...	...
	W		2.50	...	...	...	...	...	...	...	...	...	...	...	...	...	1.25	3.50
	Mo		...	0.50	0.40	1.40	0.20	...	0.75	...	...	0.40	...	...	...	0.25	...	...
	Cr		1.50	...	...	3.25	2.00	0.60	5.00	2.25	1.50	1.25	...	1.00	1.50	0.75	...	...
	V		...	...	...	...	...	...	...	...	...	...	...	0.20	0.20	...	...	...
	Ni		...	...	...	...	0.50	1.25	...	...	3.50	...	4.00	...	...	1.50	...	...
Heat-Treatment Data	Hardening Temperature	°F	1650–1750	1550–1650	1600–1700	1700–1750	1525–1550[c]	1475–1525[c]	1775–1825[c]	1550–1600[c]	1450–1500[c]	1500–1600[c]	Soln. treat.	1550–1700	1500–1600	1450–1550	1450–1600	1450–1600
		°C	899–954	843–899	871–927	927–955	829–843c	802–829c	968–996c	843–871c	788–816c	816–871c	Soln. treat.	843–927	816–871	788–843	788–871	788–871
	Tempering Temp. Range	°F	400–1200	350–800	350–800	400–1150	350–500	350–500	350–900	350–500	350–450	900–1100	Aged	350–1000	350–600	350–1000	350–500	350–500
		°C	204–649	177–427	177–427	204–621	177–260	177–260	177–482	177–260	177–232	482–593	Aged	177–538	177–316	177–538	177–260	177–260
	Approx. Tempered Hardness, RC (Rockwell C scale)		58–40	60–50	60–50	57–45	64–58[d]	64–58[d]	64–58[d]	64–58[d]	61–58[d]	37–28[d]	40–30	63–45	63–56	62–45	64–60	65–62
Relative Ratings of Properties (A = greatest to E = least)																		
Characteristics in Heat Treatment	Safety in Hardening		C	E	C	B/C	C	C	C	C	C	C	A	D	D	C	E	E
	Depth of Hardening		B	B	B	A	B[e]	B[e]	B[e]	B[e]	A[e]	B	A	B	B	B	C	C
	Resist. to Decarb.		B	C	C	B	A	A	A	A	A	A	A	A	A	A	A	A
	Stability of Shape in Heat Treatment — Quench. Med.	Air	...	...	...	A	...	...	B	...	B	C	A	...	...	...	...	...
		Oil	D	...	D	C	C	C	...	C	C	...	A	D	D	C	...	...
		Water[f]	...	E	...	...	...	...	...	E	...	...	...	E	E	...	E	E
Service Properties	Machinability		D	C/D	C/D	D	C/D	D	D/E	D	D	C/D	D	C	C	D	C	D
	Hot Hardness		D	E	E	C	E	E	D	E	E	E	D	E	E	E	E	E
	Wear Resistance		D/E	D/E	D/E	D/E	D	D	C	D	D	D/E	D	D/E	D	D	D	B/C
	Toughness		B	A	A	B	C	C	C	C	C	C	D	B	D	B	E	E

[a] Contains also about 1.20 percent A1. Solution treated in hardening.
[b] Quenched in oil.
[c] After carburizing.
[d] Carburized case.
[e] Core hardenability.
[f] Sometimes brine is used.

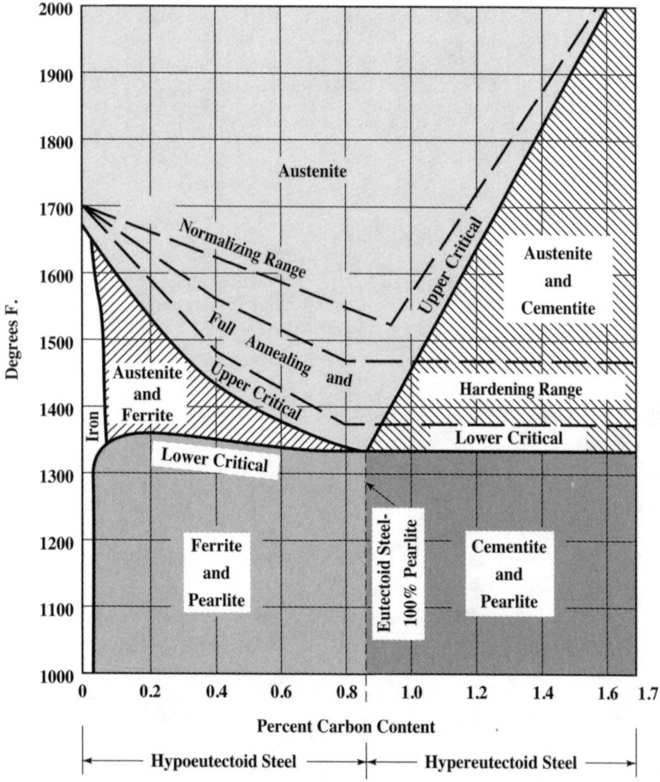

Fig. 1. Phase Diagram of Carbon Steel

Table 11. Temperature of Steel as Indicated by Color Related to Heat Treatment of Steel Cutting Tools

Temperature		Color of Steel	Processes and Tool Tempering Temperatures
°F	°C		
420	216	Faint yellow	Carbon tool steel tempering (300–1050°F), hammer heads, thin knife blades, razors, wood engraving burins, scribes.
430	221	Very pale yellow	Carbon tool steel tempering, reamers, hollow mills (solid type) for roughing on automatic screw machines, forming tools for automatic screw machines, cut off tools for automatic screw machines, formed milling cutters.
440	227	Light yellow	Carbon tool steel tempering, lathe tools, milling cutters, reamers, profile cutters for milling machines (440–450°F), drill bits (400–445°F)

Table 11. *(Continued)* **Temperature of Steel as Indicated by Color Related to Heat Treatment of Steel Cutting Tools**

Temperature		Color of Steel	Processes and Tool Tempering Temperatures
°F	°C		
450	232	Pale straw-yellow	Carbon tool steel tempering, razors, twist drills for hard service, centering tools for automatic screw machines, hatchets and axes
460	238	Straw-yellow	Carbon tool steel tempering, drill bits, dies, punches, reamers, thread rolling dies, counterbores and countersinks, snaps for pneumatic hammers (harden full length, temper to 460°F, then bring point to 520°F).
470	243	Deep straw-yellow	Carbon tool steel tempering, various kinds of wood cutting tools (470–490°F)
480	249	Dark yellow	Carbon tool steel tempering, drills, taps, knurls (485°F), cutters for tube or pipe-cutting machines.
490	254	Yellowish-brown	Carbon tool steel tempering. Thread dies for tool steel or steel tube (495°F), cold chisels, stone carving tools, punches, dies.
500	260	Brown-yellow	Carbon tool steel tempering, wood (chipping) chisels, saws, drifts, thread dies for general work, taps 1 inch or over, for use on automatic screw machines, nut taps1 inch and under.
510	266	Spotted red-brown	Carbon tool steel tempering, taps 1 inch and under, for use on automatic screw machines (515–520°F)
520	271	Brown-purple	Carbon tool steel tempering
530	277	Light purple	Carbon tool steel tempering, percussive tools, thread dies to cut thread close to shoulder (525–530°F), dies for bolt threader threading to shoulder (525–540°F)
540	282	Full purple	Carbon tool steel tempering, punches (center), wood-carving gouges
550	288	Dark purple/violet	Carbon tool steel tempering, spatulas, table knives, shear blades
560	293	Full blue	Carbon tool steel tempering, gears, screwdriver blades, springs
570	299	Dark blue	Carbon tool steel tempering, springs
600	316	Medium blue	Carbon tool steel tempering, springs, spokeshave blades, scrapers, thin knife blades (580–650°F), rivet snaps (575–600°F)
640	338	Light blue	Carbon tool steel tempering, springs (650–900°F), for hard parts
700–800	371–427	Black-red—visible in low light or dark	Carbon tool steel tempering
885	474	Red—visible at twilight	Carbon tool steel tempering
975	523.9	Red—visible in daylight	Carbon tool steel tempering
1000	538	Very dark red—visible in daylight	Carbon tool steel tempering, high-speed steel tempering (1000–1100°F)
1100	593	Dark red—visible in sunlight	High-speed steel tempering
1300	704	Dark red	
1400	760	Dark cherry-red	Carbon tool steel hardening (1350–1550°F)
1475	802	Dull cherry-red	Carbon tool steel hardening
1550	843	Light cherry-red	Alloy tool steel hardening (1500–1950°F)
1650	899	Cherry-red	Alloy tool steel hardening
1800	982	Orange-red	Alloy tool steel hardening
2000	1093	Yellow	
2300	1260	Yellow-white	High-speed steel hardening (2250–2400°F)
2400	1316	White	High-speed steel hardening
2500	1371	White	Welding
2750	1590	Brilliant white	
3000	1649	Dazzling blue-white	

Table 12. Comparative Hardness Scales for Steel

Rockwell C-Scale Hardness Number	Diamond Pyramid Hardness Number Vickers	Brinell Hardness Number 10-mm Ball, 3000-kgf Load			Rockwell Hardness Number		Rockwell Hardness Number Superficial Diam. Penetrator			Shore Sclero-scope Hard-ness Num-ber
		Standard Ball	Hultgren Ball	Tungsten Carbide Ball	A-Scale 60-kgf Load Diam. Penetrator	D-Scale 100-kgf Load Diam. Penetrator	15-N Scale 15-kgf Load	30-N Scale 30-kgf Load	45-N Scale 45-kgf Load	
68	**940**	…	…	…	**85.6**	**76.9**	**93.2**	**84.4**	**75.4**	97
67	**900**	…	…	…	**85.0**	**76.1**	**92.9**	**83.6**	**74.2**	95
66	**865**	…	…	…	**84.5**	**75.4**	**92.5**	**82.8**	**73.3**	92
65	**832**	…	…	**(739)**	**83.9**	**74.5**	**92.2**	**81.9**	**72.0**	91
64	**800**	…	…	**(722)**	**83.4**	**73.8**	**91.8**	**81.1**	**71.0**	88
63	**772**	…	…	**(705)**	**82.8**	**73.0**	**91.4**	**80.1**	**69.9**	87
62	**746**	…	…	**(688)**	**82.3**	**72.2**	**91.1**	**79.3**	**68.8**	85
61	**720**	…	…	**(670)**	**81.8**	**71.5**	**90.7**	**78.4**	**67.7**	83
60	**697**	…	**(613)**	**(654)**	**81.2**	**70.7**	**90.2**	**77.5**	**66.6**	81
59	**674**	…	**(599)**	**(634)**	**80.7**	**69.9**	**89.8**	**76.6**	**65.5**	80
58	**653**	…	**(587)**	**615**	**80.1**	**69.2**	**89.3**	**75.7**	**64.3**	78
57	**633**	…	**(575)**	**595**	**79.6**	**68.5**	**88.9**	**74.8**	**63.2**	76
56	**613**	…	**(561)**	**577**	**79.0**	**67.7**	**88.3**	**73.9**	**62.0**	75
55	**595**	…	**(546)**	**560**	**78.5**	**66.9**	**87.9**	**73.0**	**60.9**	74
54	**577**	…	**(534)**	**543**	**78.0**	**66.1**	**87.4**	**72.0**	**59.8**	72
53	**560**	…	**(519)**	**525**	**77.4**	**65.4**	**86.9**	**71.2**	**58.6**	71
52	**544**	**(500)**	**(508)**	**512**	**76.8**	**64.6**	**86.4**	**70.2**	**57.4**	69
51	**528**	**(487)**	**(494)**	**496**	**76.3**	**63.8**	**85.9**	**69.4**	**56.1**	68
50	**513**	**(475)**	**(481)**	**481**	**75.9**	**63.1**	**85.5**	**68.5**	**55.0**	67
49	**498**	**(464)**	**(469)**	**469**	**75.2**	**62.1**	**85.0**	**67.6**	**53.8**	66
48	**484**	**(451)**	**(455)**	**455**	**74.7**	**61.4**	**84.5**	**66.7**	**52.5**	64
47	**471**	**442**	**443**	**443**	**74.1**	**60.8**	**83.9**	**65.8**	**51.4**	63
46	**458**	**432**	**432**	**432**	**73.6**	**60.0**	**83.5**	**64.8**	**50.3**	62
45	**446**	**421**	**421**	**421**	**73.1**	**59.2**	**83.0**	**64.0**	**49.0**	60
44	**434**	**409**	**409**	**409**	**72.5**	**58.5**	**82.5**	**63.1**	**47.8**	58
43	**423**	**400**	**400**	**400**	**72.0**	**57.7**	**82.0**	**62.2**	**46.7**	57
42	**412**	**390**	**390**	**390**	**71.5**	**56.9**	**81.5**	**61.3**	**45.5**	56
41	**402**	**381**	**381**	**381**	**70.9**	**56.2**	**80.9**	**60.4**	**44.3**	55
40	**392**	**371**	**371**	**371**	**70.4**	**55.4**	**80.4**	**59.5**	**43.1**	54
39	**382**	**362**	**362**	**362**	**69.9**	**54.6**	**79.9**	**58.6**	**41.9**	52
38	**372**	**353**	**353**	**353**	**69.4**	**53.8**	**79.4**	**57.7**	**40.8**	51
37	**363**	**344**	**344**	**344**	**68.9**	**53.1**	**78.8**	**56.8**	**39.6**	50
36	**354**	**336**	**336**	**336**	**68.4**	**52.3**	**78.3**	**55.9**	**38.4**	49
35	**345**	**327**	**327**	**327**	**67.9**	**51.5**	**77.7**	**55.0**	**37.2**	48
34	**336**	**319**	**319**	**319**	**67.4**	**50.8**	**77.2**	**54.2**	**36.1**	47
33	**327**	**311**	**311**	**311**	**66.8**	**50.0**	**76.6**	**53.3**	**34.9**	46
32	**318**	**301**	**301**	**301**	**66.3**	**49.2**	**76.1**	**52.1**	**33.7**	44
31	**310**	**294**	**294**	**294**	**65.8**	**48.4**	**75.6**	**51.3**	**32.5**	43
30	**302**	**286**	**286**	**286**	**65.3**	**47.7**	**75.0**	**50.4**	**31.3**	42
29	**294**	**279**	**279**	**279**	**64.7**	**47.0**	**74.5**	**49.5**	**30.1**	41
28	**286**	**271**	**271**	**271**	**64.3**	**46.1**	**73.9**	**48.6**	**28.9**	41
27	**279**	**264**	**264**	**264**	**63.8**	**45.2**	**73.3**	**47.7**	**27.8**	40
26	**272**	**258**	**258**	**258**	**63.3**	**44.6**	**72.8**	**46.8**	**26.7**	38
25	**266**	**253**	**253**	**253**	**62.8**	**43.8**	**72.2**	**45.9**	**25.5**	38
24	**260**	**247**	**247**	**247**	**62.4**	**43.1**	**71.6**	**45.0**	**24.3**	37
23	**254**	**243**	**243**	**243**	**62.0**	**42.1**	**71.0**	**44.0**	**23.1**	36
22	**248**	**237**	**237**	**237**	**61.5**	**41.6**	**70.5**	**43.2**	**22.0**	35
21	**243**	**231**	**231**	**231**	**61.0**	**40.9**	**69.9**	**42.3**	**20.7**	35
20	**238**	**226**	**226**	**226**	**60.5**	**40.1**	**69.4**	**41.5**	**19.6**	34
(18)	230	219	219	219	…	…	…	…	…	33
(16)	222	212	212	212	…	…	…	…	…	32
(14)	213	203	203	203	…	…	…	…	…	31
(12)	204	194	194	194	…	…	…	…	…	29
(10)	196	187	187	187	…	…	…	…	…	28
(8)	188	179	179	179	…	…	…	…	…	27
(6)	180	171	171	171	…	…	…	…	…	26
(4)	173	165	165	165	…	…	…	…	…	25
(2)	166	158	158	158	…	…	…	…	…	24
(0)	160	152	152	152	…	…	…	…	…	24

Note: The values in this table shown in **boldface** type correspond to those shown in American Society for Testing and Materials Specification E140-67. Values in ( ) are beyond the normal range and are given for information only.

Table 13. Comparative Hardness Scales for Unhardened Steel, Soft-Temper Steel, Grey and Malleable Cast Iron, and Nonferrous Alloys

Rockwell Hardness Number			Rockwell Superficial Hardness Number			Rockwell Hardness Number			Brinell Hardness Number	
Rockwell B scale $\frac{1}{16}$" Ball Penetrator 100-kg Load	Rockwell F scale $\frac{1}{16}$" Ball Penetrator 60-kg Load	Rockwell G scale $\frac{1}{16}$" Ball Penetrator 150-kg Load	Rockwell Superficial 15-T scale $\frac{1}{16}$" Ball Penetrator 15-kg Load	Rockwell Superficial 30-T scale $\frac{1}{16}$" Ball Penetrator 30-kg Load	Rockwell Superficial 45-T scale $\frac{1}{16}$" Ball Penetrator 45-kg Load	Rockwell E scale $\frac{1}{8}$" Ball Penetrator 100-kg Load	Rockwell K scale $\frac{1}{8}$" Ball Penetrator 150-kg Load	Rockwell A scale "Brale" Penetrator 60-kg Load	Brinell Scale 10-mm Standard Ball 500-kg Load	Brinell Scale 10-mm Standard Ball 3000-kg Load
100	…	82.5	93.0	82.0	72.0	…	…	61.5	201	240
99	…	81.0	92.5	81.5	71.0	…	…	61.0	195	234
98	…	79.0	…	81.0	70.0	…	…	60.0	189	228
97	…	77.5	92.0	80.5	69.0	…	…	59.5	184	222
96	…	76.0	…	80.0	68.0	…	…	59.0	179	216
95	…	74.0	91.5	79.0	67.0	…	…	58.0	175	210
94	…	72.5	…	78.5	66.0	…	…	57.5	171	205
93	…	71.0	91.0	78.0	65.5	…	…	57.0	167	200
92	…	69.0	90.5	77.5	64.5	…	100	56.5	163	195
91	…	67.5	…	77.0	63.5	…	99.5	56.0	160	190
90	…	66.0	90.0	76.0	62.5	…	98.5	55.5	157	185
89	…	64.0	89.5	75.5	61.5	…	98.0	55.0	154	180
88	…	62.5	…	75.0	60.5	…	97.0	54.0	151	176
87	…	61.0	89.0	74.5	59.5	…	96.5	53.5	148	172
86	…	59.0	88.5	74.0	58.5	…	95.5	53.0	145	169
85	…	57.5	…	73.5	58.0	…	94.5	52.5	142	165
84	…	56.0	88.0	73.0	57.0	…	94.0	52.0	140	162
83	…	54.0	87.5	72.0	56.0	…	93.0	51.0	137	159
82	…	52.5	…	71.5	55.0	…	92.0	50.5	135	156
81	…	51.0	87.0	71.0	54.0	…	91.0	50.0	133	153
80	…	49.0	86.5	70.0	53.0	…	90.5	49.5	130	150
79	…	47.5	…	69.5	52.0	…	89.5	49.0	128	147
78	…	46.0	86.0	69.0	51.0	…	88.5	48.5	126	144
77	…	44.0	85.5	68.0	50.0	…	88.0	48.0	124	141
76	…	42.5	…	67.5	49.0	…	87.0	47.0	122	139
75	99.5	41.0	85.0	67.0	48.5	…	86.0	46.5	120	137
74	99.0	39.0	…	66.0	47.5	…	85.0	46.0	118	135
73	98.5	37.5	84.5	65.5	46.5	…	84.5	45.5	116	132
72	98.0	36.0	84.0	65.0	45.5	…	83.5	45.0	114	130
71	97.5	34.5	…	64.0	44.5	100	82.5	44.5	112	127
70	97.0	32.5	83.5	63.5	43.5	99.5	81.5	44.0	110	125
69	96.0	31.0	83.0	62.5	42.5	99.0	81.0	43.5	109	123
68	95.5	29.5	…	62.0	41.5	98.0	80.0	43.0	107	121
67	95.0	28.0	82.5	61.5	40.5	97.5	79.0	42.5	106	119
66	94.5	26.5	82.0	60.5	39.5	97.0	78.0	42.0	104	117
65	94.0	25.0	…	60.0	38.5	96.0	77.5	…	102	116
64	93.5	23.5	81.5	59.5	37.5	95.5	76.5	41.5	101	114
63	93.0	22.0	81.0	58.5	36.5	95.0	75.5	41.0	99	112
62	92.0	20.5	…	58.0	35.5	94.5	74.5	40.5	98	110
61	91.5	19.0	80.5	57.0	34.5	93.5	74.0	40.0	96	108
60	91.0	17.5	…	56.5	33.5	93.0	73.0	39.5	95	107
59	90.5	16.0	80.0	56.0	32.0	92.5	72.0	39.0	94	106
58	90.0	14.5	79.5	55.0	31.0	92.0	…	71.0	38.5	92
57	89.5	13.0	…	54.5	30.0	91.0	…	70.5	38.0	91
56	89.0	11.5	79.0	54.0	29.0	90.5	…	69.5	…	90
55	88.0	10.0	78.5	53.0	28.0	90.0	…	68.5	37.5	89
54	87.5	8.5	…	52.5	27.0	89.5	…	68.0	37.0	87
53	87.0	7.0	78.0	51.5	26.0	89.0	…	67.0	36.5	86
52	86.5	5.5	77.5	51.0	25.0	88.0	…	66.0	36.0	85
51	86.0	4.0	…	50.5	24.0	87.5	…	65.0	35.5	84
50	85.5	2.5	77.0	49.5	23.0	87.0	…	64.5	35.0	83
50	85.5	2.5	77.0	49.5	23.0	87.0	…	64.5	35.0	83
49	85.0	1.0	76.5	49.0	22.0	86.5	…	63.5	…	82

Table 13. *(Continued)* **Comparative Hardness Scales for Unhardened Steel, Soft-Temper Steel, Grey and Malleable Cast Iron, and Nonferrous Alloys**

Rockwell Hardness Number			Rockwell Superficial Hardness Number			Rockwell Hardness Number			Brinell Hardness Number	
Rockwell B scale $\frac{1}{16}$" Ball Penetrator 100-kg Load	Rockwell F scale $\frac{1}{16}$" Ball Penetrator 60-kg Load	Rockwell G scale $\frac{1}{16}$" Ball Penetrator 150-kg Load	Rockwell Superficial 15-T scale $\frac{1}{16}$" Ball Penetrator 15-kg Load	Rockwell Superficial 30-T scale $\frac{1}{16}$" Ball Penetrator 30-kg Load	Rockwell Superficial 45-T scale $\frac{1}{16}$" Ball Penetrator 45-kg Load	Rockwell E scale $\frac{1}{8}$" Ball Penetrator 100-kg Load	Rockwell K scale $\frac{1}{8}$" Ball Penetrator 150-kg Load	Rockwell A scale "Brale" Penetrator 60-kg Load	Brinell Scale 10-mm Standard Ball 500-kg Load	Brinell Scale 10-mm Standard Ball 3000-kg Load
48	84.5	…	…	48.5	20.5	85.5	…	62.5	34.5	81
47	84.0	…	76.0	47.5	19.5	85.0	…	61.5	34.0	80
46	83.0	…	75.5	47.0	18.5	84.5	…	61.0	33.5	…
45	82.5	…	…	46.0	17.5	84.0	…	60.0	33.0	79
44	82.0	…	75.0	45.5	16.5	83.5	…	59.0	32.5	78
43	81.5	…	74.5	45.0	15.5	82.5	…	58.0	32.0	77
42	81.0	…	…	44.0	14.5	82.0	…	57.5	31.5	76
41	80.5	…	74.0	43.5	13.5	81.5	…	56.5	31.0	75
40	79.5	…	73.5	43.0	12.5	81.0	…	55.5	…	…
39	79.0	…	…	42.0	11.0	80.0	…	54.5	30.5	74
38	78.5	…	73.0	41.5	10.0	79.5	…	54.0	30.0	73
37	78.0	…	72.5	40.5	9.0	79.0	…	53.0	29.5	72
36	77.5	…	…	40.0	8.0	78.5	100	52.0	29.0	…
35	77.0	…	72.0	39.5	7.0	78.0	99.5	51.5	28.5	71
34	76.5	…	71.5	38.5	6.0	77.0	99.0	50.5	28.0	70
33	75.5	…	…	38.0	5.0	76.5	…	49.5	…	69
32	75.0	…	71.0	37.5	4.0	76.0	98.5	48.5	27.5	…
31	74.5	…	…	36.5	3.0	75.5	98.0	48.0	27.0	68
30	74.0	…	70.5	36.0	2.0	75.0	…	47.0	26.5	67
29	73.5	…	70.0	35.5	1.0	74.0	97.5	46.0	26.0	…
28	73.0	…	…	34.5	…	73.5	97.0	45.0	25.5	66
27	72.5	…	69.5	34.0	…	73.0	96.5	44.5	25.0	…
26	72.0	…	69.0	33.0	…	72.5	…	43.5	24.5	65
25	71.0	…	…	32.5	…	72.0	96.0	42.5	…	64
24	70.5	…	68.5	32.0	…	71.0	95.5	41.5	24.0	…
23	70.0	…	68.0	31.0	…	70.5	…	41.0	23.5	63
22	69.5	…	…	30.5	…	70.0	95.0	40.0	23.0	…
21	69.0	…	67.5	29.5	…	69.5	94.5	39.0	22.5	62
20	68.5	…	…	29.0	…	68.5	…	38.0	22.0	…
19	68.0	…	67.0	28.5	…	68.0	94.0	37.5	21.5	61
18	67.0	…	66.5	27.5	…	67.5	93.5	36.5	…	…
17	66.5	…	…	27.0	…	67.0	93.0	35.5	21.0	60
16	66.0	…	66.0	26.0	…	66.5	…	35.0	20.5	…
15	65.5	…	65.5	25.5	…	65.5	92.5	34.0	20.0	59
14	65.0	…	…	25.0	…	65.0	92.0	33.0	…	…
13	64.5	…	65.0	24.0	…	64.5	…	32.0	…	58
12	64.0	…	64.5	23.5	…	64.0	91.5	31.5	…	…
11	63.5	…	…	23.0	…	63.5	91.0	30.5	…	…
10	63.0	…	64.0	22.0	…	62.5	90.5	29.5	…	57
9	62.0	…	…	21.5	…	62.0	…	29.0	…	…
8	61.5	…	63.5	20.5	…	61.5	90.0	28.0	…	…
7	61.0	…	63.0	20.0	…	61.0	89.5	27.0	…	56
6	60.5	…	…	19.5	…	60.5	…	26.0	…	…
5	60.0	…	62.5	18.5	…	60.0	89.0	25.5	…	55
4	59.5	…	62.0	18.0	…	59.0	88.5	24.5	…	…
3	59.0	…	…	17.0	…	58.5	88.0	23.5	…	…
2	58.0	…	61.5	16.5	…	58.0	…	23.0	…	54
1	57.5	…	61.0	16.0	…	57.5	87.5	22.0	…	…
0	57.0	…	…	15.0	…	57.0	87.0	21.0	…	53

Not applicable to annealed metals of high B-scale hardness such as austenitic stainless steels, nickel and high-nickel alloys nor to cold-worked metals of low B-scale hardness such as aluminum and the softer alloys.

(Compiled by Wilson Mechanical Instrument Co.)

Table 14. Weights of Various Metals and Shapes in Pounds per Linear Foot

Metal	Rounds	Squares	Hexagons	Octagons	Flats	Round Tubing
Steel, Carbon & Alloy	$2.673 \times D^2$	$3.403 \times D^2$	$2.947 \times D^2$	$2.819 \times D^2$	$3.403 \times T \times W$	$10.680 \times (OD-w) \times w$
Steel, Stainless 300 series	$2.700 \times D^2$	$3.437 \times D^2$	$2.977 \times D^2$	$2.847 \times D^2$	$3.437 \times T \times W$	$10.787 \times (OD-w) \times w$
Steel, Stainless 400 series	$2.673 \times D^2$	$3.403 \times D^2$	$2.947 \times D^2$	$2.819 \times D^2$	$3.403 \times T \times W$	$10.680 \times (OD-w) \times w$
Aluminum 1100	$0.925 \times D^2$	$1.180 \times D^2$	$1.020 \times D^2$	$0.976 \times D^2$	$1.180 \times T \times W$	$3.700 \times (OD-w) \times w$
Aluminum 2011	$0.963 \times D^2$	$1.227 \times D^2$	$1.062 \times D^2$	$1.016 \times D^2$	$1.227 \times T \times W$	$3.849 \times (OD-w) \times w$
Aluminum 2014	$0.954 \times D^2$	$1.214 \times D^2$	$1.052 \times D^2$	$1.006 \times D^2$	$1.214 \times T \times W$	$3.811 \times (OD-w) \times w$
Aluminum 2017	$0.954 \times D^2$	$1.214 \times D^2$	$1.052 \times D^2$	$1.006 \times D^2$	$1.214 \times T \times W$	$3.811 \times (OD-w) \times w$
Aluminum 2024	$0.954 \times D^2$	$1.214 \times D^2$	$1.052 \times D^2$	$1.006 \times D^2$	$1.214 \times T \times W$	$3.811 \times (OD-w) \times w$
Aluminum 3003	$0.935 \times D^2$	$1.190 \times D^2$	$1.031 \times D^2$	$0.986 \times D^2$	$1.190 \times T \times W$	$3.736 \times (OD-w) \times w$
Aluminum 5005	$0.925 \times D^2$	$1.178 \times D^2$	$1.020 \times D^2$	$0.976 \times D^2$	$1.178 \times T \times W$	$3.697 \times (OD-w) \times w$
Aluminum 5052	$0.916 \times D^2$	$1.166 \times D^2$	$1.010 \times D^2$	$0.966 \times D^2$	$1.166 \times T \times W$	$3.660 \times (OD-w) \times w$
Aluminum 5056	$0.897 \times D^2$	$1.142 \times D^2$	$0.989 \times D^2$	$0.946 \times D^2$	$1.142 \times T \times W$	$3.584 \times (OD-w) \times w$
Aluminum 5083	$0.907 \times D^2$	$1.154 \times D^2$	$1.000 \times D^2$	$0.956 \times D^2$	$1.154 \times T \times W$	$3.623 \times (OD-w) \times w$
Aluminum 5086	$0.907 \times D^2$	$1.154 \times D^2$	$1.000 \times D^2$	$0.956 \times D^2$	$1.154 \times T \times W$	$3.623 \times (OD-w) \times w$
Aluminum 6061	$0.925 \times D^2$	$1.178 \times D^2$	$1.020 \times D^2$	$0.976 \times D^2$	$1.178 \times T \times W$	$3.697 \times (OD-w) \times w$
Aluminum 6063	$0.916 \times D^2$	$1.166 \times D^2$	$1.010 \times D^2$	$0.966 \times D^2$	$1.166 \times T \times W$	$3.660 \times (OD-w) \times w$
Aluminum 7075	$0.954 \times D^2$	$1.214 \times D^2$	$1.052 \times D^2$	$1.006 \times D^2$	$1.214 \times T \times W$	$3.811 \times (OD-w) \times w$
Aluminum 7178	$0.963 \times D^2$	$1.227 \times D^2$	$1.062 \times D^2$	$1.016 \times D^2$	$1.227 \times T \times W$	$3.849 \times (OD-w) \times w$
Beryllium	$0.631 \times D^2$	$0.803 \times D^2$	$0.696 \times D^2$	$0.665 \times D^2$	$0.803 \times T \times W$	$2.520 \times (OD-w) \times w$
Brass	$2.897 \times D^2$	$3.689 \times D^2$	$3.195 \times D^2$	$3.056 \times D^2$	$3.689 \times T \times W$	$11.577 \times (OD-w) \times w$
Cast Iron	$2.435 \times D^2$	$3.100 \times D^2$	$2.685 \times D^2$	$2.568 \times D^2$	$3.100 \times T \times W$	$9.729 \times (OD-w) \times w$
Copper	$3.058 \times D^2$	$3.893 \times D^2$	$3.372 \times D^2$	$3.225 \times D^2$	$3.893 \times T \times W$	$12.218 \times (OD-w) \times w$
Gold	$6.591 \times D^2$	$8.392 \times D^2$	$7.268 \times D^2$	$6.950 \times D^2$	$8.392 \times T \times W$	$26.337 \times (OD-w) \times w$
Lead	$3.870 \times D^2$	$4.928 \times D^2$	$4.268 \times D^2$	$4.082 \times D^2$	$4.928 \times T \times W$	$15.465 \times (OD-w) \times w$
Magnesium	$0.612 \times D^2$	$0.779 \times D^2$	$0.675 \times D^2$	$0.646 \times D^2$	$0.779 \times T \times W$	$2.446 \times (OD-w) \times w$
Molybdenum	$3.483 \times D^2$	$4.434 \times D^2$	$3.840 \times D^2$	$3.674 \times D^2$	$4.434 \times T \times W$	$13.916 \times (OD-w) \times w$
Monel	$2.897 \times D^2$	$3.689 \times D^2$	$3.195 \times D^2$	$3.056 \times D^2$	$3.689 \times T \times W$	$11.577 \times (OD-w) \times w$
Nickel	$3.039 \times D^2$	$3.869 \times D^2$	$3.351 \times D^2$	$3.206 \times D^2$	$3.869 \times T \times W$	$12.143 \times (OD-w) \times w$
Silver	$3.579 \times D^2$	$4.557 \times D^2$	$3.946 \times D^2$	$3.775 \times D^2$	$4.557 \times T \times W$	$14.301 \times (OD-w) \times w$
Tantalum	$5.667 \times D^2$	$7.215 \times D^2$	$6.248 \times D^2$	$5.977 \times D^2$	$7.215 \times T \times W$	$22.642 \times (OD-w) \times w$
Tin	$2.491 \times D^2$	$3.172 \times D^2$	$2.747 \times D^2$	$2.628 \times D^2$	$3.172 \times T \times W$	$9.953 \times (OD-w) \times w$
Titanium	$1.537 \times D^2$	$1.575 \times D^2$	$1.695 \times D^2$	$1.621 \times D^2$	$1.957 \times T \times W$	$6.141 \times (OD-w) \times w$
Tungsten	$6.580 \times D^2$	$8.379 \times D^2$	$7.256 \times D^2$	$6.941 \times D^2$	$8.379 \times T \times W$	$26.294 \times (OD-w) \times w$
Zinc	$2.435 \times D^2$	$3.100 \times D^2$	$2.685 \times D^2$	$2.568 \times D^2$	$3.100 \times T \times W$	$9.729 \times (OD-w) \times w$
Zirconium	$2.170 \times D^2$	$0.763 \times D^2$	$0.393 \times D^2$	$0.289 \times D^2$	$0.763 \times T \times W$	$8.672 \times (OD-w) \times w$

Based on information from Steel and Aluminum Stock List and Reference Book published by Earle M. Jorgensen Co.
D = Diameter or Distance across flats, OD = Outside Diameter (0.000), T = Thickness in inches, W = Width in inches, w = wall thickness (0.000)

Table 15. Weight of Round, Square, Hexagonal, and Octagonal Carbon Bar Steel
Weight in Pounds per Linear foot, from $\frac{1}{16}$ to 3 inch Diameter

Size or diameter inches	Round	Square	Hexagonal	Octagonal
$\frac{1}{16}$	0.010	0.013	0.011	0.016
$\frac{1}{8}$	0.042	0.053	0.046	0.044
$\frac{3}{16}$	0.094	0.119	0.104	0.099
$\frac{1}{4}$	0.167	0.212	0.183	0.176
$\frac{5}{16}$	0.261	0.333	0.288	0.276
$\frac{3}{8}$	0.376	0.478	0.414	0.397
$\frac{7}{16}$	0.511	0.651	0.564	0.540
$\frac{1}{2}$	0.667	0.85	0.736	0.705
$\frac{9}{16}$	0.845	1.076	0.932	0.892
$\frac{5}{8}$	1.043	1.328	1.150	1.101
$\frac{11}{16}$	1.262	1.607	1.392	1.331
$\frac{3}{4}$	1.502	1.913	1.656	1.586
$\frac{13}{16}$	1.763	2.245	1.944	1.861
$\frac{7}{8}$	2.044	2.603	2.254	2.159
$\frac{15}{16}$	2.347	2.989	2.588	2.478
1	2.670	3.401	2.944	2.819
$1\frac{1}{16}$	3.014	3.838	3.324	3.183
$1\frac{1}{8}$	3.379	4.303	3.727	3.569
$1\frac{3}{16}$	3.766	4.795	4.152	3.976
$1\frac{1}{4}$	4.173	5.313	4.601	4.405
$1\frac{5}{16}$	4.600	5.857	5.069	4.856
$1\frac{3}{8}$	5.049	6.428	5.567	5.331
$1\frac{7}{16}$	5.518	7.026	6.075	5.826
$1\frac{1}{2}$	6.008	7.651	6.625	6.344
$1\frac{9}{16}$	6.520	8.301	7.182	6.883
$1\frac{5}{8}$	7.051	8.978	7.775	7.445
$1\frac{11}{16}$	7.604	9.682	8.378	8.028
$1\frac{3}{4}$	8.178	10.413	9.018	8.633
$1\frac{13}{16}$	8.773	11.170	9.673	9.261
$1\frac{7}{8}$	9.388	11.953	10.355	9.911
$1\frac{15}{16}$	10.024	12.763	11.053	10.574
2	10.682	13.601	11.778	11.276
$2\frac{1}{16}$	11.360	14.463	12.526	11.988
$2\frac{1}{8}$	12.059	15.353	13.296	12.724
$2\frac{3}{16}$	12.778	16.27	14.085	13.478
$2\frac{1}{4}$	13.519	17.217	14.907	14.264
$2\frac{5}{16}$	14.280	18.185	15.746	15.083
$2\frac{3}{8}$	15.068	19.178	16.609	15.893
$2\frac{7}{16}$	15.866	20.201	17.495	16.752
$2\frac{1}{2}$	16.690	21.250	18.407	17.619
$2\frac{9}{16}$	17.534	22.326	19.342	18.505
$2\frac{5}{8}$	18.401	23.428	20.294	19.436
$2\frac{11}{16}$	19.287	24.557	21.272	20.364
$2\frac{3}{4}$	20.195	25.713	22.268	21.301
$2\frac{13}{16}$	21.123	26.895	23.293	22.310

Table 15. *(Continued)* **Weight of Round, Square, Hexagonal, and Octagonal Carbon Bar Steel** Weight in Pounds per Linear foot, from $\frac{1}{16}$ to 3 inch Diameter

Size or diameter inches	Round	Square	Hexagonal	Octagonal
$2\frac{7}{8}$	22.072	28.103	24.336	23.302
$2\frac{15}{16}$	23.043	29.339	25.404	24.325
3	24.034	30.601	26.504	25.38
$3\frac{1}{16}$	25.045	31.889	…	…
$3\frac{1}{8}$	26.078	33.204	…	…
$3\frac{3}{16}$	27.132	34.545	…	…
$3\frac{1}{4}$	28.206	35.913	…	…
$3\frac{5}{16}$	28.301	37.308	…	…
$3\frac{3}{8}$	30.417	38.729	…	…
$3\frac{7}{16}$	31.554	40.176	…	…
$3\frac{1}{2}$	32.712	41.651	…	…
$3\frac{9}{16}$	33.891	43.151	…	…
$3\frac{5}{8}$	35.091	44.679	…	…
$3\frac{11}{16}$	36.311	46.233	…	…
$3\frac{3}{4}$	37.552	47.813	…	…
$3\frac{13}{16}$	38.815	49.420	…	…
$3\frac{7}{8}$	40.098	51.054	…	…
$3\frac{15}{16}$	41.401	52.714	…	…
4	42.726	54.401	…	…

Table 16. Aluminum Alloy Properties and Designations

Series Group	Alloying Elements	Comments
1 xxx	See notes below	High corrosion resistance; high thermal and electrical conductivity, low mechanical properties and good workability.
2 xxx	Copper	Require solution heat treatment to obtain optimum properties. In some cases artificial aging can further increase mechanical properties.
3 xxx	Manganese	Generally not heat treatable. 3003 is used for moderate strength applications requiring good workability.
4 xxx	Silicon	Most alloys in this series are not heat treatable.
5 xxx	Magnesium	Good welding characteristics and resistance to corrosion in marine atmospheres.
6 xxx	Magnesium and Silicon	Capable of being heat treated; may be formed in the -T4 temper and then reach full -T6 properties by artificial aging. Good formability and corrosion resistance with medium strength.
7 xxx	Zinc	When coupled with a smaller percentage of magnesium results in heat treatable alloys of very high strength. Usually other elements such as Copper and Chromium are added in small quantities.
8 xxx	Other Elements	
9 xxx		Unused Series (not currently assigned)

1000 series—1 indicates an Al of 99.00% or greater purity. The last two of the four digits indicate to the nearest hundredth the amount of Al above 99.00%.

2000–8000 series—The last two of the four digit series have no significance but are used to identify different alloys in the group.

The second digit indicates alloy modification, or special control of impurities. If the second digit is zero, it indicates no special control of impurities, or the original alloy.

Table 17. Typical Thermal Properties of Various Metals

Material and Alloy Designation[a]	Density, ρ lb/in³	Melting Point, °F		Conductivity, k, Btu/hr-ft-°F	Specific Heat, C, Btu/lb/°F	Coeff. of Expansion, α μin./in.-°F
		solidus	liquidus			
Aluminum Alloys						
2011	0.102	995	1190	82.5	0.23	12.8
2017	0.101	995	1185	99.4	0.22	13.1
2024	0.100	995	1180	109.2	0.22	12.9
3003	0.099	1190	1210	111	0.22	12.9
5052	0.097	1100	1200	80	0.22	13.2
5086	0.096	1085	1185	73	0.23	13.2
6061	0.098	1080	1200	104	0.23	13.0
7075	0.101	890	1180	70	0.23	13.1
Copper-Base Alloys						
Manganese Bronze	0.302	1590	1630	61	0.09	11.8
C11000 (Electrolytic tough pitch)	0.321	1941	1981	226	0.09	9.8
C14500 (Free-machining Cu)	0.323	1924	1967	205	0.09	9.9
C17200, C17300 (Beryllium Cu)	0.298	1590	1800	62	0.10	9.9
C18200 (Chromium Cu)	0.321	1958	1967	187	0.09	9.8
C18700 (Leaded Cu)	0.323	1750	1975	218	0.09	9.8
C22000 (Commercial bronze, 90%)	0.318	1870	1910	109	0.09	10.2
C23000 (Red brass, 85%)	0.316	1810	1880	92	0.09	10.4
C26000 (Cartridge brass, 70%)	0.313	1680	1750	70	0.09	11.1
C27000 (Yellow brass)	0.306	1660	1710	67	0.09	11.3
C28000 (Muntz metal, 60%)	0.303	1650	1660	71	0.09	11.6
C33000 (Low-leaded brass tube)	0.310	1660	1720	67	0.09	11.2
C35300 (High-leaded brass)	0.306	1630	1670	67	0.09	11.3
C35600 (Extra-high-leaded brass)	0.307	1630	1660	67	0.09	11.4
C36000 (Free-machining brass)	0.307	1630	1650	67	0.09	11.4
C36500 (Leaded Muntz metal)	0.304	1630	1650	71	0.09	11.6
C46400 (Naval brass)	0.304	1630	1650	67	0.09	11.8
C51000 (Phosphor bronze, 5% A)	0.320	1750	1920	40	0.09	9.9
C54400 (Free cutting phos. bronze)	0.321	1700	1830	50	0.09	9.6
C62300 (Aluminum bronze, 9%)	0.276	1905	1915	31.4	0.09	9.0
C62400 (Aluminum bronze, 11%)	0.269	1880	1900	33.9	0.09	9.2
C63000 (Ni-Al bronze)	0.274	1895	1930	21.8	0.09	9.0
Nickel-Silver	0.314	1870	2030	17	0.09	9.0
Nickel-Base Alloys						
Nickel 200, 201, 205	0.321	2615	2635	43.3	0.11	8.5
Hastelloy C-22	0.314	2475	2550	7.5	0.10	6.9
Hastelloy C-276	0.321	2415	2500	7.5	0.10	6.2
Inconel 718	0.296	2300	2437	6.5	0.10	7.2
Monel	0.305	2370	2460	10	0.10	8.7
Monel 400	0.319	2370	2460	12.6	0.10	7.7
Monel K500	0.306	2400	2460	10.1	0.10	7.6
Monel R405	0.319	2370	2460	10.1	0.10	7.6

Table 17. *(Continued)* **Typical Thermal Properties of Various Metals**

Material and Alloy Designation[a]	Density, ρ lb/in³	Melting Point, °F solidus	Melting Point, °F liquidus	Conductivity, *k*, Btu/hr-ft-°F	Specific Heat, *C*, Btu/lb/°F	Coeff. of Expansion, α μin./in.-°F
Stainless Steels						
S30100	0.290	2550	2590	9.4	0.12	9.4
S30200, S30300, S30323	0.290	2550	2590	9.4	0.12	9.6
S30215	0.290	2500	2550	9.2	0.12	9.0
S30400, S30500	0.290	2550	2650	9.4	0.12	9.6
S30430	0.290	2550	2650	6.5	0.12	9.6
S30800	0.290	2550	2650	8.8	0.12	9.6
S30900, S30908	0.290	2550	2650	9.0	0.12	8.3
S31000, S31008	0.290	2550	2650	8.2	0.12	8.8
S31600, S31700	0.290	2500	2550	9.4	0.12	8.8
S31703	0.290	2500	2550	8.3	0.12	9.2
S32100	0.290	2550	2600	9.3	0.12	9.2
S34700	0.290	2550	2650	9.3	0.12	9.2
S34800	0.290	2550	2650	9.3	0.12	9.3
S38400	0.290	2550	2650	9.4	0.12	9.6
S40300, S41000, S41600, S41623	0.280	2700	2790	14.4	0.11	5.5
S40500	0.280	2700	2790	15.6	0.12	6.0
S41400	0.280	2600	2700	14.4	0.11	5.8
S42000, S42020	0.280	2650	2750	14.4	0.11	5.7
S42200	0.280	2675	2700	13.8	0.11	6.2
S42900	0.280	2650	2750	14.8	0.11	5.7
S43000, S43020, S43023	0.280	2600	2750	15.1	0.11	5.8
S43600	0.280	2600	2750	13.8	0.11	5.2
S44002, S44004	0.280	2500	2700	14.0	0.11	5.7
S44003	0.280	2500	2750	14.0	0.11	5.6
S44600	0.270	2600	2750	12.1	0.12	5.8
S50100, S50200	0.280	2700	2800	21.2	0.11	6.2
Cast Iron and Steel						
Malleable Iron, A220 (50005, 60004, 80002)	0.265			29.5	0.12	7.5
Grey Cast Iron	0.25	liquidus approximately, 2100 to 2200, depending on composition		28.0	0.25	5.8
Ductile Iron, A536 (120–90–02)	0.25				0.16	5.9–6.2
Ductile Iron, A536 (100–70–03)	0.25			20.0	0.16	5.9–6.2
Ductile Iron, A536 (80–55–06)	0.25			18.0	0.15	5.9–6.2
Ductile Iron, A536 (65–45–120)	0.25			20.8	0.15	5.9–6.2
Ductile Iron, A536 (60–40–18)	0.25				0.12	5.9–6.2
Cast Steel, 3%C	0.25	liquidus, 2640		28.0	0.12	7.0
Titanium Alloys						
Commercially Pure	0.163	3000	3040	9.0	0.12	5.1
Ti-5Al-2.5Sn	0.162	2820	3000	4.5	0.13	5.3
Ti-8Mn	0.171	2730	2970	6.3	0.19	6.0

[a] Alloy designations correspond to the Aluminum Association numbers for aluminum alloys and to the unified numbering system (UNS) for copper and stainless steel alloys. A220 and A536 are ASTM specified irons.

Table 18. Characteristics of Important Plastics Families

ABS (acrylonitrile-butadiene-styrene)	Rigid, relatively low-cost thermoplastic, easily machined and thermoformed.
Acetal, POM	Engineering thermoplastic with good strength, wear resistance, and dimensional stability. More dimensionally stable than nylon under wet and humid conditions.
Acrylic, PMMA	Clear, transparent, strong, break-resistant thermoplastic with excellent chemical resistance and weatherability.
CPVC (chlorinated PVC)	Thermoplastic with properties similar to PVC, but operates to a 40–60°F (14–16°C) higher temperature.
Fiberglass	Thermosetting composite with high strength-to-weight ratio, excellent dielectric properties, and unaffected by corrosion.
Liquid crystal polymer (LCP)	Aromatic, highly inert polymer, with excellent mechanical properties, as well as chemical, fire, and temperature resistance.
Nylon	Thermoplastic with excellent impact resistance, ideal for wear applications such as bearings and gears, self-lubricating under some circumstances.
PEEK (polyetherether-ketone)	Engineering thermoplastic, excellent temperature resistance, suitable for continuous use above 500°F (260°C), excellent flexural and tensile properties.
Polyester, PET (polyethylene-terephthalate)	Dimensionally stable thermoplastic with superior machining characteristics compared to acetal.
Phenolic	Thermosetting family of plastics with minimal thermal expansion, high compressive strength, excellent wear and abrasion resistance, and a low coefficient of friction. Used for bearing applications and molded parts.
Polycarbonate, PC	Transparent tough thermoplastic with high impact strength, excellent chemical resistance and electrical properties, and good dimensional stability.
Polypropylene, PP	Good chemical resistance combined with low moisture absorption and excellent electrical properties, retains strength up to 250°F (120°C).
Polystyrene, PS	Transparent, colorless, and relatively low-cost amorphous thermoplastic. Relatively rigid with good electrical properties but brittle with poor chemical and ultraviolet properties.
Polysulfone, PSU	Durable thermoplastic, good electrical properties, operates at temperatures in excess of 300°F (150°C).
Polyurethane	Thermoplastic, excellent impact and abrasion resistance, resists sunlight and weathering.
Teflon, PTFE (polytetrafluoro-ethylene)	Thermoplastic, low coefficient of friction, withstands up to 500°F (260°C), inert to chemicals and solvents, self-lubricating with a low thermal-expansion rate.
PVC (polyvinyl chloride)	Thermoplastic, resists corrosive solutions and gases both acid and alkaline, good stiffness.
PVDF (polyvinylidene-fluoride)	Thermoplastic, outstanding chemical resistance, excellent substitute for PVC or polypropylene. Good mechanical strength and dielectric properties.

Table 19. Working with Plastics

Properties	Comments
Thermal expansion	10 times higher than metals; more heat generated. Adequate tool clearance must be provided to minimize heating. Heat must be removed by air blast or liquid coolant
Elasticity	Modulus is 10–60 times smaller than for metals; this resilience permits much greater deflection. Reduce chatter by close chucking and follow rests. Drilled or tapped holes may end up tapered or of smaller diameter than the tool.
Support	Must be firm to prevent distortion. Sharp tools are essential to keep cutting forces to a minimum.
Safety	Requires dust control, adequate ventilation, safety guards and eye protection.
Work	
Cutting Off	Speed 500–800 ft/min. Use tools with greater front and side clearance than are needed for metal. Cutting speeds: about half those used for turning operations.
Drilling	Chip flow in drilling is poor; the rake angles are insufficient and cutting speeds vary from the periphery of the drill, imposing severe loading on the workpiece. Use drills of high-speed steel or premium high-speed steel (T15, M33, or M41–M47) with low helix angles and wide, highly polished flutes. Point angles:70–120; for rigid polyvinyl chloride and acrylic use 120. Clearance angles: 9–15; for acrylic material use 12–20.
Milling	Generally use high-speed tools (M2, M3, M7, or T15). Carbide C2 is recommended for glass-reinforced nylon, silicone, polyimide, and alloy. Speeds: 800–1400 ft/min for peripheral end milling of many thermoplastics; 400–800 ft/min. for many thermosets. However, slower speeds are generally used for other milling operations: 300–500 ft/min. for some thermoplastics; 150–300 ft/min. for some thermosets.
Sawing	See *Speeds and Feeds* section of this book
Tapping and Threading	Taps should be M10, M7, or M1, molybdenum high-speed steel, with finish-ground and polished flutes. Two flute taps are recommended for holes up to 0.125 in. diameter. Speed: 50 ft/min. for through-holes in thin cast, molded or extruded thermoplastics and thermosets; 25 ft/min. for filled materials. Reduce speeds for deep or blind holes, and when the percentage of thread is 65–75%.
Turning	Use high-speed steel and carbide tools. Cutting speeds: 200–500 ft/min. Box tools are good for long, thin parts.

STANDARDS FOR DRAWINGS

Shop Prints, Reading and Interpreting

Table 1. American National Standard Lines for Engineering Drawings
ANSI/ASME Y14.2-2014

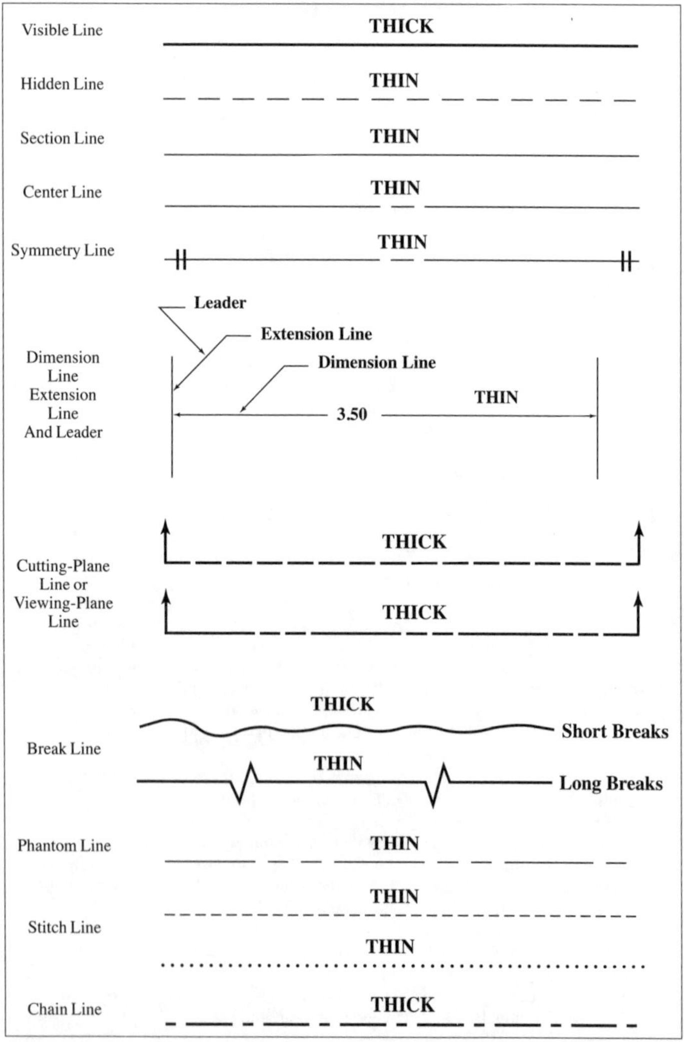

Table 2. American National Standard Symbols for Section Lining
ANSI Y14.2M-1979, R1987[a]

Symbol	Material	Symbol	Material
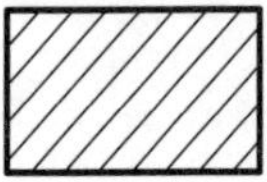	Cast and malleable iron (Also for general use of all materials)	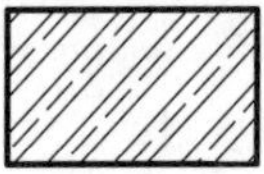	Titanium and refractory material
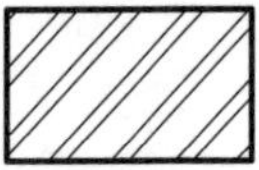	Steel	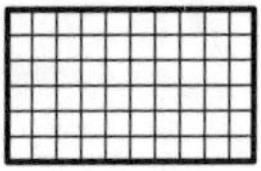	Electric windings, electro magnets, resistance, etc.
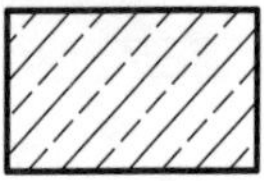	Bronze, brass, copper, and compositions	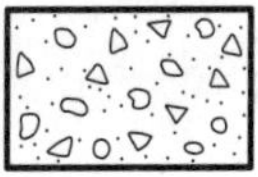	Concrete
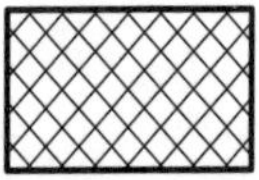	White metal, zinc, lead, babbitt, and alloys	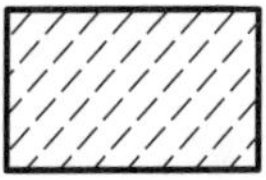	Marble, slate, glass, porcelain, etc.
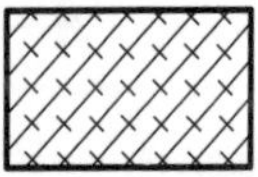	Magnesium, aluminum, and aluminum alloys	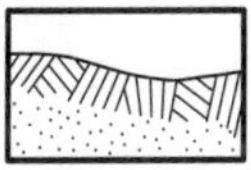	Earth
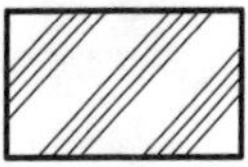	Rubber, plastic electrical insulation		Rock
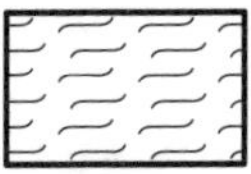	Cork, felt, fabric, leather, fiber	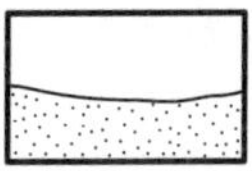	Sand
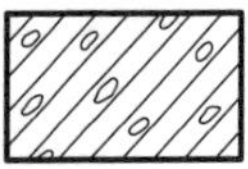	Sound insulation	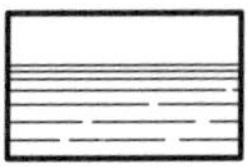	Water and other liquids
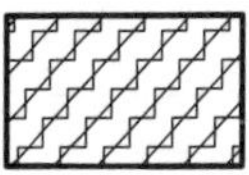	Thermal insulation	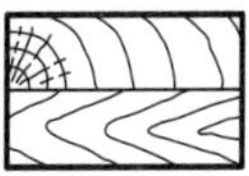	Wood—across grain Wood—with grain

[a] This table has been removed from the current version of standard and is retained here for reference.

Table 3. ASME Geometric Symbols *ASME Y14.5-2018*

Symbol for		Symbol for	
Straightness	—	Diameter	Ø
Flatness	⏥	Basic Dimension	50
Circularity	○	Reference Dimension	(50)
Cylindricity	⌭	Datum Feature (triangle may be filled in or not)	A
Angularity	∠	Dimension Origin	⌖→
Perpendicularity	⊥	Feature Control Frame	⌖ Ø0.5Ⓜ A B C
Parallelism	//	Conical Taper	⌲
Position	⌖	Slope	⌳
Profile of a Line	⌒	Counterbore	⌴
Profile of a Surface	⌓	Spotface	SF
Circular Runout (arrowhead may be filled in or not)	↗	Countersink	⌵
Total Runout (arrowheads may be filled in or not)	⌰	Depth/Deep	↧
All Around	○	Square	□
All Over	◎	Dimension Not to Scale	15
At Maximum Material Condition	Ⓜ	Number of Places	8X
At Maximum Material Boundary	Ⓜ	Arc Length	⌒105
At Least Material Condition	Ⓛ	Radius	R
At Least Material Boundary	Ⓛ	Spherical Radius	SR
Projected Tolerance Zone	Ⓟ	Spherical Diameter	SØ
Tangent Plane	Ⓣ	Controlled Radius	CR
Free State	Ⓕ	Between (arrowheads may be filled in or not)	↔
Unequally Disposed Profile	Ⓤ	Statistical Tolerance	⟨ST⟩
Envelope Principle	Default	Continuous Feature	⟨CF⟩
Independency	Ⓘ	Datum Target	Ø6 A1 or Ø6 A1
Dynamic Profile Tolerance	Δ	Movable Datum Target	A1
Translation	▷	Target Point	X
From / To	→		

Table 4. ISO Geometric Symbols *ISO 1101:2017*

Symbol for		Symbol for	
Straightness	⏤	Independency	Default
Flatness	⏥	Dynamic Profile Tolerance	Δ
Circularity	○	From / To	→
Cylindricity	⌭	Diameter	ϕ
Angularity	∠	Theoretically Exact Dimension	[50]
Perpendicularity	⟂	Auxiliary Dimension	(50)
Parallelism	//	Datum Feature (triangle may be filled in or not)	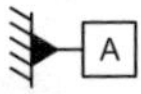
Position	⌖	Dimension Origin	
Concentricity and Coaxiality	◎	Feature Control Frame	⌖ ϕ 0.5 Ⓜ A B C
Symmetry	⌯	Conical Taper	⌲
Profile of a Line	⌒	Slope	⌳
Profile of a Surface	⌓	Square	□
Circular Runout	↗	Dimension Not to Scale	15
Total Runout	⌰	Number of Places	8x
All Around		Arc Length	⌒105
All Over		Radius	R
At Maximum Material Condition	Ⓜ	Spherical Radius	SR
At Least Material Condition	Ⓛ	Spherical Diameter	Sϕ
Projected Tolerance Zone	Ⓟ	Between (arrowheads may be filled in or not)	↔
Tangent Plane	Ⓣ	Datum Target	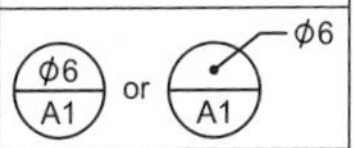
Free State	Ⓕ	Movable Datum Target	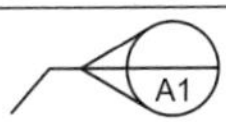
Unequally Disposed Profile	UZ	Target Point	✕
Envelope Principle	Ⓔ		

Note: This table includes the most commonly used symbols; for additional symbols, refer to the standard.

Table 5. American National Standard Symbols for Datum Referencing in Engineering Drawing *ASME Y14.5-2018*

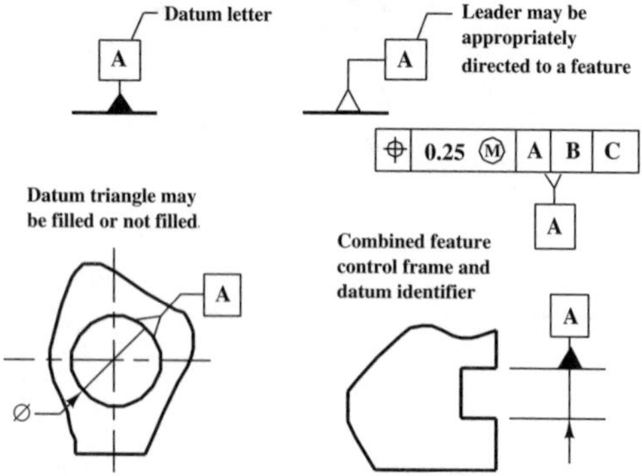

Fig. 1. Datum Feature Symbol

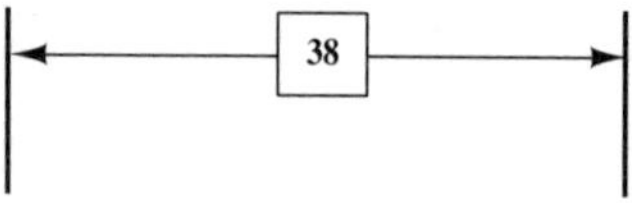

Fig. 2. Basic Dimensions

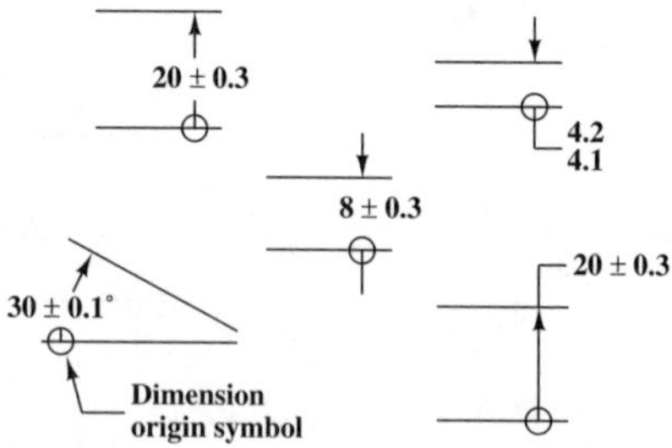

Fig. 3. Dimension Origin Symbol

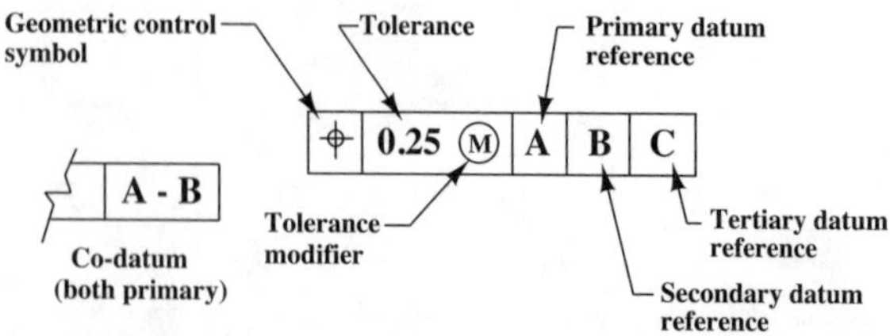

Fig. 4. Feature Control Frame and Datum Order of Precedence

Table 5. *(Continued)* **American National Standard Symbols for Datum Referencing in Engineering Drawing** *ASME Y14.5-2018*

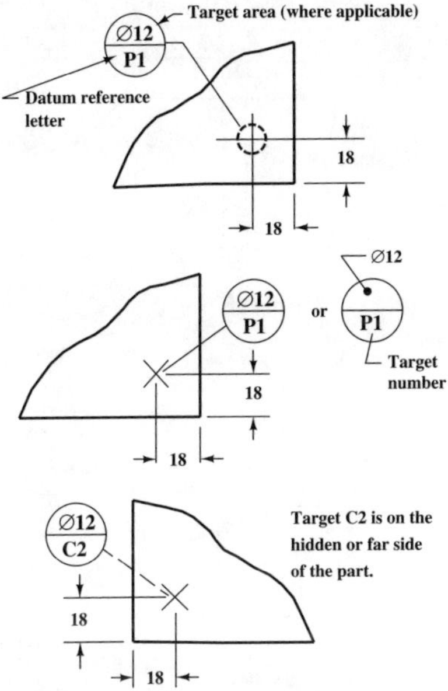

Fig. 5. Datum Target Symbols

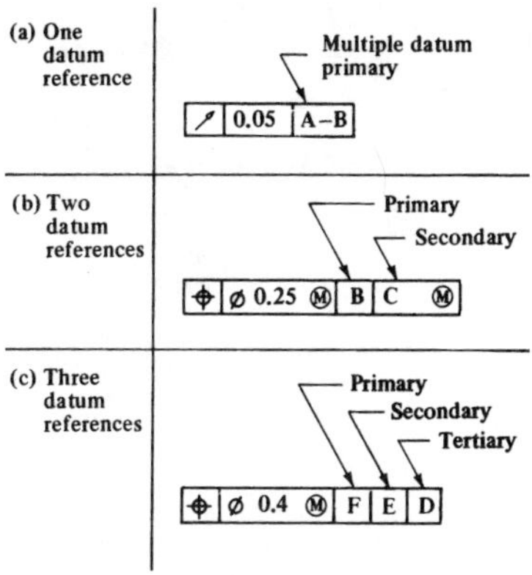

Fig. 6. Order of Precedence of Datum References

Table 5. *(Continued)* **American National Standard Symbols for Datum Referencing in Engineering Drawing** *ANSI/ASME Y14.5-2018*

This on the drawing

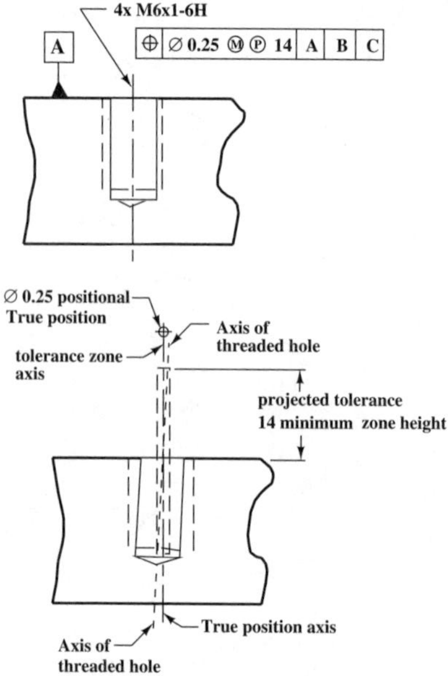

Means this

Fig. 7. Projected Tolerance Zone Application

This on the drawing

Means this

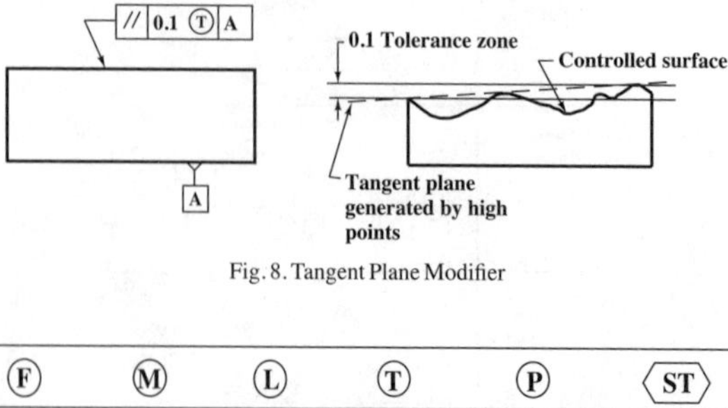

Fig. 8. Tangent Plane Modifier

Ⓕ	Ⓜ	Ⓛ	Ⓣ	Ⓟ	⟨ST⟩
Free State	MMC	LMC	Tangent Plane	Projected Tolerance Zone	Statistical Tolerance

Fig. 9. Tolerance Modifiers

SURFACE TEXTURE

Surface Texture Symbols.—The symbol used to designate control of surface irregularities is shown in Fig. 1a. Where surface texture values other than roughness average are specified, the symbol must be drawn with the horizontal extension as shown in Fig. 1e.

Table 1. Surface Texture Symbols and Construction

Symbol	Meaning
 Fig. 1a.	Basic Surface Texture Symbol. Surface may be produced by any method except when the bar or circle (Fig. 1b or 1d) is specified.
 Fig. 1b.	Material Removal By Machining Is Required. The horizontal bar indicates that material removal by machining is required to produce the surface and that material must be provided for that purpose.
 Fig. 1c.	Material Removal Allowance. The number indicates the amount of stock to be removed by machining in millimeters (or inches). Tolerances may be added to the basic value shown or in a general note.
 Fig. 1d.	Material Removal Prohibited. The circle in the V-shape indicates that the surface must be produced by processes such as casting, forging, hot finishing, cold finishing, die casting, powder metallurgy or injection molding without subsequent removal of material.
 Fig. 1e.	Surface Texture Symbol. To be used when any surface characteristics are specified above the horizontal line or the right of the symbol. Surface may be produced by any method except when the bar or circle (Fig. 1b and 1d) is specified.

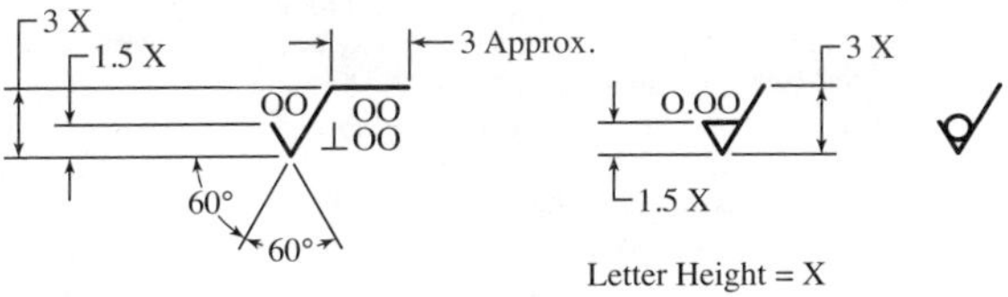

Fig. 1f.

Use of Surface Texture Symbols: When required from a functional standpoint, the desired surface characteristics should be specified. Where no surface texture control is specified, the surface produced by normal manufacturing methods is satisfactory provided it is within the limits of size (and form) specified in accordance with ANSI/ASME Y14.5-2018, "Dimensioning and Tolerancing." This is not viewed as good practice; there should always be some maximum value, either specifically or by default (for example, in the manner of the note shown in Fig. 2).

Material Removal Required or Prohibited: The surface texture symbol is modified when necessary to require or prohibit removal of material. When it is necessary to indicate that a surface must be produced by removal of material by machining, apply the symbol shown in Fig. 1b. When required, the amount of material to be removed is specified as shown in Fig. 1c, in millimeters for metric drawings and in inches for nonmetric drawings. Tolerance for material removal may be added to the basic value shown or specified in a general note. When it is necessary to indicate that a surface must be produced without material removal, use the machining prohibited symbol as shown in Fig. 1d.

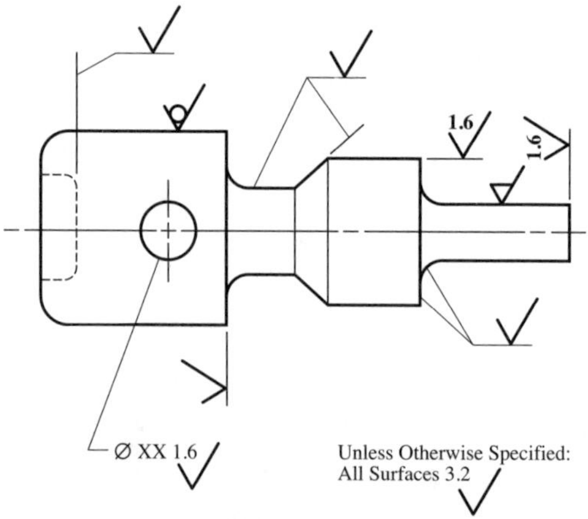

Fig. 2. Application of Surface Texture Symbols

Proportions of Surface Texture Symbols: The recommended proportions for drawing the surface texture symbol are shown in Fig. 1f. The letter height and line width should be the same as that for dimensions and dimension lines.

Applying Surface Texture Symbols.—The point of the symbol should be on a line representing the surface, an extension line of the surface, or to a leader line directed to the surface. The symbol may be specified following a diameter dimension. The long leg (and extension) shall be to the right as the drawing is read. For parts requiring extensive and uniform surface roughness control, a general note may be added to the drawing which applies to each surface texture symbol specified without values as shown in Fig. 2.

When the symbol is used with a dimension it affects the entire surface defined by the dimension. Areas of transition, such as chamfers and fillets, shall conform with the roughest adjacent finished area unless otherwise indicated.

Surface texture values, unless otherwise specified, apply to the complete surface. Drawings or specifications for plated or coated parts shall indicate whether the surface texture values apply before plating, after plating, or both before and after plating.

Include in the symbol only those values required to specify and verify the required texture characteristics. Values should be in metric units for metric drawings and nonmetric units for nonmetric drawings.

Roughness and waviness measurements, unless otherwise specified, apply in the direction that gives the maximum reading; generally across the lay.

Cutoff or Roughness Sampling Length: Standard values are listed in Table 2. When no value is specified, the value 0.8 mm (0.030 in.) applies.

Table 2. Standard Roughness Sampling Length (Cutoff) Values

mm	in.	mm	in.
0.08	0.003	2.5	0.1
0.25	0.010	8.0	0.3
0.80	0.030	25.0	1.0

Roughness Average (Ra): The preferred series of specified roughness average values is given in Table 3.

Table 3. Preferred Series Roughness Average (Ra) Values

μm	μin	μm	μin
0.012	0.5	1.25	50
0.025[a]	1[a]	1.60[a]	63[a]
0.050[a]	2[a]	2.0	80
0.075[a]	3	2.5	100
0.10[a]	4[a]	3.2[a]	125[a]
0.125	5	4.0	160
0.15	6	5.0	200
0.20[a]	8[a]	6.3[a]	250[a]
0.25	10	8.0	320
0.32	13	10.0	400
0.40[a]	16[a]	12.5[a]	500[a]
0.50	20	15	600
0.63	25	20	800
0.80[a]	32[a]	25[a]	1000[a]
1.00	40	…	…

[a] Recommended

Waviness Height: The preferred series of maximum waviness height values is listed in Table 4. Waviness is not currently shown in ISO Standards. It is included here to follow present industry practice in the United States.

Table 4. Preferred Series Maximum Waviness Height Values

mm	in.	mm	in.	mm	in.
0.00002	0.00002	0.008	0.0003	0.12	0.005
0.00003	0.00003	0.012	0.0005	0.20	0.008
0.00005	0.00005	0.020	0.0008	0.25	0.010
0.00008	0.00008	0.025	0.001	0.38	0.015
0.0001	0.0001	0.05	0.002	0.50	0.020
0.0002	0.0002	0.08	0.003	0.80	0.030

Lay: Symbols for designating the direction of lay are shown and interpreted in Table 5.

Metric Dimensions on Drawings: The length units of the metric system that are most generally used in connection with any work relating to mechanical engineering are the meter (39.37 inches) and the millimeter (0.03937 inch). One meter equals 1000 millimeters. On mechanical drawings, all dimensions are generally given in millimeters, no matter how large the dimensions may be. In fact, dimensions of such machines as locomotives and large electrical apparatus are given exclusively in millimeters. This practice is adopted to avoid mistakes due to misplacing decimal points, or misreading dimensions when other units are used as well. When dimensions are given in millimeters, many of them can be given without resorting to decimal points, as a millimeter is only a little more than $^1/_{32}$ inch. Only dimensions of precision need be given in decimals of a millimeter; such dimensions are generally given in hundredths of a millimeter—for example, 0.02 millimeter, which is equal to 0.0008 inch. As 0.01 millimeter is equal to 0.0004 inch, dimensions are seldom given with greater accuracy than to hundredths of a millimeter.

Scales of Metric Drawings: Drawings made to the metric system are not made to scales of $^1/_2$, $^1/_4$, $^1/_8$, etc., as with drawings made to the English system. If the object cannot be drawn full size, it may be drawn $^1/_2$, $^1/_5$, $^1/_{10}$, $^1/_{20}$, $^1/_{50}$, $^1/_{100}$, $^1/_{200}$, $^1/_{500}$, or $^1/_{1000}$ size. If the object is too small and has to be drawn larger, it is drawn 2, 5, or 10 times its actual size.

Table 5. Lay Symbols

Lay Symbol	Meaning	Example Showing Direction of Tool Marks
=	Lay approximately parallel to the line representing the surface to which the symbol is applied.	
⊥	Lay approximately perpendicular to the line representing the surface to which the symbol is applied.	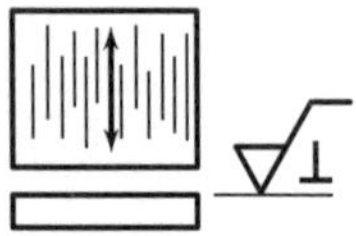
X	Lay angular in both directions to line representing the surface to which the symbol is applied.	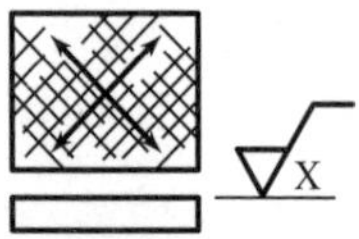
M	Lay multidirectional	
C	Lay approximately circular relative to the center of the surface to which the symbol is applied.	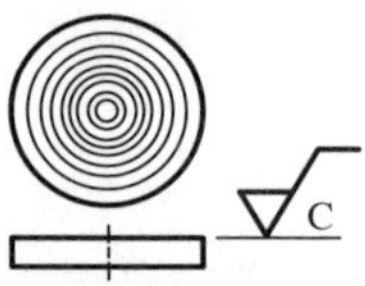
R	Lay approximately radial relative to the center of the surface to which the symbol is applied.	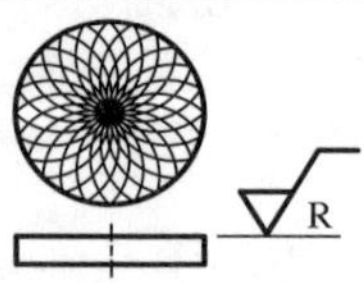
P	Lay particulate, non-directional, or protuberant.	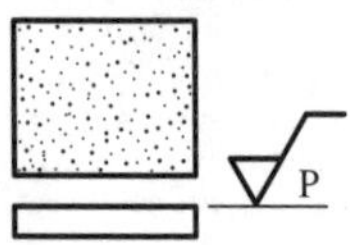

Example Designations.—Table 6 illustrates examples of designations of roughness, waviness, and lay by insertion of values in appropriate positions relative to the symbol.

Table 6. Application of Surface Texture Values to Symbol

Symbol	Description	Symbol	Description
1.6	Roughness average rating is placed at the left of the long leg. The specification of only one rating shall indicate the maximum value and any lesser value shall be acceptable. Specify in micrometers (microinch).	1.6 3.5	Material removal by machining is required to produce the surface. The basic amount of stock provided for material removal is specified at the left of the short leg of the symbol. Specify in millimeters (inch).
1.6 0.8	The specification of maximum and minimum roughness average values indicates permissible range of roughness. Specify in micrometers (microinch).	1.6	Removal of material is prohibited.
0.8 0.005-5	Maximum waviness height rating is the first rating place above the horizontal extension. Any lesser rating shall be acceptable. Specify in millimeters (inch).	0.8 ⊥	Lay designation is indicated by the lay symbol placed at the right of the long leg.
	Maximum waviness spacing rating is the second rating placed above the horizontal extension and to the right of the waviness height rating. Any lesser rating shall be acceptable. Specify in millimeters (inch).	0.8 2.5	Roughness sampling length or cutoff rating is placed below the horizontal extension. When no value is shown, 0.80 mm (0.030 inch) applies. Specify in millimeters (inch).
		0.8 ⊥ 0.5	Where required maximum roughness spacing shall be placed at the right of the lay symbol. Any lesser rating shall be acceptable. Specify in millimeters (inch).

Examples of Special Designations

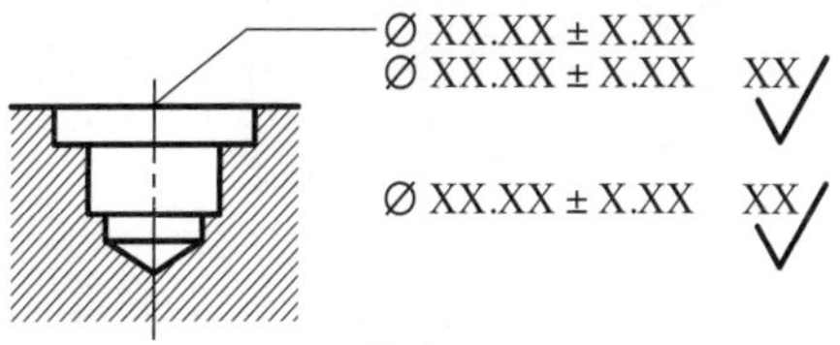

Fig. 3a.

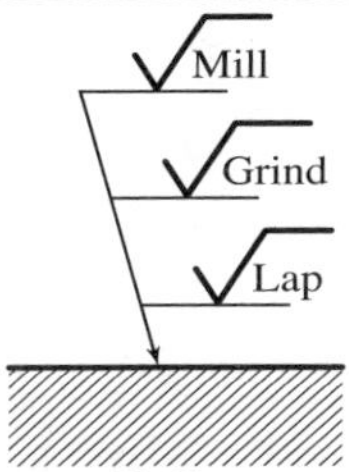

Fig. 3b.

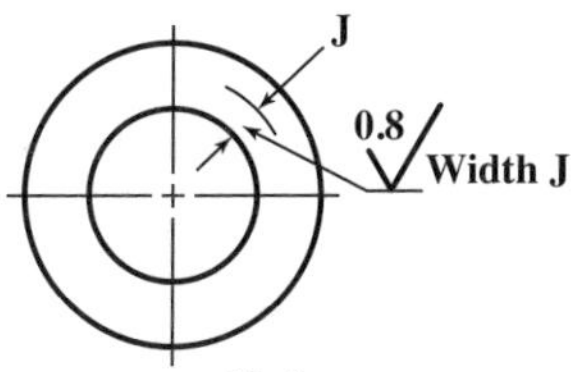

Fig. 3c.

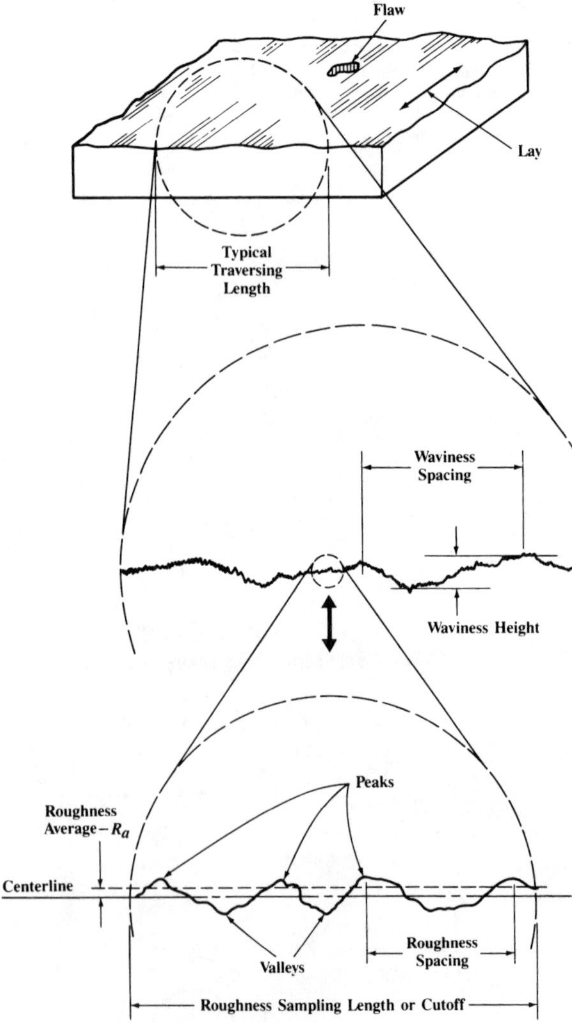

Fig. 4. Pictorial Display of Surface Characteristics

Table 7. Surface Roughness Produced by Common Production Methods

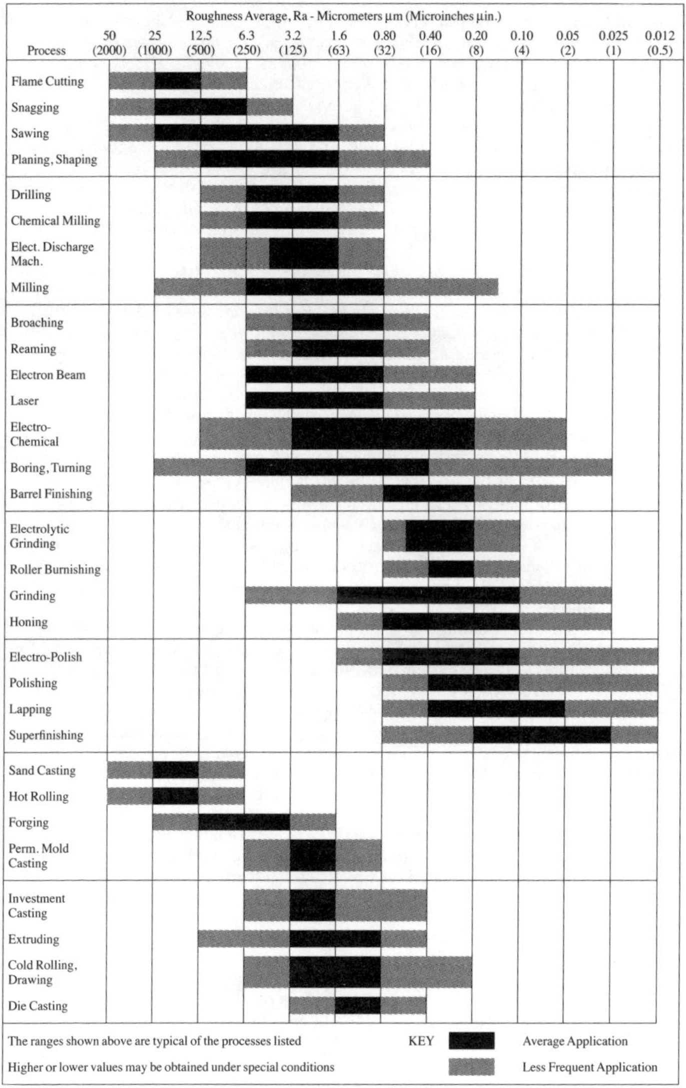

CORROSION

Types of Corrosion.—Corrosion is a process by which a material and/or its properties deteriorate due to interaction with one or more external substances. Environmental conditions, surface conditions, and stresses can initiate or accelerate effects of corrosion, ranging from aesthetic changes to interference with electrical connections and weakening of a structural or mechanical part that can lead to failure.

Uniform (General) Corrosion: This process describes a chemical or electrochemical attack that affects an entire exposed surface. The corroding part will thin; corrosive substances may accumulate on the surface.

Chemical Corrosion: In most cases, chemical damage is the destructive reaction of a contacting substance acting directly on and degrading a material, such as acid affecting a metal. Chemical corrosion also may result in oxide formation or deposition of other surface coatings.

Dry and High-Temperature Corrosion: Also called *scaling*, this is caused by chemicals (gases, molten salts, or solids) on a surface in a dry atmosphere; particulate abrasion may be involved. High temperatures accelerate the process and cause oxidation, carburization, chlorination, and sulfidation.

Electrochemical Corrosion: Electrically active material exposed to an electrolyte can form an electrical cell where ionization occurs and electrons move from an anodic (active) material to a cathodic (noble) material. The anode corrodes through oxidation faster or differently than when alone; oxygen deprivation may result in pockets of increased corrosion. The cathode experiences reduction and may develop a protective oxide layer. *Wet and damp/atmospheric corrosion* occurs when water and contaminants in the environment form an electrolyte liquid or damp film on the surface of a part.

Methods of Protection.—These include material choice, passivation, polishing, coatings and other barriers, environmental controls, reducing stresses on components, separating incompatible materials from each other, or electrolyte solutions. To minimize corrosion, fully research interactions between dissimilar metals before using them in an assembly, match potentials of part(s) to the environment, use barriers, and add auxiliary anodes or cathodes.

Cathodic Protection: Adding one or more sacrificial anodes (a part or a coating) to a system can protect a part that would be the anode in an electrochemical reaction. Adding a DC power supply improves system performance and extends anode life. Cathodic protection is not for all corrosive environments; it can accelerate hydrogen embrittlement.

Anodic Protection: This newer method employs additional cathodes and an applied current to shift the target material's potential into the passive range. It works in more corrosive environments and can be achieved with much lower current density than cathodic protection but is limited to materials that exhibit active-passive behavior (surfaces that can change from active to passive when exposed to oxidizers or applied current).

Galvanic Corrosion: Sometimes called *bi-metallic corrosion*, in this process dissimilar metals or alloys interact when in contact with or electrically coupled by a conductive fluid. Corrosion depends on metal composition, differences in electrochemical potential and distance between them, wetted surface areas, conductivity and pH of fluid, oxygenation, films or deposits, passive film stability, and exposure of metals to processes such as welding. The US military published empirical compatibility information in MIL-STD-889B, "Dissimilar Metals" (Table 1). Many factors influence galvanic corrosion, so use all such data only as a general guide.

A useful method of comparing electrochemical potentials of metals is referencing a galvanic series. A useful standard is ASTM G82-98 (2014), "Standard Guide for Development and Use of a Galvanic Series for Predicting Galvanic Corrosion Performance." Table 2 is based on Army Missile Command Report S-T-67-11, "Practical Galvanic Series." Materials closer in the series have less corrosion inducing potential; polarity reversals can occur. Another galvanic compatibility guide is an anodic index (Table 3). Table 3 is an example based on data in US military specification MIL-14072F, "Finishes for Ground Based Electronic Equipment," relative to a gold reference.

Table 1. Galvanic Compatibility of Metals in Select Environments

Metal or Alloy (Active (Anodic) → Noble (Cathodic))		A	B	C	D	E	F	G	H	I	J	K	L	M	N	O	P	Q	R	S	T
Magnesium	A	C,C,C	C,C,I	C,C,I	I,I,I	I,I,I	I,I,I	I,I,I	I,I,I	I,I,I	I,I,I	I,I,I	I,I,I	I,I,I	I,I,I	I,I,I	I,I,I	I,I,I	I,I,I	I,I,I	I,I,I
Zinc	B		C,C,C	C,C,C	C,C,I	C,C,I	I,I,I	I,I,I	I,I,I	I,I,I	I,I,I	I,I,I	I,I,I	I,I,I	I,I,I	I,I,I	I,I,I	I,I,I	I,I,I	I,I,I	I,I,I
Cadmium, Beryllium	C			C,C,C	C,C,I	C,C,I	I,I,I	I,I,I	I,I,I	I,I,I	I,I,I	I,I,I	I,I,I	I,I,I	I,I,I	I,I,I	I,I,I	I,I,I	I,I,I	I,I,I	I,I,I
Aluminum, Al-Zn, Al-Mg	D				C,C,C	C,C,C	I,I,I	I,I,I	I,I,I	I,I,I	I,I,I	I,I,I	I,I,I	I,I,I	I,I,I	I,I,I	I,I,I	I,I,I	I,I,I	I,I,I	I,I,I
Aluminum-Copper	E					C,C,C	I,I,I	I,I,I	I,I,I	I,I,I	I,I,I	I,I,I	I,I,I	I,I,I	I,I,I	I,I,I	I,I,I	I,I,I	I,I,I	I,I,I	I,I,I
Carbon Steels, Low-Alloy Steels	F						C,C,C	C,I,I	I,I,I	I,I,I	I,I,I	I,I,I	I,I,I	I,I,I	I,I,I	I,I,I	C,C,I	I,I,I	I,I,I	I,I,I	I,I,I
Lead	G							C,C,C	C,C,C	C,C,I	C,C,I	C,C,I	C,C,I	I,I,I	C,I,I	C,I,I	C,I,I	C,I,I	C,C,I	C,I,I	C,I,I
Tin, Tin-Lead, Indium	H								C,C,C	I,I,I	C,I,I	C,I,I	C,C,I	C,C,I	C,C,I	C,C,I	C,C,C	C,I,C	C,C,C	I,I,I	I,I,I
Martensitic Stainless Steels (includes 420), Ferritic Stainless Steels	I									C,C,C	C,I,I	I,I,I	I,I,I	C,I,I	I,I,I	I,I,I	I,I,I	I,I,I	I,I,I	I,I,I	I,I,I
Chromium, Molybdenum, Tungsten	J										C,C,C	C,C,C	C,C,I	C,I,I	I,I,I	C,C,I	C,C,I	C,C,I	C,C,I	C,C,I	C,C,I
Austenitic Stainless Steels (includes 200 and 300 Series), PH Stainless Steels, Super Strength Stainless Steels, Heat-Resistant Stainless Steel	K											C,C,C	I,I,I	I,I,I	I,I,I	C,C,I	C,C,I	C,C,I	C,C,I	C,C,I	C,C,I
Lead Brass, Bronze	L												C,C,C	C,C,I	C,C,I	C,I,I	C,C I	C,I,I	C,C,I	C,C,I	C,C,I
Low Copper Brass, Low Copper Bronze	M													C,C,C	C,C,I	C,C,I	C,C,I	C,I,I	C,C,I	C,I,I	C,I,I
High Copper Brass, High Copper Bronze	N														C,C,C	C,I,I	C,C,I	C,I,I	C,C,I	C,I,I	C,I,I
Copper-High Nickel, Monel	O															C,C,C	C,C,I	C,C,I	C,C,I	C,C,I	C,C,I
Nickel, Cobalt	P																C,C,C	C,C,I	C,C,I	C,C,I	C,C,I
Titanium	Q																	C,C,C	C,C,I	C,C,I	C,C,I
Silver	R																		C,C,C	C,C,C	C,C,I
Palladium, Rhodium, Gold, Platinum	S																			C,C,C	C,C,C
Graphite	T																				C,C,C

Values
I = Incompatible (risk of significant galvanic interaction)
C = Compatible (negligible galvanic interaction likely)

Position of Compatibility Values (1,2,3)
1 = Industrial atmosphere
2 = Marine atmosphere
3 = Seawater immersion

Example: C,C,I = Compatible in industrial atmosphere (1)
Compatible in marine atmosphere (2)
Incompatible in seawater immersion (3)

Table 2. Sample Galvanic Series, General Seawater Environment

Active (Anodic)

Magnesium
Mg alloy AZ-31B
Mg alloy HK-31A
Zinc (hot-dip, die cast, or plated)
Beryllium (hot-pressed)
Aluminum 7072 clad on 7075
Aluminum 2014-T3
Aluminum 1160-H14
Aluminum 7079-T6
Cadmium (plated)
Uranium
Aluminum 218 (die cast)
Aluminum 5052-0
Aluminum 5052-H12
Aluminum 5456-0, H353
Aluminum 5052-H32
Aluminum 1100-0
Aluminum 3003-H25
Aluminum 6061-T6
Aluminum A360 (die cast)
Aluminum 7075-T6
Aluminum 6061-0
Indium
Aluminum 2014-0
Aluminum 2024-T4
Aluminum 5052-H16
Tin (plated)
Stainless Steel 430 (active)
Lead
Steel 1010
Iron (cast)
Stainless Steel 410 (active)
Copper (plated, cast, or wrought)
Nickel (plated)
Chromium (plated)
Tantalum
AM350 (active)
Stainless Steel 310 (active)
Stainless Steel 301 (active)
Stainless Steel 304 (active)
Stainless Steel 430 (active)
Stainless Steel 410 (active)
Stainless Steel 17-7PH (active)
Tungsten
Niobium (columbium) 1% Zr
Brass, Yellow, 268
Uranium 8% Mo
Brass, Naval, 464
Yellow Brass
Muntz Metal 280
Brass (plated)
Nickel-Silver (18% Ni)
Stainless Steel 316L (active)
Bronze 220
Copper 110
Red Brass
Stainless Steel 347 (active)
Molybdenum (commercial pure)
Copper-Nickel 715
Admiralty Brass
Stainless Steel 202 (active)
Bronze, Phosphor 534 (B-1)
Monel 400
Stainless Steel 201 (active)
Carpenter 20 (active)
Stainless Steel 321 (active)
Stainless Steel 316 (active)
Stainless Steel 309 (active)
Stainless Steel 17-7PH (passive)
Silicon Bronze 655
Stainless Steel 304 (passive)
Stainless Steel 301 (passive)
Stainless Steel 321 (passive)
Stainless Steel 201 (passive)
Stainless Steel 286 (passive)
Stainless Steel 316L (passive)
AM355 (active)
Stainless Steel 202 (passive)
Carpenter 20 (passive)
AM355 (passive)
A286 (passive)
Titanium 5A1, 2.5 Sn
Titanium 13V, 11Cr, 3Al (annealed)
Titanium 6Al, 4V
(solution treated and aged)
Titanium 6Al, 4V (anneal)
Titanium 8Mn
Titanium 13V, 11Cr 3Al
(solution heat treated and aged)
Titanium 75A
AM350 (passive)
Silver
Gold
Graphite

Noble (Cathodic)

Table 3. Anodic Index (Gold Reference)

Metals and Alloys	Anodic Index (V)	
Gold, Gold-Platinum Alloys, Wrought Platinum	−0.00	*Cathodic (Noble)* ↑
Rhodium Plated on Silver-Plated Copper	−0.10	
Silver, High Silver Alloys	−0.15	
Nickel, Monel, High Nickel-Copper Alloys, Titanium Alloys	−0.30	
Copper, Low Brasses or Bronzes, Silver Solder, High Copper-Nickel-Zinc Alloys, Nickel-Chromium Alloys	−0.35	
Yellow Brasses and Bronzes	−0.40	
High Brasses and Bronzes, Naval Brass, Muntz Metal	−0.45	
18% Chromium-type Stainless Steels (includes 300 Series)	−0.50	
Chromium Plated, Tin Plated, 12% Chromium-type Stainless Steels (some 400 Series Steels)	−0.60	
Tin Plate, Tin-Lead Solder, Terneplate (Lead-Tin Alloy)	−0.65	
Lead, High Lead Alloys	−0.70	
2000 Series Wrought Aluminum	−0.75	
Plain Carbon Steels, Low Alloy Steels, Wrought Iron, Gray Malleable Iron	−0.85	
Wrought Aluminum Alloys other than 2000 Series, Cast Aluminum-Silicon Alloys	−0.90	
Cast Aluminum Alloys other than Silicon Type, Cadmium, Cadmium Chromate	−0.95	
Hot-Dip Zinc Plate, Galvanized Steel	−1.20	
Wrought Zinc, Zinc Die-Casting Alloys, Zinc Plate	−1.25	
Magnesium, Magnesium Alloys	−1.75	
Beryllium	−1.85	*Anodic (Active)*

Evaluation of Galvanic Couples in Various Environments

Environmental Conditions	Maximum Differential (V)
Controlled: Temperature and humidity are controlled, as in laboratory or office spaces.	0.50
Normal: Industrial environments like warehouses where temperature and humidity are not controlled.	0.25
Harsh: Outdoors, high humidity, or salt exposure environments.	0.15

Subtract one index value from another to determine the electrochemical potential difference between the metals.

ALLOWANCES AND TOLERANCES

ANSI Standard Limits and Fits (ANSI/ASME B4.1-1967 [2009; out of print]).—This American National Standard for Preferred Limits and Fits for Cylindrical Parts presents definitions of terms applying to fits between plain (non-threaded) cylindrical parts and makes recommendations on preferred sizes, allowances, tolerances, and fits for use wherever they are applicable. This standard is in accord with the recommendations of American-British-Canadian (ABC) conferences for diameters of up to 20 inches.They should have application for a wide range of products.

Preferred Basic Sizes.—In specifying fits, the basic size of mating parts may be chosen from the decimal series or the fractional series in the following Table 1.

Table 1. Preferred Basic Sizes *ANSI/ASME B4.1-1967 (2009; out of print)*

Decimal			Fractional					
0.010	2.00	8.50	$\frac{1}{64}$	0.015625	$2\frac{1}{4}$	2.2500	$9\frac{1}{2}$	9.5000
0.012	2.20	9.00	$\frac{1}{32}$	0.03125	$2\frac{1}{2}$	2.5000	10	10.0000
0.016	2.40	9.50	$\frac{1}{16}$	0.0625	$2\frac{3}{4}$	2.7500	$10\frac{1}{2}$	10.5000
0.020	2.60	10.00	$\frac{3}{32}$	0.09375	3	3.0000	11	11.0000
0.025	2.80	10.50	$\frac{1}{8}$	0.1250	$3\frac{1}{4}$	3.2500	$11\frac{1}{2}$	11.5000
0.032	3.00	11.00	$\frac{5}{32}$	0.15625	$3\frac{1}{2}$	3.5000	12	12.0000
0.040	3.20	11.50	$\frac{3}{16}$	0.1875	$3\frac{3}{4}$	3.7500	$12\frac{1}{2}$	12.5000
0.05	3.40	12.00	$\frac{1}{4}$	0.2500	4	4.0000	13	13.0000
0.06	3.60	12.50	$\frac{5}{16}$	0.3125	$4\frac{1}{4}$	4.2500	$13\frac{1}{2}$	13.5000
0.08	3.80	13.00	$\frac{3}{8}$	0.3750	$4\frac{1}{2}$	4.5000	14	14.0000
0.10	4.00	13.50	$\frac{7}{16}$	0.4375	$4\frac{3}{4}$	4.7500	$14\frac{1}{2}$	14.5000
0.12	4.20	14.00	$\frac{1}{2}$	0.5000	5	5.0000	15	15.0000
0.16	4.40	14.50	$\frac{9}{16}$	0.5625	$5\frac{1}{4}$	5.2500	$15\frac{1}{2}$	15.5000
0.20	4.60	15.00	$\frac{5}{8}$	0.6250	$5\frac{1}{2}$	5.5000	16	16.0000
0.24	4.80	15.50	$\frac{11}{16}$	0.6875	$5\frac{3}{4}$	5.7500	$16\frac{1}{2}$	16.5000
0.30	5.00	16.00	$\frac{3}{4}$	0.7500	6	6.0000	17	17.0000
0.40	5.20	16.50	$\frac{7}{8}$	0.8750	$6\frac{1}{2}$	6.5000	$17\frac{1}{2}$	17.5000
0.50	5.40	17.00	1	1.0000	7	7.0000	18	18.0000
0.60	5.60	17.50	$1\frac{1}{4}$	1.2500	$7\frac{1}{2}$	7.5000	$18\frac{1}{2}$	18.5000
0.80	5.80	18.00	$1\frac{1}{2}$	1.5000	8	8.0000	19	19.0000
1.00	6.00	18.50	$1\frac{3}{4}$	1.7500	$8\frac{1}{2}$	8.5000	$19\frac{1}{2}$	19.5000
1.20	6.50	19.00	2	2.0000	9	9.0000	20	20.0000
1.40	7.00	19.50	…	…	…	…	…	…
1.60	7.50	20.00	All dimensions are in inches.					
1.80	8.00	…						

Table 2. Preferred Series of Tolerances and Allowances (In thousandths of an inch)
ANSI/ASME B4.1-1967 (2009; out of print)

0.1	1	10	100	0.3	3	30	…
…	1.2	12	125	…	3.5	35	…
0.15	1.4	14	…	0.4	4	40	…
…	1.6	16	160	…	4.5	45	…
…	1.8	18	…	0.5	5	50	…
0.2	2	20	200	0.6	6	60	…
…	2.2	22	…	0.7	7	70	…
0.25	2.5	25	250	0.8	8	80	…
…	2.8	28	…	0.9	9	…	…

Standard Tolerances.—The series of standard tolerances shown in Table 3 are so arranged that for any one grade they represent approximately similar production difficulties throughout the range of sizes. This table provides a suitable range from which appropriate tolerances for holes and shafts can be selected and enables standard gages to be used. The tolerances shown in Table 3 have been used in the succeeding tables for different classes of fits.

Table 3a. ANSI/ASME Standard Tolerances
ANSI/ASME B4.1-1967 (2009; out of print)

Nominal Size, Inches		Grade									
		4	5	6	7	8	9	10	11	12	13
Over	To	Tolerances in thousandths of an inch[a]									
0	0.12	0.12	0.15	0.25	0.4	0.6	1.0	1.6	2.5	4	6
0.12	0.24	0.15	0.20	0.3	0.5	0.7	1.2	1.8	3.0	5	7
0.24	0.40	0.15	0.25	0.4	0.6	0.9	1.4	2.2	3.5	6	9
0.40	0.71	0.2	0.3	0.4	0.7	1.0	1.6	2.8	4.0	7	10
0.71	1.19	0.25	0.4	0.5	0.8	1.2	2.0	3.5	5.0	8	12
1.19	1.97	0.3	0.4	0.6	1.0	1.6	2.5	4.0	6	10	16
1.97	3.15	0.3	0.5	0.7	1.2	1.8	3.0	4.5	7	12	18
3.15	4.73	0.4	0.6	0.9	1.4	2.2	3.5	5	9	14	22
4.73	7.09	0.5	0.7	1.0	1.6	2.5	4.0	6	10	16	25
7.09	9.85	0.6	0.8	1.2	1.8	2.8	4.5	7	12	18	28
9.85	12.41	0.6	0.9	1.2	2.0	3.0	5.0	8	12	20	30
12.41	15.75	0.7	1.0	1.4	2.2	3.5	6	9	14	22	35
15.75	19.69	0.8	1.0	1.6	2.5	4	6	10	16	25	40
19.69	30.09	0.9	1.2	2.0	3	5	8	12	20	30	50
30.09	41.49	1.0	1.6	2.5	4	6	10	16	25	40	60
41.49	56.19	1.2	2.0	3	5	8	12	20	30	50	80
56.19	76.39	1.6	2.5	4	6	10	16	25	40	60	100
76.39	100.9	2.0	3	5	8	12	20	30	50	80	125
100.9	131.9	2.5	4	6	10	16	25	40	60	100	160
131.9	171.9	3	5	8	12	20	30	50	80	125	200
171.9	200	4	6	10	16	25	40	60	100	160	250

[a] All tolerances above the heavy line are in accordance with American-British-Canadian (ABC) agreements.

Table 3b. Relation of Machining Processes to Tolerance Grades
ANSI/ASME B4.1-1967 (2009; out of print)

This chart may be used as a general guide to determine the machining processes that will, under normal conditions, produce work within the tolerance grades indicated.

MACHINING OPERATION	TOLERANCE GRADES 4	5	6	7	8	9	10	11	12	13
Lapping & Honing	X	X								
Cylindrical Grinding		X	X	X						
Surface Grinding		X	X	X	X					
Diamond Turning		X	X	X						
Diamond Boring		X	X	X						
Broaching		X	X	X	X					
Reaming			X	X	X	X	X			
Turning				X	X	X	X	X	X	X
Boring					X	X	X	X	X	X
Milling							X	X	X	X
Planing & Shaping							X	X	X	X
Drilling							X	X	X	X

Table 4. Designation of Standard Fits

Letter Symbols	Definition of Fits	Description
RC	Running and Sliding Clearance	Intended to provide a similar running performance, with suitable lubrication allowance, throughout the range of sizes.
RC 1	Close Sliding	Intended for the accurate location of parts that must be assembled without perceptible play.[a]
RC 2	Sliding	Intended for accurate location, but with greater maximum clearance than class RC 1. Parts made to this fit move and turn easily but are not intended to run freely, and in large sizes may seize with small temperature changes.[a]
RC 3	Precision Running	The closest fits that can be expected to run freely, intended for precision work at slow speeds and light journal pressures, but are not suitable where appreciable temperature differences are likely to be encountered.
RC 4	Close Running	Intended chiefly for running fits on accurate machinery with moderate surface speeds and journal pressures, where accurate location and minimum play are desired.
RC 5 & RC 6	Medium Running	Intended for higher running speeds, or heavy journal pressures, or both.
RC 7	Free Running	Intended for use where accuracy is not essential, or where large temperature variations are likely to be encountered, or under both conditions.
RC 8 & RC 9	Loose Running	Intended for use where wide commercial tolerances may be necessary, together with an allowance on the external member.
LC	Locational Clearance	Intended to determine only the location of the mating parts; they may provide rigid or accurate location, as with interference fits, or provide some freedom of location, as with clearance fits.
LC	Locational Clearance	Intended for parts that are normally stationary, but that can be freely assembled. They range from snug fits for parts requiring accuracy of location, through the medium clearance fits for parts such as spigots, to the looser fastener fits where freedom of assembly is of importance.
LT	Locational Transitional	A compromise between clearance and interference fits, for applications where accuracy of location is important, but a small amount of clearance or interference is permissible.

Table 4. *(Continued)* **Designation of Standard Fits**

Letter Symbols	Definition of Fits	Description
LN	Locational Interference	Used where accuracy of location is of prime importance, and for parts requiring rigidity and alignment with no special requirements for bore pressure. Not intended for parts designed to transmit frictional loads from one part to another by virtue of the tightness of fit. These conditions are covered by force fits.
FN	Force or Shrink	A special type of interference fit, normally characterized by maintenance of constant bore pressures throughout the range of sizes. The interference therefore varies almost directly with the diameter, and the difference between its minimum and maximum value is small, to maintain the resulting pressures within reasonable limits.
FN 1	Light Drive	Requiring light assembly pressures, and producing more or less permanent assemblies. They are suitable for the thin sections or long fits, or in cast-iron external members.
FN 2	Medium Drive	Suitable for ordinary steel parts, or for shrink fits on light sections. The tightest fits that can be used with high-grade cast iron external members.
FN 3	Heavy Drive	Suitable for heavier steel parts or for shrink fits in medium sections.
FN 4 & FN 5	Force	Suitable for parts that can be highly stressed, or for shrink fits where the heavy pressing forces required are impractical.
B	Bilateral Hole (Modified Standard Fit)	The symbols used for these fits are identical with those used for standard fits; thus, LC 4 B is a clearance locational fit, Class 4, except it is produced with a bilateral hole.
S	Basic Shaft (Modified Standard Fit)	The symbols used for these fits are identical with those used for standard fits; thus, LC 4 S is a clearance locational fit, Class 4, except it is produced on a basic shaft basis.

[a] Note: The clearances, used chiefly as slide fits, increase more slowly with the diameter than for other classes, so that accurate location is maintained even at the expense of free relative motion.

Graphical Representation of Limits and Fits.—A visual comparison of the hole and shaft tolerances and the clearances or interferences provided by the various types and classes of fits can be obtained from the following diagrams. These show disposition of hole and shaft tolerances (in thousandths of an inch) with respect to basic size (0) for a nominal diameter of 1 inch, per ANSI/ASME B4.1-1967 (R2009; out of print).

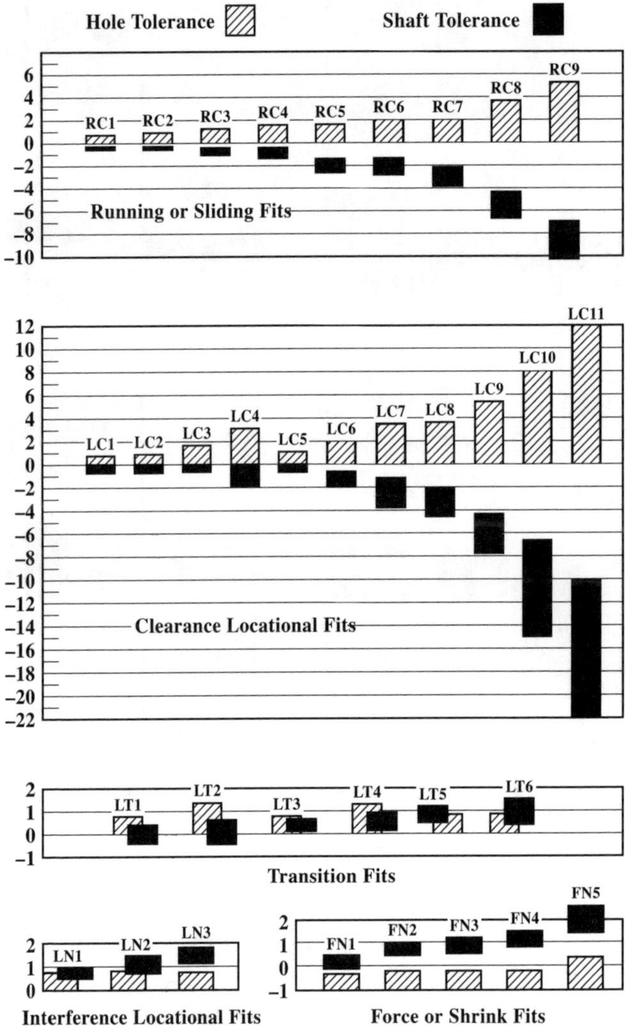

Table 5. American National Standard Running and Sliding Fits *ANSI/ASME B4.1-1967 (2009; out of print)*

Nominal Size Range, Inches		Class RC 1			Class RC 2			Class RC 3			Class RC 4		
			Standard Tolerance Limits			Standard Tolerance Limits			Standard Tolerance Limits			Standard Tolerance Limits	
		Clearance[a]	Hole H5	Shaft g4	Clearance[a]	Hole H6	Shaft g5	Clearance[a]	Hole H7	Shaft f6	Clearance[a]	Hole H8	Shaft f7
Over	To	Values shown below are in thousandths of an inch											
0	0.12	0.1 0.45	+0.2 0	−0.1 −0.25	0.1 0.55	+0.25 0	−0.1 −0.3	0.3 0.95	+0.4 0	−0.3 −0.55	0.3 1.3	+0.6 0	−0.3 −0.7
0.12	0.24	0.15 0.5	+0.2 0	−0.15 −0.3	0.15 0.65	+0.3 0	−0.15 −0.35	0.4 1.12	+0.5 0	−0.4 −0.7	0.4 1.6	+0.7 0	−0.4 −0.9
0.24	0.40	0.2 0.6	+0.25 0	−0.2 −0.35	0.2 0.85	+0.4 0	−0.2 −0.45	0.5 1.5	+0.6 0	−0.5 −0.9	0.5 2.0	+0.9 0	−0.5 −1.1
0.40	0.71	0.25 0.75	+0.3 0	−0.25 −0.45	0.25 0.95	+0.4 0	−0.25 −0.55	0.6 1.7	+0.7 0	−0.6 −1.0	0.6 2.3	+1.0 0	−0.6 −1.3
0.71	1.19	0.3 0.95	+0.4 0	−0.3 −0.55	0.3 1.2	+0.5 0	−0.3 −0.7	0.8 2.1	+0.8 0	−0.8 −1.3	0.8 2.8	+1.2 0	−0.8 −1.6
1.19	1.97	0.4 1.1	+0.4 0	−0.4 −0.7	0.4 1.4	+0.6 0	−0.4 −0.8	1.0 2.6	+1.0 0	−1.0 −1.6	1.0 3.6	+1.6 0	−1.0 −2.0
1.97	3.15	0.4 1.2	+0.5 0	−0.4 −0.7	0.4 1.6	+0.7 0	−0.4 −0.9	1.2 3.1	+1.2 0	−1.2 −1.9	1.2 4.2	+1.8 0	−1.2 −2.4
3.15	4.73	0.5 1.5	+0.6 0	−0.5 −0.9	0.5 2.0	+0.9 0	−0.5 −1.1	1.4 3.7	+1.4 0	−1.4 −2.3	1.4 5.0	+2.2 0	−1.4 −2.8
4.73	7.09	0.6 1.8	+0.7 0	−0.6 −1.1	0.6 2.3	+1.0 0	−0.6 −1.3	1.6 4.2	+1.6 0	−1.6 −2.6	1.6 5.7	+2.5 0	−1.6 −3.2
7.09	9.85	0.6 2.0	+0.8 0	−0.6 −1.2	0.6 2.6	+1.2 0	−0.6 −1.4	2.0 5.0	+1.8 0	−2.0 −3.2	2.0 6.6	+2.8 0	−2.0 −3.8
9.85	12.41	0.8 2.3	+0.9 0	−0.8 −1.4	0.8 2.9	+1.2 0	−0.8 −1.7	2.5 5.7	+2.0 0	−2.5 −3.7	2.5 7.5	+3.0 0	−2.5 −4.5
12.41	15.75	1.0 2.7	+1.0 0	−1.0 −1.7	1.0 3.4	+1.4 0	−1.0 −2.0	3.0 6.6	+2.2 0	−3.0 −4.4	3.0 8.7	+3.5 0	−3.0 −5.2
15.75	19.69	1.2 3.0	+1.0 0	−1.2 −2.0	1.2 3.8	+1.6 0	−1.2 −2.2	4.0 8.1	+2.5 0	−4.0 −5.6	4.0 10.5	+4.0 0	−4.0 −6.5

[a] Pairs of values shown represent minimum and maximum amounts of clearance resulting from application of standard tolerance limits.

Table 6. American National Standard Running and Sliding Fits *ANSI/ASME B4.1-1967 (2009; out of print)*

Nominal Size Range, Inches		Class RC 5			Class RC 6			Class RC 7			Class RC 8			Class RC 9		
			Standard Tolerance Limits			Standard Tolerance Limits			Standard Tolerance Limits			Standard Tolerance Limits			Standard Tolerance Limits	
Over	To	Clearance[a]	Hole H8	Shaft e7	Clearance[a]	Hole H9	Shaft e8	Clearance[a]	Hole H9	Shaft d8	Clearance[a]	Hole H10	Shaft c9	Clearance[a]	Hole H11	Shaft
		Values shown below are in thousandths of an inch														
0	0.12	0.6 1.6	+0.6 0	−0.6 −1.0	0.6 2.2	+1.0 0	−0.6 −1.2	1.0 2.6	+1.0 0	−1.0 −1.6	2.5 5.1	+1.6 0	−2.5 −3.5	4.0 8.1	+2.5 0	−4.0 −5.6
0.12	0.24	0.8 2.0	+0.7 0	−0.8 −1.3	0.8 2.7	+1.2 0	−0.8 −1.5	1.2 3.1	+1.2 0	−1.2 −1.9	2.8 5.8	+1.8 0	−2.8 −4.0	4.5 9.0	+3.0 0	−4.5 −6.0
0.24	0.40	1.0 2.5	+0.9 0	−1.0 −1.6	1.0 3.3	+1.4 0	−1.0 −1.9	1.6 3.9	+1.4 0	−1.6 −2.5	3.0 6.6	+2.2 0	−3.0 −4.4	5.0 10.7	+3.5 0	−5.0 −7.2
0.40	0.71	1.2 2.9	+1.0 0	−1.2 −1.9	1.2 3.8	+1.6 0	−1.2 −2.2	2.0 4.6	+1.6 0	−2.0 −3.0	3.5 7.9	+2.8 0	−3.5 −5.1	6.0 12.8	+4.0 0	−6.0 −8.8
0.71	1.19	1.6 3.6	+1.2 0	−1.6 −2.4	1.6 4.8	+2.0 0	−1.6 −2.8	2.5 5.7	+2.0 0	−2.5 −3.7	4.5 10.0	+3.5 0	−4.5 −6.5	7.0 15.5	+5.0 0	−7.0 −10.5
1.19	1.97	2.0 4.6	+1.6 0	−2.0 −3.0	2.0 6.1	+2.5 0	−2.0 −3.6	3.0 7.1	+2.5 0	−3.0 −4.6	5.0 11.5	+4.0 0	−5.0 −7.5	8.0 18.0	+6.0 0	−8.0 −12.0
1.97	3.15	2.5 5.5	+1.8 0	−2.5 −3.7	2.5 7.3	+3.0 0	−2.5 −4.3	4.0 8.8	+3.0 0	−4.0 −5.8	6.0 13.5	+4.5 0	−6.0 −9.0	9.0 20.5	+7.0 0	−9.0 −13.5
3.15	4.73	3.0 6.6	+2.2 0	−3.0 −4.4	3.0 8.7	+3.5 0	−3.0 −5.2	5.0 10.7	+3.5 0	−5.0 −7.2	7.0 15.5	+5.0 0	−7.0 −10.5	10.0 24.0	+9.0 0	−10.0 −15.0
4.73	7.09	3.5 7.6	+2.5 0	−3.5 −5.1	3.5 10.0	+4.0 0	−3.5 −6.0	6.0 12.5	+4.0 0	−6.0 −8.5	8.0 18.0	+6.0 0	−8.0 −12.0	12.0 28.0	+10.0 0	−12.0 −18.0
7.09	9.85	4.0 8.6	+2.8 0	−4.0 −5.8	4.0 11.3	+4.5 0	−4.0 −6.8	7.0 14.3	+4.5 0	−7.0 −9.8	10.0 21.5	+7.0 0	−10.0 −14.5	15.0 34.0	+12.0 0	−15.0 −22.0
9.85	12.41	5.0 10.0	+3.0 0	−5.0 −7.0	5.0 13.0	+5.0 0	−5.0 −8.0	8.0 16.0	+5.0 0	−8.0 −11.0	12.0 25.0	+8.0 0	−12.0 −17.0	18.0 38.0	+12.0 0	−18.0 −26.0
12.41	15.75	6.0 11.7	+3.5 0	−6.0 −8.2	6.0 15.5	+6.0 0	−6.0 −9.5	10.0 19.5	+6.0 0	−10.0 −13.5	14.0 29.0	+9.0 0	−14.0 −20.0	22.0 45.0	+14.0 0	−22.0 −31.0
15.75	19.69	8.0 14.5	+4.0 0	−8.0 −10.5	8.0 18.0	+6.0 0	−8.0 −12.0	12.0 22.0	+6.0 0	−12.0 −16.0	16.0 32.0	+10.0 0	−16.0 −22.0	25.0 51.0	+16.0 0	−25.0 −35.0

Tolerance limits given in body of table are added to or subtracted from basic size (as indicated by + or − sign) to obtain maximum and minimum sizes of mating parts. All data above heavy lines are in accord with ABC agreements. Symbols H5, g4, etc., are hole and shaft designations in ABC system. Limits for sizes above 19.69 inches are also given in the ANSI Standard.

Table 7. American National Standard Clearance Locational Fits *ANSI/ASME B4.1-1967 (2009; out of print)*

Nominal Size Range, Inches		Class LC 1			Class LC 2			Class LC 3			Class LC 4			Class LC 5		
			Standard Tolerance Limits			Standard Tolerance Limits			Standard Tolerance Limits			Standard Tolerance Limits			Standard Tolerance Limits	
		Clearance[a]	Hole H6	Shaft h5	Clearance[a]	Hole H7	Shaft h6	Clearance[a]	Hole H8	Shaft h7	Clearance[a]	Hole H10	Shaft h9	Clearance[a]	Hole H7	Shaft g6
Over	To	Values shown below are in thousandths of an inch														
0	0.12	0	+0.25	0	0	+0.4	0	0	+0.6	0	0	+1.6	0	0.1	+0.4	−0.1
		0.45	0	−0.2	0.65	0	−0.25	1	0	−0.4	2.6	0	−1.0	0.75	0	−0.35
0.12	0.24	0	+0.3	0	0	+0.5	0	0	+0.7	0	0	+1.8	0	0.15	+0.5	−0.15
		0.5	0	−0.2	0.8	0	−0.3	1.2	0	−0.5	3.0	0	−1.2	0.95	0	−0.45
0.24	0.40	0	+0.4	0	0	+0.6	0	0	+0.9	0	0	+2.2	0	0.2	+0.6	−0.2
		0.65	0	−0.25	1.0	0	−0.4	1.5	0	−0.6	3.6	0	−1.4	1.2	0	−0.6
0.40	0.71	0	+0.4	0	0	+0.7	0	0	+1.0	0	0	+2.8	0	0.25	+0.7	−0.25
		0.7	0	−0.3	1.1	0	−0.4	1.7	0	−0.7	4.4	0	−1.6	1.35	0	−0.65
0.71	1.19	0	+0.5	0	0	+0.8	0	0	+1.2	0	0	+3.5	0	0.3	+0.8	−0.3
		0.9	0	−0.4	1.3	0	−0.5	2	0	−0.8	5.5	0	−2.0	1.6	0	−0.8
1.19	1.97	0	+0.6	0	0	+1.0	0	0	+1.6	0	0	+4.0	0	0.4	+1.0	−0.4
		1.0	0	−0.4	1.6	0	−0.6	2.6	0	−1	6.5	0	−2.5	2.0	0	−1.0
1.97	3.15	0	+0.7	0	0	+1.2	0	0	+1.8	0	0	+4.5	0	0.4	+1.2	−0.4
		1.2	0	−0.5	1.9	0	−0.7	3	0	−1.2	7.5	0	−3	2.3	0	−1.1
3.15	4.73	0	+0.9	0	0	+1.4	0	0	+2.2	0	0	+5.0	0	0.5	+1.4	−0.5
		1.5	0	−0.6	2.3	0	−0.9	3.6	0	−1.4	8.5	0	−3.5	2.8	0	−1.4
4.73	7.09	0	+1.0	0	0	+1.6	0	0	+2.5	0	0	+6.0	0	0.6	+1.6	−0.6
		1.7	0	−0.7	2.6	0	−1.0	4.1	0	−1.6	10.0	0	−4	3.2	0	−1.6
7.09	9.85	0	+1.2	0	0	+1.8	0	0	+2.8	0	0	+7.0	0	0.6	+1.8	−0.6
		2.0	0	−0.8	3.0	0	−1.2	4.6	0	−1.8	11.5	0	−4.5	3.6	0	−1.8
9.85	12.41	0	+1.2	0	0	+2.0	0	0	+3.0	0	0	+8.0	0	0.7	+2.0	−0.7
		2.1	0	−0.9	3.2	0	−1.2	5	0	−2.0	13.0	0	−5	3.9	0	−1.9
12.41	15.75	0	+1.4	0	0	+2.2	0	0	+3.5	0	0	+9.0	0	0.7	+2.2	−0.7
		2.4	0	−1.0	3.6	0	−1.4	5.7	0	−2.2	15.0	0	−6	4.3	0	−2.1
15.75	19.69	0	+1.6	0	0	+2.5	0	0	+4	0	0	+10.0	0	0.8	+2.5	−0.8
		2.6	0	−1.0	4.1	0	−1.6	6.5	0	−2.5	16.0	0	−6	4.9	0	−2.4

[a] Pairs of values shown represent minimum and maximum amounts of interference resulting from application of standard tolerance limits.

Table 8. American National Standard Clearance Locational Fits *ANSI/ASME B4.1-1967 (2009; out of print)*

Nominal Size Range, Inches		Class LC 6			Class LC 7			Class LC 8			Class LC 9			Class LC 10			Class LC 11		
			Std. Tolerance Limits			Std. Tolerance Limits			Std. Tolerance Limits			Std. Tolerance Limits			Std. Tolerance Limits			Std. Tolerance Limits	
Over	To	Clearance[a]	Hole H9	Shaft f8	Clearance[a]	Hole H10	Shaft e9	Clearance[a]	Hole H10	Shaft d9	Clearance[a]	Hole H11	Shaft c10	Clearance[a]	Hole H12	Shaft	Clearance[a]	Hole H13	Shaft
		Values shown below are in thousandths of an inch																	
0	0.12	0.3	+1.0	−0.3	0.6	+1.6	−0.6	1.0	+1.6	−1.0	2.5	+2.5	−2.5	4	+4	−4	5	+6	−5
		1.9	0	−0.9	3.2	0	−1.6	2.0	0	−2.0	6.6	0	−4.1	12	0	−8	17	0	−11
0.12	0.24	0.4	+1.2	−0.4	0.8	+1.8	−0.8	1.2	+1.8	−1.2	2.8	+3.0	−2.8	4.5	+5	−4.5	6	+7	−6
		2.3	0	−1.1	3.8	0	−2.0	4.2	0	−2.4	7.6	0	−4.6	14.5	0	−9.5	20	0	−13
0.24	0.40	0.5	+1.4	−0.5	1.0	+2.2	−1.0	1.6	+2.2	−1.6	3.0	+3.5	−3.0	5	+6	−5	7	+9	−7
		2.8	0	−1.4	4.6	0	−2.4	5.2	0	−3.0	8.7	0	−5.2	17	0	−11	25	0	−16
0.40	0.71	0.6	+1.6	−0.6	1.2	+2.8	−1.2	2.0	+2.8	−2.0	3.5	+4.0	−3.5	6	+7	−6	8	+10	−8
		3.2	0	−1.6	5.6	0	−2.8	6.4	0	−3.6	10.3	0	−6.3	20	0	−13	28	0	−18
0.71	1.19	0.8	+2.0	−0.8	1.6	+3.5	−1.6	2.5	+3.5	−2.5	4.5	+5.0	−4.5	7	+8	−7	10	+12	−10
		4.0	0	−2.0	7.1	0	−3.6	8.0	0	−4.5	13.0	0	−8.0	23	0	−15	34	0	−22
1.19	1.97	1.0	+2.5	−1.0	2.0	+4.0	−2.0	3.6	+4.0	−3.0	5.0	+6	−5.0	8	+10	−8	12	+16	−12
		5.1	0	−2.6	8.5	0	−4.5	9.5	0	−5.5	15.0	0	−9.0	28	0	−18	44	0	−28
1.97	3.15	1.2	+3.0	−1.0	2.5	+4.5	−2.5	4.0	+4.5	−4.0	6.0	+7	−6.0	10	+12	−10	14	+18	−14
		6.0	0	−3.0	10.0	0	−5.5	11.5	0	−7.0	17.5	0	−10.5	34	0	−22	50	0	−32
3.15	4.73	1.4	+3.5	−1.4	3.0	+5.0	−3.0	5.0	+5.0	−5.0	7	+9	−7	11	+14	−11	16	+22	−16
		7.1	0	−3.6	11.5	0	−6.5	13.5	0	−8.5	21	0	−12	39	0	−25	60	0	−38
4.73	7.09	1.6	+4.0	−1.6	3.5	+6.0	−3.5	6	+6	−6	8	+10	−8	12	+16	−12	18	+25	−18
		8.1	0	−4.1	13.5	0	−7.5	16	0	−10	24	0	−14	44	0	−28	68	0	−43
7.09	9.85	2.0	+4.5	−2.0	4.0	+7.0	−4.0	7	+7	−7	10	+12	−10	16	+18	−16	22	+28	−22
		9.3	0	−4.8	15.5	0	−8.5	18.5	0	−11.5	29	0	−17	52	0	−34	78	0	−50
9.85	12.41	2.2	+5.0	−2.2	4.5	+8.0	−4.5	7	+8	−7	12	+12	−12	20	+20	−20	28	+30	−28
		10.2	0	−5.2	17.5	0	−9.5	20	0	−12	32	0	−20	60	0	−40	88	0	−58
12.41	15.75	2.5	+6.0	−2.5	5.0	+9.0	−5	8	+9	−8	14	+14	−14	22	+22	−22	30	+35	−30
		12.0	0	−6.0	20.0	0	−11	23	0	−14	37	0	−23	66	0	−44	100	0	−65
15.75	19.69	2.8	+6.0	−2.8	5.0	+10.0	−5	9	+10	−9	16	+16	−16	25	+25	−25	35	+40	−35
		12.8	0	−6.8	21.0	0	−11	25	0	−15	42	0	−26	75	0	−50	115	0	−75

Tolerance limits given in body of table are added or subtracted to basic size (as indicated by + or − sign) to obtain maximum and minimum sizes of mating parts. All data above heavy lines are in accordance with American-British-Canadian (ABC) agreements. Symbols H6, H7, s6, etc., are hole and shaft designations in ABC system. Limits for sizes above 19.69 inches are not covered by ABC agreements but are given in the ANSI Standard.

Table 9. ANSI/ASME Standard Transition Locational Fits *ANSI/ASME B4.1-1967 (2009; out of print)*

Nominal Size Range, Inches		Class LT 1			Class LT 2			Class LT 3			Class LT 4			Class LT 5			Class LT 6		
			Std. Tolerance Limits			Std. Tolerance Limits			Std. Tolerance Limits			Std. Tolerance Limits			Std. Tolerance Limits			Std. Tolerance Limits	
Over	To	Fit[a]	Hole H7	Shaft js6	Fit[a]	Hole H8	Shaft js7	Fit[a]	Hole H7	Shaft k6	Fit[a]	Hole H8	Shaft k7	Fit[a]	Hole H7	Shaft n6	Fit[a]	Hole H7	Shaft n7
		Values shown below are in thousandths of an inch																	
0	0.12	−0.12 +0.52	+0.4 0	+0.12 −0.12	−0.2 +0.8	+0.6 0	+0.2 −0.2	… …	… …	… …	… …	… …	… …	−0.5 +0.15	+0.4 0	+0.5 +0.25	−0.65 +0.15	+0.4 0	+0.65 +0.25
0.12	0.24	−0.15 +0.65	+0.5 0	+0.15 −0.15	−0.25 +0.95	+0.7 0	+0.25 −0.25	… …	… …	… …	… …	… …	… …	−0.6 +0.2	+0.5 0	+0.6 +0.3	−0.8 +0.2	+0.5 0	+0.8 +0.3
0.24	0.40	−0.2 +0.8	+0.6 0	+0.2 −0.2	−0.3 +1.2	+0.9 0	+0.3 −0.3	−0.5 +0.5	+0.6 0	+0.5 +0.1	−0.7 +0.8	+0.9 0	+0.7 +0.1	−0.8 +0.2	+0.6 0	+0.8 +0.4	−1.0 +0.2	+0.6 0	+1.0 +0.4
0.40	0.71	−0.2 +0.9	+0.7 0	+0.2 −0.2	−0.35 +1.35	+1.0 0	+0.35 −0.35	−0.5 +0.6	+0.7 0	+0.5 +0.1	−0.8 +0.9	+1.0 0	+0.8 +0.1	−0.9 +0.2	+0.7 0	+0.9 +0.5	−1.2 +0.2	+0.7 0	+1.2 +0.5
0.71	1.19	−0.25 +1.05	+0.8 0	+0.25 −0.25	−0.4 +1.6	+1.2 0	+0.4 −0.4	−0.6 +0.7	+0.8 0	+0.6 +0.1	−0.9 +1.1	+1.2 0	+0.9 +0.1	−1.1 +0.2	+0.8 0	+1.1 +0.6	−1.4 +0.2	+0.8 0	+1.4 +0.6
1.19	1.97	−0.3 +1.3	+1.0 0	+0.3 −0.3	−0.5 +2.1	+1.6 0	+0.5 −0.5	−0.7 +0.9	+1.0 0	+0.7 +0.1	−1.1 +1.5	+1.6 0	+1.1 +0.1	−1.3 +0.3	+1.0 0	+1.3 +0.7	−1.7 +0.3	+1.0 0	+1.7 +0.7
1.97	3.15	−0.3 +1.5	+1.2 0	+0.3 −0.3	−0.6 +2.4	+1.8 0	+0.6 −0.6	−0.8 +1.1	+1.2 0	+0.8 +0.1	−1.3 +1.7	+1.8 0	+1.3 +0.1	−1.5 +0.4	+1.2 0	+1.5 +0.8	−2.0 +0.4	+1.2 0	+2.0 +0.8
3.15	4.73	−0.4 +1.8	+1.4 0	+0.4 −0.4	−0.7 +2.9	+2.2 0	+0.7 −0.7	−1.0 +1.3	+1.4 0	+1.0 +0.1	−1.5 +2.1	+2.2 0	+1.5 +0.1	−1.9 +0.4	+1.4 0	+1.9 +1.0	−2.4 +0.4	+1.4 0	+2.4 +1.0
4.73	7.09	−0.5 +2.1	+1.6 0	+0.5 −0.5	−0.8 +3.3	+2.5 0	+0.8 −0.8	−1.1 +1.5	+1.6 0	+1.1 +0.1	−1.7 +2.4	+2.5 0	+1.7 +0.1	−2.2 +0.4	+1.6 0	+2.2 +1.2	−2.8 +0.4	+1.6 0	+2.8 +1.2
7.09	9.85	−0.6 +2.4	+1.8 0	+0.6 −0.6	−0.9 +3.7	+2.8 0	+0.9 −0.9	−1.4 +1.6	+1.8 0	+1.4 +0.2	−2.0 +2.6	+2.8 0	+2.0 +0.2	−2.6 +0.4	+1.8 0	+2.6 +1.4	−3.2 +0.4	+1.8 0	+3.2 +1.4
9.85	12.41	−0.6 +2.6	+2.0 0	+0.6 −6.6	−1.0 +4.0	+3.0 0	+1.0 −1.0	−1.4 +1.8	+2.0 0	+1.4 +0.2	−2.2 +2.8	+3.0 0	+2.2 +0.2	−2.6 +0.6	+2.0 0	+2.6 +1.4	−3.4 +0.6	+2.0 0	+3.4 +1.4
12.41	15.75	−0.7 +2.9	+2.2 0	+0.7 −0.7	−1.0 +4.5	+3.5 0	+1.0 −1.0	−1.6 +2.0	+2.2 0	+1.6 +0.2	−2.4 +3.3	+3.5 0	+2.4 +0.2	−3.0 +0.6	+2.2 0	+3.0 +1.6	−3.8 +0.6	+2.2 0	+3.8 +1.6
15.75	19.69	−0.8 +3.3	+2.5 0	+0.8 −0.8	−1.2 +5.2	+4.0 0	+1.2 −1.2	−1.8 +2.3	+2.5 0	+1.8 +0.2	−2.7 +3.8	+4.0 0	+2.7 +0.2	−3.4 +0.7	+2.5 0	+3.4 +1.8	−4.3 +0.7	+2.5 0	+4.3 +1.8

[a] Pairs of values shown represent maximum amount of interference (−) and maximum amount of clearance (+) resulting from application of standard tolerance limits.

All data above heavy lines are in accord with ABC agreements. Symbols H7, js6, etc., are hole and shaft designations in the ABC system.

Table 10. ANSI/ASME Standard Interference Locational Fits
ANSI/ASME B4.1-1967 (2009; out of print)

Nominal Size Range, Inches	Class LN 1			Class LN 2			Class LN 3		
	Limits of Interference	Standard Limits		Limits of Interference	Standard Limits		Limits of Interference	Standard Limits	
		Hole H6	Shaft n5		Hole H7	Shaft p6		Hole H7	Shaft r6
Over To	Values shown below are in thousandths of an inch								
0 – 0.12	0	+0.25	+0.45	0	+0.4	+0.65	0.1	+0.4	+0.75
	0.45	0	+0.25	0.65	0	+0.4	0.75	0	+0.5
0.12 – 0.24	0	+0.3	+0.5	0	+0.5	+0.8	0.1	+0.5	+0.9
	0.5	0	+0.3	0.8	0	+0.5	0.9	0	+0.6
0.24 – 0.40	0	+0.4	+0.65	0	+0.6	+1.0	0.2	+0.6	+1.2
	0.65	0	+0.4	1.0	0	+0.6	1.2	0	+0.8
0.40 – 0.71	0	+0.4	+0.8	0	+0.7	+1.1	0.3	+0.7	+1.4
	0.8	0	+0.4	1.1	0	+0.7	1.4	0	+1.0
0.71 – 1.19	0	+0.5	+1.0	0	+0.8	+1.3	0.4	+0.8	+1.7
	1.0	0	+0.5	1.3	0	+0.8	1.7	0	+1.2
1.19 – 1.97	0	+0.6	+1.1	0	+1.0	+1.6	0.4	+1.0	+2.0
	1.1	0	+0.6	1.6	0	+1.0	2.0	0	+1.4
1.97 – 3.15	0.1	+0.7	+1.3	0.2	+1.2	+2.1	0.4	+1.2	+2.3
	1.3	0	+0.8	2.1	0	+1.4	2.3	0	+1.6
3.15 – 4.73	0.1	+0.9	+1.6	0.2	+1.4	+2.5	0.6	+1.4	+2.9
	1.6	0	+1.0	2.5	0	+1.6	2.9	0	+2.0
4.73 – 7.09	0.2	+1.0	+1.9	0.2	+1.6	+2.8	0.9	+1.6	+3.5
	1.9	0	+1.2	2.8	0	+1.8	3.5	0	+2.5
7.09 – 9.85	0.2	+1.2	+2.2	0.2	+1.8	+3.2	1.2	+1.8	+4.2
	2.2	0	+1.4	3.2	0	+2.0	4.2	0	+3.0
9.85 – 12.41	0.2	+1.2	+2.3	0.2	+2.0	+3.4	1.5	+2.0	+4.7
	2.3	0	+1.4	3.4	0	+2.2	4.7	0	+3.5
12.41 – 15.75	0.2	+1.4	+2.6	0.3	+2.2	+3.9	2.3	+2.2	+5.9
	2.6	0	+1.6	3.9	0	+2.5	5.9	0	+4.5
15.75 – 19.69	0.2	+1.6	+2.8	0.3	+2.5	+4.4	2.5	+2.5	+6.6
	2.8	0	+1.8	4.4	0	+2.8	6.6	0	+5.0

Tolerance limits given in body of table are added or subtracted to basic size (as indicated by + or – sign) to obtain maximum and minimum sizes of mating parts.

All data in this table are in accordance with American-British-Canadian (ABC) agreements.

Limits for sizes above 19.69 inches are not covered by ABC agreements but are given in the ANSI Standard.

Symbols H7, p6, etc., are hole and shaft designations in the ABC system.

Table 11. ANSI/ASME Standard Force and Shrink Fits *ANSI/ASME B4.1-1967 (2009; out of print)*

Nominal Size Range, Inches		Class FN 1			Class FN 2			Class FN 3			Class FN 4			Class FN 5		
			Standard Tolerance Limits			Standard Tolerance Limits			Standard Tolerance Limits			Standard Tolerance Limits			Standard Tolerance Limits	
Over	To	Interference[a]	Hole H6	Shaft	Interference[a]	Hole H7	Shaft s6	Interference[a]	Hole H7	Shaft t6	Interference[a]	Hole H7	Shaft u6	Interference[a]	Hole H8	Shaft x7
		Values shown below are in thousandths of an inch														
0	0.12	0.05	+0.25	+0.5	0.2	+0.4	+0.85	…	…	…	0.3	+0.4	+0.95	0.3	+0.6	+1.3
		0.5	0	+0.3	0.85	0	+0.6				0.95	0	+0.7	1.3	0	+0.9
0.12	0.24	0.1	+0.3	+0.6	0.2	+0.5	+1.0	…	…	…	0.4	+0.5	+1.2	0.5	+0.7	+1.7
		0.6	0	+0.4	1.0	0	+0.7				1.2	0	+0.9	1.7	0	+1.2
0.24	0.40	0.1	+0.4	+0.75	0.4	+0.6	+1.4	…	…	…	0.6	+0.6	+1.6	0.5	+0.9	+2.0
		0.75	0	+0.5	1.4	0	+1.0				1.6	0	+1.2	2.0	0	+1.4
0.40	0.56	0.1	+0.4	+0.8	0.5	+0.7	+1.6	…	…	…	0.7	+0.7	+1.8	0.6	+1.0	+2.3
		0.8	0	+0.5	1.6	0	+1.2				1.8	0	+1.4	2.3	0	+1.6
0.56	0.71	0.2	+0.4	+0.9	0.5	+0.7	+1.6	…	…	…	0.7	+0.7	+1.8	0.8	+1.0	+2.5
		0.9	0	+0.6	1.6	0	+1.2				1.8	0	+1.4	2.5	0	+1.8
0.71	0.95	0.2	+0.5	+1.1	0.6	+0.8	+1.9	…	…	…	0.8	+0.8	+2.1	1.0	+1.2	+3.0
		1.1	0	+0.7	1.9	0	+1.4				2.1	0	+1.6	3.0	0	+2.2
0.95	1.19	0.3	+0.5	+1.2	0.6	+0.8	+1.9	0.8	+0.8	+2.1	+1.0	+0.8	+2.3	1.3	+1.2	+3.3
		1.2	0	+0.8	1.9	0	+1.4	2.1	0	+1.6	2.3	0	+1.8	3.3	0	+2.5
1.19	1.58	0.3	+0.6	+1.3	0.8	+1.0	+2.4	1.0	+1.0	+2.6	1.5	+1.0	+3.1	1.4	+1.6	+4.0
		1.3	0	+0.9	2.4	0	+1.8	2.6	0	+2.0	3.1	0	+2.5	4.0	0	+3.0
1.58	1.97	0.4	+0.6	+1.4	0.8	+1.0	+2.4	1.2	+1.0	+2.8	1.8	+1.0	+3.4	2.4	+1.6	+5.0
		1.4	0	+1.0	2.4	0	+1.8	2.8	0	+2.2	3.4	0	+2.8	5.0	0	+4.0
1.97	2.56	0.6	+0.7	+1.8	0.8	+1.2	+2.7	1.3	+1.2	+3.2	2.3	+1.2	+4.2	3.2	+1.8	+6.2
		1.8	0	+1.3	2.7	0	+2.0	3.2	0	+2.5	4.2	0	+3.5	6.2	0	+5.0
2.56	3.15	0.7	+0.7	+1.9	1.0	+1.2	+2.9	1.8	+1.2	+3.7	2.8	+1.2	+4.7	4.2	+1.8	+7.2
		1.9	0	+1.4	2.9	0	+2.2	3.7	0	+3.0	4.7	0	+4.0	7.2	0	+6.0
3.15	3.94	0.9	+0.9	+2.4	1.4	+1.4	+3.7	2.1	+1.4	+4.4	3.6	+1.4	+5.9	4.8	+2.2	+8.4
		2.4	0	+1.8	3.7	0	+2.8	4.4	0	+3.5	5.9	0	+5.0	8.4	0	+7.0
3.94	4.73	1.1	+0.9	+2.6	1.6	+1.4	+3.9	2.6	+1.4	+4.9	4.6	+1.4	+6.9	5.8	+2.2	+9.4
		2.6	0	+2.0	3.9	0	+3.0	4.9	0	+4.0	6.9	0	+6.0	9.4	0	+8.0

Table 11. *(Continued)* **ANSI/ASME Standard Force and Shrink Fits** *ANSI/ASME B4.1-1967 (2009; out of print)*

Nominal Size Range, Inches		Class FN 1			Class FN 2			Class FN 3			Class FN 4			Class FN 5		
			Standard Tolerance Limits			Standard Tolerance Limits			Standard Tolerance Limits			Standard Tolerance Limits			Standard Tolerance Limits	
Over	To	Interference[a]	Hole H6	Shaft	Interference[a]	Hole H7	Shaft s6	Interference[a]	Hole H7	Shaft t6	Interference[a]	Hole H7	Shaft u6	Interference[a]	Hole H8	Shaft x7
		Values shown below are in thousandths of an inch														
4.73	5.52	1.2	+1.0	+2.9	1.9	+1.6	+4.5	3.4	+1.6	+6.0	5.4	+1.6	+8.0	7.5	+2.5	+11.6
		2.9	0	+2.2	4.5	0	+3.5	6.0	0	+5.0	8.0	0	+7.0	11.6	0	+10.0
5.52	6.30	1.5	+1.0	+3.2	2.4	+1.6	+5.0	3.4	+1.6	+6.0	5.4	+1.6	+8.0	9.5	+2.5	+13.6
		3.2	0	+2.5	5.0	0	+4.0	6.0	0	+5.0	8.0	0	+7.0	13.6	0	+12.0
6.30	7.09	1.8	+1.0	+3.5	2.9	+1.6	+5.5	4.4	+1.6	+7.0	6.4	+1.6	+9.0	9.5	+2.5	+13.6
		3.5	0	+2.8	5.5	0	+4.5	7.0	0	+6.0	9.0	0	+8.0	13.6	0	+12.0
7.09	7.88	1.8	+1.2	+3.8	3.2	+1.8	+6.2	5.2	+1.8	+8.2	7.2	+1.8	+10.2	11.2	+2.8	+15.8
		3.8	0	+3.0	6.2	0	+5.0	8.2	0	+7.0	10.2	0	+9.0	15.8	0	+14.0
7.88	8.86	2.3	+1.2	+4.3	3.2	+1.8	+6.2	5.2	+1.8	+8.2	8.2	+1.8	+11.2	13.2	+2.8	+17.8
		4.3	0	+3.5	6.2	0	+5.0	8.2	0	+7.0	11.2	0	+10.0	17.8	0	+16.0
8.86	9.85	2.3	+1.2	+4.3	4.2	+1.8	+7.2	6.2	+1.8	+9.2	10.2	+1.8	+13.2	13.2	+2.8	+17.8
		4.3	0	+3.5	7.2	0	+6.0	9.2	0	+8.0	13.2	0	+12.0	17.8	0	+16.0
9.85	11.03	2.8	+1.2	+4.9	4.0	+2.0	+7.2	7.0	+2.0	+10.2	10.0	+2.0	+13.2	15.0	+3.0	+20.0
		4.9	0	+4.0	7.2	0	+6.0	10.2	0	+9.0	13.2	0	+12.0	20.0	0	+18.0
11.03	12.41	2.8	+1.2	+4.9	5.0	+2.0	+8.2	7.0	+2.0	+10.2	12.0	+2.0	+15.2	17.0	+3.0	+22.0
		4.9	0	+4.0	8.2	0	+7.0	10.2	0	+9.0	15.2	0	+14.0	22.0	0	+20.0
12.41	13.98	3.1	+1.4	+5.5	5.8	+2.2	+9.4	7.8	+2.2	+11.4	13.8	+2.2	+17.4	18.5	+3.5	+24.2
		5.5	0	+4.5	9.4	0	+8.0	11.4	0	+10.0	17.4	0	+16.0	24.2	0	+22.0
13.98	15.75	3.6	+1.4	+6.1	5.8	+2.2	+9.4	9.8	+2.2	+13.4	15.8	+2.2	+19.4	21.5	+3.5	+27.2
		6.1	0	+5.0	9.4	0	+8.0	13.4	0	+12.0	19.4	0	+18.0	27.2	0	+25.0
15.75	17.72	4.4	+1.6	+7.0	6.5	+2.5	+10.6	+9.5	+2.5	+13.6	17.5	+2.5	+21.6	24.0	+4.0	+30.5
		7.0	0	+6.0	10.6	0	+9.0	13.6	0	+12.0	21.6	0	+20.0	30.5	0	+28.0
17.72	19.69	4.4	+1.6	+7.0	7.5	+2.5	+11.6	11.5	+2.5	+15.6	19.5	+2.5	+23.6	26.0	+4.0	+32.5
		7.0	0	+6.0	11.6	0	+10.0	15.6	0	+14.0	23.6	0	+22.0	32.5	0	+30.0

[a] Pairs of values shown represent minimum and maximum amounts of interference resulting from application of standard tolerance limits.

All data above heavy lines are in accordance with American-British-Canadian (ABC) agreements. Symbols H6, H7, s6, etc., are hole and shaft designations in the ABC system. Limits for sizes above 19.69 inches are not covered by ABC agreements but are given in the ANSI standard.

American National Standard Preferred Metric Limits and Fits.—This standard ANSI/ASME B4.2-1978 (R2009) describes the ISO system of metric limits and fits for mating parts as approved for general engineering usage in the United States. It establishes: (1) the designation symbols used to define dimensional limits on drawings, material stock, related tools, gages, etc.; (2) the preferred basic sizes (first and second choices); (3) the preferred tolerance zones (first, second, and third choices); (4) the definitions of related terms; and (5) the preferred limits and fits for sizes (first choice only) up to and including 500 millimeters.

The general terms "hole" and "shaft" can also be taken to refer to the space containing or contained by two parallel faces of any part, such as the width of a slot, or the thickness of a key.

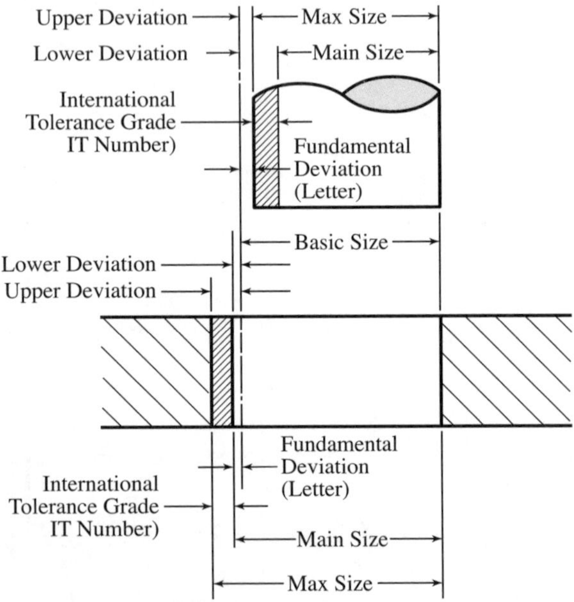

Fig. 1. Illustration of Definitions

Definitions.—The most important terms relating to limits and fits are shown in Fig. 1 and are defined as follows:

Basic Size: The size to which limits of deviation are assigned. The basic size is the same for both members of a fit. For example, it is designated by the numbers 40 in 40H7.

Deviation: The algebraic difference between a size and the corresponding basic size.

Upper Deviation: The algebraic difference between the maximum limit of size and the corresponding basic size.

Table 12. American National Standard Preferred Metric Sizes
ANSI/ASME B4.2-1978 (R2009)

Basic Size, mm		Basic Size, mm		Basic Size, mm		Basic Size, mm	
1st Choice	2nd Choice	1st Choice	2nd Choice	1st Choice	2nd Choice	1st Choice	2nd Choice
1	…	6	…	40	…	250	…
…	1.1	…	7	…	45	…	280
1.2	…	8	…	50	…	300	…
…	1.4	…	9	…	55	…	350
1.6	…	10	…	60	…	400	…
…	1.8	…	11	…	70	…	450
2	…	12	…	80	…	500	…
…	2.2	…	14	…	90	…	550
2.5	…	16	…	100	…	600	…
…	2.8	…	18	…	110	…	700
3	…	20	…	120	…	800	…
…	3.5	…	22	…	140	…	900
4	…	25	…	160	…	1000	…
…	4.5	…	28	…	180	…	…
5	…	30	…	200	…	…	…
…	5.5	…	35	…	220	…	…

Preferred Fits.—First-choice tolerance zones are used to establish preferred fits in the Standard for Preferred Metric Limits and Fits, ANSI/ASME B4.2, as shown in Figs. 2 and 3. A complete listing of first-, second-, and third-choice tolerance zones is given in the Standard.

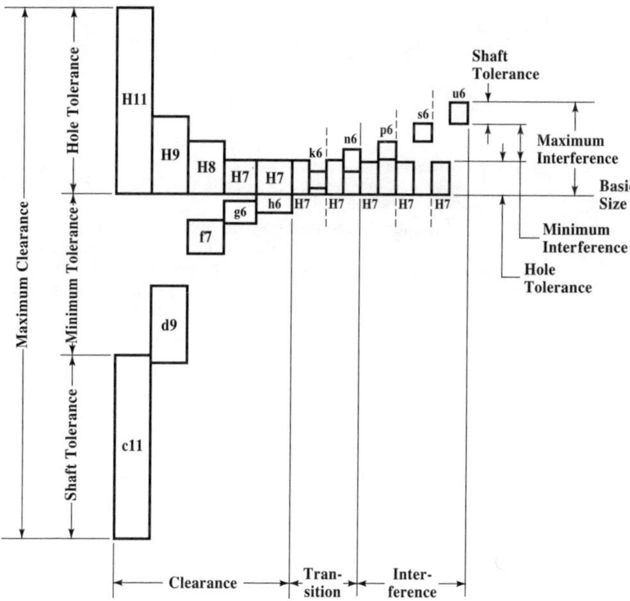

Fig. 2. Preferred Hole Basis Fits

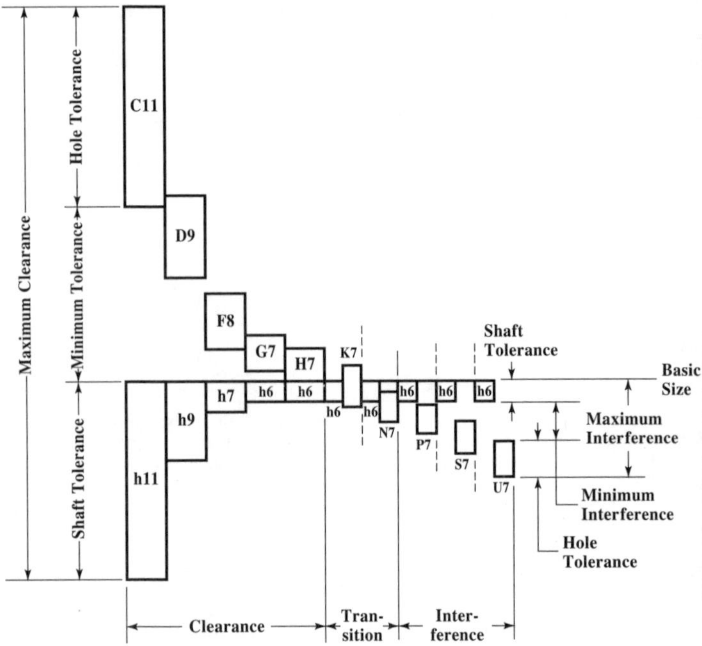

Fig. 3. Preferred Shaft Basis Fits

Hole basis fits have a fundamental deviation of H on the hole, and shaft basis fits have a fundamental deviation of h on the shaft and are shown in Fig. 2 for hole basis and Fig. 3 for shaft basis fits. A description of both types of fits, which have the same relative fit condition, is given in Table 13. Normally, the hole basis system is preferred; however, when a common shaft mates with several holes, the shaft basis system should be used.

The hole basis and shaft basis fits shown in Table 13 are combined with the first-choice sizes shown in Table 12 to form Tables 14, 15, 16, and 17, where specific limits as well as the resultant fits are tabulated.

If the required size is not tabulated in Tables 14 through 17, then the preferred fit can be calculated from numerical values given in an appendix of ANSI/ASME B4.2-1978 (R2009). It is anticipated that other fit conditions may be necessary to meet special requirements, and a preferred fit can be loosened or tightened simply by selecting a standard tolerance zone as given in the Standard. Information on how to calculate limit dimensions, clearances, and interferences for nonpreferred fits and sizes can also be found in an appendix of this Standard.

By combining the IT grade number and the tolerance position letter, the tolerance symbol is established that identifies the actual maximum and minimum limits of the part. The tolerances size is thus defined by the basic size of the part followed by a symbol composed of a letter and a number, such as 40H7, 40f7, etc.

Table 13. Description of Preferred Fits

	ISO SYMBOL		DESCRIPTION	
	Hole Basis	**Shaft Basis**		
Clearance Fits	H11/c11	C11/h11	*Loose running* fit for wide commercial tolerances or allowances on external members.	↑ More Clearance
	H9/d9	D9/h9	*Free running* fit not for use where accuracy is essential, but good for large temperature variations, high running speeds, or heavy journal pressures.	
	H8/f7	F8/h7	*Close Running* fit for running on accurate machines and for accurate moderate speeds and journal pressures.	
	H7/g6	G7/h6	*Sliding fit* not intended to run freely, but to move and turn freely and locate accurately.	
	H7/h6	H7/h6	*Locational clearance* fit provides snug fit for locating stationary parts; but can be freely assembled and disassembled.	
Transition Fits	H7/k6	K7/h6	*Locational transition* fit for accurate location, a compromise between clearance and interference.	
	H7/n6	N7/h6	*Locational transition* fit for more accurate location where greater interference is permissible.	
Interference Fits	H7/p6[a]	P7/h6	*Locational interference* fit for parts requiring rigidity and alignment with prime accuracy of location but without special bore pressure requirements.	More Interference ↓
	H7/s6	S7/h6	*Medium drive* fit for ordinary steel parts or shrink fits on light sections; the tightest fit usable with cast iron.	
	H7/u6	U7/h6	*Force* fit suitable for parts that can be highly stressed or for shrink fits where the heavy pressing forces required are impractical.	

[a] Transition fit for basic sizes in range from 0 through 3 mm.

Table 14. American National Standard Preferred Hole Basis Metric Clearance Fits *ANSI/ASME B4.2-1978 (R2009)*

Basic Size[a]		Loose Running			Free Running			Close Running			Sliding			Locational Clearance		
		Hole H11	Shaft c11	Fit[b]	Hole H9	Shaft d9	Fit[b]	Hole H8	Shaft f7	Fit[b]	Hole H7	Shaft g6	Fit[b]	Hole H7	Shaft h6	Fit[b]
1	Max	1.060	0.940	0.180	1.025	0.980	0.070	1.014	0.994	0.030	1.010	0.998	0.018	1.010	1.000	0.016
	Min	1.000	0.880	0.060	1.000	0.995	0.020	1.000	0.984	0.006	1.000	0.992	0.002	1.000	0.994	0.000
1.2	Max	1.260	1.140	0.180	1.225	1.180	0.070	1.214	1.194	0.030	1.210	1.198	0.018	1.210	1.200	0.016
	Min	1.200	1.080	0.060	1.200	1.155	0.020	1.200	1.184	0.006	1.200	1.192	0.002	1.200	1.194	0.000
1.6	Max	1.660	1.540	0.180	1.625	1.580	0.070	1.614	1.594	0.030	1.610	1.598	0.018	1.610	1.600	0.016
	Min	1.600	1.480	0.060	1.600	1.555	0.020	1.600	1.584	0.006	1.600	1.592	0.002	1.600	1.594	0.000
2	Max	2.060	1.940	0.180	2.025	1.980	0.070	2.014	1.994	0.030	2.010	1.998	0.018	2.010	2.000	0.016
	Min	2.000	1.880	0.060	2.000	1.955	0.020	2.000	1.984	0.006	2.000	1.992	0.002	2.000	1.994	0.000
2.5	Max	2.560	2.440	0.180	2.525	2.480	0.070	2.514	2.494	0.030	2.510	2.498	0.018	2.510	2.500	0.016
	Min	2.500	2.380	0.060	2.500	2.455	0.020	2.500	2.484	0.006	2.500	2.492	0.002	2.500	2.494	0.000
3	Max	3.060	2.940	0.180	3.025	2.980	0.070	3.014	2.994	0.030	3.010	2.998	0.018	3.010	3.000	0.016
	Min	3.000	2.880	0.060	3.000	2.955	0.020	3.000	2.984	0.006	3.000	2.992	0.002	3.000	2.994	0.000
4	Max	4.075	3.930	0.220	4.030	3.970	0.090	4.018	3.990	0.040	4.012	3.996	0.024	4.012	4.000	0.020
	Min	4.000	3.855	0.070	4.000	3.940	0.030	4.000	3.978	0.010	4.000	3.988	0.004	4.000	3.992	0.000
5	Max	5.075	4.930	0.220	5.030	4.970	0.090	5.018	4.990	0.040	5.012	4.996	0.024	5.012	5.000	0.020
	Min	5.000	4.855	0.070	5.000	4.940	0.030	5.000	4.978	0.010	5.000	4.988	0.004	5.000	4.992	0.000
6	Max	6.075	5.930	0.220	6.030	5.970	0.090	6.018	5.990	0.040	6.012	5.996	0.024	6.012	6.000	0.020
	Min	6.000	5.855	0.070	6.000	5.940	0.030	6.000	5.978	0.010	6.000	5.988	0.004	6.000	5.992	0.000
8	Max	8.090	7.920	0.260	8.036	7.960	0.112	8.022	7.987	0.050	8.015	7.995	0.029	8.015	8.000	0.024
	Min	8.000	7.830	0.080	8.000	7.924	0.040	8.000	7.972	0.013	8.000	7.986	0.005	8.000	7.991	0.000
10	Max	10.090	9.920	0.260	10.036	9.960	0.112	10.022	9.987	0.050	10.015	9.995	0.029	10.015	10.000	0.024
	Min	10.000	9.830	0.080	10.000	9.924	0.040	10.000	9.972	0.013	10.000	9.986	0.005	10.000	9.991	0.000
12	Max	12.110	11.905	0.315	12.043	11.956	0.136	12.027	11.984	0.061	12.018	11.994	0.035	12.018	12.000	0.029
	Min	12.000	11.795	0.095	12.000	11.907	0.050	12.000	11.966	0.016	12.000	11.983	0.006	12.000	11.989	0.000
16	Max	16.110	15.905	0.315	16.043	15.950	0.136	16.027	15.984	0.061	16.018	15.994	0.035	16.018	16.000	0.029
	Min	16.000	15.795	0.095	16.000	15.907	0.050	16.000	15.966	0.016	16.000	15.983	0.006	16.000	15.989	0.000
20	Max	20.130	19.890	0.370	20.052	19.935	0.169	20.033	19.980	0.074	20.021	19.993	0.041	20.021	20.000	0.034
	Min	20.000	19.760	0.110	20.000	19.883	0.065	20.000	19.959	0.020	20.000	19.980	0.007	20.000	19.987	0.000
25	Max	25.130	24.890	0.370	25.052	24.935	0.169	25.033	24.980	0.074	25.021	24.993	0.041	25.021	25.000	0.034
	Min	25.000	24.760	0.110	25.000	24.883	0.065	25.000	24.959	0.020	25.000	24.980	0.007	25.000	24.987	0.000

Table 14. *(Continued)* **American National Standard Preferred Hole Basis Metric Clearance Fits** *ANSI/ASME B4.2-1978 (R2009)*

Basic Size[a]		Loose Running			Free Running			Close Running			Sliding			Locational Clearance		
		Hole H11	Shaft c11	Fit[b]	Hole H9	Shaft d9	Fit[b]	Hole H8	Shaft f7	Fit[b]	Hole H7	Shaft g6	Fit[b]	Hole H7	Shaft h6	Fit[b]
30	Max	30.130	29.890	0.370	30.052	29.935	0.169	30.033	29.980	0.074	30.021	29.993	0.041	30.021	30.000	0.034
	Min	30.000	29.760	0.110	30.000	29.883	0.065	30.000	29.959	0.020	30.000	29.980	0.007	30.000	29.987	0.000
40	Max	40.160	39.880	0.440	40.062	39.920	0.204	40.039	39.975	0.089	40.025	39.991	0.050	40.025	40.000	0.041
	Min	40.000	39.720	0.120	40.000	39.858	0.080	40.000	39.950	0.025	40.000	39.975	0.009	40.000	39.984	0.000
50	Max	50.160	49.870	0.450	50.062	49.920	0.204	50.039	49.975	0.089	50.025	49.991	0.050	50.025	50.000	0.041
	Min	50.000	49.710	0.130	50.000	49.858	0.080	50.000	49.950	0.025	50.000	49.975	0.009	50.000	49.984	0.000
60	Max	60.190	59.860	0.520	60.074	59.900	0.248	60.046	59.970	0.106	60.030	59.990	0.059	60.030	60.000	0.049
	Min	60.000	59.670	0.140	60.000	59.826	0.100	60.000	59.940	0.030	60.000	59.971	0.010	60.000	59.981	0.000
80	Max	80.190	79.850	0.530	80.074	79.900	0.248	80.046	79.970	0.106	80.030	79.990	0.059	80.030	80.000	0.049
	Min	80.000	79.660	0.150	80.000	79.826	0.100	80.000	79.940	0.030	80.000	79.971	0.010	80.000	79.981	0.000
100	Max	100.220	99.830	0.610	100.087	99.880	0.294	100.054	99.964	0.125	100.035	99.988	0.069	100.035	100.000	0.057
	Min	100.000	99.610	0.170	100.000	99.793	0.120	100.000	99.929	0.036	100.000	99.966	0.012	100.000	99.978	0.000
120	Max	120.220	119.820	0.620	120.087	119.880	0.294	120.054	119.964	0.125	120.035	119.988	0.069	120.035	120.000	0.057
	Min	120.000	119.600	0.180	120.000	119.793	0.120	120.000	119.929	0.036	120.000	119.966	0.012	120.000	119.978	0.000
160	Max	160.250	159.790	0.710	160.100	159.855	0.345	160.063	159.957	0.146	160.040	159.986	0.079	160.040	160.000	0.065
	Min	160.000	159.540	0.210	160.000	159.755	0.145	160.000	159.917	0.043	160.000	159.961	0.014	160.000	159.975	0.000
200	Max	200.290	199.760	0.820	200.115	199.830	0.400	200.072	199.950	0.168	200.046	199.985	0.090	200.046	200.000	0.075
	Min	200.000	199.470	0.240	200.000	199.715	0.170	200.000	199.904	0.050	200.000	199.956	0.015	200.000	199.971	0.000
250	Max	250.290	249.720	0.860	250.115	249.830	0.400	250.072	249.950	0.168	250.046	249.985	0.090	250.046	250.000	0.075
	Min	250.000	249.430	0.280	250.000	249.715	0.170	250.000	249.904	0.050	250.000	249.956	0.015	250.000	249.971	0.000
300	Max	300.320	299.670	0.970	300.130	299.810	0.450	300.081	299.944	0.189	300.052	299.983	0.101	300.052	300.000	0.084
	Min	300.000	299.350	0.330	300.000	299.680	0.190	300.000	299.892	0.056	300.000	299.951	0.017	300.000	299.968	0.000
400	Max	400.360	399.600	1.120	400.140	399.790	0.490	400.089	399.938	0.208	400.057	399.982	0.111	400.057	400.000	0.093
	Min	400.000	399.240	0.400	400.000	399.650	0.210	400.000	399.881	0.062	400.000	399.946	0.018	400.000	399.964	0.000
500	Max	500.400	499.520	1.280	500.155	499.770	0.540	500.097	499.932	0.228	500.063	499.980	0.123	500.063	500.000	0.103
	Min	500.000	499.120	0.480	500.000	499.615	0.230	500.000	499.869	0.068	500.000	499.940	0.020	500.000	499.960	0.000

[a] The sizes shown are first-choice basic sizes (see Table 12). Preferred fits for other sizes can be calculated from data given in ANSI B4.2-1978, R2004.
[b] All fits shown in this table have clearance.
All dimensions are in millimeters.

Table 15. American National Standard Preferred Hole Basis Metric Transition and Interference Fits *ANSI/ASME B4.2-1978 (R2009)*

Basic Size[a]		Locational Transition			Locational Transition			Locational Interference			Medium Drive			Force		
		Hole H7	Shaft k6	Fit[b]	Hole H7	Shaft n6	Fit[b]	Hole H7	Shaft p6	Fit[b]	Hole H7	Shaft s6	Fit[b]	Hole H7	Shaft u6	Fit[b]
1	Max	1.010	1.006	+0.010	1.010	1.010	+0.006	1.010	1.012	+0.004	1.010	1.020	−0.004	1.010	1.024	−0.008
	Min	1.000	1.000	−0.006	1.000	1.004	−0.010	1.000	1.006	−0.012	1.000	1.014	−0.020	1.000	1.018	−0.024
1.2	Max	1.210	1.206	+0.010	1.210	1.210	+0.006	1.210	1.212	+0.004	1.210	1.220	−0.004	1.210	1.224	−0.008
	Min	1.200	1.200	−0.006	1.200	1.204	−0.010	1.200	1.206	−0.012	1.200	1.214	−0.020	1.200	1.218	−0.024
1.6	Max	1.610	1.606	+0.010	1.610	1.610	+0.006	1.610	1.612	+0.004	1.610	1.620	−0.004	1.610	1.624	−0.008
	Min	1.600	1.600	−0.006	1.600	1.604	−0.010	1.600	1.606	−0.012	1.600	1.614	−0.020	1.600	1.618	−0.024
2	Max	2.010	2.006	+0.010	2.010	2.010	+0.006	2.010	2.012	+0.004	2.010	2.020	−0.004	2.010	2.024	−0.008
	Min	2.000	2.000	−0.006	2.000	2.004	−0.010	2.000	2.006	−0.012	2.000	2.014	−0.020	2.000	2.018	−0.024
2.5	Max	2.510	2.506	+0.010	2.510	2.510	+0.006	2.510	2.512	+0.004	2.510	2.520	−0.004	2.510	2.524	−0.008
	Min	2.500	2.500	−0.006	2.500	2.504	−0.010	2.500	2.506	−0.012	2.500	2.514	−0.020	2.500	2.518	−0.024
3	Max	3.010	3.006	+0.010	3.010	3.010	+0.006	3.010	3.012	+0.004	3.010	3.020	−0.004	3.010	3.024	−0.008
	Min	3.000	3.000	−0.006	3.000	3.004	−0.010	3.000	3.006	−0.012	3.000	3.014	−0.020	3.000	3.018	−0.024
4	Max	4.012	4.009	+0.011	4.012	4.016	+0.004	4.012	4.020	0.000	4.012	4.027	−0.007	4.012	4.031	−0.011
	Min	4.000	4.001	−0.009	4.000	4.008	−0.016	4.000	4.012	−0.020	4.000	4.019	−0.027	4.000	4.023	−0.031
5	Max	5.012	5.009	+0.011	5.012	5.016	+0.004	5.012	5.020	0.000	5.012	5.027	−0.007	5.012	5.031	−0.011
	Min	5.000	5.001	−0.009	5.000	5.008	−0.016	5.000	5.012	−0.020	5.000	5.019	−0.027	5.000	5.023	−0.031
6	Max	6.012	6.009	+0.011	6.012	6.016	+0.004	6.012	6.020	0.000	6.012	6.027	−0.007	6.012	6.031	−0.011
	Min	6.000	6.001	−0.009	6.000	6.008	−0.016	6.000	6.012	−0.020	6.000	6.019	−0.027	6.000	6.023	−0.031
8	Max	8.015	8.010	+0.014	8.015	8.019	+0.005	8.015	8.024	0.000	8.015	8.032	−0.008	8.015	8.037	−0.013
	Min	8.000	8.001	−0.010	8.000	8.010	−0.019	8.000	8.015	−0.024	8.000	8.023	−0.032	8.000	8.028	−0.037
10	Max	10.015	10.010	+0.014	10.015	10.019	+0.005	10.015	10.024	0.000	10.015	10.032	−0.008	10.015	10.034	−0.013
	Min	10.000	10.001	−0.010	10.000	10.010	−0.019	10.000	10.015	−0.024	10.000	10.023	−0.032	10.000	10.028	−0.037
12	Max	12.018	12.012	+0.017	12.018	12.023	+0.006	12.018	12.029	0.000	12.018	12.039	−0.010	12.018	12.044	−0.015
	Min	12.000	12.001	−0.012	12.000	12.012	−0.023	12.000	12.018	−0.029	12.000	12.028	−0.039	12.000	12.033	−0.044
16	Max	16.018	16.012	+0.017	16.018	16.023	+0.006	16.018	16.029	0.000	16.018	16.039	−0.010	16.018	16.044	−0.015
	Min	16.000	16.001	−0.012	16.000	16.012	−0.023	16.000	16.018	−0.029	16.000	16.028	−0.039	16.000	16.033	−0.044
20	Max	20.021	20.015	+0.019	20.021	20.028	+0.006	20.021	20.035	−0.001	20.021	20.048	−0.014	20.021	20.054	−0.020
	Min	20.000	20.002	−0.015	20.000	20.015	−0.028	20.000	20.022	−0.035	20.000	20.035	−0.048	20.000	20.041	−0.054
25	Max	25.021	25.015	+0.019	25.021	25.028	+0.006	25.021	25.035	−0.001	25.021	25.048	−0.014	25.021	25.061	−0.027
	Min	25.000	25.002	−0.015	25.000	25.015	−0.028	25.000	25.022	−0.035	25.000	25.035	−0.048	25.000	25.048	−0.061

Table 15. *(Continued)* **American National Standard Preferred Hole Basis Metric Transition and Interference Fits** *ANSI/ASME B4.2-1978 (R2009)*

Basic Size[a]		Locational Transition			Locational Transition			Locational Interference			Medium Drive			Force		
		Hole H7	Shaft k6	Fit[b]	Hole H7	Shaft n6	Fit[b]	Hole H7	Shaft p6	Fit[b]	Hole H7	Shaft s6	Fit[b]	Hole H7	Shaft u6	Fit[b]
30	Max	30.021	30.015	+0.019	30.021	30.028	+0.006	30.021	30.035	−0.001	30.021	30.048	−0.014	30.021	30.061	−0.027
	Min	30.000	30.002	−0.015	30.000	30.015	−0.028	30.000	30.022	−0.035	30.000	30.035	−0.048	30.000	30.048	−0.061
40	Max	40.025	40.018	+0.023	40.025	40.033	+0.008	40.025	40.042	−0.001	40.025	40.059	−0.018	40.025	40.076	−0.035
	Min	40.000	40.002	−0.018	40.000	40.017	−0.033	40.000	40.026	−0.042	40.000	40.043	−0.059	40.000	40.060	−0.076
50	Max	50.025	50.018	+0.023	50.025	50.033	+0.008	50.025	50.042	−0.001	50.025	50.059	−0.018	50.025	50.086	−0.045
	Min	50.000	50.002	−0.018	50.000	50.017	−0.033	50.000	50.026	−0.042	50.000	50.043	−0.059	50.000	50.070	−0.086
60	Max	60.030	60.021	+0.028	60.030	60.039	+0.010	60.030	60.051	−0.002	60.030	60.072	−0.023	60.030	60.106	−0.057
	Min	60.000	60.002	−0.021	60.000	60.020	−0.039	60.000	60.032	−0.051	60.000	60.053	−0.072	60.000	60.087	−0.106
80	Max	80.030	80.021	+0.028	80.030	80.039	+0.010	80.030	80.051	−0.002	80.030	80.078	−0.029	80.030	80.121	−0.072
	Min	80.000	80.002	−0.021	80.000	80.020	−0.039	80.000	80.032	−0.051	80.000	80.059	−0.078	80.000	80.102	−0.121
100	Max	100.035	100.025	+0.032	100.035	100.045	+0.012	100.035	100.059	−0.002	100.035	100.093	−0.036	100.035	100.146	−0.089
	Min	100.000	100.003	−0.025	100.000	100.023	−0.045	100.000	100.037	−0.059	100.000	100.071	−0.093	100.000	100.124	−0.146
120	Max	120.035	120.025	+0.032	120.035	120.045	+0.012	120.035	120.059	−0.002	120.035	120.101	−0.044	120.035	120.166	−0.109
	Min	120.000	120.003	−0.025	120.000	120.023	−0.045	120.000	120.037	−0.059	120.000	120.079	−0.101	120.000	120.144	−0.166
160	Max	160.040	160.028	+0.037	160.040	160.052	+0.013	160.040	160.068	−0.003	160.040	160.125	−0.060	160.040	160.215	−0.150
	Min	160.000	160.003	−0.028	160.000	160.027	−0.052	160.000	160.043	−0.068	160.000	160.100	−0.125	160.000	160.190	−0.215
200	Max	200.046	200.033	+0.042	200.046	200.060	+0.015	200.046	200.079	−0.004	200.046	200.151	−0.076	200.046	200.265	−0.190
	Min	200.000	200.004	−0.033	200.000	200.031	−0.060	200.000	200.050	−0.079	200.000	200.122	−0.151	200.000	200.236	−0.265
250	Max	250.046	250.033	+0.042	250.046	250.060	+0.015	250.046	250.079	−0.004	250.046	250.169	−0.094	250.046	250.313	−0.238
	Min	250.000	250.004	−0.033	250.000	250.031	−0.060	250.000	250.050	−0.079	250.000	250.140	−0.169	250.000	250.284	−0.313
300	Max	300.052	300.036	+0.048	300.052	300.066	+0.018	300.052	300.088	−0.004	300.052	300.202	−0.118	300.052	300.382	−0.298
	Min	300.000	300.004	−0.036	300.000	300.034	−0.066	300.000	300.056	−0.088	300.000	300.170	−0.202	300.000	300.350	−0.382
400	Max	400.057	400.040	+0.053	400.057	400.073	+0.020	400.057	400.098	−0.005	400.057	400.244	−0.151	400.057	400.471	−0.378
	Min	400.000	400.004	−0.040	400.000	400.037	−0.073	400.000	400.062	−0.098	400.000	400.208	−0.244	400.000	400.435	−0.471
500	Max	500.063	500.045	+0.058	500.063	500.080	+0.023	500.063	500.108	−0.005	500.063	500.292	−0.189	500.063	500.580	−0.477
	Min	500.000	500.005	−0.045	500.000	500.040	−0.080	500.000	500.068	−0.108	500.000	500.252	−0.292	500.000	500.540	−0.580

[a] The sizes shown are first-choice basic sizes (see Table 12). Preferred fits for other sizes can be calculated from data given in ANSI B4.2-1978, R2004.

[b] A plus sign indicates clearance; a minus sign indicates interference.

All dimensions are in millimeters.

Table 16. American National Standard Preferred Shaft Basis Metric Clearance Fits *ANSI/ASME B4.2-1978 (R2009)*

Basic Size[a]		Loose Running			Free Running			Close Running			Sliding			Locational Clearance		
		Hole C11	Shaft h11	Fit[b]	Hole D9	Shaft h9	Fit[b]	Hole F8	Shaft h7	Fit[b]	Hole G7	Shaft h6	Fit[b]	Hole H7	Shaft h6	Fit[b]
1	Max	1.120	1.000	0.180	1.045	1.000	0.070	1.020	1.000	0.030	1.012	1.000	0.018	1.010	1.000	0.016
	Min	1.060	0.940	0.060	1.020	0.975	0.020	1.006	0.990	0.006	1.002	0.994	0.002	1.000	0.994	0.000
1.2	Max	1.320	1.200	0.180	1.245	1.200	0.070	1.220	1.200	0.030	1.212	1.200	0.018	1.210	1.200	0.016
	Min	1.260	1.140	0.060	1.220	1.175	0.020	1.206	1.190	0.006	1.202	1.194	0.002	1.200	1.194	0.000
1.6	Max	1.720	1.600	0.180	1.645	1.600	0.070	1.620	1.600	0.030	1.612	1.600	0.018	1.610	1.600	0.016
	Min	1.660	1.540	0.060	1.620	1.575	0.020	1.606	1.590	0.006	1.602	1.594	0.002	1.600	1.594	0.000
2	Max	2.120	2.000	0.180	2.045	2.000	0.070	2.020	2.000	0.030	2.012	2.000	0.018	2.010	2.000	0.016
	Min	2.060	1.940	0.060	2.020	1.975	0.020	2.006	1.990	0.006	2.002	1.994	0.002	2.000	1.994	0.000
2.5	Max	2.620	2.500	0.180	2.545	2.500	0.070	2.520	2.500	0.030	2.512	2.500	0.018	2.510	2.500	0.016
	Min	2.560	2.440	0.060	2.520	2.475	0.020	2.506	2.490	0.006	2.502	2.494	0.002	2.500	2.494	0.000
3	Max	3.120	3.000	0.180	3.045	3.000	0.070	3.020	3.000	0.030	3.012	3.000	0.018	3.010	3.000	0.016
	Min	3.060	2.940	0.060	3.020	2.975	0.020	3.006	2.990	0.006	3.002	2.994	0.002	3.000	2.994	0.000
4	Max	4.145	4.000	0.220	4.060	4.000	0.090	4.028	4.000	0.040	4.016	4.000	0.024	4.012	4.000	0.020
	Min	4.070	3.925	0.070	4.030	3.970	0.030	4.010	3.988	0.010	4.004	3.992	0.004	4.000	3.992	0.000
5	Max	5.145	5.000	0.220	5.060	5.000	0.090	5.028	5.000	0.040	5.016	5.000	0.024	5.012	5.000	0.020
	Min	5.070	4.925	0.070	5.030	4.970	0.030	5.010	4.988	0.010	5.004	4.992	0.004	5.000	4.992	0.000
6	Max	6.145	6.000	0.220	6.060	6.000	0.090	6.028	6.000	0.040	6.016	6.000	0.024	6.012	6.000	0.020
	Min	6.070	5.925	0.070	6.030	5.970	0.030	6.010	5.988	0.010	6.004	5.992	0.004	6.000	5.992	0.000
8	Max	8.170	8.000	0.260	8.076	8.000	0.112	8.035	8.000	0.050	8.020	8.000	0.029	8.015	8.000	0.024
	Min	8.080	7.910	0.080	8.040	7.964	0.040	8.013	7.985	0.013	8.005	7.991	0.005	8.000	7.991	0.000
10	Max	10.170	10.000	0.260	10.076	10.000	0.112	10.035	10.000	0.050	10.020	10.000	0.029	10.015	10.000	0.024
	Min	10.080	9.910	0.080	10.040	9.964	0.040	10.013	9.985	0.013	10.005	9.991	0.005	10.000	9.991	0.000
12	Max	12.205	12.000	0.315	12.093	12.000	0.136	12.043	12.000	0.061	12.024	12.000	0.035	12.018	12.000	0.029
	Min	12.095	11.890	0.095	12.050	11.957	0.050	12.016	11.982	0.016	12.006	11.989	0.006	12.000	11.989	0.000
16	Max	16.205	16.000	0.315	16.093	16.000	0.136	16.043	16.000	0.061	16.024	16.000	0.035	16.018	16.000	0.029
	Min	16.095	15.890	0.095	16.050	15.957	0.050	16.016	15.982	0.016	16.006	15.989	0.006	16.000	15.989	0.000
20	Max	20.240	20.000	0.370	20.117	20.000	0.169	20.053	20.000	0.074	20.028	20.000	0.041	20.021	20.000	0.034
	Min	20.110	19.870	0.110	20.065	19.948	0.065	20.020	19.979	0.020	20.007	19.987	0.007	20.000	19.987	0.000
25	Max	25.240	25.000	0.370	25.117	25.000	0.169	25.053	25.000	0.074	25.028	25.000	0.041	25.021	25.000	0.034
25	Min	25.110	24.870	0.110	25.065	24.948	0.065	25.020	24.979	0.020	25.007	24.987	0.007	25.000	24.987	0.000

Table 16. *(Continued)* **American National Standard Preferred Shaft Basis Metric Clearance Fits** *ANSI/ASME B4.2-1978 (R2009)*

Basic Size[a]		Loose Running			Free Running			Close Running			Sliding			Locational Clearance		
		Hole C11	Shaft h11	Fit[b]	Hole D9	Shaft h9	Fit[b]	Hole F8	Shaft h7	Fit[b]	Hole G7	Shaft h6	Fit[b]	Hole H7	Shaft h6	Fit[b]
30	Max	30.240	30.000	0.370	30.117	30.000	0.169	30.053	30.000	0.074	30.028	30.000	0.041	30.021	30.000	0.034
	Min	30.110	29.870	0.110	30.065	29.948	0.065	30.020	29.979	0.020	30.007	29.987	0.007	30.000	29.987	0.000
40	Max	40.280	40.000	0.440	40.142	40.000	0.204	40.064	40.000	0.089	40.034	40.000	0.050	40.025	40.000	0.041
	Min	40.120	39.840	0.120	40.080	39.938	0.080	40.025	39.975	0.025	40.009	39.984	0.009	40.000	39.984	0.000
50	Max	50.290	50.000	0.450	50.142	50.000	0.204	50.064	50.000	0.089	50.034	50.000	0.050	50.025	50.000	0.041
	Min	50.130	49.840	0.130	50.080	49.938	0.080	50.025	49.975	0.025	50.009	49.984	0.009	50.000	49.984	0.000
60	Max	60.330	60.000	0.520	60.174	60.000	0.248	60.076	60.000	0.106	60.040	60.000	0.059	60.030	60.000	0.049
	Min	60.140	59.810	0.140	60.100	59.926	0.100	60.030	59.970	0.030	60.010	59.981	0.010	60.000	59.981	0.000
80	Max	80.340	80.000	0.530	80.174	80.000	0.248	80.076	80.000	0.106	80.040	80.000	0.059	80.030	80.000	0.049
	Min	80.150	79.810	0.150	80.100	79.926	0.100	80.030	79.970	0.030	80.010	79.981	0.010	80.000	79.981	0.000
100	Max	100.390	100.000	0.610	100.207	100.000	0.294	100.090	100.000	0.125	100.047	100.000	0.069	100.035	100.000	0.057
	Min	100.170	99.780	0.170	100.120	99.913	0.120	100.036	99.965	0.036	100.012	99.978	0.012	100.000	99.978	0.000
120	Max	120.400	120.000	0.620	120.207	120.000	0.294	120.090	120.000	0.125	120.047	120.000	0.069	120.035	120.000	0.057
	Min	120.180	119.780	0.180	120.120	119.913	0.120	120.036	119.965	0.036	120.012	119.978	0.012	120.000	119.978	0.000
160	Max	160.460	160.000	0.710	160.245	160.000	0.345	160.106	160.000	0.146	160.054	160.000	0.079	160.040	160.000	0.065
	Min	160.210	159.750	0.210	160.145	159.900	0.145	160.043	159.960	0.043	160.014	159.975	0.014	160.000	159.975	0.000
200	Max	200.530	200.000	0.820	200.285	200.000	0.400	200.122	200.000	0.168	200.061	200.000	0.090	200.046	200.000	0.075
	Min	200.240	199.710	0.240	200.170	199.885	0.170	200.050	199.954	0.050	200.015	199.971	0.015	200.000	199.971	0.000
250	Max	250.570	250.000	0.860	250.285	250.000	0.400	250.122	250.000	0.168	250.061	250.000	0.090	250.046	250.000	0.075
	Min	250.280	249.710	0.280	250.170	249.885	0.170	250.050	249.954	0.050	250.015	249.971	0.015	250.000	249.971	0.000
300	Max	300.650	300.000	0.970	300.320	300.000	0.450	300.137	300.000	0.189	300.069	300.000	0.101	300.052	300.000	0.084
	Min	300.330	299.680	0.330	300.190	299.870	0.190	300.056	299.948	0.056	300.017	299.968	0.017	300.000	299.968	0.000
400	Max	400.760	400.000	1.120	400.350	400.000	0.490	400.151	400.000	0.208	400.075	400.000	0.111	400.057	400.000	0.093
	Min	400.400	399.640	0.400	400.210	399.860	0.210	400.062	399.943	0.062	400.018	399.964	0.018	400.000	399.964	0.000
500	Max	500.880	500.000	1.280	500.385	500.000	0.540	500.165	500.000	0.228	500.083	500.000	0.123	500.063	500.000	0.103
	Min	500.480	499.600	0.480	500.230	499.845	0.230	500.068	499.937	0.068	500.020	499.960	0.020	500.000	499.960	0.000

[a] The sizes shown are first-choice basic sizes (see Table 12). Preferred fits for other sizes can be calculated from data given in ANSI/ASME B4.2-1978 (R2009).
[b] All fits shown in this table have clearance.
All dimensions are in millimeters.

Table 17. American National Standard Preferred Shaft Basis Metric Transition and Interference Fits *ANSI/ASME B4.2-1978 (R2009)*

Basic Size[a]		Locational Transition			Locational Transition			Locational Interference			Medium Drive			Force		
		Hole K7	Shaft h6	Fit[b]	Hole N7	Shaft h6	Fit[b]	Hole P7	Shaft h6	Fit[b]	Hole S7	Shaft h6	Fit[b]	Hole U7	Shaft h6	Fit[b]
1	Max	1.000	1.000	+0.006	0.996	1.000	+0.002	0.994	1.000	0.000	0.986	1.000	−0.008	0.982	1.000	−0.012
	Min	0.990	0.994	−0.010	0.986	0.954	−0.014	0.984	0.994	−0.016	0.976	0.994	−0.024	0.972	0.994	−0.028
1.2	Max	1.200	1.200	+0.006	1.196	1.200	+0.002	1.194	1.200	0.000	1.186	1.200	−0.008	1.182	1.200	−0.012
	Min	1.190	1.194	−0.010	1.186	1.194	−0.014	1.184	1.194	−0.016	1.176	1.194	−0.024	1.172	1.194	−0.028
1.6	Max	1.600	1.600	+0.006	1.596	1.600	+0.002	1.594	1.600	0.000	1.586	1.600	−0.008	1.582	1.600	−0.012
	Min	1.590	1.594	−0.010	1.586	1.594	−0.014	1.584	1.594	−0.016	1.576	1.594	−0.024	1.572	1.594	−0.028
2	Max	2.000	2.000	+0.006	1.996	2.000	+0.002	1.994	2.000	0.000	1.986	2.000	−0.008	1.982	2.000	−0.012
	Min	1.990	1.994	−0.010	1.986	1.994	−0.014	1.984	1.994	−0.016	1.976	1.994	−0.024	1.972	1.994	−0.028
2.5	Max	2.500	2.500	+0.006	2.496	2.500	+0.002	2.494	2.500	0.000	2.486	2.500	−0.008	2.482	2.500	−0.012
	Min	2.490	2.494	−0.010	2.486	2.494	−0.014	2.484	2.494	−0.016	2.476	2.494	−0.024	2.472	2.494	−0.028
3	Max	3.000	3.000	+0.006	2.996	3.000	+0.002	2.994	3.000	0.000	2.986	3.000	−0.008	2.982	3.000	−0.012
	Min	2.990	2.994	−0.010	2.986	2.994	−0.014	2.984	2.994	−0.016	2.976	2.994	−0.024	2.972	2.994	−0.028
4	Max	4.003	4.000	+0.011	3.996	4.000	+0.004	3.992	4.000	0.000	3.985	4.000	−0.007	3.981	4.000	−0.011
	Min	3.991	3.992	−0.009	3.984	3.992	−0.016	3.980	3.992	−0.020	3.973	3.992	−0.027	3.969	3.992	−0.031
5	Max	5.003	5.000	+0.011	4.996	5.000	+0.004	4.992	5.000	0.000	4.985	5.000	−0.007	4.981	5.000	−0.011
	Min	4.991	4.992	−0.009	4.984	4.992	−0.016	4.980	4.992	−0.020	4.973	4.992	−0.027	4.969	4.992	−0.031
6	Max	6.003	6.000	+0.011	5.996	6.000	+0.004	5.992	6.000	0.000	5.985	6.000	−0.007	5.981	6.000	−0.011
	Min	5.991	5.992	−0.009	5.984	5.992	−0.016	5.980	5.992	−0.020	5.973	5.992	−0.027	5.969	5.992	−0.031
8	Max	8.005	8.000	+0.014	7.996	8.000	+0.005	7.991	8.000	0.000	7.983	8.000	−0.008	7.978	8.000	−0.013
	Min	7.990	7.991	−0.010	7.981	7.991	−0.019	7.976	7.991	−0.024	7.968	7.991	−0.032	7.963	7.991	−0.037
10	Max	10.005	10.000	+0.014	9.996	10.000	+0.005	9.991	10.000	0.000	9.983	10.000	−0.008	9.978	10.000	−0.013
	Min	9.990	9.991	−0.010	9.981	9.991	−0.019	9.976	9.991	−0.024	9.968	9.991	−0.032	9.963	9.991	−0.037
12	Max	12.006	12.000	+0.017	11.995	12.000	+0.006	11.989	12.000	0.000	11.979	12.000	−0.010	11.974	12.000	−0.015
	Min	11.988	11.989	−0.012	11.977	11.989	−0.023	11.971	11.989	−0.029	11.961	11.989	−0.039	11.956	11.989	−0.044
16	Max	16.006	16.000	+0.017	15.995	16.000	+0.006	15.989	16.000	0.000	15.979	16.000	−0.010	15.974	16.000	−0.015
	Min	15.988	15.989	−0.012	15.977	15.989	−0.023	15.971	15.989	−0.029	15.961	15.989	−0.039	15.956	15.989	−0.044
20	Max	20.006	20.000	+0.019	19.993	20.000	+0.006	19.986	20.000	−0.001	19.973	20.000	−0.014	19.967	20.000	−0.020
	Min	19.985	19.987	−0.015	19.972	19.987	−0.028	19.965	19.987	−0.035	19.952	19.987	−0.048	19.946	19.987	−0.054
25	Max	25.006	25.000	+0.019	24.993	25.000	+0.006	24.986	25.000	−0.001	24.973	25.000	−0.014	24.960	25.000	−0.027
	Min	24.985	24.987	−0.015	24.972	24.987	−0.028	24.965	24.987	−0.035	24.952	24.987	−0.048	24.939	24.987	−0.061

Table 17. *(Continued)* **American National Standard Preferred Shaft Basis Metric Transition and Interference Fits** *ANSI/ASME B4.2-1978 (R2009)*

Basic Size[a]		Locational Transition			Locational Transition			Locational Interference			Medium Drive			Force		
		Hole K7	Shaft h6	Fit[b]	Hole N7	Shaft h6	Fit[b]	Hole P7	Shaft h6	Fit[b]	Hole S7	Shaft h6	Fit[b]	Hole U7	Shaft h6	Fit[b]
30	Max	30.006	30.000	+0.019	29.993	30.000	+0.006	29.986	30.000	−0.001	29.973	30.000	−0.014	29.960	30.000	−0.027
	Min	29.985	29.987	−0.015	29.972	29.987	−0.028	29.965	29.987	−0.035	29.952	29.987	−0.048	29.939	29.987	−0.061
40	Max	40.007	40.000	+0.023	39.992	40.000	+0.008	39.983	40.000	−0.001	39.966	40.000	−0.018	39.949	40.000	−0.035
	Min	39.982	39.984	−0.018	39.967	39.984	−0.033	39.958	39.984	−0.042	39.941	39.984	−0.059	39.924	39.984	−0.076
50	Max	50.007	50.000	+0.023	49.992	50.000	+0.008	49.983	50.000	−0.001	49.966	50.000	−0.018	49.939	50.000	−0.045
	Min	49.982	49.984	−0.018	49.967	49.984	−0.033	49.958	49.984	−0.042	49.941	49.984	−0.059	49.914	49.984	−0.086
60	Max	60.009	60.000	+0.028	59.991	60.000	+0.010	59.979	60.000	−0.002	59.958	60.000	−0.023	59.924	60.000	−0.087
	Min	59.979	59.981	−0.021	59.961	59.981	−0.039	59.949	59.981	−0.051	59.928	59.981	−0.072	59.894	59.981	−0.106
80	Max	80.009	80.000	+0.028	79.991	80.000	+0.010	79.979	80.000	−0.002	79.952	80.000	−0.029	79.909	80.000	−0.072
	Min	79.979	79.981	−0.021	79.961	79.981	−0.039	79.949	79.981	−0.051	79.922	79.981	−0.078	79.879	79.981	−0.121
100	Max	100.010	100.000	+0.032	99.990	100.000	+0.012	99.976	100.000	−0.002	99.942	100.000	−0.036	99.889	100.000	−0.089
	Min	99.975	99.978	−0.025	99.955	99.978	−0.045	99.941	99.978	−0.059	99.907	99.978	−0.093	99.854	99.978	−0.146
120	Max	120.010	120.000	+0.032	119.990	120.000	+0.012	119.976	120.000	−0.002	119.934	120.000	−0.044	119.869	120.000	−0.109
	Min	119.975	119.978	−0.025	119.955	119.978	−0.045	119.941	119.978	−0.059	119.899	119.978	−0.101	119.834	119.978	−0.166
160	Max	160.012	160.000	+0.037	159.988	160.000	+0.013	159.972	160.000	−0.003	159.915	160.000	−0.060	159.825	160.000	−0.150
	Min	159.972	159.975	−0.028	159.948	159.975	−0.052	159.932	159.975	−0.068	159.875	159.975	−0.125	159.785	159.975	−0.215
200	Max	200.013	200.00	+0.042	199.986	200.000	+0.015	199.967	200.000	−0.004	199.895	200.000	−0.076	199.781	200.000	−0.190
	Min	199.967	199.971	−0.033	199.940	199.971	−0.060	199.921	199.971	−0.079	199.849	199.971	−0.151	199.735	199.971	−0.265
250	Max	250.013	250.000	+0.042	249.986	250.000	+0.015	249.967	250.000	−0.004	249.877	250.000	−0.094	249.733	250.000	−0.238
	Min	249.967	249.971	−0.033	249.940	249.971	−0.060	249.921	249.971	−0.079	249.831	249.971	−0.169	249.687	249.971	−0.313
300	Max	300.016	300.000	+0.048	299.986	300.000	+0.018	299.964	300.000	−0.004	299.850	300.000	−0.118	299.670	300.000	−0.298
	Min	299.964	299.968	−0.036	299.934	299.968	−0.066	299.912	299.968	−0.088	299.798	299.968	−0.202	299.618	299.968	−0.382
400	Max	400.017	400.000	+0.053	399.984	400.000	+0.020	399.959	400.000	−0.005	399.813	400.000	−0.151	399.586	400.000	−0.378
	Min	399.960	399.964	−0.040	399.927	399.964	−0.073	399.902	399.964	−0.098	399.756	399.964	−0.244	399.529	399.964	−0.471
500	Max	500.018	500.000	+0.058	499.983	500.000	+0.023	499.955	500.000	−0.005	499.771	500.000	−0.189	499.483	500.000	−0.477
	Min	499.955	499.960	−0.045	499.920	499.960	−0.080	499.892	499.960	−0.108	499.708	499.960	−0.292	499.420	499.960	−0.580

[a] The sizes shown are first-choice basic sizes (see Table 12). Preferred fits for other sizes can be calculated from data given in ANSI/ASME B4.2-1978 (R2009).
[b] A plus sign indicates clearance; a minus sign indicates interference.
All dimensions are in millimeters.

Table 18. American National Standard Gagemakers Tolerances
ANSI/ASME B4.4M-1981 (R2014)

Gagemakers Tolerance			Workpiece Tolerance	
	Class	ISO Symbol[a]	IT Grade	Recommended Gage Usage
Rejection of Good Parts Increase ↑	ZM	0.05 IT11	IT11	Low-precision gages recommended to be used to inspect workpieces held to internal (hole) tolerances C11 and H11 and to external (shaft) tolerances c11 and h11.
	YM	0.05 IT9	IT9	Gages recommended to be used to inspect workpieces held to internal (hole) tolerances D9 and H9 and to external (shaft) tolerances d9 and h9.
	XM	0.05 IT8	IT8	Precision gages recommended to be used to inspect workpieces held to internal (hole) tolerances F8 and H8.
Gage Cost Increase ↓	XXM	0.05 IT7	IT7	Recommended to be used for gages to inspect workpieces held to internal (hole) tolerances G7, H7, K7, N7, P7, S7, and U7, and to external (shaft) tolerances f7 and h7.
	XXXM	0.05 IT6	IT6	High-precision gages recommended to be used to inspect workpieces held to external (shaft) tolerances g6, h6, k6, n6, p6, s6, and u6.

[a] Gagemakers tolerance is equal to 5 percent of workpiece tolerance or 5 percent of applicable IT grade value. See table *American National Standard Gagemakers Tolerances ANSI/ASME B4.4M-1981 (R2014).*

For workpiece tolerance class values, see previous Tables 14, 15, 16, and 17.

Table 19. American National Standard Gagemakers Tolerances
ANSI/ASME B4.4M-1981 (R2014)

Basic Size		Class ZM (0.05 IT11)	Class YM (0.05 IT9)	Class XM (0.05 IT8)	Class XXM (0.05 IT7)	Class XXXM (0.05 IT6)
Over	To					
0	3	0.0030	0.0012	0.0007	0.0005	0.0003
3	6	0.0037	0.0015	0.0009	0.0006	0.0004
6	10	0.0045	0.0018	0.0011	0.0007	0.0005
10	18	0.0055	0.0021	0.0013	0.0009	0.0006
18	30	0.0065	0.0026	0.0016	0.0010	0.0007
30	50	0.0080	0.0031	0.0019	0.0012	0.0008
50	80	0.0095	0.0037	0.0023	0.0015	0.0010
80	120	0.0110	0.0043	0.0027	0.0017	0.0011
120	180	0.0125	0.0050	0.0031	0.0020	0.0013
180	250	0.0145	0.0057	0.0036	0.0023	0.0015
250	315	0.0160	0.0065	0.0040	0.0026	0.0016
315	400	0.0180	0.0070	0.0044	0.0028	0.0018
400	500	0.0200	0.0077	0.0048	0.0031	0.0020

All dimensions are in millimeters. For closer gagemakers tolerance classes than Class XXXM, specify 5 percent of IT5, IT4, or IT3 and use the designation 0.05 IT5, 0.05 IT4, etc.

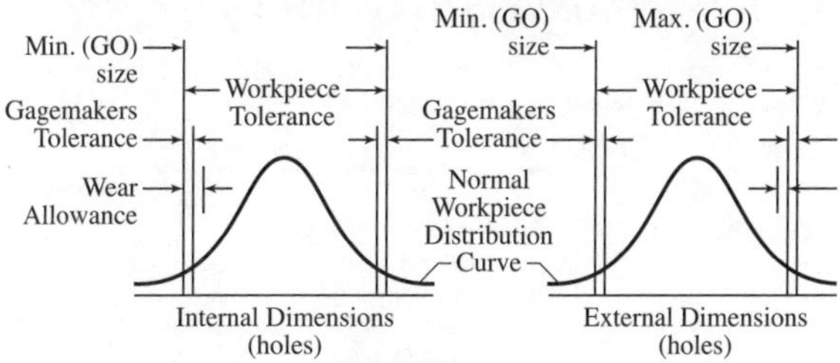

Fig. 4. Relationship between Gagemakers Tolerance, Wear Allowance, and Workpiece Tolerance

Relation of Machining Processes to IT Tolerance Grades

	IT Grades							
	4	5	6	7	8	9	10	11
Lapping & Honing	■	■						
Cylindrical Grinding		■	■	■				
Surface Grinding		■	■	■	■			
Diamond Turning		■	■	■				
Diamond Boring		■	■	■				
Broaching		■	■	■	■			
Powder Metal sizes			■	■	■			
Reaming			■	■	■	■	■	
Turning				■	■	■	■	■
Powder Metal sintered				■	■	■	■	
Boring				■	■	■	■	■
Milling							■	■
Planing & Shaping							■	■
Drilling							■	■
Punching							■	■
Die Casting								■

Practical Usage of International Tolerance Grades

	IT Grades																	
	01	0	1	2	3	4	5	6	7	8	9	10	11	12	13	14	15	16
Measuring Tools	■	■	■	■	■	■	■	■	■									
Fits							■	■	■	■	■	■	■					
Material										■	■	■	■	■	■	■		
Large Manufacturing Tolerances														■	■	■	■	■

CONVERSION FACTORS

In the table of conversion factors that follows, the symbols for SI units, multiples, and submultiples are given in parentheses in the right hand column.

Table 1. Metric Conversion Factors

Multiply	By	To Obtain
Length		
centimeter	0.03280840	foot
centimeter	0.3937008	inch
fathom	1.8288[a]	meter (m)
foot	0.3048[a]	meter (m)
foot	30.48[a]	centimeter (cm)
foot	304.8[a]	millimeter (mm)
inch	0.0254[a]	meter (m)
inch	2.54[a]	centimeter (cm)
inch	25.4[a]	millimeter (mm)
kilometer	0.6213712	mile [US statute]
meter	39.37008	inch
meter	0.5468066	fathom
meter	3.280840	foot
meter	0.1988388	rod
meter	1.093613	yard
meter	0.0006213712	mile [US statute]
microinch	0.0254[a]	micrometer [micron] (μm)
micrometer [micron]	39.37008	microinch
mile [US statute]	1609.344[a]	meter (m)
mile [US statute]	1.609344[a]	kilometer (km)
millimeter	0.003280840	foot
millimeter	0.03937008	inch
rod	5.0292[a]	meter (m)
yard	0.9144[a]	meter (m)
Area		
acre	4046.856	meter^2 (m^2)
acre	0.4046856	hectare
centimeter^2	0.1550003	inch^2
centimeter^2	0.001076391	foot^2
foot^2	0.09290304[a]	meter^2 (m^2)
foot^2	929.0304[a]	centimeter^2 (cm^2)
foot^2	92,903.04[a]	millimeter^2 (mm^2)
hectare	2.471054	acre
inch^2	645.16[a]	millimeter^2 (mm^2)
inch^2	6.4516[a]	centimeter^2 (cm^2)
inch^2	0.00064516[a]	meter^2 (m^2)
meter^2	1550.003	inch^2
meter^2	10.763910	foot^2
meter^2	1.195990	yard^2
meter^2	0.0002471054	acre

Table 1. *(Continued)* **Metric Conversion Factors**

Multiply	By	To Obtain
$mile^2$	2.5900	$kilometer^2$
$millimeter^2$	0.00001076391	$foot^2$
$millimeter^2$	0.001550003	$inch^2$
$yard^2$	0.8361274	$meter^2$ (m^2)
Volume (including Capacity)		
$centimeter^3$	0.06102376	$inch^3$
$foot^3$	28.31685	liter
$foot^3$	28.31685	liter
gallon [UK liquid]	0.004546092	$meter^3$ (m^3)
gallon [UK liquid]	4.546092	liter
gallon [US liquid]	0.003785412	$meter^3$ (m^3)
gallon [US liquid]	3.785412	liter
$inch^3$	16,387.06	$millimeter^3$ (mm^3)
$inch^3$	16.38706	$centimeter^3$ (cm^3)
$inch^3$	0.00001638706	$meter^3$ (m^3)
liter	0.001[a]	$meter^3$ (m^3)
liter	0.2199692	gallon [UK liquid]
liter	0.2641720	gallon [US liquid]
liter	0.03531466	$foot^3$
$meter^3$	219.9692	gallon [UK liquid]
$meter^3$	264.1720	gallon [US liquid]
$meter^3$	35.31466	$foot^3$
$meter^3$	1.307951	$yard^3$
$meter^3$	1000.[a]	liter
$meter^3$	61,023.76	$inch^3$
$millimeter^3$	0.00006102376	$inch^3$
quart [US liquid]	0.946	liter
quart [UK liquid]	1.136	liter
$yard^3$	0.7645549	$meter^3$ (m^3)
Velocity, Acceleration, and Flow		
centimeter / second	1.968504	foot / minute
centimeter / second	0.03280840	foot / second
centimeter / minute	0.3937008	inch / minute
foot / hour	0.00008466667	meter / second (m / s)
foot / hour	0.00508[a]	meter / minute
foot / hour	0.3048[a]	meter / hour
foot / minute	0.508[a]	centimeter / second
foot / minute	18.288[a]	meter / hour
foot / minute	0.3048[a]	meter / minute
foot / minute	0.00508[a]	meter / second (m / s)
foot / second	30.48[a]	centimeter / second
foot / second	18.288[a]	meter / minute
foot / second	0.3048[a]	meter / second (m / s)
foot / $second^2$	0.3048[a]	meter / $second^2$ (m / s^2)
$foot^3$ / minute	28.31685	liter / minute

Table 1. *(Continued)* **Metric Conversion Factors**

Multiply	By	To Obtain
foot^3 / minute	0.0004719474	meter^3 / second (m^3 / s)
gallon [US liquid] / min.	0.003785412	meter^3 / minute
gallon [US liquid] / min.	0.00006309020	meter^3 / second (m^3 / s)
gallon [US liquid] / min.	0.06309020	liter / second
gallon [US liquid] / min.	3.785412	liter / minute
gallon [UK liquid] / min.	0.004546092	meter^3 / minute
gallon [UK liquid] / min.	0.00007576820	meter^3 / second (m^3 / s)
inch / minute	25.4[a]	millimeter / minute
inch / minute	2.54[a]	centimeter / minute
inch / minute	0.0254[a]	meter / minute
inch / second^2	0.0254[a]	meter / second^2 (m / s^2)
kilometer / hour	0.6213712	mile / hour [US statute]
liter / minute	0.03531466	foot^3 / minute
liter / minute	0.2641720	gallon [US liquid] / minute
liter / second	15.85032	gallon [US liquid] / minute
mile / hour	1.609344[a]	kilometer / hour
millimeter / minute	0.03937008	inch / minute
meter / second	11,811.02	foot / hour
meter / second	196.8504	foot / minute
meter / second	3.280840	foot / second
meter / second^2	3.280840	foot / second^2
meter / second^2	39.37008	inch / second^2
meter / minute	3.280840	foot / minute
meter / minute	0.05468067	foot / second
meter / minute	39.37008	inch / minute
meter / hour	3.280840	foot / hour
meter / hour	0.05468067	foot / minute
meter^3 / second	2118.880	foot^3 / minute
meter^3 / second	13,198.15	gallon [UK liquid] / minute
meter^3 / second	15,850.32	gallon [US liquid] / minute
meter^3 / minute	219.9692	gallon [UK liquid] / minute
meter^3 / minute	264.1720	gallon [US liquid] / minute
Mass and Density		
grain [$\frac{1}{7000}$ 1b avoirdupois]	0.06479891	gram (g)
gram	15.43236	grain
gram	0.001[a]	kilogram (kg)
gram	0.03527397	ounce [avoirdupois]
gram	0.03215074	ounce [troy]
gram / centimeter^3	0.03612730	pound / inch^3
hundredweight [long]	50.80235	kilogram (kg)
hundredweight [short]	45.35924	kilogram (kg)
kilogram	1000.[a]	gram (g)
kilogram	35.27397	ounce [avoirdupois]
kilogram	32.15074	ounce [troy]

Table 1. *(Continued)* **Metric Conversion Factors**

Multiply	By	To Obtain
kilogram	2.204622	pound [avoirdupois]
kilogram	0.06852178	slug
kilogram	0.0009842064	ton [long]
kilogram	0.001102311	ton [short]
kilogram	0.001[a]	ton [metric]
kilogram	0.001[a]	tonne
kilogram	0.01968413	hundredweight [long]
kilogram	0.02204622	hundredweight [short]
kilogram / meter^3	0.06242797	pound / foot^3
kilogram / meter^3	0.01002242	pound / gallon [UK liquid]
kilogram / meter^3	0.008345406	pound / gallon [US liquid]
ounce [avoirdupois]	28.34952	gram (g)
ounce [avoirdupois]	0.02834952	kilogram (kg)
ounce [troy]	31.10348	gram (g)
ounce [troy]	0.03110348	kilogram (kg)
pound [avoirdupois]	0.4535924	kilogram (kg)
pound / foot^3	16.01846	kilogram / meter^3 (kg / m^3)
pound / inch^3	27.67990	gram / centimeter^3 (g / cm^3)
pound / gal [US liquid]	119.8264	kilogram / meter^3 (kg / m^3)
pound / gal [UK liquid]	99.77633	kilogram / meter^3 (kg / m^3)
slug	14.59390	kilogram (kg)
ton [long 2240 lb]	1016.047	kilogram (kg)
ton [short 2000 lb]	907.1847	kilogram (kg)
ton [metric]	1000.[a]	kilogram (kg)
ton [Metric]	0.9842	ton [long 2240 lb]
ton [Metric]	1.1023	ton [short 2000 lb]
tonne	1000.[a]	kilogram (kg)
Force and Force / Length		
dyne	0.00001[a]	newton (N)
kilogram-force	9.806650[a]	newton (N)
kilopound	9.806650[a]	newton (N)
newton	0.1019716	kilogram-force
newton	0.1019716	kilopound
newton	0.2248089	pound-force
newton	100,000.[a]	dyne
newton	7.23301	poundal
newton	3.596942	ounce-force
newton / meter	0.005710148	pound / inch
newton / meter	0.06852178	pound / foot
ounce-force	0.2780139	newton (N)
pound-force	4.448222	newton (N)
poundal	0.1382550	newton (N)
pound / inch	175.1268	newton / meter (N / m)
pound / foot	14.59390	newton / meter (N / m)

Table 1. *(Continued)* **Metric Conversion Factors**

Multiply	By	To Obtain
Bending Moment or Torque		
dyne-centimeter	0.0000001[a]	newton-meter (N · m)
kilogram-meter	9.806650[a]	newton-meter (N · m)
ounce-inch	7.061552	newton-millimeter
ounce-inch	0.007061552	newton-meter (N · m)
newton-meter	0.7375621	pound-foot
newton-meter	10,000,000.[a]	dyne-centimeter
newton-meter	0.1019716	kilogram-meter
newton-meter	141.6119	ounce-inch
newton-millimeter	0.1416119	ounce-inch
pound-foot	1.355818	newton-meter (N · m)
Moment Of Inertia and Section Modulus		
moment of inertia [kg · m^2]	23.73036	pound-foot^2
moment of inertia [kg · m^2]	3417.171	pound-inch^2
moment of inertia [lb · ft^2]	0.04214011	kilogram-meter^2 (kg · m^2)
moment of inertia [lb · inch^2]	0.0002926397	kilogram-meter^2 (kg · m^2)
moment of section [foot^4]	0.008630975	meter^4 (m^4)
moment of section [inch^4]	41.62314	centimeter^4
moment of section [meter^4]	115.8618	foot^4
moment of section [centimeter^4]	0.02402510	inch^4
section modulus [foot^3]	0.02831685	meter^3 (m^3)
section modulus [inch^3]	0.00001638706	meter^3 (m^3)
section modulus [meter^3]	35.31466	foot^3
section modulus [meter^3]	61,023.76	inch^3
Momentum		
kilogram-meter / second	7.233011	pound-foot / second
kilogram-meter / second	86.79614	pound-inch / second
pound-foot / second	0.1382550	kilogram-meter / second (kg · m / s)
pound-inch / second	0.01152125	kilogram-meter / second (kg · m / s)
Pressure and Stress		
atmosphere [14.6959 lb / inch^2]	101,325.	pascal (Pa)
bar	100,000.[a]	pascal (Pa)
bar	14.50377	pound / inch^2
bar	100,000.[a]	newton / meter^2 (N / m^2)
hectobar	0.6474898	ton [long] / inch^2
kilogram / centimeter^2	14.22334	pound / inch^2
kilogram / meter^2	9.806650[a]	newton / meter^2 (N / m^2)
kilogram / meter^2	9.806650[a]	pascal (Pa)
kilogram / meter^2	0.2048161	pound / foot^2
kilonewton / meter^2	0.1450377	pound / inch^2
newton / centimeter^2	1.450377	pound / inch^2
newton / meter^2	0.00001[a]	bar
newton / meter^2	1.0[a]	pascal (Pa)

Table 1. *(Continued)* **Metric Conversion Factors**

Multiply	By	To Obtain
newton / meter^2	0.0001450377	pound / inch^2
newton / meter^2	0.1019716	kilogram / meter^2
newton / millimeter^2	145.0377	pound / inch^2
pascal	0.00000986923	atmosphere
pascal	0.00001[a]	bar
pascal	0.1019716	kilogram / meter^2
pascal	1.0[a]	newton / meter^2 (N / m^2)
pascal	0.02088543	pound / foot^2
pascal	0.0001450377	pound / inch^2
pound / foot^2	4.882429	kilogram / meter^2
pound / foot^2	47.88026	pascal (Pa)
pound / inch^2	0.06894757	bar
pound / inch^2	0.07030697	kilogram / centimeter^2
pound / inch^2	0.6894757	newton / centimeter^2
pound / inch^2	6.894757	kilonewton / meter^2
pound / inch^2	6894.757	newton / meter^2 (N / m^2)
pound / inch^2	0.006894757	newton / millimeter^2 (N / mm^2)
pound / inch^2	6894.757	pascal (Pa)
ton [long] / inch^2	1.544426	hectobar
	Energy and Work	
Btu [International Table]	1055.056	joule (J)
Btu [mean]	1055.87	joule (J)
calorie [mean]	4.19002	joule (J)
foot-pound	1.355818	joule (J)
foot-poundal	0.04214011	joule (J)
joule	0.0009478170	Btu [International Table]
joule	0.0009470863	Btu [mean]
joule	0.2386623	calorie [mean]
joule	0.7375621	foot-pound
joule	23.73036	foot-poundal
joule	0.9998180	joule [International US]
joule	0.9999830	joule [US legal, 1948]
joule [International US]	1.000182	joule (J)
joule [US legal, 1948]	1.000017	joule (J)
joule	.0002777778	watt-hour
watt-hour	3600.[a]	joule (J)
	Power	
Btu [International Table] / hour	0.2930711	watt (W)
foot-pound / hour	0.0003766161	watt (W)
foot-pound / minute	0.02259697	watt (W)
horsepower [550 ft-lb / s]	0.7456999	kilowatt (kW)
horsepower [550 ft-lb / s]	745.6999	watt (W)
horsepower [electric]	746.[a]	watt (W)
horsepower [metric]	735.499	watt (W)
horsepower [UK]	745.70	watt (W)
kilowatt	1.341022	horsepower [550 ft-lb / s]

Table 1. *(Continued)* **Metric Conversion Factors**

Multiply	By	To Obtain
watt	2655.224	foot-pound / hour
watt	44.25372	foot-pound / minute
watt	0.001341022	horsepower [550 ft-lb / s]
watt	0.001340483	horsepower [electric]
watt	0.001359621	horsepower [metric]
watt	0.001341022	horsepower [UK]
watt	3.412141	Btu [International Table] / hour
	Viscosity	
centipoise	0.001[a]	pascal-second (Pa · s)
centistoke	0.000001[a]	meter^2 / second (m^2 / s)
meter^2 / second	1,000,000.[a]	centistoke
meter^2 / second	10,000.[a]	stoke
pascal-second	1000.[a]	centipoise
pascal-second	10.[a]	poise
poise	0.1[a]	pascal-second (Pa · s)
stoke	0.0001[a]	meter^2 / second (m^2 / s)
	Temperature	
To Convert From	**To**	**Use Formula**
temperature Celsius, t_C	temperature Kelvin, t_K	$t_K = t_C + 273.15$
temperature Fahrenheit, t_F	temperature Kelvin, t_K	$t_K = (t_F + 459.67) / 1.8$
temperature Celsius, t_C	temperature Fahrenheit, t_F	$t_F = 1.8\, t_C + 32$
temperature Fahrenheit, t_F	temperature Celsius, t_C	$t_C = (t_F - 32) / 1.8$
temperature Kelvin, t_K	temperature Celsius, t_C	$t_C = t_K - 273.15$
temperature Kelvin, t_K	temperature Fahrenheit, t_F	$t_F = 1.8\, t_K - 459.67$
temperature Kelvin, t_K	temperature Rankine, t_R	$t_R = 9/5\, t_K$
temperature Rankine, t_R	temperature Kelvin, t_K	$t_K = 5/9\, t_R$

[a] The figure is exact.
The right-hand column shows symbols of SI units, multiples, and sub-multiples in parentheses.

Table 2. Factors and Prefixes for Forming Decimal Multiples and Sub-Multiples of the SI Units

Number	Power	Prefix	Subdivisions or Multiples	Example based on meter	Symbol
1E–06	10^{-6}	micro	one-millionth	micrometer	μ
0.001	10^{-3}	milli	one-thousandth	millimeter	m
0.01	10^{-2}	centi	one-hundredth	centimeter	c
0.1	10^{-1}	deci	one-tenth	decimeter	d
1	10^{0}	…	one	meter	…
10	10^{1}	deka	ten	dekameter	da
100	10^{2}	hecto	one hundred	hectometer	h
1,000	10^{3}	kilo	one thousand	kilometer	k
1,000,000	10^{6}	mega	one million	megameter	M
1,000,000,000	10^{9}	giga	one billion	gigameter	G
1,000,000,000,000	10^{12}	tera	one trillion	terameter	T

Use of Conversion Tables.—On this and following pages, tables are given that permit conversion between English and metric units over a wide range of values. Where the desired value cannot be obtained directly from these tables, a simple addition of two or more values taken directly from the table will suffice as shown in the following examples:

Example 1: Find the millimeter equivalent of 0.4476 inch.

4	in	=	10.1600	mm
.04	in	=	1.01600	mm
.007	in	=	0.17780	mm
.0006	in	=	0.01524	mm
.4476	in	=	11.36904	mm

Example 2: Find the inch equivalent of 84.9 mm.

80.	mm	=	3.14961	in
4.	mm	=	0.15748	in
0.9	mm	=	0.03543	in
84.9	mm	=	3.34252	in

Table 3. Inch–Millimeter and Inch–Centimeter Conversion Table
(Based on 1 inch = 25.4 millimeters, exactly)

Inches To Millimeters											
in.	mm	in.	mm	in.	mm	in.	mm	in.	mm	in.	mm
10	254.00000	1	25.40000	0.1	2.54000	.01	0.25400	0.001	0.02540	0.0001	0.00254
20	508.00000	2	50.80000	0.2	5.08000	.02	0.50800	0.002	0.05080	0.0002	0.00508
30	762.00000	3	76.20000	0.3	7.62000	.03	0.76200	0.003	0.07620	0.0003	0.00762
40	1,016.00000	4	101.60000	0.4	10.16000	.04	1.01600	0.004	0.10160	0.0004	0.01016
50	1,270.00000	5	127.00000	0.5	12.70000	.05	1.27000	0.005	0.12700	0.0005	0.01270
60	1,524.00000	6	152.40000	0.6	15.24000	.06	1.52400	0.006	0.15240	0.0006	0.01524
70	1,778.00000	7	177.80000	0.7	17.78000	.07	1.77800	0.007	0.17780	0.0007	0.01778
80	2,032.00000	8	203.20000	0.8	20.32000	.08	2.03200	0.008	0.20320	0.0008	0.02032
90	2,286.00000	9	228.60000	0.9	22.86000	.09	2.2860	0.009	0.22860	0.0009	0.02286
100	2,540.00000	10	254.00000	1.0	25.40000	.10	2.54000	0.010	0.25400	0.0010	0.02540

Millimeters To Inches											
mm	in.	mm	in.	mm	in	mm	in.	mm	in.	mm	in.
100	3.93701	10	0.39370	1	0.03937	0.1	0.00394	0.01	.000039	0.001	0.00004
200	7.87402	20	0.78740	2	0.07874	0.2	0.00787	0.02	.00079	0.002	0.00008
300	11.81102	30	1.18110	3	0.11811	0.3	0.01181	0.03	.00118	0.003	0.00012
400	15.74803	40	1.57480	4	0.15748	0.4	0.01575	0.04	.00157	0.004	0.00016
500	19.68504	50	1.96850	5	0.19685	0.5	0.01969	0.05	.00197	0.005	0.00020
600	23.62205	60	2.36220	6	0.23622	0.6	0.02362	0.06	.00236	0.006	0.00024
700	27.55906	70	2.75591	7	0.27559	0.7	0.02756	0.07	.00276	0.007	0.00028
800	31.49606	80	3.14961	8	0.31496	0.8	0.03150	0.08	.00315	0.008	0.00031
900	35.43307	90	3.54331	9	0.35433	0.9	0.03543	0.09	.00354	0.009	0.00035
1,000	39.37008	100	3.93701	10	0.39370	1.0	0.03937	0.10	.00394	0.010	0.00039

For inches to centimeters, shift decimal point in mm column one place to left and read centimeters, thus:

40 in. = 1016 mm = 101.6 cm

For centimeters to inches, shift decimal point of centimeter value one place to right and enter mm column, thus:

70 cm = 700 mm = 27.55906 inches

Table 4. Decimals of an Inch to Millimeters
(Based on 1 inch = 25.4 millimeters, exactly)

Inches	0.000	0.001	0.002	0.003	0.004	0.005	0.006	0.007	0.008	0.009
	Millimeters									
0.000	…	0.0254	0.0508	0.0762	0.1016	0.1270	0.1524	0.1778	0.2032	0.2286
0.010	0.2540	0.2794	0.3048	0.3302	0.3556	0.3810	0.4064	0.4318	0.4572	0.4826
0.020	0.5080	0.5334	0.5588	0.5842	0.6096	0.6350	0.6604	0.6858	0.7112	0.7366
0.030	0.7620	0.7874	0.8128	0.8382	0.8636	0.8890	0.9144	0.9398	0.9652	0.9906
0.040	1.0160	1.0414	1.0668	1.0922	1.1176	1.1430	1.1684	1.1938	1.2192	1.2446
0.050	1.2700	1.2954	1.3208	1.3462	1.3716	1.3970	1.4224	1.4478	1.4732	1.4986
0.060	1.5240	1.5494	1.5748	1.6002	1.6256	1.6510	1.6764	1.7018	1.7272	1.7526
0.070	1.7780	1.8034	1.8288	1.8542	1.8796	1.9050	1.9304	1.9558	1.9812	2.0066
0.080	2.0320	2.0574	2.0828	2.1082	2.1336	2.1590	2.1844	2.2098	2.2352	2.2606
0.090	2.2860	2.3114	2.3368	2.3622	2.3876	2.4130	2.4384	2.4638	2.4892	2.5146
0.100	2.5400	2.5654	2.5908	2.6162	2.6416	2.6670	2.6924	2.7178	2.7432	2.7686
0.110	2.7940	2.8194	2.8448	2.8702	2.8956	2.9210	2.9464	2.9718	2.9972	3.0226
0.120	3.0480	3.0734	3.0988	3.1242	3.1496	3.1750	3.2004	3.2258	3.2512	3.2766
0.130	3.3020	3.3274	3.3528	3.3782	3.4036	3.4290	3.4544	3.4798	3.5052	3.5306
0.140	3.5560	3.5814	3.6068	3.6322	3.6576	3.6830	3.7084	3.7338	3.7592	3.7846
0.150	3.8100	3.8354	3.8608	3.8862	3.9116	3.9370	3.9624	3.9878	4.0132	4.0386
0.160	4.0640	4.0894	4.1148	4.1402	4.1656	4.1910	4.2164	4.2418	4.2672	4.2926
0.170	4.3180	4.3434	4.3688	4.3942	4.4196	4.4450	4.4704	4.4958	4.5212	4.5466
0.180	4.5720	4.5974	4.6228	4.6482	4.6736	4.6990	4.7244	4.7498	4.7752	4.8006
0.190	4.8260	4.8514	4.8768	4.9022	4.9276	4.9530	4.9784	5.0038	5.0292	5.0546
0.200	5.0800	5.1054	5.1308	5.1562	5.1816	5.2070	5.2324	5.2578	5.2832	5.3086
0.210	5.3340	5.3594	5.3848	5.4102	5.4356	5.4610	5.4864	5.5118	5.5372	5.5626
0.220	5.5880	5.6134	5.6388	5.6642	5.6896	5.7150	5.7404	5.7658	5.7912	5.8166
0.230	5.8420	5.8674	5.8928	5.9182	5.9436	5.9690	5.9944	6.0198	6.0452	6.0706
0.240	6.0960	6.1214	6.1468	6.1722	6.1976	6.2230	6.2484	6.2738	6.2992	6.3246
0.250	6.3500	6.3754	6.4008	6.4262	6.4516	6.4770	6.5024	6.5278	6.5532	6.5786
0.260	6.6040	6.6294	6.6548	6.6802	6.7056	6.7310	6.7564	6.7818	6.8072	6.8326
0.270	6.8580	6.8834	6.9088	6.9342	6.9596	6.9850	7.0104	7.0358	7.0612	7.0866
0.280	7.1120	7.1374	7.1628	7.1882	7.2136	7.2390	7.2644	7.2898	7.3152	7.3406
0.290	7.3660	7.3914	7.4168	7.4422	7.4676	7.4930	7.5184	7.5438	7.5692	7.5946
0.300	7.6200	7.6454	7.6708	7.6962	7.7216	7.7470	7.7724	7.7978	7.8232	7.8486
0.310	7.8740	7.8994	7.9248	7.9502	7.9756	8.0010	8.0264	8.0518	8.0772	8.1026
0.320	8.1280	8.1534	8.1788	8.2042	8.2296	8.2550	8.2804	8.3058	8.3312	8.3566
0.330	8.3820	8.4074	8.4328	8.4582	8.4836	8.5090	8.5344	8.5598	8.5852	8.6106
0.340	8.6360	8.6614	8.6868	8.7122	8.7376	8.7630	8.7884	8.8138	8.8392	8.8646
0.350	8.8900	8.9154	8.9408	8.9662	8.9916	9.0170	9.0424	9.0678	9.0932	9.1186
0.360	9.1440	9.1694	9.1948	9.2202	9.2456	9.2710	9.2964	9.3218	9.3472	9.3726
0.370	9.3980	9.4234	9.4488	9.4742	9.4996	9.5250	9.5504	9.5758	9.6012	9.6266
0.380	9.6520	9.6774	9.7028	9.7282	9.7536	9.7790	9.8044	9.8298	9.8552	9.8806
0.390	9.9060	9.9314	9.9568	9.9822	10.0076	10.0330	10.0584	10.0838	10.1092	10.1346
0.400	10.1600	10.1854	10.2108	10.2362	10.2616	10.2870	10.3124	10.3378	10.3632	10.3886
0.410	10.4140	10.4394	10.4648	10.4902	10.5156	10.5410	10.5664	10.5918	10.6172	10.6426
0.420	10.6680	10.6934	10.7188	10.7442	10.7696	10.7950	10.8204	10.8458	10.8712	10.8966
0.430	10.9220	10.9474	10.9728	10.9982	11.0236	11.0490	11.0744	11.0998	11.1252	11.1506
0.440	11.1760	11.2014	11.2268	11.2522	11.2776	11.3030	11.3284	11.3538	11.3792	11.4046
0.450	11.4300	11.4554	11.4808	11.5062	11.5316	11.5570	11.5824	11.6078	11.6332	11.6586
0.460	11.6840	11.7094	11.7348	11.7602	11.7856	11.8110	11.8364	11.8618	11.8872	11.9126
0.470	11.9380	11.9634	11.9888	12.0142	12.0396	12.0650	12.0904	12.1158	12.1412	12.1666
0.480	12.1920	12.2174	12.2428	12.2682	12.2936	12.3190	12.3444	12.3698	12.3952	12.4206
0.490	12.4460	12.4714	12.4968	12.5222	12.5476	12.5730	12.5984	12.6238	12.6492	12.6746
0.500	12.7000	12.7254	12.7508	12.7762	12.8016	12.8270	12.8524	12.8778	12.9032	12.9286
0.510	12.9540	12.9794	13.0048	13.0302	13.0556	13.0810	13.1064	13.1318	13.1572	13.1826

Table 4. *(Continued)* **Decimals of an Inch to Millimeters**
(Based on 1 inch = 25.4 millimeters, exactly)

Inches	0.000	0.001	0.002	0.003	0.004	0.005	0.006	0.007	0.008	0.009
	Millimeters									
0.520	13.2080	13.2334	13.2588	13.2842	13.3096	13.3350	13.3604	13.3858	13.4112	13.4366
0.530	13.4620	13.4874	13.5128	13.5382	13.5636	13.5890	13.6144	13.6398	13.6652	13.6906
0.540	13.7160	13.7414	13.7668	13.7922	13.8176	13.8430	13.8684	13.8938	13.9192	13.9446
0.550	13.9700	13.9954	14.0208	14.0462	14.0716	14.0970	14.1224	14.1478	14.1732	14.1986
0.560	14.2240	14.2494	14.2748	14.3002	14.3256	14.3510	14.3764	14.4018	14.4272	14.4526
0.570	14.4780	14.5034	14.5288	14.5542	14.5796	14.6050	14.6304	14.6558	14.6812	14.7066
0.580	14.7320	14.7574	14.7828	14.8082	14.8336	14.8590	14.8844	14.9098	14.9352	14.9606
0.590	14.9860	15.0114	15.0368	15.0622	15.0876	15.1130	15.1384	15.1638	15.1892	15.2146
0.600	15.2400	15.2654	15.2908	15.3162	15.3416	15.3670	15.3924	15.4178	15.4432	15.4686
0.610	15.4940	15.5194	15.5448	15.5702	15.5956	15.6210	15.6464	15.6718	15.6972	15.7226
0.620	15.7480	15.7734	15.7988	15.8242	15.8496	15.8750	15.9004	15.9258	15.9512	15.9766
0.630	16.0020	16.0274	16.0528	16.0782	16.1036	16.1290	16.1544	16.1798	16.2052	16.2306
0.640	16.2560	16.2814	16.3068	16.3322	16.3576	16.3830	16.4084	16.4338	16.4592	16.4846
0.650	16.5100	16.5354	16.5608	16.5862	16.6116	16.6370	16.6624	16.6878	16.7132	16.7386
0.660	16.7640	16.7894	16.8148	16.8402	16.8656	16.8910	16.9164	16.9418	16.9672	16.9926
0.670	17.0180	17.0434	17.0688	17.0942	17.1196	17.1450	17.1704	17.1958	17.2212	17.2466
0.680	17.2720	17.2974	17.3228	17.3482	17.3736	17.3990	17.4244	17.4498	17.4752	17.5006
0.690	17.5260	17.5514	17.5768	17.6022	17.6276	17.6530	17.6784	17.7038	17.7292	17.7546
0.700	17.7800	17.8054	17.8308	17.8562	17.8816	17.9070	17.9324	17.9578	17.9832	18.0086
0.710	18.0340	18.0594	18.0848	18.1102	18.1356	18.1610	18.1864	18.2118	18.2372	18.2626
0.720	18.2880	18.3134	18.3388	18.3642	18.3896	18.4150	18.4404	18.4658	18.4912	18.5166
0.730	18.5420	18.5674	18.5928	18.6182	18.6436	18.6690	18.6944	18.7198	18.7452	18.7706
0.740	18.7960	18.8214	18.8468	18.8722	18.8976	18.9230	18.9484	18.9738	18.9992	19.0246
0.750	19.0500	19.0754	19.1008	19.1262	19.1516	19.1770	19.2024	19.2278	19.2532	19.2786
0.760	19.3040	19.3294	19.3548	19.3802	19.4056	19.4310	19.4564	19.4818	19.5072	19.5326
0.770	19.5580	19.5834	19.6088	19.6342	19.6596	19.6850	19.7104	19.7358	19.7612	19.7866
0.780	19.8120	19.8374	19.8628	19.8882	19.9136	19.9390	19.9644	19.9898	20.0152	20.0406
0.790	20.0660	20.0914	20.1168	20.1422	20.1676	20.1930	20.2184	20.2438	20.2692	20.2946
0.800	20.3200	20.3454	20.3708	20.3962	20.4216	20.4470	20.4724	20.4978	20.5232	20.5486
0.810	20.5740	20.5994	20.6248	20.6502	20.6756	20.7010	20.7264	20.7518	20.7772	20.8026
0.820	20.8280	20.8534	20.8788	20.9042	20.9296	20.9550	20.9804	21.0058	21.0312	21.0566
0.830	21.0820	21.1074	21.1328	21.1582	21.1836	21.2090	21.2344	21.2598	21.2852	21.3106
0.840	21.3360	21.3614	21.3868	21.4122	21.4376	21.4630	21.4884	21.5138	21.5392	21.5646
0.850	21.5900	21.6154	21.6408	21.6662	21.6916	21.7170	21.7424	21.7678	21.7932	21.8186
0.860	21.8440	21.8694	21.8948	21.9202	21.9456	21.9710	21.9964	22.0218	22.0472	22.0726
0.870	22.0980	22.1234	22.1488	22.1742	22.1996	22.2250	22.2504	22.2758	22.3012	22.3266
0.880	22.3520	22.3774	22.4028	22.4282	22.4536	22.4790	22.5044	22.5298	22.5552	22.5806
0.890	22.6060	22.6314	22.6568	22.6822	22.7076	22.7330	22.7584	22.7838	22.8092	22.8346
0.900	22.8600	22.8854	22.9108	22.9362	22.9616	22.9870	23.0124	23.0378	23.0632	23.0886
0.910	23.1140	23.1394	23.1648	23.1902	23.2156	23.2410	23.2664	23.2918	23.3172	23.3426
0.920	23.3680	23.3934	23.4188	23.4442	23.4696	23.4950	23.5204	23.5458	23.5712	23.5966
0.930	23.6220	23.6474	23.6728	23.6982	23.7236	23.7490	23.7744	23.7998	23.8252	23.8506
0.940	23.8760	23.9014	23.9268	23.9522	23.9776	24.0030	24.0284	24.0538	24.0792	24.1046
0.950	24.1300	24.1554	24.1808	24.2062	24.2316	24.2570	24.2824	24.3078	24.3332	24.3586
0.960	24.3840	24.4094	24.4348	24.4602	24.4856	24.5110	24.5364	24.5618	24.5872	24.6126
0.970	24.6380	24.6634	24.6888	24.7142	24.7396	24.7650	24.7904	24.8158	24.8412	24.8666
0.980	24.8920	24.9174	24.9428	24.9682	24.9936	25.0190	25.0444	25.0698	25.0952	25.1206
0.990	25.1460	25.1714	25.1968	25.2222	25.2476	25.2730	25.2984	25.3238	25.3492	25.3746
1.000	25.4000	…	…	…	…	…	…	…	…	…

Use Table 3 to obtain whole inch equivalents to add to decimal equivalents above. All values given in this table are exact; figures to the right of the last place figures are all zeros.

Table 5. Millimeters to Inches
(Based on 1 inch = 25.4 millimeters, exactly)

Millimeters	0	1	2	3	4	5	6	7	8	9
	Inches									
0	…	0.03937	0.07874	0.11811	0.15748	0.19685	0.23622	0.27559	0.31496	0.35433
10	0.39370	0.43307	0.47244	0.51181	0.55118	0.59055	0.62992	0.66929	0.70866	0.74803
20	0.78740	0.82677	0.86614	0.90551	0.94488	0.98425	1.02362	1.06299	1.10236	1.14173
30	1.18110	1.22047	1.25984	1.29921	1.33858	1.37795	1.41732	1.45669	1.49606	1.53543
40	1.57480	1.61417	1.65354	1.69291	1.73228	1.77165	1.81102	1.85039	1.88976	1.92913
50	1.96850	2.00787	2.04724	2.08661	2.12598	2.16535	2.20472	2.24409	2.28346	2.32283
60	2.36220	2.40157	2.44094	2.48031	2.51969	2.55906	2.59843	2.63780	2.67717	2.71654
70	2.75591	2.79528	2.83465	2.87402	2.91339	2.95276	2.99213	3.03150	3.07087	3.11024
80	3.14961	3.18898	3.22835	3.26772	3.30709	3.34646	3.38583	3.42520	3.46457	3.50394
90	3.54331	3.58268	3.62205	3.66142	3.70079	3.74016	3.77953	3.81890	3.85827	3.89764
100	3.93701	3.97638	4.01575	4.05512	4.09449	4.13386	4.17323	4.21260	4.25197	4.29134
110	4.33071	4.37008	4.40945	4.44882	4.48819	4.52756	4.56693	4.60630	4.64567	4.68504
120	4.72441	4.76378	4.80315	4.84252	4.88189	4.92126	4.96063	5.00000	5.03937	5.07874
130	5.11811	5.15748	5.19685	5.23622	5.27559	5.31496	5.35433	5.39370	5.43307	5.47244
140	5.51181	5.55118	5.59055	5.62992	5.66929	5.70866	5.74803	5.78740	5.82677	5.86614
150	5.90551	5.94488	5.98425	6.02362	6.06299	6.10236	6.14173	6.18110	6.22047	6.25984
160	6.29921	6.33858	6.37795	6.41732	6.45669	6.49606	6.53543	6.57480	6.61417	6.65354
170	6.69291	6.73228	6.77165	6.81102	6.85039	6.88976	6.92913	6.96850	7.00787	7.04724
180	7.08661	7.12598	7.16535	7.20472	7.24409	7.28346	7.32283	7.36220	7.40157	7.44094
190	7.48031	7.51969	7.55906	7.59843	7.63780	7.67717	7.71654	7.75591	7.79528	7.83465
200	7.87402	7.91339	7.95276	7.99213	8.03150	8.07087	8.11024	8.14961	8.18898	8.22835
210	8.26772	8.30709	8.34646	8.38583	8.42520	8.46457	8.50394	8.54331	8.58268	8.62205
220	8.66142	8.70079	8.74016	8.77953	8.81890	8.85827	8.89764	8.93701	8.97638	9.01575
230	9.05512	9.09449	9.13386	9.17323	9.21260	9.25197	9.29134	9.33071	9.37008	9.40945
240	9.44882	9.48819	9.52756	9.56693	9.60630	9.64567	9.68504	9.72441	9.76378	9.80315
250	9.84252	9.88189	9.92126	9.96063	10.0000	10.0394	10.0787	10.1181	10.1575	10.1969
260	10.2362	10.2756	10.3150	10.3543	10.3937	10.4331	10.4724	10.5118	10.5512	10.5906
270	10.6299	10.6693	10.7087	10.7480	10.7874	10.8268	10.8661	10.9055	10.9449	10.9843
280	11.0236	11.0630	11.1024	11.1417	11.1811	11.2205	11.2598	11.2992	11.3386	11.3780
290	11.4173	11.4567	11.4961	11.5354	11.5748	11.6142	11.6535	11.6929	11.7323	11.7717
300	11.8110	11.8504	11.8898	11.9291	11.9685	12.0079	12.0472	12.0866	12.1260	12.1654
310	12.2047	12.2441	12.2835	12.3228	12.3622	12.4016	12.4409	12.4803	12.5197	12.5591
320	12.5984	12.6378	12.6772	12.7165	12.7559	12.7953	12.8346	12.8740	12.9134	12.9528
330	12.9921	13.0315	13.0709	13.1102	13.1496	13.1890	13.2283	13.2677	13.3071	13.3465
340	13.3858	13.4252	13.4646	13.5039	13.5433	13.5827	13.6220	13.6614	13.7008	13.7402
350	13.7795	13.8189	13.8583	13.8976	13.9370	13.9764	14.0157	14.0551	14.0945	14.1339
360	14.1732	14.2126	14.2520	14.2913	14.3307	14.3701	14.4094	14.4488	14.4882	14.5276
370	14.5669	14.6063	14.6457	14.6850	14.7244	14.7638	14.8031	14.8425	14.8819	14.9213
380	14.9606	15.0000	15.0394	15.0787	15.1181	15.1575	15.1969	15.2362	15.2756	15.3150
390	15.3543	15.3937	15.4331	15.4724	15.5118	15.5512	15.5906	15.6299	15.6693	15.7087
400	15.7480	15.7874	15.8268	15.8661	15.9055	15.9449	15.9843	16.0236	16.0630	16.1024
410	16.1417	16.1811	16.2205	16.2598	16.2992	16.3386	16.3780	16.4173	16.4567	16.4961
420	16.5354	16.5748	16.6142	16.6535	16.6929	16.7323	16.7717	16.8110	16.8504	16.8898
430	16.9291	16.9685	17.0079	17.0472	17.0866	17.1260	17.1654	17.2047	17.2441	17.2835
440	17.3228	17.3622	17.4016	17.4409	17.4803	17.5197	17.5591	17.5984	17.6378	17.6772
450	17.7165	17.7559	17.7953	17.8346	17.8740	17.9134	17.9528	17.9921	18.0315	18.0709
460	18.1102	18.1496	18.1890	18.2283	18.2677	18.3071	18.3465	18.3858	18.4252	18.4646
470	18.5039	18.5433	18.5827	18.6220	18.6614	18.7008	18.7402	18.7795	18.8189	18.8583
480	18.8976	18.9370	18.9764	19.0157	19.0551	19.0945	19.1339	19.1732	19.2126	19.2520
490	19.2913	19.3307	19.3701	19.4094	19.4488	19.4882	19.5276	19.5669	19.6063	19.6457

Table 5. *(Continued)* **Millimeters to Inches**
(Based on 1 inch = 25.4 millimeters, exactly)

Millimeters	0	1	2	3	4	5	6	7	8	9
	Inches									
500	19.6850	19.7244	19.7638	19.8031	19.8425	19.8819	19.9213	19.9606	20.0000	20.0394
510	20.0787	20.1181	20.1575	20.1969	20.2362	20.2756	20.3150	20.3543	20.3937	20.4331
520	20.4724	20.5118	20.5512	20.5906	20.6299	20.6693	20.7087	20.7480	20.7874	20.8268
530	20.8661	20.9055	20.9449	20.9843	21.0236	21.0630	21.1024	21.1417	21.1811	21.2205
540	21.2598	21.2992	21.3386	21.3780	21.4173	21.4567	21.4961	21.5354	21.5748	21.6142
550	21.6535	21.6929	21.7323	21.7717	21.8110	21.8504	21.8898	21.9291	21.9685	22.0079
560	22.0472	22.0866	22.1260	22.1654	22.2047	22.2441	22.2835	22.3228	22.3622	22.4016
570	22.4409	22.4803	22.5197	22.5591	22.5984	22.6378	22.6772	22.7165	22.7559	22.7953
580	22.8346	22.8740	22.9134	22.9528	22.9921	23.0315	23.0709	23.1102	23.1496	23.1890
590	23.2283	23.2677	23.3071	23.3465	23.3858	23.4252	23.4646	23.5039	23.5433	23.5827
600	23.6220	23.6614	23.7008	23.7402	23.7795	23.8189	23.8583	23.8976	23.9370	23.9764
610	24.0157	24.0551	24.0945	24.1339	24.1732	24.2126	24.2520	24.2913	24.3307	24.3701
620	24.4094	24.4488	24.4882	24.5276	24.5669	24.6063	24.6457	24.6850	24.7244	24.7638
630	24.8031	24.8425	24.8819	24.9213	24.9606	25.0000	25.0394	25.0787	25.1181	25.1575
640	25.1969	25.2362	25.2756	25.3150	25.3543	25.3937	25.4331	25.4724	25.5118	25.5512
650	25.5906	25.6299	25.6693	25.7087	25.7480	25.7874	25.8268	25.8661	25.9055	25.9449
660	25.9843	26.0236	26.0630	26.1024	26.1417	26.1811	26.2205	26.2598	26.2992	26.3386
670	26.3780	26.4173	26.4567	26.4961	26.5354	26.5748	26.6142	26.6535	26.6929	26.7323
680	26.7717	26.8110	26.8504	26.8898	26.9291	26.9685	27.0079	27.0472	27.0866	27.1260
690	27.1654	27.2047	27.2441	27.2835	27.3228	27.3622	27.4016	27.4409	27.4803	27.5197
700	27.5591	27.5984	27.6378	27.6772	27.7165	27.7559	27.7953	27.8346	27.8740	27.9134
710	27.9528	27.9921	28.0315	28.0709	28.1102	28.1496	28.1890	28.2283	28.2677	28.3071
720	28.3465	28.3858	28.4252	28.4646	28.5039	28.5433	28.5827	28.6220	28.6614	28.7008
730	28.7402	28.7795	28.8189	28.8583	28.8976	28.9370	28.9764	29.0157	29.0551	29.0945
740	29.1339	29.1732	29.2126	29.2520	29.2913	29.3307	29.3701	29.4094	29.4488	29.4882
750	29.5276	29.5669	29.6063	29.6457	29.6850	29.7244	29.7638	29.8031	29.8425	29.8819
760	29.9213	29.9606	30.0000	30.0394	30.0787	30.1181	30.1575	30.1969	30.2362	30.2756
770	30.3150	30.3543	30.3937	30.4331	30.4724	30.5118	30.5512	30.5906	30.6299	30.6693
780	30.7087	30.7480	30.7874	30.8268	30.8661	30.9055	30.949	30.9843	31.0236	31.0630
790	31.1024	31.1417	31.1811	31.2205	31.2598	31.2992	31.3386	31.3780	31.4173	31.4567
800	31.4961	31.5354	31.5748	31.6142	31.6535	31.6929	31.7323	31.7717	31.8110	31.8504
810	31.8898	31.9291	31.9685	32.0079	32.0472	32.0866	32.1260	32.1654	32.2047	32.2441
820	32.2835	32.3228	32.3622	32.4016	32.4409	32.4803	32.5197	32.5591	32.5984	32.6378
830	32.6772	32.7165	32.7559	32.7953	32.8346	32.8740	32.9134	32.9528	32.9921	33.0315
840	33.0709	33.1102	33.1496	33.1890	33.2283	33.2677	33.3071	33.3465	33.3858	33.4252
850	33.4646	33.5039	33.5433	33.5827	33.6220	33.6614	33.7008	33.7402	33.7795	33.8189
860	33.8583	33.8976	33.9370	33.9764	34.0157	34.0551	34.0945	34.1339	34.1732	34.2126
870	34.2520	34.2913	34.3307	34.3701	34.4094	34.4488	34.4882	34.5276	34.5669	34.6063
880	34.6457	34.6850	34.7244	34.7638	34.8031	34.8425	34.8819	34.9213	34.9606	35.0000
890	35.0394	35.0787	35.1181	35.1575	35.1969	35.2362	35.2756	35.3150	35.3543	35.3937
900	35.4331	35.4724	35.5118	35.5512	35.5906	35.6299	35.6693	35.7087	35.7480	35.7874
910	35.8268	35.8661	35.9055	35.9449	35.9843	36.0236	36.0630	36.1024	36.1417	36.1811
920	36.2205	36.2598	36.2992	36.3386	36.3780	36.4173	36.4567	36.4961	36.5354	36.5748
930	36.6142	36.6535	36.6929	36.7323	36.7717	36.8110	36.8504	36.8898	36.9291	36.9685
940	37.0079	37.0472	37.0866	37.1260	37.1654	37.2047	37.2441	37.2835	37.3228	37.3622
950	37.4016	37.409	37.4803	37.5197	37.5591	37.5984	37.6378	37.6772	37.7165	37.7559
960	37.7953	37.8346	37.8740	37.9134	37.9528	37.9921	38.0315	38.0709	38.1102	38.1496
970	38.1800	38.2283	38.2677	38.3071	38.3465	38.3858	38.4252	38.4646	38.5039	38.5433
980	38.5827	38.6220	38.6614	38.7008	38.7402	38.7795	38.8189	38.8583	38.8976	38.9370
990	38.9764	39.0157	39.0551	39.0945	39.1339	39.1732	39.2126	39.2520	39.2913	39.3307
1000	39.3701	…	…	…	…	…	…	…	…	…

Table 6. Fractional Inch–Millimeter and Feet–Millimeter Conversions
(Based on 1 inch = 25.4 millimeters, exactly)

Fractional Inch to Millimeters

in.	mm	in.	mm	in.	mm	in.	mm
1/64	0.397	17/64	6.747	33/64	13.097	49/64	19.447
1/32	0.794	9/32	7.144	17/32	13.494	25/32	19.844
3/64	1.191	19/64	7.541	35/64	13.891	51/64	20.241
1/16	1.588	5/16	7.938	9/16	14.288	13/16	20.638
5/64	1.984	21/64	8.334	37/64	14.684	53/64	21.034
3/32	2.381	11/32	8.731	19/32	15.081	27/32	21.431
7/64	2.778	23/64	9.128	39/64	15.478	55/64	21.828
1/8	3.175	3/8	9.525	5/8	15.875	7/8	22.225
9/64	3.572	25/64	9.922	41/64	16.272	57/64	22.622
5/32	3.969	13/32	10.319	21/32	16.669	29/32	23.019
11/64	4.366	27/64	10.716	43/64	17.066	59/64	23.416
3/16	4.762	7/16	11.112	11/16	17.462	15/16	23.812
13/64	5.159	29/64	11.509	45/64	17.859	61/64	24.209
7/32	5.556	15/32	11.906	23/32	18.256	31/32	24.606
15/64	5.953	31/64	12.303	47/64	18.653	63/64	25.003
1/4	6.350	1/2	12.700	3/4	19.050	1	25.400

Inches to Millimeters

in.	mm	in.	mm	in.	mm	in.	mm	in.	mm	in.	mm
1	25.4	3	76.2	5	127.0	7	177.8	9	228.6	11	279.4
2	50.8	4	101.6	6	152.4	8	203.2	10	254.0	12	304.8

Feet to Millimeters

ft	mm	ft	mm	ft	mm	ft	mm	ft	mm
100	30,480	10	3,048	1	304.8	0.1	30.48	0.01	3.048
200	60,960	20	6,096	2	609.6	0.2	60.96	0.02	6.096
300	91,440	30	9,144	3	914.4	0.3	91.44	0.03	9.144
400	121,920	40	12,192	4	1,219.2	0.4	121.92	0.04	12.192
500	152,400	50	15,240	5	1,524.0	0.5	152.40	0.05	15.240
600	182,880	60	18,288	6	1,828.8	0.6	182.88	0.06	18.288
700	213,360	70	21,336	7	2,133.6	0.7	213.36	0.07	21.336
800	243,840	80	24,384	8	2,438.4	0.8	243.84	0.08	24.384
900	274,320	90	27,432	9	2,743.2	0.9	274.32	0.09	27.432
1,000	304,800	100	30,480	10	3,048.0	1.0	304.80	0.10	30.480

Example 1: Find millimeter equivalent of 293 feet, 5 47/64 inches.

200 ft	=	60,960.	mm
90 ft	=	27,432.	mm
3 ft	=	914.4	mm
5 in.	=	127.0	mm
47/64 in.	=	18.653	mm
293 ft 5 47/64 in.	=	89,452.053	mm

Example 2: Find millimeter equivalent of 71.86 feet.

70. ft	=	21,336.	mm
1. ft	=	304.8	mm
.80 ft	=	243.84	mm
.06 ft	=	18.288	mm
71.86 ft	=	21,902.928	mm

Table 7. Thousandths of an Inch to Millimeters Conversion Table

Inch	Millimeters									
	0	1	2	3	4	5	6	7	8	9
0.001	0.02540	0.02794	0.03048	0.03302	0.03556	0.03810	0.04064	0.04318	0.04572	0.04826
0.002	0.05080	0.05334	0.05588	0.05842	0.06096	0.06350	0.06604	0.06858	0.07112	0.07366
0.003	0.07620	0.07874	0.08128	0.08382	0.08636	0.08890	0.09144	0.09398	0.09652	0.09906
0.004	0.10160	0.10414	0.10668	0.10922	0.11176	0.11430	0.11684	0.11938	0.12192	0.12446
0.005	0.12700	0.12954	0.13208	0.13462	0.13716	0.13970	0.14224	0.14478	0.14732	0.14986
0.006	0.15240	0.15494	0.15748	0.16002	0.16256	0.16510	0.16764	0.17018	0.17272	0.17526
0.007	0.17780	0.18034	0.18288	0.18542	0.18796	0.19050	0.19304	0.19558	0.19812	0.20066
0.008	0.20320	0.20574	0.20828	0.21082	0.21336	0.21590	0.21844	0.22098	0.22352	0.22606
0.009	0.22860	0.23114	0.23368	0.23622	0.23876	0.24130	0.24384	0.24638	0.24892	0.25146
0.01	0.25400	0.25654	0.25908	0.26162	0.26416	0.26670	0.26924	0.27178	0.27432	0.27686
0.02	0.50800	0.53340	0.55880	0.58420	0.60960	0.63500	0.66040	0.68580	0.71120	0.73660
0.03	0.76200	0.78740	0.81280	0.83820	0.86360	0.88900	0.91440	0.93980	0.96520	0.99060
0.04	1.01600	1.04140	1.06680	1.09220	1.11760	1.14300	1.16840	1.19380	1.21920	1.24460
0.05	1.27000	1.29540	1.32080	1.34620	1.37160	1.39700	1.42240	1.44780	1.47320	1.49860
0.06	1.52400	1.54940	1.57480	1.60020	1.62560	1.65100	1.67640	1.70180	1.72720	1.75260
0.07	1.77800	1.80340	1.82880	1.85420	1.87960	1.90500	1.93040	1.95580	1.98120	2.00660
0.08	2.03200	2.05740	2.08280	2.10820	2.13360	2.15900	2.18440	2.20980	2.23520	2.26060
0.09	2.28600	2.31140	2.33680	2.36220	2.38760	2.41300	2.43840	2.46380	2.48920	2.51460
0.1	2.54000	2.56540	2.59080	2.61620	2.64160	2.66700	2.69240	2.71780	2.74320	2.76860
0.2	5.08000	5.10540	5.13080	5.15620	5.18160	5.20700	5.23240	5.25780	5.28320	5.30860
0.3	7.62000	7.64540	7.67080	7.69620	7.72160	7.74700	7.77240	7.79780	7.82320	7.84860
0.4	10.16000	10.18540	10.21080	10.23620	10.26160	10.28700	10.31240	10.33780	10.36320	10.38860
0.5	12.70000	12.72540	12.75080	12.77620	12.80160	12.82700	12.85240	12.87780	12.90320	12.92860
0.6	15.24000	15.26540	15.29080	15.31620	15.34160	15.36700	15.39240	15.41780	15.44320	15.46860
0.7	17.78000	17.80540	17.83080	17.85620	17.88160	17.90700	17.93240	17.95780	17.98320	18.00860
0.8	20.32000	20.34540	20.37080	20.39620	20.42160	20.44700	20.47240	20.49780	20.52320	20.54860
0.9	22.86000	22.88540	22.91080	22.93620	22.96160	22.98700	23.01240	23.03780	23.06320	23.08860
1.0	25.40000	…	…	…	…	…	…	…	…	…

Rounding Off Numbers

Rules	Examples		
When the last digit is followed by a 0,1,2,3,4, it is retained and unchanged. This is known as rounding down.	3.60040	=	3.600
	3.60027	=	3.600
When the last digit is followed by a 5,6,7,8 or 9, it is increased by 1.This is known as rounding up.	3.60070	=	3.601
	3.60056	=	3.601
When the first digit neglected is a 5 followed by zeros the rounding is exactly equal to half a unit of the last digit retained. However, the last digit retained is then the closest even number.	0.12500	=	0.12
	0.15500	=	0.16
	3.60350	=	3.604
	3.60450	=	3.604

Table 8. Fundamental Constants

Constant	Symbol	Value	
		SI Units	Other Units
Electronic Charge	e	1.60210×10^{-19} C	4.80298×10^{-10} e.s.u.
Electronic rest mass	m_e	9.1091×10^{-31} kilogram	5.48597×10^{-4} a.m.u
Electronic radius	r_e	2.81777×10^{-15} meter	...
Proton rest mass	m_p	1.67252×10^{-27} kilogram	1.00727663 a.m.u
Neutron rest mass	m_n	1.67482×10^{-27} kilogram	1.0086654 a.m.u
Planck's constant	h	6.62559×10^{-34} joule-second	6.62559×10^{-27} erg second
Velocity of light	c	2.997925×10^{8} meters/s	186281 miles/s
Avogadro's constant	L, N_A	6.02252×10^{23} per mole	...
Loschmidt's constant	N_l	2.68719×10^{25} m^{-3}	2.68719×10^{19} cm^{-3}
Gas constant	R	8.3143 J K^{-1} mol^{-1}	1.9858 calories $°C^{-1}mol^{-1}$
Boltzmann's constant	$\kappa = \frac{R}{N_A}$	1.38054×10^{-23} J K^{-1}	3.29729×10^{-24} calories $°C^{-1}$
Faraday's constant	F	9.64870×10^{4} Coulomb mol^{-1}	2.89261×10^{14} e.s.u mol^{-1}
Stefan-Boltzman Constant	σ	5.6697×10^{-8} W $m^{-2}K^{4}$	5.6697×10^{-5} e.s.u. mol^{-1}
Gravitational constant	G	6.670×10^{-11} N m^2 kg^{-2}	6.670×10^{-11} dyne cm^2 g^{-2}
Electrical permeability	μ_o	$4\pi \times 10^{-7}$ H m^{-1}	...
Magnetic permeability	ε_o	8.85418×10^{-12} F m^{-1}	...
Euler's constant	γ	0.5772	...
Golden ratio	Φ	1.6180	...

Table 9. Functions of π

Constant	Numerical Value	Logarithm	Constant	Numerical Value	Logarithm
π	3.141593	0.49715	2π	6.283185	0.79818
3π	9.424778	0.97427	4π	12.566370	1.09921
$2\pi/3$	2.094395	0.32105	$4\pi/3$	4.188790	0.62209
$\pi \div 2$	1.570796	0.19611	$\pi \div 3$	1.047197	0.02003
$\pi \div 4$	0.785398	-0.10491	$\pi \div 6$	0.523598	-0.28100
$\pi\sqrt{2}$	4.442882	0.64766	$\pi\sqrt{3}$	5.441398	0.73571
$\pi/\sqrt{2}$	2.221441	0.34663	$\pi/\sqrt{3}$	1.813799	0.25859
π^2	9.869604	0.99430	$1 \div \pi^2$	0.101321	-0.99430
$1 \div \pi$	0.318310	-0.49715	$1 \div 2\pi$	0.159155	-0.79818
$1 \div \pi^3$	0.032252	-1.49145	π^3	31.006277	1.49145
$\sqrt{\pi}$	1.772454	0.24858	$\sqrt[3]{\pi}$	1.464592	0.16572

Table 10. Functions of *g*

Constant	Numerical Value, ft/s²	Numerical Value, m/s²	Constant	Numerical Value	Numerical Value, m/s²
g	32.16	9.81	g^2	1034.266	96.2361
$2g$	64.32	19.62	$1 \div 2g$	0.01555	0.101936
$\sqrt{2g}$	8.01998	4.43	$1 \div \sqrt{g}$	0.17634	0.319275
$\pi \div \sqrt{g}$	0.55398	1.00	$\pi \div (2\sqrt{g})$	0.39172	0.70916

Table 11. Functions of *e*

Constants	Numerical value	Constants	Numerical value
e	2.71828	$1/e$	0.3679
$1/e^2$	0.13534	e^{π}	23.141

Table 12. Weights and Volumes

Constant	Numerical Value	Logarithm
Weight in pounds of:		
Water column, 1″ × 1″ × 1 ft.	0.4335	-0.36301
1 US gallon of water, 39.1°F.	8.34	0.92117
1 cu. ft. of water, 39.1° F.	62.4245	1.79536
1 cu. in. of water, 39.1°F.	0.0361	-1.44249
1 cu. ft. of air, 32°F., atmospheric pressure	0.08073	-1.09297
Volume in cu. ft. of:		
1 pound of water, 39.1°F	0.01602	-1.79534
1 pound of air, 32°F., atmospheric pressure	12.387	1.09297
Volume in gallons of 1 pound of water, 39.1°F	0.1199	-0.92118
Volume in cu. in. of 1 pound of water, 39.1 °F	27.70	1.44248
Gallons in one cu. ft.	7.4805	0.87393
Atmospheric pressure in pounds per sq. inch	14.696	1.16720

Roman Numerals

I	1	VI	6	XX	20	LX	60	CC	200	DC	600	MM	2000
II	2	VII	7	XXX	30	LXX	70	CCC	300	DCC	700	IMM	1999
III	3	VIII	8	XL	40	LXXX	80	CD	400	DCCC	800	IL	49
IV	4	IX	9	XLV	45	XC	90	ID	499	CM	900	IC	99
V	5	X	10	L	50	C	100	D	500	M	1000		

Greek Letters and Standard Abbreviations

The Greek letters are frequently used in mathematical expressions and formulas. The Greek alphabet is given below.

Α	α	Alpha	Η	η	Eta	Ν	ν	Nu	Τ	τ	Tau
Β	β	Beta	Θ	ϑ θ	Theta	Ξ	ξ	Xi	Υ	υ	Upsilon
Γ	γ	Gamma	Ι	ι	Iota	Ο	ο	Omicron	Φ	φ	Phi
Δ	δ	Delta	Κ	κ	Kappa	Π	π	Pi	Ξ	χ	Chi
Ε	ε	Epsilon	Λ	λ	Lambda	Ρ	ρ	Rho	Ψ	ψ	Psi
Ζ	ζ	Zeta	Μ	μ	Mu	Σ	σ ς	Sigma	Ω	ω	Omega

Table 13. Conversion Factors

Multiply	By	To Obtain
Celsius	$C \times 1.8 + 32$	Fahrenheit
Celsius	$C + 273.15$	Kelvin
Circumference	6.2832	radians
Degrees/second (angular)	0.002778	revolutions/sec
Degrees (angular)	60	minutes
Degrees (angular)	0.01111	quadrants
Degrees/second (angular)	0.01745	radians/sec
Degrees (angular)	3600	seconds
Degrees (angular)	0.01745	radians
Degrees/second (angular)	0.1667	revolutions/min (rpm)
Fahrenheit	$F + 459.67$	Rankine
Fahrenheit	$[F - 32] \times 5/9$	Celsius
Horsepower	0.7457	kilowatts
Horsepower	33,000	foot-pounds/min
Horsepower	550	foot-pounds/sec
Horsepower	745.7	watts
Horsepower-hr	2.6845×10^{13}	ergs
Horsepower-hr	0.7457	kilowatt-hrs
Horsepower-hr	1.98×10^{6}	foot-pounds
Minutes (angular)	60	seconds
Minutes (angular)	2.909×10^{-4}	radians
Minutes (angular)	1.852×10^{-4}	quadrants
Minutes (angular)	0.01667	degrees
Quadrants (angular)	5400	minutes
Quadrants (angular)	1.571	radians
Quadrants (angular)	90	degrees
Radians/sec	57.3	degrees/sec
Radians/sec	9.549	revolutions/min
Radians/sec	0.1592	revolutions/sec
Radians	0.6366	quadrants
Radians	57.3	degrees
Radians	3438	minutes
Rankine	R-459.67	Fahrenheit
Revolutions/min	0.1047	radians/sec
Revolutions/min	0.01667	revolutions/sec
Revolutions/sec	360	degrees/sec
Revolutions/sec	6.283	radians/sec
Revolutions/min	6	degrees/sec
Revolutions/sec	60	revolutions/sec
Revolutions	6.283	radians
Revolutions	4	quadrants
Revolutions	360	degrees
Seconds (angular)	4.848×10^{-6}	radians
Seconds (angular)	3.087×10^{-6}	quadrants
Seconds (angular)	0.01667	minutes
Seconds (angular)	2.778×10^{-4}	degrees

Table 14. °C → °F and °R Temperature Conversion °F → °C and K

K	°C		°F	°R	K	°C		°F	°R	K	°C		°F	°R
0.0	−273.2	**−459.7**	…	…	261.5	−11.7	**11**	51.8	511.5	293.7	20.6	**69**	156.2	615.9
5.4	−267.8	**−450**	…	…	262.0	−11.1	**12**	53.6	513.3	294.3	21.1	**70**	158.0	617.7
10.9	−262.2	**−440**	…	…	262.6	−10.6	**13**	55.4	515.1	294.8	21.7	**71**	159.8	619.5
16.5	−256.7	**−430**	…	…	263.2	−10.0	**14**	57.2	516.9	295.4	22.2	**72**	161.6	621.3
22.0	−251.1	**−420**	…	…	263.7	−9.4	**15**	59.0	518.7	295.9	22.8	**73**	163.4	623.1
27.6	−245.6	**−410**	…	…	264.3	−8.9	**16**	60.8	520.5	296.5	23.3	**74**	165.2	624.9
33.2	−240.0	**−400**	…	…	264.8	−8.3	**17**	62.6	522.3	297.0	23.9	**75**	167.0	626.7
38.7	−234.4	**−390**	…	…	265.4	−7.8	**18**	64.4	524.1	297.6	24.4	**76**	168.8	628.5
44.3	−228.9	**−380**	…	…	265.9	−7.2	**19**	66.2	525.9	298.2	25.0	**77**	170.6	630.3
49.8	−223.3	**−370**	…	…	266.5	−6.7	**20**	68.0	527.7	298.7	25.6	**78**	172.4	632.1
55.4	−217.8	**−360**	…	…	267.0	−6.1	**21**	69.8	529.5	299.3	26.1	**79**	174.2	633.9
60.9	−212.2	**−350**	…	…	267.6	−5.6	**22**	71.6	531.3	299.8	26.7	**80**	176.0	635.7
66.5	−206.7	**−340**	…	…	268.2	−5.0	**23**	73.4	533.1	300.4	27.2	**81**	177.8	637.5
72.0	−201.1	**−330**	…	…	268.7	−4.4	**24**	75.2	534.9	300.9	27.8	**82**	179.6	639.3
77.6	−195.6	**−320**	…	…	269.3	−3.9	**25**	77.0	536.7	301.5	28.3	**83**	181.4	641.1
83.2	−190.0	**−310**	…	…	269.8	−3.3	**26**	78.8	538.5	302.0	28.9	**84**	183.2	642.9
88.7	−184.4	**−300**	…	…	270.4	−2.8	**27**	80.6	540.3	302.6	29.4	**85**	185.0	644.7
94.3	−178.9	**−290**	…	…	270.9	−2.2	**28**	82.4	542.1	303.2	30.0	**86**	186.8	646.5
99.8	−173.3	**−280**	…	…	271.5	−1.7	**29**	84.2	543.9	303.7	30.6	**87**	188.6	648.3
103.6	−169.5	**−273.2**	−459.7	0.0	272.0	−1.1	**30**	86.0	545.7	304.3	31.1	**88**	190.4	650.1
105.4	−167.8	**−270**	−454.0	5.7	272.6	−0.6	**31**	87.8	547.5	304.8	31.7	**89**	192.2	651.9
110.9	−162.2	**−260**	−436.0	23.7	273.2	0.0	**32**	89.6	549.3	305.4	32.2	**90**	194.0	653.7
116.5	−156.7	**−250**	−418.0	41.7	273.7	0.6	**33**	91.4	551.1	305.9	32.8	**91**	195.8	655.5
122.0	−151.1	**−240**	−400.0	59.7	274.3	1.1	**34**	93.2	552.9	306.5	33.3	**92**	197.6	657.3
127.6	−145.6	**−230**	−382.0	77.7	274.8	1.7	**35**	95.0	554.7	307.0	33.9	**93**	199.4	659.1
133.2	−140.0	**−220**	−364.0	95.7	275.4	2.2	**36**	96.8	556.5	307.6	34.4	**94**	201.2	660.9
138.7	−134.4	**−210**	−346.0	113.7	275.9	2.8	**37**	98.6	558.3	308.2	35.0	**95**	203.0	662.7
144.3	−128.9	**−200**	−328.0	131.7	276.5	3.3	**38**	100.4	560.1	308.7	35.6	**96**	204.8	664.5
149.8	−123.3	**−190**	−310.0	149.7	277.0	3.9	**39**	102.2	561.9	309.3	36.1	**97**	206.6	666.3
155.4	−117.8	**−180**	−292.0	167.7	277.6	4.4	**40**	104.0	563.7	309.8	36.7	**98**	208.4	668.1
160.9	−112.2	**−170**	−274.0	185.7	278.2	5.0	**41**	105.8	565.5	310.4	37.2	**99**	210.2	669.9
166.5	−106.7	**−160**	−256.0	203.7	278.7	5.6	**42**	107.6	567.3	310.9	37.8	**100**	212.0	671.7
172.0	−101.1	**−150**	−238.0	221.7	279.3	6.1	**43**	109.4	569.1	311.5	38.3	**101**	213.8	673.5
177.6	−95.6	**−140**	−220.0	239.7	279.8	6.7	**44**	111.2	570.9	312.0	38.9	**102**	215.6	675.3
183.2	−90.0	**−130**	−202.0	257.7	280.4	7.2	**45**	113.0	572.7	312.6	39.4	**103**	217.4	677.1
188.7	−84.4	**−120**	−184.0	275.7	280.9	7.8	**46**	114.8	574.5	313.2	40.0	**104**	219.2	678.9
194.3	−78.9	**−110**	−166.0	293.7	281.5	8.3	**47**	116.6	576.3	313.7	40.6	**105**	221.0	680.7
199.8	−73.3	**−100**	−148.0	311.7	282.0	8.9	**48**	118.4	578.1	314.3	41.1	**106**	222.8	682.5
205.4	−67.8	**−90**	−130.0	329.7	282.6	9.4	**49**	120.2	579.9	314.8	41.7	**107**	224.6	684.3
210.9	−62.2	**−80**	−112.0	347.7	283.2	10.0	**50**	122.0	581.7	315.4	42.2	**108**	226.4	686.1
216.5	−56.7	**−70**	−94.0	365.7	283.7	10.6	**51**	123.8	583.5	315.9	42.8	**109**	228.2	687.9
222.0	−51.1	**−60**	−76.0	383.7	284.3	11.1	**52**	125.6	585.3	316.5	43.3	**110**	230.0	689.7
227.6	−45.6	**−50**	−58.0	401.7	284.8	11.7	**53**	127.4	587.1	317.0	43.9	**111**	231.8	691.5
233.2	−40.0	**−40**	−40.0	419.7	285.4	12.2	**54**	129.2	588.9	317.6	44.4	**112**	233.6	693.3
238.7	−34.4	**−30**	−22.0	437.7	285.9	12.8	**55**	131.0	590.7	318.2	45.0	**113**	235.4	695.1
244.3	−28.9	**−20**	−4.0	455.7	286.5	13.3	**56**	132.8	592.5	318.7	45.6	**114**	237.2	696.9
249.8	−23.3	**−10**	14.0	473.7	287.0	13.9	**57**	134.6	594.3	319.3	46.1	**115**	239.0	698.7
255.4	−17.8	**0**	32.0	491.7	287.6	14.4	**58**	136.4	596.1	319.8	46.7	**116**	240.8	700.5
255.9	−17.2	**1**	33.8	493.5	288.2	15.0	**59**	138.2	597.9	320.4	47.2	**117**	242.6	702.3
256.5	−16.7	**2**	35.6	495.3	288.7	15.6	**60**	140.0	599.7	320.9	47.8	**118**	244.4	704.1
257.0	−16.1	**3**	37.4	497.1	289.3	16.1	**61**	141.8	601.5	321.5	48.3	**119**	246.2	705.9
257.6	−15.6	**4**	39.2	498.9	289.8	16.7	**62**	143.6	603.3	322.0	48.9	**120**	248.0	707.7
258.2	−15.0	**5**	41.0	500.7	290.4	17.2	**63**	145.4	605.1	322.6	49.4	**121**	249.8	709.5
258.7	−14.4	**6**	42.8	502.5	290.9	17.8	**64**	147.2	606.9	323.2	50.0	**122**	251.6	711.3
259.3	−13.9	**7**	44.6	504.3	291.5	18.3	**65**	149.0	608.7	323.7	50.6	**123**	253.4	713.1
259.8	−13.3	**8**	46.4	506.1	292.0	18.9	**66**	150.8	610.5	324.3	51.1	**124**	255.2	714.9
260.4	−12.8	**9**	48.2	507.9	292.6	19.4	**67**	152.6	612.3	324.8	51.7	**125**	257.0	716.7
260.9	−12.2	**10**	50.0	509.7	293.2	20.0	**68**	154.4	614.1	325.4	52.2	**126**	258.8	718.5

Table 14. *(Continued)* **°C → °F and °R Temperature Conversion °F → °C and K**

K	°C		°F	°R	K	°C		°F	°R	K	°C		°F	°R
325.9	52.8	**127**	260.6	720.3	357.6	84.4	**184**	363.2	822.9	741.5	468.3	**875**	1607.0	2066.7
326.5	53.3	**128**	262.4	722.1	358.2	85.0	**185**	365.0	824.7	755.4	482.2	**900**	1652.0	2111.7
327.0	53.9	**129**	264.2	723.9	358.7	85.6	**186**	366.8	826.5	769.3	496.1	**925**	1697.0	2156.7
327.6	54.4	**130**	266.0	725.7	359.3	86.1	**187**	368.6	828.3	783.2	510.0	**950**	1742.0	2201.7
328.2	55.0	**131**	267.8	727.5	359.8	86.7	**188**	370.4	830.1	797.0	523.9	**975**	1787.0	2246.7
328.7	55.6	**132**	269.6	729.3	360.4	87.2	**189**	372.2	831.9	810.9	537.8	**1000**	1832.0	2291.7
329.3	56.1	**133**	271.4	731.1	360.9	87.8	**190**	374.0	833.7	838.7	565.6	**1050**	1922.0	2381.7
329.8	56.7	**134**	273.2	732.9	361.5	88.3	**191**	375.8	835.5	866.5	593.3	**1100**	2012.0	2471.7
330.4	57.2	**135**	275.0	734.7	362.0	88.9	**192**	377.6	837.3	894.3	621.1	**1150**	2102.0	2561.7
330.9	57.8	**136**	276.8	736.5	362.6	89.4	**193**	379.4	839.1	922.0	648.9	**1200**	2192.0	2651.7
331.5	58.3	**137**	278.6	738.3	363.2	90.0	**194**	381.2	840.9	949.8	676.7	**1250**	2282.0	2741.7
332.0	58.9	**138**	280.4	740.1	363.7	90.6	**195**	383.0	842.7	977.6	704.4	**1300**	2372.0	2831.7
332.6	59.4	**139**	282.2	741.9	364.3	91.1	**196**	384.8	844.5	1005.4	732.2	**1350**	2462.0	2921.7
333.2	60.0	**140**	284.0	743.7	364.8	91.7	**197**	386.6	846.3	1033.2	760.0	**1400**	2552.0	3011.7
333.7	60.6	**141**	285.8	745.5	365.4	92.2	**198**	388.4	848.1	1060.9	787.8	**1450**	2642.0	3101.7
334.3	61.1	**142**	287.6	747.3	365.9	92.8	**199**	390.2	849.9	1088.7	815.6	**1500**	2732.0	3191.7
334.8	61.7	**143**	289.4	749.1	366.5	93.3	**200**	392.0	851.7	1116.5	843.3	**1550**	2822.0	3281.7
335.4	62.2	**144**	291.2	750.9	367.0	93.9	**201**	393.8	853.5	1144.3	871.1	**1600**	2912.0	3371.7
335.9	62.8	**145**	293.0	752.7	367.6	94.4	**202**	395.6	855.3	1172.0	898.9	**1650**	3002.0	3461.7
336.5	63.3	**146**	294.8	754.5	368.2	95.0	**203**	397.4	857.1	1199.8	926.7	**1700**	3092.0	3551.7
337.0	63.9	**147**	296.6	756.3	368.7	95.6	**204**	399.2	858.9	1227.6	954.4	**1750**	3182.0	3641.7
337.6	64.4	**148**	298.4	758.1	369.3	96.1	**205**	401.0	860.7	1255.4	982.2	**1800**	3272.0	3731.7
338.2	65.0	**149**	300.2	759.9	369.8	96.7	**206**	402.8	862.5	1283.2	1010.0	**1850**	3362.0	3821.7
338.7	65.6	**150**	302.0	761.7	370.4	97.2	**207**	404.6	864.3	1310.9	1037.8	**1900**	3452.0	3911.7
339.3	66.1	**151**	303.8	763.5	370.9	97.8	**208**	406.4	866.1	1338.7	1065.6	**1950**	3542.0	4001.7
339.8	66.7	**152**	305.6	765.3	371.5	98.3	**209**	408.2	867.9	1366.5	1093.3	**2000**	3632.0	4091.7
340.4	67.2	**153**	307.4	767.1	372.0	98.9	**210**	410.0	869.7	1394.3	1121.1	**2050**	3722.0	4181.7
340.9	67.8	**154**	309.2	768.9	372.6	99.4	**211**	411.8	871.5	1422.0	1148.9	**2100**	3812.0	4271.7
341.5	68.3	**155**	311.0	770.7	373.2	100.0	**212**	413.6	873.3	1449.8	1176.7	**2150**	3902.0	4361.7
342.0	68.9	**156**	312.8	772.5	377.6	104.4	**220**	428.0	887.7	1477.6	1204.4	**2200**	3992.0	4451.7
342.6	69.4	**157**	314.6	774.3	383.2	110.0	**230**	446.0	905.7	1505.4	1232.2	**2250**	4082.0	4541.7
343.2	70.0	**158**	316.4	776.1	388.7	115.6	**240**	464.0	923.7	1533.2	1260.0	**2300**	4172.0	4631.7
343.7	70.6	**159**	318.2	777.9	394.3	121.1	**250**	482.0	941.7	1560.9	1287.8	**2350**	4262.0	4721.7
344.3	71.1	**160**	320.0	779.7	408.2	135.0	**275**	527.0	986.7	1588.7	1315.6	**2400**	4352.0	4811.7
344.8	71.7	**161**	321.8	781.5	422.0	148.9	**300**	572.0	1031.7	1616.5	1343.3	**2450**	4442.0	4901.7
345.4	72.2	**162**	323.6	783.3	435.9	162.8	**325**	617.0	1076.7	1644.3	1371.1	**2500**	4532.0	4991.7
345.9	72.8	**163**	325.4	785.1	449.8	176.7	**350**	662.0	1121.7	1672.0	1398.9	**2550**	4622.0	5081.7
346.5	73.3	**164**	327.2	786.9	463.7	190.6	**375**	707.0	1166.7	1699.8	1426.7	**2600**	4712.0	5171.7
347.0	73.9	**165**	329.0	788.7	477.6	204.4	**400**	752.0	1211.7	1727.6	1454.4	**2650**	4802.0	5261.7
347.6	74.4	**166**	330.8	790.5	491.5	218.3	**425**	797.0	1256.7	1755.4	1482.2	**2700**	4892.0	5351.7
348.2	75.0	**167**	332.6	792.3	505.4	232.2	**450**	842.0	1301.7	1783.2	1510.0	**2750**	4982.0	5441.7
348.7	75.6	**168**	334.4	794.1	519.3	246.1	**475**	887.0	1346.7	1810.9	1537.8	**2800**	5072.0	5531.7
349.3	76.1	**169**	336.2	795.9	533.2	260.0	**500**	932.0	1391.7	1838.7	1565.6	**2850**	5162.0	5621.7
349.8	76.7	**170**	338.0	797.7	547.0	273.9	**525**	977.0	1436.7	1866.5	1593.3	**2900**	5252.0	5711.7
350.4	77.2	**171**	339.8	799.5	560.9	287.8	**550**	1022.0	1481.7	1894.3	1621.1	**2950**	5342.0	5801.7
350.9	77.8	**172**	341.6	801.3	574.8	301.7	**575**	1067.0	1526.7	1922.0	1648.9	**3000**	5432.0	5891.7
351.5	78.3	**173**	343.4	803.1	588.7	315.6	**600**	1112.0	1571.7	2033.2	1760.0	**3200**	5792.0	6251.7
352.0	78.9	**174**	345.2	804.9	602.6	329.4	**625**	1157.0	1616.7	2144.3	1871.1	**3400**	6152.0	6611.7
352.6	79.4	**175**	347.0	806.7	616.5	343.3	**650**	1202.0	1661.7	2255.4	1982.2	**3600**	6512.0	6971.7
353.2	80.0	**176**	348.8	808.5	630.4	357.2	**675**	1247.0	1706.7	2366.5	2093.3	**3800**	6872.0	7331.7
353.7	80.6	**177**	350.6	810.3	644.3	371.1	**700**	1292.0	1751.7	2477.6	2204.4	**4000**	7232.0	7691.7
354.3	81.1	**178**	352.4	812.1	658.2	385.0	**725**	1337.0	1796.7	2588.7	2315.6	**4200**	7592.0	8051.7
354.8	81.7	**179**	354.2	813.9	672.0	398.9	**750**	1382.0	1841.7	2699.8	2426.7	**4400**	7952.0	8411.7
355.4	82.2	**180**	356.0	815.7	685.9	412.8	**775**	1427.0	1886.7	2810.9	2537.8	**4600**	8312.0	8771.7
355.9	82.8	**181**	357.8	817.5	699.8	426.7	**800**	1472.0	1931.7	2922.0	2648.9	**4800**	8672.0	9131.7
356.5	83.3	**182**	359.6	819.3	713.7	440.6	**825**	1517.0	1976.7	3033.2	2760.0	**5000**	9032.0	9491.7
357.0	83.9	**183**	361.4	821.1	727.6	454.4	**850**	1562.0	2021.7	...	...	...	...	...

Table converts °C → °F and °R, or °F → °C and K. Find "convert from" temperature in **bold** column and read result from °F and °R or °C and K columns. *Example 1:* 183 °C = 361.4 °F and 821.1 °R. *Example 2:* 183 °F = 83.9 °C and 357.0 K.

USEFUL ONLINE RESOURCES

Professional Organizations and Government Agencies	Websites
Aerospace Industries Association (AIA)	www.aia-aerospace.org
America Makes	www.americamakes.us
American Bearing Manufacturers Association (ABMA)	www.americanbearings.org
American Bureau of Shipping (ABS)	ww2.eagle.org
American Coatings Association (ACA)	www.paint.org
American Composites Manufacturers Association (ACMA)	acmanet.org
American Institute of Aeronautics and Astronautics (AIAA)	www.aiaa.org
American Institute of Architects (AIA)	www.aia.org
American Institute of Steel Construction (AISC)	www.aisc.org
American Iron and Steel Institute	www.steel.org
American Mold Builders Association (AMBA)	amba.org
American National Standards Institute (ANSI)	www.ansi.org
American Petroleum Institute (API)	www.api.org
American Society for Engineering Education (ASEE)	www.asee.org
American Society for Metals (ASM)	www.asminternational.org
American Society for Testing and Materials (ASTM)	www.astm.org
American Society of Civil Engineers (ASCE)	www.asce.org
American Society of Mechanical Engineers (ASME)	www.asme.org
American Welding Society (AWS)	www.aws.org
American Wire Producers Association (AWPA)	www.awpa.org
The Association for Manufacturing Technology (AMT)	www.amtonline.org
Association of Professional Model Makers (APMM)	www.modelmakers.org
British Engineering Manufacturers Association (BEMA)	www.bema.co.uk
British Plastics Federation (BPF)	www.bpf.co.uk

Professional Organizations and Government Agencies	Websites
British Standards Institution (BSI)	www.bsigroup.com
Canadian Tooling & Machining Association (CTMA)	www.ctma.com
Deutsches Institut für Normung (DIN)	www.din.de
European Power Metallurgy Association (EPMA)	www.epma.com
Federal Aviation Administration (FAA)	www.faa.gov
Industrial Fasteners Institute (IFI)	www.indfast.org
Institute of Electrical and Electronics Engineers (IEEE)	www.ieee.org
Institute of Measurement and Control (INSTMC)	www.instmc.org
Institution of Engineering and Technology (IET)	www.theiet.org
International Association of Machinists and Aerospace Workers (IAM)	www.goiam.org
International Organization for Standardization (ISO)	www.iso.org
Laser Institute of America (LIA)	www.lia.org
Machinery and Allied Products Institute (MAPI)	www.mapi.net
The Manufacturers' Association (MA)	www.mascpa.org
Mechanical Contractors Association of America (MCAA)	www.mcaa.org
National Aeronautics and Space Administration (NASA)	www.nasa.gov
National Association for Surface Finishing (NASF)	nasf.org
National Association of Manufacturers (NAM)	www.nam.org
National Center for Manufacturing Sciences (NCMS)	www.ncms.org
National Electrical Manufacturers Association (NEMA)	www.nema.org
National Fire Protection Association (NFPA)	www.nfpa.org
National Institute for Metalworking Skills (NIMS)	www.nims-skills.org
National Institute of Standards and Technology (NIST)	www.nist.gov
National Math Foundation (NMF)	nationalmathfoundation.org

Professional Organizations and Government Agencies	Websites
The National Network for Manufacturing Innovation (NNMI)	www.energy.gov/eere/amo/national-network-manufacturing-innovation
National Science Foundation (NSF)	www.nsf.gov
National Society of Professional Engineers (NSPE)	www.nspe.org
The National STEM Foundation (N-STEM)	n-stem.org
National Tooling & Machining Association (NTMA)	ntma.org
Naval Facilities Engineering Command (NAVFAC)	www.navfac.navy.mil
Plant Engineering and Maintenance Association of Canada (PEMAC)	www.pemac.org
Precision Machined Products Association (PMPA)	www.pmpa.org
Precision Metalforming Association (PMA)	www.pma.org
Product Development & Management Association (PDMA)	www.pdma.org
Robotics Industries Association (RIA)	www.robotics.org
SAE International	www.sae.org
SkillsUSA	www.skillsusa.org
Society for Mining, Metallurgy & Exploration (SME)	www.smenet.org
Society of Manufacturing Engineers (SME)	www.sme.org
Society of Plastic Engineers (SPE)	www.4spe.org
The World Foundry Organization (WFO)	www.thewfo.com
Uni-Bell PVC Pipe Association	www.uni-bell.org
United Abrasives Manufacturers' Association	uama.org
US Air Force	www.af.mil
US Army Corps of Engineers	www.usa.gov/federal-agencies/u-s-army-corps-of-engineers
US Department of Education, Office of Educational Technology	tech.ed.gov
US Government Printing Office	www.gpo.gov
US National Laboratories	www.usa.gov/federal-agencies/national-laboratories
US STEM Foundation (US STEM)	www.usstem.org
World Steel Association	www.worldsteel.org

Other Industry Sources and Publications	Websites
AutoDesk	www.autodesk.com
Business Industrial Network	bin95.com
CAD Innovation	cadinnovation.com
CADdigest	www.caddigest.com
Cutting Tool Engineering	www.ctemag.com
Demand Driven Institute	www.demanddriveninstitute.com
Digital Machinist	www.digitalmachinist.net
Eastec: A Manufacturing Technology Series Event	easteconline.com
Engineers Edge	www.engineersedge.com
Haas Technical Education Center	www.haascnc.com/htec.html
The Home Shop Machinist	www.homeshopmachinist.net
Industrial Press, home of *Machinery's Handbook*	books.industrialpress.com
Industry Week	www.industryweek.com
International Manufacturing Technology Show (MTS)	www.imts.com
Lifetime Reliability Solutions	www.lifetime-reliability.com
Lloyd's Register North America	www.lr.org/en-us
MachineTools.com	www.machinetools.com
Machinio.com	www.machinio.com
Make	makezine.com
Manufacturing.net	www.manufacturing.net
Modern Machine Shop	www.mmsonline.com
MSC Direct	www.mscdirect.com
New Equipment Digest	www.newequipment.com
Photonics	www.photonics.com
Plastics Machinery Magazine	www.plasticsmachinerymagazine.com
PlumbingSupply.com	www.plumbingsupply.com
Tech Briefs	www.techbriefs.com
Tech Directions	www.techdirections.com
Travers Tools	www.travers.com
Westec: A Manufacturing Technology Series Event	westeconline.com

Note: The above list includes selected entities and websites that the *Machinery's Handbook* team deem potentially useful to readers at the time of publication. Suggestions for additions to this list can be emailed to us at MHT@industrialpress.com. For further related resources and information, please visit us at books.industrialpress.com.

INDEX

A

D

E–F

G

H

I–K

L

M

N

O–P

R

S

T

U–V

W

NOTES

Machinery's Handbook 31
Digital Edition

Since the first edition published more than 100 years ago, *Machinery's Handbook* has been an authoritative and comprehensive, yet highly practical, and easy-to-use tool. The *Machinery's Handbook 31 Digital Edition* makes access to this vast collection of information even easier and includes:

- The complete contents of the printed *Machinery's Handbook, 31st Edition*, with thousands of changes and almost 100 pages added since the last edition.
- Nearly 800 pages of additional archival content extracted over time from previous editions.
- Table of contents and indexes for material only available in the *Digital Edition*.
- Useful indexes of standards and materials covered throughout this expanded edition.
- The complete contents of the companion volume *Guide to the Use of Tables and Formulas in Machinery's Handbook, 31st Edition*, with handy links to *Digital Edition* pages.

With the *Machinery's Handbook 31 Digital Edition*, you can:

- View and print text, tables, and graphics.
- Zoom to magnify pages.
- Search the complete contents.
- Efficiently access information you need with thousands of clickable links.
- Choose online and offline viewing options on your PC, Mac, iPad, iPhone, and Android devices.
- For more information on the *Digital Edition* and *Digital Edition Upgrade* (available to previous edition *Machinery's Handbook* CD-ROM owners and eBook subscribers), visit us at books.industrialpress.com and ebooks.industrialpress.com.